AF368962

Antimicrobial Materials with Medical Applications

Antimicrobial Materials with Medical Applications

Editors

Sotiris K. Hadjikakou
Christina N. Banti
Andreas K. Rossos

MDPI • Basel • Beijing • Wuhan • Barcelona • Belgrade • Manchester • Tokyo • Cluj • Tianjin

Editors
Sotiris K. Hadjikakou
Department of Chemistry
University of Ioannina
Ioannina
Greece

Christina N. Banti
Department of Chemistry
Univerisity of Ioannina
Ioannina
Greece

Andreas K. Rossos
Department of Chemistry
University of Ioannina
IΩANNINA
Greece

Editorial Office
MDPI
St. Alban-Anlage 66
4052 Basel, Switzerland

This is a reprint of articles from the Special Issue published online in the open access journal *International Journal of Molecular Sciences* (ISSN 1422-0067) (available at: www.mdpi.com/journal/ijms/special_issues/materials_medical_applications).

For citation purposes, cite each article independently as indicated on the article page online and as indicated below:

LastName, A.A.; LastName, B.B.; LastName, C.C. Article Title. *Journal Name* **Year**, *Volume Number*, Page Range.

ISBN 978-3-0365-4004-7 (Hbk)
ISBN 978-3-0365-4003-0 (PDF)

Contents

About the Editors

Sotiris K. Hadjikakou

Dr Sotiris K. Hadjikakou (Professor): He is Professor of Biological Inorganic Chemistry at the University of Ioannina Greece and Coordinator of the Graduate Program in Biological Inorganic Chemistry. He graduated from the Chemistry Department of the Aristotle University of Thessaloniki, Greece, from where he was awarded a Ph.D. degree with "honors". He completed his post-doctoral research on Organometallic Chemistry, in the Department of Chemistry, University of Dortmund, Germany, and also worked as a visiting researcher at the Department of Chemistry and Biology, University of ESSEX, Colchester UK. His research interests in the field of Biological Inorganic Chemistry and Bio-Organometallic Chemistry include Drug activation (NSAIDs, Antibiotics or Antithyroid) with main group elements, the development of new therapeutic Anticancer or Antibacterial agents, and the development of new active medical devices. Prof. Hadjikakou is a member of the editorial board of the *Journal of Inorganic Biochemistry*, *Antibiotics*, and the *International Journal of Molecular Science*.

Christina N. Banti

Dr Christina N Banti (Adjunct Lecturer): Dr Banti received her bachelor's degree in Biological Applications and Technologies, University of Ioannina, Greece in 2009. She was awarded an M.Sc. in Bioinorganic Chemistry, University of Ioannina, and obtained her PhD from the Department of Physiology, School of Medicine, University of Ioannina, Greece. She is now a Postdoctoral Researcher and Adjunct Lecturer of the Department of Chemistry, University of Ioannina. Her research interests are mainly focused on bioinorganic chemistry and particularly on the evaluation of new anticancer or antibacterial metallodrugs and the study of their interactions with DNA or enzymes.

Andreas K. Rossos

Dr. Andreas K. Rossos: Dr Rossos graduated from the School of Chemical Engineering of the National Technical University of Athens, Greece, from where he was also awarded an M.Sc. degree in Materials Science and Technology. He obtained his PhD from the Department of Materials Science Engineering, University of Ioannina, Greece. Dr Rossos worked as Postdoctoral Researcher in the team of Professor Dwayne Miller in the field of Atomically Resolved Dynamics, Center for Free-Electron Laser Science (CFEL) of the Max Plank Institute for the Structure and Dynamic of Matter, Hamburg, Germany.

Preface to "Antimicrobial Materials with Medical Applications"

This Special Issue of the journal *IJMS* entitled "Antimicrobial Materials with Medical Applications"covers a selection of recent research and review articles in the field of antimicrobial materials and their medical applications.

Infectious diseases are a continuous threat to human health. New methods for the appropriate use of disinfectants and antibiotics have been developed to reduce the microbial activity, associated infections and the increase in antimicrobial resistance. Thus, the healthcare sector is facing a totally new challenge. Potential and promising weapons against bacterial growth and the development of multi-drug-resistant bacteria have been found in new antimicrobial materials.

Research on the design and development of new antimicrobial materials and their medical applications (such as antimicrobial surfaces, medical devices, contact lens, package materials, etc.), bring together stakeholders from different disciplines. The reader of this Special Issue will gain an appreciation of the real role of antimicrobial materials and their medical applications.

Sotiris K. Hadjikakou, Christina N. Banti, and Andreas K. Rossos
Editors

International Journal of
Molecular Sciences

MDPI

Editorial

Antimicrobial Materials with Medical Applications

Christina N. Banti and Sotiris K. Hadjikakou *

Laboratory of Biological Inorganic Chemistry, Department of Chemistry, University of Ioannina,
45110 Ioannina, Greece; cbanti@uoi.gr
* Correspondence: shadjika@uoi.gr

Citation: Banti, C.N.; Hadjikakou, S.K. Antimicrobial Materials with Medical Applications. *Int. J. Mol. Sci.* **2022**, 23, 1890. https://doi.org/10.3390/ijms23031890

Received: 12 January 2022
Accepted: 3 February 2022
Published: 8 February 2022

Publisher's Note: MDPI stays neutral with regard to jurisdictional claims in published maps and institutional affiliations.

This Special Issue of the *International Journal of Molecular Sciences*, entitled "Antimicrobial Materials with Medical Applications", covers a selection of recent research and review articles in the field of antimicrobial materials, as well as their medical applications. Moreover, it provides an overview of recent developments and the latest research in this increasingly diverse field. It also presents, with particular emphasis, the applications of new antimicrobial surfaces, medical devices, contact lens, packaging materials, etc.

Infectious diseases are a continuous threat to human health. New methods for the appropriate use of disinfectants and antibiotics have been developed to reduce microbial activity, associated infections, and increases in antimicrobial resistance. To overcome microbial infections and antimicrobial resistance, various antimicrobial materials, including small molecules and macromolecules, and inorganic and organic agents, have been developed and evaluated. Thus, the healthcare sector is facing totally new challenges. Potential and promising weapons against bacterial growth and the development of multi-drug resistant bacteria have been found in new antimicrobial materials. The development of new long-term or permanent antimicrobial materials, which go beyond the resistance of microbes to modern antibiotics is a research, technological, and financial issue of great importance. During the preparation of this Special Issue, the current worldwide public health crisis of COVID-19 particularly highlighted the emergent need for materials that inactivate on contact, not only microbes, but also viruses, further emphasizing the importance of the development of new antimicrobial and antiviral materials.

This Special Issue is composed of thirteen articles that are briefly reviewed below.

Jain et al. reviewed recent advances in green synthesis, in the context of the physico-chemical and biological properties of green silver nanoparticles [1]. Coelho et al. reviewed the effects of different cavity disinfectants on bond strength and the clinical success of composites and glass ionomer restorations of primary teeth [2]. van Hengel et al. presented a review on the biomaterial properties, antibacterial behavior, and biocompatibility of titanium implants that were biofunctionalized by plasma electrolytic oxidation (PEO) using Ag, Cu, and Zn [3]. Coelho et al. conducted a review on the effects of different cavity disinfectants on restoration adhesion and clinical success [4]. Meretoudi et al. dispersed silver nanoparticles (AgNPs(ORLE)) of oregano leaf extract (ORLE) in polymer hydrogels (pHEMA@ORLE_2 and pHEMA@AgNPs(ORLE)_2) using hydroxyethyl–methacrylate (HEMA). The materials were characterized and the antimicrobial activity of the materials was investigated against Gram-negative or Gram-positive bacteria strains [5]. Hung et al. synthesized curcumin analogs, and tested their antibacterial activity against Gram-positive aerobic bacteria [6]. Bidossi et al. investigated the in vitro ability of antibiotic-eluting hydroxyapatite/calcium sulfate bone graft substitute to prevent bacterial adhesion and biofilm formation by clinically relevant microorganisms [7]. Piszczek et al. assessed the microbiocidal activity of tri- and tetranuclear oxo-titanium(IV) complexes, which were dispersed in a poly(methyl methacrylate) matrix [8]. Marinas et al. extracted and characterized cellulose from *Gleditsia triacanthos* pods, and used it to fabricate a wound dressing. Moreover, the antioxidant properties and the antimicrobial activities of these materials were evaluated [9]. Li et al. proposed a simple and eco-friendly strategy to efficiently assemble zinc oxide nanoparticles and silver nanoparticles on sericin–agarose composite film to impart superior antimicrobial activity [10]. Khan et al. presented the green synthesis of chromium oxide nanoparticles using a leaf extract of *Abutilon indicum (L.) Sweet* as a reducing and capping agent. The biological activities were also evaluated [11]. Gouyau et al. prepared, synthesized, and tested the antibacterial activity of 12 nm gold and silver nanoparticles [12]. Tuñón-Molina et al. developed a single-use transparent antimicrobial face shield composed of polyethylene terephthalate and an antimicrobial coating of benzalkonium chloride for facial protective equipment [13].

Conflicts of Interest: The authors declare no conflict of interest.

References

1. Jain, A.S.; Pawar, P.S.; Sarkar, A.; Junnuthula, V.; Dyawanapelly, S. Bionanofactories for Green Synthesis of Silver Nanoparticles: Toward Antimicrobial Applications. *Int. J. Mol. Sci.* **2021**, *22*, 11993. [CrossRef] [PubMed]
2. Coelho, A.; Amaro, I.; Apolónio, A.; Paula, A.; Saraiva, J.; Ferreira, M.M.; Marto, C.M.; Carrilho, E. Effect of Cavity Disinfectants on Adhesion to Primary Teeth—A Systematic Review. *Int. J. Mol. Sci.* **2021**, *22*, 4398. [CrossRef] [PubMed]
3. Van Hengel, I.A.J.; Tierolf, M.W.A.M.; Fratila-Apachitei, L.E.; Apachitei, I.; Zadpoor, A.A. Antibacterial Titanium Implants Biofunctionalized by Plasma Electrolytic Oxidation with Silver, Zinc, and Copper: A Systematic Review. *Int. J. Mol. Sci.* **2021**, *22*, 3800. [CrossRef] [PubMed]
4. Coelho, A.; Amaro, I.; Rascão, B.; Marcelino, I.; Paula, A.; Saraiva, J.; Spagnuolo, G.; Marques Ferreira, M.; Miguel Marto, C.; Carrilho, E. Effect of Cavity Disinfectants on Dentin Bond Strength and Clinical Success of Composite Restorations—A Systematic Review of In Vitro, In Situ and Clinical Studies. *Int. J. Mol. Sci.* **2021**, *22*, 353. [CrossRef] [PubMed]
5. Meretoudi, A.; Banti, C.N.; Raptis, P.K.; Papachristodoulou, C.; Kourkoumelis, N.; Ikiades, A.A.; Zoumpoulakis, P.; Mavromoustakos, T.; Hadjikakou, S.K. Silver Nanoparticles from Oregano Leaves' Extracts as Antimicrobial Components for Non-Infected Hydrogel Contact Lenses. *Int. J. Mol. Sci.* **2021**, *22*, 3539. [CrossRef] [PubMed]
6. Hung, S.-J.; Hong, Y.-A.; Lin, K.-Y.; Hua, Y.-W.; Kuo, C.-J.; Hu, A.; Shih, T.-L.; Chen, H.-P. Efficient Photodynamic Killing of Gram-Positive Bacteria by Synthetic Curcuminoids. *Int. J. Mol. Sci.* **2020**, *21*, 9024. [CrossRef] [PubMed]
7. Bidossi, A.; Bottagisio, M.; Logoluso, N.; De Vecchi, E. In Vitro Evaluation of Gentamicin or Vancomycin Containing Bone Graft Substitute in the Prevention of Orthopedic Implant-Related Infections. *Int. J. Mol. Sci.* **2020**, *21*, 9250. [CrossRef] [PubMed]
8. Piszczek, P.; Kubiak, B.; Golińska, P.; Radtke, A. Oxo-Titanium(IV) Complex/Polymer Composites—Synthesis, Spectroscopic Characterization and Antimicrobial Activity Test. *Int. J. Mol. Sci.* **2020**, *21*, 9663. [CrossRef] [PubMed]
9. Marinas, I.C.; Oprea, E.; Geana, E.-I.; Tutunaru, O.; Pircalabioru, G.G.; Zgura, I.; Chifiriuc, M.C. Valorization of Gleditsia triacanthos Invasive Plant Cellulose Microfibers and Phenolic Compounds for Obtaining Multi-Functional Wound Dressings with Antimicrobial and Antioxidant Properties. *Int. J. Mol. Sci.* **2021**, *22*, 33. [CrossRef] [PubMed]
10. Li, W.; Huang, Z.; Cai, R.; Yang, W.; He, H.; Wang, Y. Rational Design of Ag/ZnO Hybrid Nanoparticles on Sericin/Agarose Composite Film for Enhanced Antimicrobial Applications. *Int. J. Mol. Sci.* **2021**, *22*, 105. [CrossRef] [PubMed]
11. Khan, S.A.; Shahid, S.; Hanif, S.; Almoallim, H.S.; Alharbi, S.A.; Sellami, H. Green Synthesis of Chromium Oxide Nanoparticles for Antibacterial, Antioxidant Anticancer, and Biocompatibility Activities. *Int. J. Mol. Sci.* **2021**, *22*, 502. [CrossRef] [PubMed]
12. Gouyau, J.; Duval, R.E.; Boudier, A.; Lamouroux, E. Investigation of Nanoparticle Metallic Core Antibacterial Activity: Gold and Silver Nanoparticles against Escherichia coli and Staphylococcus aureus. *Int. J. Mol. Sci.* **2021**, *22*, 1905. [CrossRef] [PubMed]
13. Tuñón-Molina, A.; Martí, M.; Muramoto, Y.; Noda, T.; Takayama, K.; Serrano-Aroca, Á. Antimicrobial Face Shield: Next Generation of Facial Protective Equipment against SARS-CoV-2 and Multidrug-Resistant Bacteria. *Int. J. Mol. Sci.* **2021**, *22*, 9518. [CrossRef] [PubMed]

International Journal of
Molecular Sciences

Review

Bionanofactories for Green Synthesis of Silver Nanoparticles: Toward Antimicrobial Applications

Ashvi Sanjay Jain [1,†], Pranita Subhash Pawar [1,†], Aira Sarkar [2], Vijayabhaskarreddy Junnuthula [3,*] and Sathish Dyawanapelly [1,*]

[1] Department of Pharmaceutical Sciences & Technology, Institute of Chemical Technology, Nathalal Parekh Marg, Matunga, Mumbai 400019, India; ashvisanjay001@gmail.com (A.S.J.); 18phtps.pawar@ug.ictmumbai.edu.in (P.S.P.)
[2] Chemical and Biomolecular Engineering, Johns Hopkins University, Baltimore, MD 21218, USA; asarka11@jhu.edu
[3] Drug Research Program, Faculty of Pharmacy, University of Helsinki, Viikinkaari 5 E, 00790 Helsinki, Finland
* Correspondence: junnuthula.vijayabhaskarreddy@helsinki.fi (V.J.); sa.dyawanapelly@ictmumbai.edu.in (S.D.)
† These authors are contributed equally to this work.

Abstract: Among the various types of nanoparticles and their strategy for synthesis, the green synthesis of silver nanoparticles has gained much attention in the biomedical, cellular imaging, cosmetics, drug delivery, food, and agrochemical industries due to their unique physicochemical and biological properties. The green synthesis strategies incorporate the use of plant extracts, living organisms, or biomolecules as bioreducing and biocapping agents, also known as bionanofactories for the synthesis of nanoparticles. The use of green chemistry is ecofriendly, biocompatible, nontoxic, and cost-effective. We shed light on the recent advances in green synthesis and physicochemical properties of green silver nanoparticles by considering the outcomes from recent studies applying SEM, TEM, AFM, UV/Vis spectrophotometry, FTIR, and XRD techniques. Furthermore, we cover the antibacterial, antifungal, and antiparasitic activities of silver nanoparticles.

Keywords: bioreduction; biocapping agent; bionanofactories; biomedical; green synthesis; silver nanoparticles

Citation: Jain, A.S.; Pawar, P.S.; Sarkar, A.; Junnuthula, V.; Dyawanapelly, S. Bionanofactories for Green Synthesis of Silver Nanoparticles: Toward Antimicrobial Applications. *Int. J. Mol. Sci.* **2021**, *22*, 11993. https://doi.org/10.3390/ijms222111993

Academic Editors: Sotiris Hadjikakou, Christina N. Banti and Andreas K. Rossos

Received: 16 September 2021
Accepted: 3 November 2021
Published: 5 November 2021

Publisher's Note: MDPI stays neutral with regard to jurisdictional claims in published maps and institutional affiliations.

1. Introduction

Nanotechnology is coming into focus owing to its plethora of applications that can be elucidated as the manipulation of a material using several procedures to create matters with some desired specific properties. It usually involves particles possessing dimensions from 1–100 nm [1]. Their appreciable surface-area-to-volume ratio is the most significant attribute responsible for their extensive use in electronics, nanomedicine, biomaterials, and food [2]. Various physicochemical pathways are employed for the fabrication of nanoparticles (NPs) in the industry. These involve chemical reduction, chemical solution deposition, the sol–gel process, photochemical reduction, and electrochemical reduction. Other methods include laser desorption, sputter deposition, lithographic techniques, layer-by-layer growth, the Langmuir–Blodgett method, the hydrolysis coprecipitation method, the wet chemical method, and catalytic routes [3]. These synthetic approaches compel one to move toward the application of reactive and toxic reducing agents and stabilizing agents, as well as high radiation. However, due to the toxicity associated with them, they adversely affect living organisms, as well as the environment, thus posing various limitations to their use [4]. Hence, to accomplish our needs, over the past decade, efforts have been globally made to reduce the generation of hazardous waste. Green synthesis can be accomplished by adopting the 12 fundamental postulates of "green chemistry" proposed by Anastas and Warner and integrating them with modern developments. This can reduce the use of harmful chemicals and elevate the efficiency of the process [3,5].

Hence, green synthesis is a possible alternative to harsh physical and chemical operations. The use of nontoxic solvents [6–9] and sustainable materials represents crucial components that need attention in this ecofriendly approach [10]. Several factors influence the choice of the green approach over conventional methods. The term "green" is not the color, but the concept of synthesizing nanoparticles from metal salts by exploiting the reducing property of biologically active compounds. These biologically active compounds may be obtained from microorganisms (both live and dead), herbal extracts (from leaf, root, whole body, flower, fruit, bark, latex, etc.), and animal extracts [11]. Nanoparticles derived from biological materials are known as biogenic nanoparticles, and the involved synthesis process is known as the green synthesis of nanoparticles [2]. This concept of NP synthesis was earliest proposed by Raveendran et al. by employing β-D-glucose as a reducing agent and starch as the capping agent in the preparation of silver nanoparticles (AgNPs) [5]. The specific antimicrobial mechanisms of AgNPs still remain unknown. According to research investigations, the expected antimicrobial activity of silver nanoparticles are proposed in Figure 1. AgNPs releases Ag$^+$ ions, which can accumulate on the cell wall and cell membranes of microorganisms and further enter into cytoplasm. Inside the cell, Ag$^+$ ions generate reactive oxygen species (ROS), which are the key agent for antimicrobial activity, involving (1) inhibition of DNA synthesis, (2) inhibition of mRNA synthesis, (3) cell membrane destruction and the leakage of the cell constituents, (4) inhibition of protein synthesis, (5) inhibition of cell-wall synthesis (6) mitochondrial damage, and (7) inhibition of the electron transport chain. These effects eventually lead to cell death. In addition to being able to release silver ions, AgNPs can themselves kill bacteria.

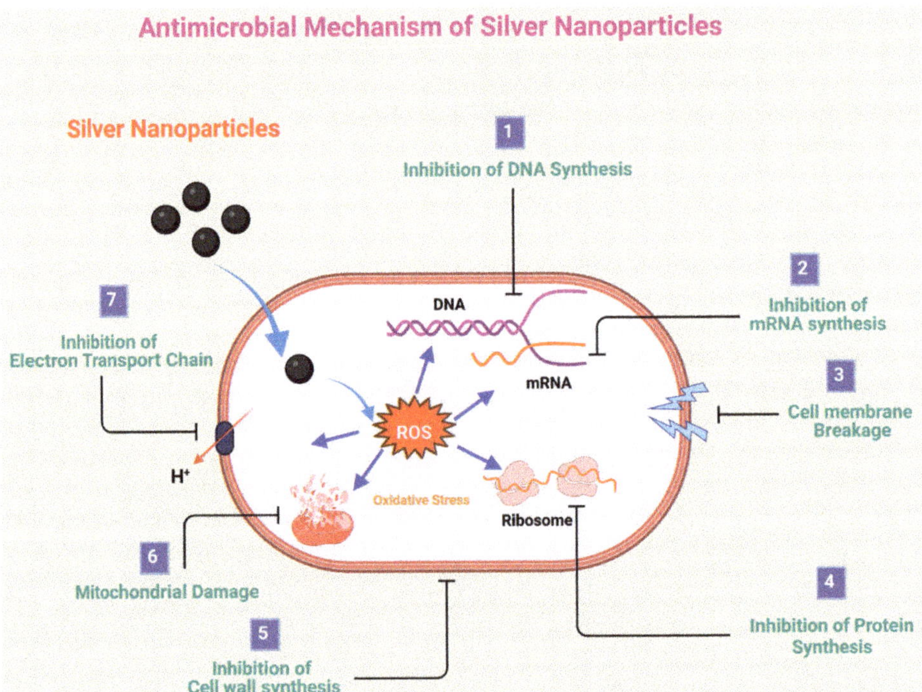

Figure 1. Antimicrobial mechanism of silver nanoparticles: (1) inhibition of DNA synthesis, (2) inhibition of mRNA synthesis, (3) cell membrane destruction and the leakage of the cell constituents, (4) inhibition of protein synthesis, (5) inhibition of cell-wall synthesis, (6) mitochondrial damage, and (7) inhibition of electron transport chain.

The antimicrobial properties of AgNPs are well reported in the literature; however, the safety of these particles and surface cytotoxicity issues in living cells are of a concern for their use in biomedical applications. Recently, Barbalinardo at al. studied the surface cytotoxicity of AgNPs after surface modulation using an oligo (ethylene glycol)-based ligand (11-mercaptoundecyl)hexa(ethylene glycol(EG6OH)) and concluded that the cytotoxicity of nanoparticles was reduced with an increase in ligand density; they further stated that rational design and engineering could potentially minimize the side-effects of the AgNPs [12]. Furthermore, EG6OH-coated AgNPs were not internalized and did not show any cytotoxicity in mouse embryonic fibroblast (NIH-3T3) cells [13]. These studies revealed that the relationship among surface coatings, ligand density, and protein corona formation characteristics are useful in modulating surface properties for better use of AgNPs.

The green chemistry perspective involves the three main steps generally involved in the preparation of nanoparticles. These include the selection of a solvent medium, focusing on the selection of a "green" alternative for the reducing agent and a harmless substance to stabilize the NPs, as a vast number of conventional methods depend on organic solvents, majorly contributing to the hydrophobic properties of the capping materials involved in the process. Another concern is the selection of a reducing agent. Most processes reported until now mostly employ hydrazine, dimethyl formamide (DMF), and sodium borohydride (NaBH$_4$) as reducing agents, although they are highly reactive and pose various environmental issues [5]. Hence, the chemical reduction methods can be replaced by biogenic reduction, which is a "bottom-up" technique, wherein the extract of a natural product that possesses innate properties of stabilizing, growth-terminating and capping of NPs replaces the harmful and toxic reducing agents [11]. Furthermore, this method seems to be more atom-efficient as the particles are built atom-by-atom in the process and do not require the use of protection/deprotection processes used in a traditional organic approach [10]. The final challenge to be solved is to select an appropriate capping material that can be employed to make the nanoparticle surface unreactive (passivation). The choice of capping agent is influenced by various challenges that exist in the process of synthesis and vary with numerous factors. However, those substances utilized as reducing and stabilizing agents include proteins, enzymes, sugars, and certain phytochemicals such as flavonoids, terpenoids, and cofactors in these green synthesis methods [5]. This helps to produce NPs which are environmentally friendly, low-cost, and nonpolluting especially for healthcare and biology applications that demand high-grade purity [11]. Hence, green synthesis is advantageous as it is economical, environmentally friendly, and uncomplicated for large-scale synthesis as the plants and their extracts employed represent promising solutions due to their availability, suitability for mass production, and environmentally benign nature. Moreover, this approach does not demand employing various industrial processes for maintaining high temperature, pressure, or energy requirements [14]. Green NPs are synthesized in a one-step procedure that is advantageous for controlling and manipulating the crystal growth, stabilization, and particle size and shape. This single-step reduction technique requires a lower amount of energy for synthesis as the processes are operated at near-ambient temperature, pressure, and pH, which again follows the principles of green chemistry [3]. This also helps to prevent particle toxicity and reactivity toward our health and the environment, as a lack of predictability and composition ambiguity of the nanoparticles are not experienced. Moreover, expensive metal salts such as gold and silver can be recycled from the waste generated by applying these green fabrication strategies, thereby controlling the issue of limited reserves and high prices of these metals. Thus, green chemistry is aimed at thwarting waste, using sustainable materials, and employing techniques that lower the risk to living beings, which is accomplished via the green synthesis of nanoparticles [10].

2. Strategies for Green Synthesis of Silver Nanoparticles

Numerous classes of microorganisms and plants have been employed to successfully accomplish the green synthesis of NPs. These methods of synthesis have also garnered importance as an alternative strategy for the development of gold, silver, zinc, titanium, and palladium NPs. Several reviews have been reported on the methods for biosynthesis of NPs [15] that majorly deal with plants [16,17] microbes [18], marine organisms [19], and phototrophic eukaryotes. The potential microbial bionanofactories that can be utilized to synthesize NPs intra- or extracellularly include bacteria, algae, yeast, and fungi [20]. Different parts of the plant including leaves, stem, bark, root, fruit, and flower can be selected for this new ecofriendly approach of synthesis. This vast range of phytochemicals includes *Aloe vera* plant extract [20], *Mangifera indica* fruit extract [21], *Murraya koenigii* leaf [22], and others carbohydrates [23]. Glucose was also utilized to synthesize AgNPs along with the appearance of stabilizing agents such as soluble starch [5,24,25], sucrose, and maltose [26]. One of the very well-known plants is *Eucalyptus* [27], whereas *Bacillus methylotrophicus* [28] has also attracted interest in green synthesis. Ubiquitous plant materials such as *Coffee arabica* seeds [29] and *Azadirachta indica* [30] have also played a significant role in extensively utilizing plant materials in this process of green synthesis. Even peanut shells have been trialed for green silver NP synthesis [31]. Spices such as pepper leaf extract were reported in [32] as an example of the green synthesis of nanoparticles. Other synthesis methods utilizing various parts of plants are summarized throughout this review. The higher degree of safety and stability, as well as biocompatibility, offered by these green NPs is due to their surface being capped with nontoxic biomolecules. Moulton et al. noted that polyphenols as capping agents can potentially impart superior antioxidant effects to the synthesized AgNPs [33]. Furthermore, studies have found that the metabolites present in plant extracts, such as proteins [34] and chlorophyll [35], act as the capping agents for synthesized AgNPs. Figure 2 represents the strategies of green synthesis and the use of bionanofactories in the synthesis of silver nanoparticles, and their physicochemical properties are briefly outlined in Table 1.

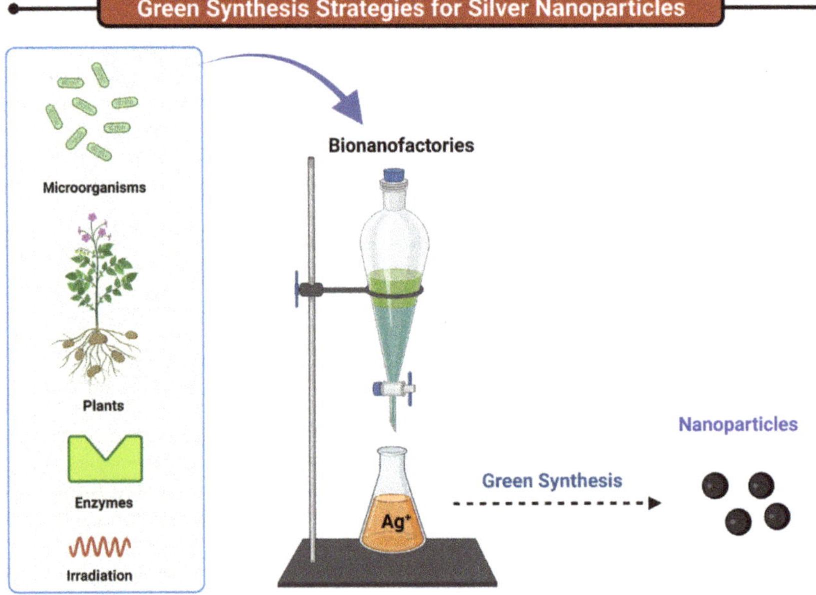

Figure 2. Green synthesis of nanoparticles of silver nanoparticles.

Table 1. Use of bionanofactories in the synthesis of silver nanoparticles, along with a brief outline of their physicochemical and antimicrobial properties.

Category of Biofactories	Reducing Agent	UV/Vis	FTIR (cm^{-1})	Structural Analysis	In Vitro Activity Studies	Ref.
Green Algae	*Ulva compressa* *Cladophora glomerata*	403 nm (<6 h), 443 nm (after 24 h)	3278, 1634, 1539	**AFM** • For AgNPs (*U. compressa*): Sa 1.01 nm; Sq 1.48 nm; Sz 9.09 nm. • For AgNPs (*C. glomerata*): Sa 0.471 nm; Sq 0.848 nm; Sz 5.90 nm. **XRD** • Crystalline • Structure—face-centered cubic (FCC) Particle size • AgNPs (*U. compressa*)—66.3 nm • AgNPs (*C. glomerata*)—81.8	• Antimicrobial efficacy against species such as *K. pneumoniae, P. aeruginosa, E. coli, E. faecium,* and *S. aureus*	[36]
	Spirogyra varians	420–430 nm	3423, 2927, 1645, 1515, 1429	**XRD** • Face-centered cubic (FCC) **SEM** • Average size ~35 nm • Shape—uniform and quasi-spherical	• Antibacterial *activity against* Staphylococcus aureus, Bacillus cereus, *Salmonella typhimurium,* Escherichia coli, and *Listeria monocytogenes*	[37]
Brown Algae	*Spatoglossum asperum*	440 nm	1638, 1034, 3447, 1034.95, 1384.14	**SEM** • Shape—spherical to oval • Size—32 to 51 nm **TEM** • Shape—mostly spherical • Size—20 to 46 nm **XRD** • Crystalline • Structure—face-centered cubic (FCC) **EDS** • Strong signal indicating presence of metallic silver	• Agar bioassay method showed reduction of bacterial colonies of *K. pneumoniae* with increasing concentration.	[38]

Table 1. *Cont.*

Category of Biofactories	Reducing Agent	UV/Vis	FTIR (cm^{-1})	Structural Analysis	In Vitro Activity Studies	Ref.
Red Algae	*Spyridia fusiformis*	450 nm	3907, 3779, 3410, 2927, 2853, 2593, 1644, 1416, 1170, 749	HR-TEM ▪ Shape—variable shapes such as spherical, triangle, pseudo-spherical, and some rounded rectangle shapes ▪ Most observed shape and size—spherical and 5 to 50 nm ▪ Average size of NPs—32.70 nm XRD ▪ Crystalline ▪ Structure—face-centered cubic (FCC)	▪ Antibacterial activity of AgNPs (concentration = 100 µg/mL) Maximum zone of inhibition: ▪ In *K. pneumoniae* = 26 mm ▪ In *S. aureus* = 24 mm	[39]
	Gracilaria corticata	424 nm; another peak at 220 nm maybe due to the presence of amide bond	2921, 1630, 1455	EDS ▪ Optical absorption peak observed at 3 keV is typical for the absorption of metallic silver XRD ▪ Crystalline ▪ Structure—Face-centered cubic (FCC) TEM ▪ Size range = 10–35 nm	▪ Cytotoxic activity—on Hep2 cell lines, IC$_{50}$ = 62.5 µg/mL	[40]

Table 1. *Cont.*

Category of Biofactories	Reducing Agent	UV/Vis	FTIR (cm^{-1})	Structural Analysis	In Vitro Activity Studies	Ref.
Blue-Green Algae	*Nostoc* sp.	419 nm	3443.96, 3385.61, 2923.83, 2853.32 1644.73	SEM ■ Shape—spherical ■ Average size 51–100 nm TEM ■ Spherical and well-dispersed. TEM-SAED pattern ■ Structure—face-centered cubic ■ Nature—crystalline XRD ■ Nature—crystalline	■ No significant cytotoxicity against MCF-7 breast cancer cells at lower concentration; cytotoxicity increased with increasing concentration from 0 µL/mL to 50 µL/mL ■ Good antibacterial and antifungal activity	[41]
	Anabaena sp.	420 nm	3383.72, 2930.60, 1651.13, 1076.58	TEM ■ Size: 10–50 nm XRD ■ Nature—crystalline ■ Structure—face-centered cubic (FCC) ■ Calculated particle size—20 nm	■ Ag-CNP treatment inhibited growth of tumor and cancer cells and induced apoptosis	[42]
Actinomycetes	*Streptacidiphilus durhamensis*	-	3421, 1384.4, 1623, 1480	TEM ■ Spherical shape EDS ■ Indicated Ag as the major element with a ~3 keV signal ■ Heterogeneous particle size distribution observed in the range of ~8–48 nm.	■ Antimicrobial activity against the tested strains such as *S. aureus*, *B. subtilis*, *E. coli*, *P. aeruginosa*, *K. pneumoniae*, and *P. mirabilis* but not *S. infantis* ■ Synergistic effects of bio(AgNPs) with various standard antibiotics	[43]

Table 1. *Cont.*

Category of Biofactories	Reducing Agent	UV/Vis	FTIR (cm^{-1})	Structural Analysis	In Vitro Activity Studies	Ref.
	Streptomyces rochei	410 nm	3420.14, 2932.23, 2362.37, 1639.20, 1430.92, 1115.62, 613.252	EDS ▪ Optical absorption peak for AgNPs observed at 3.5 keV SEM ▪ Size: 22 to 85 nm	▪ AgNPs exhibited synergistic effects with antibiotics such as ciprofloxacin, ampicillin, streptomycin, gentamicin, tetracycline and lincomycin ▪ AgNPs reduced the density of bacterial cells and acted as an antibiofouling agent	[44]
	Streptomyces sp.	425 nm	3695.61, 1585.49, 1398.39, 1151.50, 1068.56	XRD ▪ Nature—crystalline ▪ Structure—face-centered cubic (FCC) SEM ▪ Size range = 21–45 nm	▪ As compared to the cell-free supernatant, synthesized AgNPs showed high anticandidal activity	[45]
	Nocardiopsis sp. MBRC-1	420 nm	3440, 2923, 2853, 1655, 1460, 685	SEM ▪ Average particle size ~45 ± 0.05 nm FE-SEM and EDS ▪ The optical absorption peak was observed at 3 keV TEM ▪ Size range—30–90 nm ▪ Average particle size—45 ± 0.15 nm ▪ Shape—spherical	▪ Excellent antimicrobial activity by AgNPs observed against *Bacillus subtilis*, *Pseudomonas aeruginosa*, and *Candida albicans* ▪ Cytotoxicity studies against HeLa cancer cell lines, IC$_{50}$ = 200 μg/mL	[46]

Table 1. *Cont.*

Category of Biofactories	Reducing Agent	UV/Vis	FTIR (cm^{-1})	Structural Analysis	In Vitro Activity Studies	Ref.
Fungi	*Arthroderma fulvum*	420 nm	-	XRD ■ Nature—crystalline ■ Structure—face-centered cubic (FCC) TEM ■ Shape—spherical ■ Average diameter—15.5 ± 2.5 nm Particle size analysis ■ Average diameter = 20.56 nm	■ Potential antifungal activity against fungi such as *Candida* spp., *Aspergillus* spp., and *Fusarium* spp. observed ■ Compared to antifungal drugs such as itraconazole and fluconazole, the biosynthesized AgNPs at concentrations near to 1 mg/mL showed a broader antifungal spectrum	[47]
	Mushroom *Pleurotus ostreatus*	400–470 nm	3318, 2944, 1612, 1411	HR-TEM and FE-SEM ■ Shape—spherical in shape ■ Average size range—10–40 nm Size distribution analysis ■ Average size—28 nm EDS analysis ■ 13% of Ag and rest presence of C & O recorded at 3 keV	■ Antibacterial activity against *B. subtilis, B. cereus, S. aureus, E. coli,* and *P. aeruginosa* ■ Bactericidal activity observed against *B. cereus, E. coli,* and *P. aeruginosa*	[48]

Table 1. *Cont.*

Category of Biofactories	Reducing Agent	UV/Vis	FTIR (cm^{-1})	Structural Analysis	In Vitro Activity Studies	Ref.
	Raphanus sativus	426 nm	3145, 1597, 1402, 1109, 1213, 995, 911, 699, 504	XRD • Structure—face-centered cubic (FCC) • Calculated mean size—~25 nm. TEM • Shape—spherical • Size range—10–30 nm SAED • Nature—crystalline EDS • Strong silver peaks at 3 keV AFM • Monodispersed AgNPs, • Average particle size 4 to 28 nm	• Antibacterial activity against human pathogenic bacteria such as *Bacillus subtilis, Staphylococcus aureus, Escherichia coli,* and *Serratia marcescens*	[49]
	Endophytic fungus *Curvularialunata*	422 nm	3430.86, 1573.16, 1483.37, 1402.84, 1260.95, 1123.70	SEM • Shape—spherical • Size (diameter) range—10 to 50 nm • Average size—26 nm EDS • Strong metal signal peak of Ag observed XRD • Structure—face-centered cubic (FCC)	• Synthesized AgNPs, along with antibiotics, exhibited inhibitory activity against Gram-negative and Gram-positive bacterial pathogens such as *E. coli, Pseudomonas aeruginosa, Salmonella paratyphi, Bacillus subtillis, Staphylococcus aureus,* and *Bacillus cereus*	[50]

Table 1. *Cont.*

Category of Biofactories	Reducing Agent	UV/Vis	FTIR (cm^{-1})	Structural Analysis	In Vitro Activity Studies	Ref.
	Penicillium polonicum	430 nm	2920.23, 2850.79, 1747.51, 1508.33, 1473.62, 1338.60, 1361.74, 1240.16, 1035.77, 760.88	TEM ■ Common shape—spherical ■ Size range—10 to 15 nm ■ Above size of 30 nm, hexagonal NPs HR-TEM ■ Nature—crystalline SAED ■ Nature—crystalline Particle size analysis ■ Polydisperse AgNPs in the size range of 10–15 nm observed EDS ■ Absorption peak recorded at 3 keV	■ Killing kinetic assay depicted that complete killing of *A. baumanii* bacterial cells occurred within 6 h of exposure time to AgNPs	[51]
Bacteria	*Pseudomonas deceptionensis* DC5	428 nm	-	XRD ■ Nature—crystalline FE-TEM ■ Shape—spherical ■ Size range—10 to 30 nm EDS ■ A peak recorded at 3 keV Particle size analysis ■ Average particles size—127 nm	■ Activity efficiency observed in descending order against pathogens *V. parahaemolyticus*, *C. albicans*, *S. aureus*, *S. enterica*, and *B. anthracis* ■ AgNPs at concentration of 5 μg/L found to inhibit biofilm formed by *S. aureus* and *P. aeruginosa*	[52]

Table 1. *Cont.*

Category of Biofactories	Reducing Agent	UV/Vis	FTIR (cm^{-1})	Structural Analysis	In Vitro Activity Studies	Ref.
	Weissellaoryzae DC6-	432 nm	-	FE-TEM ▪ Shape—spherical ▪ Size range—10 to 30 nm EDS ▪ Highest peak recorded at 3 keV XRD ▪ Nature—crystalline Particle size analysis ▪ Average particle size—150.2 nm ▪ Polydispersity index (PDI)—0.176	▪ Descending order of antimicrobial potential observed against *S. aureus*, *C. albicans*, *B. cereus*, *V. parahaemolyticus*, *E. coli*, and *B. anthracis* ▪ AgNPs at a concentration of about 5–6 μg found to inhibit the biofilm formed by *S. aureus* and *P. aeruginosa*	[53]
	Bacillus thuringiensis	413 nm	1644, 1549, 1520, 1114, 564, 550, 546, 523	FE-SEM ▪ Shape—spherical ▪ Average diameter range—10 to 30 nm TEM ▪ Size range 10 to 30 nm	▪ Purified AgNPs showed relatively stronger antibacterial activity against *E. coli* than the commercially available AgNPs	[54]
	Halotolerant *Bacillus endophyticus* SCU-L	420 nm	3400, 2969, 1650, 1560, 1453, 1401, 1227, 1083	XRD ▪ Structure—face-centered cubic TEM ▪ Shape—spherical ▪ Average size ~5.1 nm	▪ Antimicrobial activity observed against *C. albicans*, *E. coli*, *S. typhi*, and *S. aureus* ▪ AgNPs showed broad-spectrum antimicrobial activity against both Gram-positive and Gram-negative pathogens, as well as a fungus strain	[55]
	Phenerochaete chrysosporium (MTCC-787)	430 nm	767, 1642, 2137,3400	TEM ▪ Shape—different shapes, such as spherical and oval ▪ Size range—34 to 90 nmAFM ▪ Agglomerated silver nanostructures	▪ Gram-negative clinical pathogens showed a higher susceptibility to AgNPs than Gram-positive pathogens	[56]

Table 1. *Cont.*

Category of Biofactories	Reducing Agent	UV/Vis	FTIR (cm^{-1})	Structural Analysis	In Vitro Activity Studies	Ref.
Plants (Roots)	*Delphinium denudatum*	416 nm	3354, 2952, 2063, 1651, 1419, 1383, 1354, 1171, 1093, 780, 672, 605	XRD ■ Structure—face-centered cubic (FCC) ■ Nature—crystalline FE SEM ■ Spherical shape and size not more than 85 nm	■ Antibacterial activity observed against *S. aureus*, *B. cereus* NCIM 2106, *E. coli*, and *P. aeruginosa* ATCC ■ Susceptibility of dengue vector *aegypti* larvae to AgNPs increased when exposure time extended to 48 h; the LC$_{50}$ values of AgNPs were 96 ppm (24 h) and 9.6 ppm (48 h) against second-instar larvae of *A. aegypti*	[56]
	Alpinia katsumadai	417 nm and during reaction 436 nm	3371–3377, 2980–2978, 1649–1653, 1389–1385, 1045–1049, 888-881	FE-TEM ■ Shape—quasi-spherical ■ Nature—well-dispersed and scattered EDS ■ Strong absorption peak recorded at 3 keV XRD ■ Structure—face-centered cubic ■ Nature—crystalline	■ Antibacterial activity observed against *S. aureus*, *E. coli*, and *P. aeruginosa*; thus, it was concluded that so-prepared AgNPs exhibited effective antioxidant, antibacterial, and anticancer activities	[57]
	Aloe vera leaves	420 nm	-	SEM ■ Shape—spherical ■ Size range—70.7 to 192.02 nm (size varied with temperature) XRD ■ Structure—face-centered cubic (FCC)	■ Minimal cytotoxicity to human PBMCs	[58]

Table 1. *Cont.*

Category of Biofactories	Reducing Agent	UV/Vis	FTIR (cm^{-1})	Structural Analysis	In Vitro Activity Studies	Ref.
	Piper nigrum leaf and stem	460 nm	3697, 3313, 3195, 2298, 1670, 1456, 1336, 1193, 1118, 811, 750, 651, 601	XRD ■ Structure—face-centered cubic ■ Nature—crystalline SEM ■ Shape—spherical EDS ■ Intense signal at 3 keV TEM ■ Stem extracts of *P. nigrum* 9 to 30 nm ■ Leaf extracts of *P. nigrum* - Small-sized AgNPs: 4 to 14 nm - Large-sized AgNPs: 20 to 50 nm	■ Antibacterial activity—Stem and leaf synthesized AgNPs (at 50 µL) showed activity against *Citrobacter freundii* and *Erwinia cacticida*	[59]
Plants (Fruit and Peel)	Banana peel	430nm	2353–2351, 1732–1755, 1640–1643, 1532–1537, 1445–1454	SEM–EDS ■ A distinct signal and high atomic percent values for silver were obtained XRD ■ Structure—face-centered cubic (FCC) ■ Nature—crystalline	■ AgNPs exhibited potent antifungal activity against the tested pathogenic strains of *C. albicans* and *C. lipolytica* ■ The antibacterial activity of AgNPs was observed against *E. coli*, E. aerogenes, *Klebsiella* sp., and *Shigella* sp.	[60]

Table 1. *Cont.*

Category of Biofactories	Reducing Agent	UV/Vis	FTIR (cm^{-1})	Structural Analysis	In Vitro Activity Studies	Ref.
	Tribulus terrestris dried fruit	435 nm	-	XRD ■ Nature—crystalline AFM ■ Shape—spherical ■ Particle size ~24.631 nm TEM ■ Shape—spherical shape ■ Average size—22 nm	■ Antimicrobial activity against *S. pyogens*, *S. aureus*, *B. subtilis*, *P. aeruginosa*, and *E. coli*.	[61]
	Lemon	400–430 nm	-	AFM ■ Particle dimensions—height 12 nm, width 100 nm ■ SEM—NPs consisted of agglomerates of small grains with diameter of approximately 75 nm	■ Disc diffusion method showed that NPs reduced the growth of both *E. coli* and *Bacillus subtilis*	[62]

Table 1. *Cont.*

Category of Biofactories	Reducing Agent	UV/Vis	FTIR (cm^{-1})	Structural Analysis	In Vitro Activity Studies	Ref.
Other biosynthesizing Agents	Gum kondagogu (*Cochlospermum gossypium*)	416 nm	3443, 2916, 2850, 1727, 1630, 1597, 1384, 1351, 1254, 1148, 1043	TEM ■ Shape—anisotropic nanostructures such as nanotriangles, a few nanorods, hexagonal and polygonal nanoprisms, and abundant unevenly shaped nanoparticles were observed ■ Nature—polydisperse ■ For 30 min of reaction time, size of 55.0 nm; for 60 min of reaction time, size of 18.9 nm SAED ■ Nature—crystalline ■ Structure—face-centered cubic (FCC) TEM ■ Shape—spherical ■ Average particle size: for 30 min of reaction time, 11.2 nm; for 60 min of reaction time, 4.5nm XRD ■ Nature—crystalline ■ Structure—face-centered cubic	■ Antibacterial activity was observed against *S. aureus* ATCC 25923, *E. coli* ATCC 25922, *E. coli* ATCC 35218, and *P. aeruginosa* ATCC 27853	[63]

Table 1. *Cont.*

Category of Biofactories	Reducing Agent	UV/Vis	FTIR (cm^{-1})	Structural Analysis	In Vitro Activity Studies	Ref.
	Dextran T40	423 nm	-	AFM • Particle size range—10 to 60 nm TEM • Shape—spherical • Size ~5–10 nm SAED • Nature—crystalline. EDS • Optical absorption peak at 3 keV XRD • Structure—face-centered cubic (FCC)	• Antimicrobial activity observed against *B. subtilis*, *B. cereus*, *E. coli*, *S. aureus*, and *P. aeruginosa*.	[64]
	Casein (milk protein)	400 to 500 nm	1644, 1514	SEM and TEM • Shape—spherical agglomerates formed upon carefully decreasing the pH to 3.32 • Size—average diameter of about 60 to 80 nm	• AgNPs at a dose of 0.025 µg/mL, i.e., below LD$_{50}$ value, was observed to be fairly distributed in cytoplasm of living cells imaged by CLSM	[65]

Sa—average roughness, Sq—root-mean-square roughness, Sz—ten-point height.

2.1. Phytosynthesis

Plants have always been exploited by humans since the Stone Age for their metabolites, which have proven to be a pillar of human survival. Similarly, there have been numerous experiments for this emerging method of NP synthesis that can enhance the potential applications of the plants and their extracts in this field. Many studies have shown that phytoconstituents such as flavonoids, terpenoids, pectin, sugars, ascorbic acid, and carotenoids present in powders or extracts of roots, shoots, bark, leaves, peel, flowers, and fruits can function as reducing and capping agents to develop NPs [66].

For in vitro green synthesis, the required chemical metabolites are first extracted from plant organs and then suitably incubated with NP precursors to produce NPs. The obtained NPs are subject to centrifugation and washing, allowing them to be collected. Furthermore, the NPs are characterized by employing various methods, and studies for analyzing the release of Ag^+ from the AgNPs are also conducted. In one such study mentioned [67], it was observed that, after entering the aquatic environment, AgNPs would release silver ions, which would decrease the stability of the AgNPs. Furthermore, Lee et al. suggested that the release of Ag ions follows first-order kinetics [68]. There are several factors affecting the release rates of Ag ions that must be considered while evaluating these synthesized NPs; they mainly include particle size, environmental factors, e.g., pH, temperature, and dissolved oxygen [69,70], and capping agents [71]. However, it has been observed that silver ions exhibit different physiochemical properties and biological toxicity from the synthesized AgNPs; hence, detailed studies are necessary before these AgNPs are put into real-life application [69]. In another interesting study, AgNP synthesis was reported by Forough et al. utilizing two plants, wherein an aqueous extract of soap-root (*Acanthe phylum bracteatum*) and an aqueous manna extract of *Hedysarum* were employed as the stabilizing agent and reducing agent, respectively. Manna has been widely used in Asia as it possesses laxative properties [72].

2.1.1. Extracts of Roots

The synthesis of metallic nanoparticles using *Medicago sativa* [73] is perhaps one of the earliest records on the generation of AgNPs utilizing a plant part as a source. Alfalfa roots absorb the reduced silver (Ag^+ to Ag^0) from agar medium and transmit it to the shoots in the identical oxidation state (Ag^0). Then, Ag atoms in the shoots organize themselves by joining together and forming larger arrangements to produce NPs. TEM/STEM analysis displayed the aggregation of Ag atoms in the interior of the plant tissue, whereby they underwent nucleation and NP formation. An aqueous root extract of *Parthenium hysterophorus* has been employed to reduce silver ions and synthesize stable green NPs, which further showed larvicidal activity toward *Culex quinquefasciatus* in mosquito control [74]. Figure 3 shows the bioreduction of silver ions into silver nanoparticles.

2.1.2. Extracts of Seeds

To illustrate AgNP synthesis using plant seeds, Bar et al. described the fabrication of green silver NPs by utilizing the seed extract from *Jatropha curcas* [75], wherein it was noted that the major phytoconstituents including curcain (an enzyme), curcacycline A (a cyclic octapeptide), and curcacycline B (a cyclic nonapeptide) could be employed as reducing and capping agents. The resultant NPs were further evaluated and characterized by HR-TEM, XRD, and UV–Vis spectroscopy. The analytical results showed two broad distributions of AgNPs, among which those having a diameter from 20 to 40 nm possessed a spherical shape, whereas the other particles were found to be larger and uneven in shape. It was further observed that the cavity of the cyclic peptides (curcacycline A and curcacycline B) stabilized the smaller NPs, while the sizeable ones were stabilized by the enzyme curcain. This interpretation was based on the demonstration that the peptides of proteins or carbonyl groups of amino-acid residues have strong metal-binding affinity [76]. Hence, the protein can protect the NPs by preventing their agglomeration, thus working as an encapsulating agent. It was believed that the cyclic proteins, curcacycline A or curcacycline B, first entrapped the Ag ions in their core structure. The subsequent reduction and stabilization of AgNPs happened in situ by the amide groups of the host peptide under suitable process conditions. Since the radius of most AgNPs obtained was comparable to the cavity of cyclic peptides, it was considered that the cyclic peptide cavity stabilized the smaller AgNPs, whereas the irregularly sized AgNPs were stabilized by the enzyme curcain, owing to its large, folded protein structure. Studies also concluded that the AgNPs synthesized by curcain latex were stable even after 1 month [77].

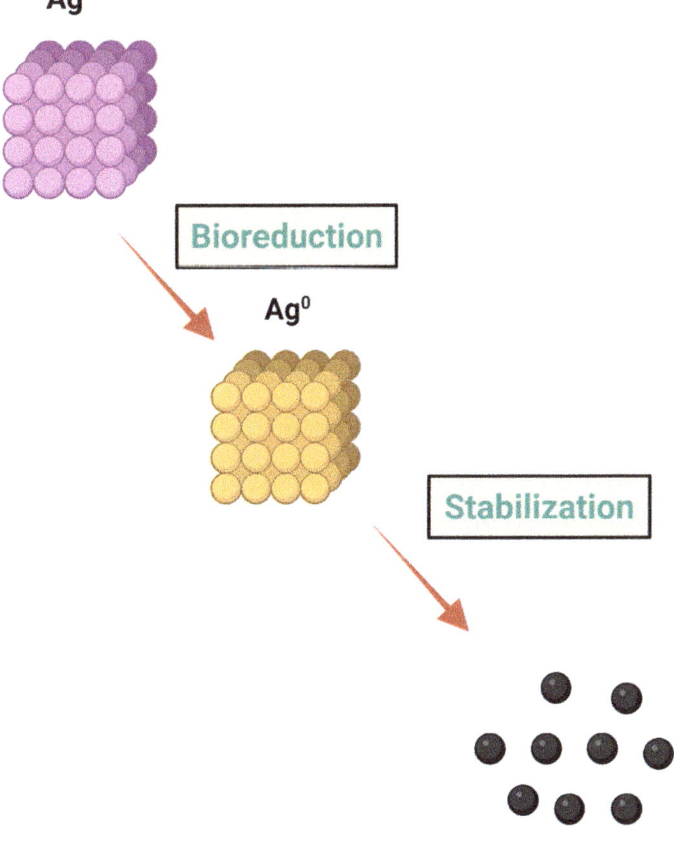

Figure 3. Green synthesis of silver nanoparticles from bioreduction of silver ions. The first step involves the bioreduction of positive Ag^+ into the zero-valent Ag^0 metal, while the last step involves the stabilization of metal NPs.

Vidhu et al. demonstrated the application of *Macrotyloma uniflorum* for AgNP synthesis. The plant is commonly known as horse gram, which is an herbaceous type of plant having a numerous pharmaceutical properties [77]. Several plant parts such as seeds and leaves are used for the treatment of asthma, heart conditions, bronchitis, urinary discharges, leukoderma, etc. Seeds of horse gram are a good source of molybdenum, iron, phenolic compounds, and antioxidants. The phytochemical analysis of the constituent seed indicated the presence of different phenolic acids such as 3,4-dihydroxy benzoic acid, *p*-hydroxy benzoic acid, vanillic acid, sinapic acid, syringic acid, caffeic acid, ferulic acid, and *p*-coumaric acid. The carbonyl and hydroxyl groups present in phenolic compounds were found to be capable of binding to metals [77]. Such compounds help in the inactivation of ions via the process of chelation. One of the constituents of the seed, i.e., caffeic acid, is thought to possess great antioxidant activity as it possesses an added conjugation in its propanoic side-chain that favors the delocalization of electrons via resonance. Here, it is synthesized from 4-hydroxy cinnamic acid in plants and is converted to ferulic acid after the release of a hydrogen. It is believed that this active hydrogen mediates the reduction of silver ions that form the resultant AgNPs. Hence, it is believed that the presence of proteins and phenolic compounds may act as the strategic factors for the synthesis of AgNPs. Hence, it is evident that this NP synthesis method utilizing the biochemical approach, by employing several plant extracts to play a key role as reducing and capping agents, is potentially a promising method for future developments on similar lines.

2.1.3. Extracts of Fruits

As per the literature, fruits have played a major role in ecofriendly AgNP synthesis. Amin et al. employed fruit extract from the *Solanum xanthocarpum* plant for the reduction and capping of AgNPs. It is a thorny plant known as Indian nightshade or yellow-berried nightshade, which grows in various terrains of the Indo-Pakistan subcontinent. These fruits are a rich source of apigenin glycosides, quercitrin, and flavonoids, and their extract displays antimicrobial, antioxidant, and anthelmintic properties [78]. This study concluded that pH, temperature, and the molar ratio of $AgNO_3$ to *S. xanthocarpum* extract (SXE) influence the reduction of Ag^+ and size of AgNPs. These fabricated particles exhibited urease-inhibitory and anti-*H. pylori* activities; accordingly, the study hinted at the potential antibacterial and urease-inhibitory activities of the AgNPs. This synthesis route for the AgNPs utilized SXE extract with $AgNO_3$ at 45 °C for 25 min, resulting in a band centered at 406 nm with surface plasmon resonance (SPR). These synthesized particles were observed to be spherical and monodispersed in nature with a size of around 10 nm. These NPs displayed appreciable effectiveness against the antibiotic-susceptible and antibiotic-resistant strains of *H. pylori*.

Indian gooseberry (*Emblica officinalis*) fruit extract was employed as a reducing agent to fabricate AgNPs [76]. In the study, treatment of aqueous chloroauric acid solution and silver sulfate with *Emblica officinalis* fruit extract, resulting in the reduction of Ag^+ ions into highly stable NPs. TEM analysis of the AgNPs reported here demonstrated that they were approximately 10–20 nm in size.

Li et al. reported the synthesis of AgNPs using *Capsicum annuum* extract. Studies suggested that the relationship between recognition–reduction–limited nucleation and growth is essential to elucidate the mechanism of formation of AgNPs. The first step was the recognition step wherein the proteins present in the *Capsicum annuum* extract interacted with the Ag ions through electrostatic interactions [79]. Then, Ag ions were reduced by proteins in the extract, which resulted in the generation of silver nuclei, as well as caused variations in the secondary structures of proteins. Moreover, the reduction of silver ions and their further accretion on these nuclei resulted in their subsequent growth. Larger AgNPs were formed with an increase in time. The polycrystalline phase turned into single crystalline phase via Ostwald ripening due to the increase in aging time, resulting in large-sized AgNPs.

In another study, a plant from the Bromeliaceae family, *Ananas comosus L.* (pineapple) [80], which has several beneficial properties including antioxidant activity, was used to produce AgNPs using the juice of the pulpy fruit. Phenolic bioactive constituents, present in vegetables and fruits, have been found to be majorly responsible for health benefits [81]. Ferulic acid in pineapples is believed to act as a reducing agent, which is oxidized by $AgNO_3$, further leading to the formation of the AgNPs. Another phenol known to be present in pineapple extract is chlorogenic acid. Such antioxidants present in pineapple juice act synergistically as reducing agents and stabilizing agents for silver metal ions.

2.1.4. Extracts of Leaves

Medicinal herbs such as *Hibiscus rosa sinensis* are effectively utilized in the treatment of hypertension, pyrexia, liver disorder etc. Philip et al. successfully used the above for AgNPs synthesis, wherein a quick change of the solution color to golden yellow indicated the formation of AgNPs. Its leaf extract contains antioxidant compounds and certain organic acids (essentially malic acid), proteins, flavonoids, anthocyanins, and vitamin C.

Interestingly, in another examination, Singh et al. [82] demonstrated the reduction of Ag ions present in a silver nitrate solution with the application of an aqueous extract of *Argemone mexicana* leaf. The color of the aqueous solution of the Ag ions changed from watery to yellowish brown due to the reduction of silver ions, indicating nanoparticle formation after the mixing of the *Argemone* leaf in the complex [74]. Studies suggested that the biosynthesized NPs are extremely toxic against various pathogenic fungi and bacteria at a concentration of 30 ppm for their growth control. The results from SEM and XRD studies displayed that the particle size range was 25–50 nm, and they were cubic in structure. The fact that the bioreduction of Ag^+ ions to AgNPs was due to the capping action of the plant extract was further confirmed by FTIR analysis [83].

The literature suggests that the plant extract from *Ocimum sanctum* (Tulsi) can be a good source of stabilizers, as well as bioreducing agents. Studies have shown that the glycosides, alkaloids, saponins, and tannins contained in the extract can be used to treat diarrhea, headaches, worms, and cough. Jain et al. employed green synthesis strategies to develop stable AgNPs using a leaf extract of quercetin and tulsi. TEM micrographs of AgNPs indicated the spherical and unform size of NPs with quercetin (11.35 nm) and tulsi (14.6 nm) as reducing and capping agents. In the case of quercetin, the size of nanoparticles was increased from 11.35 nm to 18 nm upon increasing the pH to 10 [84].

Mallikarjuna et al. developed nanoparticles 3–20 nm in size, which were characterized using TEM, XRD, UV/Vis spectroscopy, and FTIR techniques [85]. Moreover, these reduced AgNPs were covered with proteins and metabolites, e.g., terpenoids, with the functional groups of carboxylic acids, amines, ketones, alcohols, and aldehydes. The FTIR studies showed that the carbonyl groups from the amino-acid residues and proteins possess significant potential to bind metal, suggesting that the proteins (perhaps from the metal NPs i.e., capping of AgNPs) could put a stop to the aggregation that stabilizes the medium. This experiment hinted at the ability of the biological molecules to display binary actions in aiding the formation and stabilization of AgNPs in the aqueous medium.

In one study, a leaf extract of *Parthenium hysterophorus* was employed for the optimized green synthesis of AgNPs with an average particle size of 187.87 ± 4.89 nm and zeta potential of −34 ± 3.12 mV (shown Figure 4). A significant anti-inflammatory activity of NPs was observed. In addition, the in vitro cytotoxicity of AgNPs displayed potential anticancer activity after treatment of B16F10 and HepG2 cell lines at 24 h and 48 h. This leaf extract of *Parthenium hysterophorus*-based AgNPs could be a promising *Candida*te as an antimicrobial, antioxidant, anti-inflammatory, and antitumor agent for treatment [86].

Jha et al. reported another strategy involving the extract of *Cycas revoluta* (family Cycadaceae) [87], the source of sago. It is considered to be a rich source of fatty acids such as palmitic, stearic, oleic, and behenic acids, as well as flavonoids. Flavonoids are basically phenolic compounds which are found in almost all vascular plants. Amentiflavone and hinokiflavone are present in *Cycas* leaves as characteristic biflavonyls, which again act as reducing agents. $AgNO_3$ solution was treated with this leaf broth for 4 h, and their images were recorded by TEM, displaying discrete spherical nanoparticles possessing a diameter of around 2–6 nm. To ascertain the crystal structure of AgNPs, XRD was applied, with the lattice parameter showing good agreement with previous publications. There was an immediate change in color wherein the extract turned yellowish brown after adding the *Cycas* ethanol extract to the $AgNO_3$ solution. Here, the generation of AgNPs occurred as a result of reduction, after which UV/Vis spectroscopy at 449 nm was conducted on the resultant AgNPs [87].

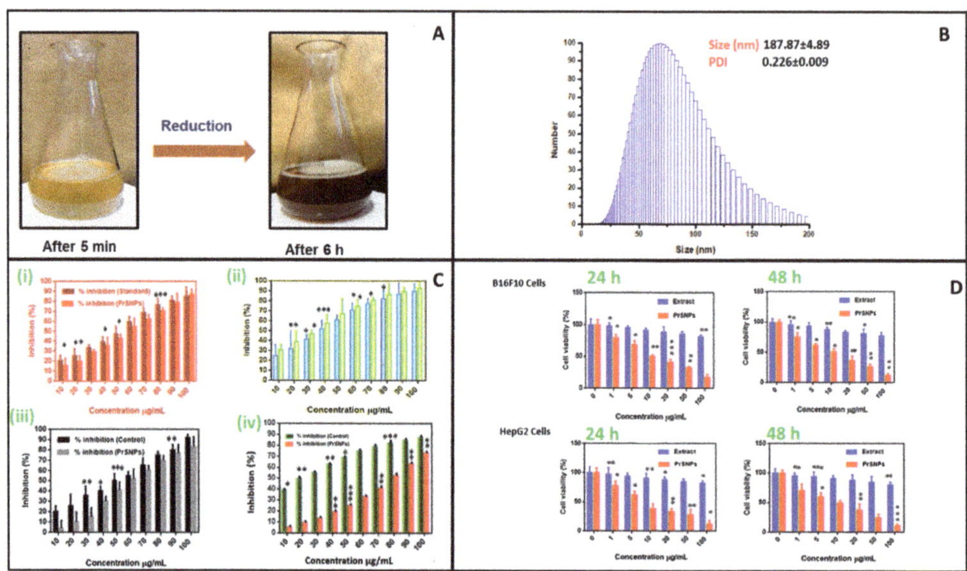

Figure 4. Green synthesis of AgNPs using leaf extract of *P. hysterophorus*. (**A**) The reduction of AgNO$_3$ by *P. hysterophorus* extract indicated a color change after 5 min and 6 h. (**B**) Particle size distribution. (**C**) Anti-inflammatory activity of AgNPs according to (i) DPPH assay, (ii) H$_2$O$_2$ assay, (iii) NO free radical-scavenging assay, and (iv) nitric oxide radical-scavenging assay. (**D**) In vitro cytotoxicity test on B16F10 and HepG2 cell lines after 24 h and 48 h treatment; * $p < 0.05$, ** $p < 0.01$ and *** $p < 0.001$, unpaired Student's *t*-test. CC-BY License [86].

Another study designed a cost-effective, simple, and green synthesis method of AgNPs using an extract of mulberry leaves as a reducing and stabilizing agent. The generated NPs had a mean size of 20 nm and possessed a face-centered cubic (FCC) structure. It is quite evident that plants are "biofactories", as the rate of synthesis of NPs with the application of plant products/extracts is faster than that when using microorganisms, while the resultant NPs are also more stable [88].

Other recent green methods to obtain AgNPs include those given by [89] employing *Olea europaea* leaf extract. These were synthesized by utilizing hot-water olive leaf extracts (OLE) that function as a reducing and a stabilizing agent; they were further tested for activity against drug-resistant bacteria. The spherical AgNPs synthesized possessed an average size of 20–25 nm. The AgNPs at 0.03–0.07 mg/mL concentration appreciably inhibited bacterial growth of multidrug resistant *Pseudomonas aeruginosa (P. aeruginosa)*, *Staphylococcus aureus (S. aureus)*, and *Escherichia coli (E. coli)*. This study also highlighted that the aqueous olive leaf extract exhibited no notable effect at the concentrations utilized for the preparation of these NPs. Elavazhagan et al. [90] developed AuNPs and AgNPs using *Memecylon edule* leaf extracts, which is a shrub also known as iron wood tree. According to various studies, saponins are the most favorable phytoconstituents for NP synthesis. Some studies have also demonstrated that the application of plant extracts for AgNP synthesis is faster than fungi- or bacteria-mediated synthesis [91]. For instance, Shankar et al. demonstrated the synthesis of NPs utilizing *Pelargonium* leaf in 9 h, while synthesis took about 24 to 124 h for the previously mentioned reactions [35].

The literature suggests that one of the major advantages of phytosynthesis is the easy handling of plant compounds or extracts and ready accessibility. Furthermore, the plants possess several active agents that facilitate Ag ion reduction, augmenting the synthetic procedure. It has also drawn attention since this route is capable of providing an economical ecofriendly protocol that occurs in a single step via a nonpathogenic synthesis route [92].

It has also been reported that centrifugation [83] can be used to obtain the AgNPs in pellet or powder form. In the case of the formulation of AgNPs suspensions, the product can be obtained in powder form via oven-drying [93]. Today, most parts of the plant such as roots, latex, stem, leaves, flowers, and seeds are used for NP synthesis. The most important factor to be kept in mind is the presence of bioactive agents in these parts responsible for the reduction and stabilization of nanoparticles. The medicinal plants used for AgNP creation are useful for control over the size and shape of the particles; in addition, such plants also offer their antimicrobial properties to the synthesized NPs. The utilization of plant extracts for the green synthesis of NPs could also be beneficial compared to other green synthesis processes, as these do not demand the intricate process of nurturing cell cultures as required in cases of microbial synthesis. A group of researchers also developed AgNPs using various plant leaf extracts such as *Aloe vera* [85], *Cinnamomum camphora* [85], *Camellia sinensis* [94], *Diopyros kaki* leaf, *Magnolia kobus* [95], *Geranium* leaf [35], *Acalypha indica* leaf [96], *Coriandrum sativum* [97], *Sorbus aucuparia* leaf [98], *Gliricidia sepium*, and rose leaf [99].

2.2. Microbial Synthesis of Silver NPs

Various types of microorganisms have been explored as biofactories for the synthesis of green NPs, and these strategies are extensively discussed in the literature [18,100]. Ahluwalia et al. demonstrated the utility of the fungus *Trichoderma harzianum* for AgNP synthesis, in addition to its extensive use as an agricultural fungicide [101]. The efficiency of the method was proven by the formation of NPs that are stable beyond 3 months of manufacturing. In addition, various types of broths such as lysogeny broth, peptone broth, nutrient broth, yeast extract, yeast mold broth, and tryptic soy broth have also been investigated for synthesis [102]. The formation of NPs is majorly impacted by two critical parameters, broth pH and light condition [102]. There are two mechanisms involved in the microbial synthesis of AgNPs [103]: intracellular synthesis and extracellular synthesis. In the intracellular method, the enzymes and the other biomolecules present inside the microbial cells are accountable for Ag ion reduction to NPs [2], and nucleation of the synthesized AgNPs is caused by the accumulation of Ag inside the cell, wherein the process continues with the growth of microbes. Once the optimum growth of cells is achieved, the live cells are harvested. Furthermore, special treatment procedures are employed for release of the synthesized NPs from the harvested cells [103]. In a study, Otari et al. demonstrated the intracellular synthesis of AgNPs using *Rhodococcus* spp. When tested against pathogenic microorganisms such as *Pseudomonas arugenosa*, *Stapylococcus aureus*, *Klebsiella pneumoniae*, *Enterococcus faecalis*, and *Escherichia coli*, fabricated AgNPs were found to exhibit great bacteriostatic and bactericidal activity. In contrast to the conventional physical and chemical methods of synthesis of AgNPs, this approach offered a cheaper and greener route of synthesis with scope for bioremediation [104]. In the extracellular method, the synthesis is done using extracellular secretions of the bacterial cells, and this method offers added advantages over the intracellular method, such as ease of separation along with the absence of downstream processing protocols [2]. In another study, Singh et al. (2014) revealed the extracellular biosynthesis of AgNPs using *Penicillium* spp., isolated from *Curcuma longa* (turmeric) leaves. The synthesized silver NPs showed appreciable activity against multidrug-resistant bacteria such as *Staphylococcus aureus* and *Escherichia coli* [105]. As discussed earlier, biosynthetic methods mediated by microbes can be categorized into extracellular and intracellular synthesis according to the location of NP production. Of these methods, the extracellular synthesis of NPs is still under study to comprehend the mechanisms employed for synthesis and a rapid scale-up. The intracellular mechanism for the green synthesis of metallic NPs has also been investigated [15] by using various types of plant and microbial species [16,17]. This microbial synthesis includes a range of reactions which involve trapping, bioreduction, and capping. The enzymes present in the cell wall of microbial species reduce the metal ions [18]. The intracellular synthesis of nanoparticles has several limitations such as low production and difficult purification [17].

However, there are several drawbacks of microbe-mediated NP synthesis. These include the expenses associated with upstream and downstream processing, making it an expensive resource-intensive synthetic route. Additionally, this method seems less feasible for industrial application because microbial cells demand an extremely specific environment for optimal growth. It was also found that, although microorganisms possess resistance mechanism against Ag ions, which is beneficial for the effective production of AgNPs, they show varying degrees of such resistance depending on the organism. Such resistance aids in the synthesis of AgNPs at a high concentration without killing the microbial cells. Generally, with increasing concentration, the rate of cell death increases [103].

2.2.1. Bacteria-Mediated Green Synthesis of Silver NPs

Among the various classes of microbes [106], the use of bacteria is gaining importance and is prevalent because of its easy and extensively studied genetic modification protocols, simple handling, and rising accomplishments [107]. Bacteria are regarded as promising *Candidate*s for this ecofriendly route of synthesis, which is attributed to their intrinsic potential to reduce heavy metals. Several factors such as organic functional groups present in the bacterial cell wall work synergistically to carry out the reduction [108]. In one such study [109], *Enterococcus* species isolated from fermented foods and further extracts of various strains ($n = 6$) were employed for the generation of nanoparticles. The prepared NPs displayed antimicrobial activity against multidrug-resistant species including *E. coli*, *K. pneumoniae*, and *P. vulgaris*. In addition, these NPs showed synergistic antimicrobial activity with ampicillin, ciprofloxacin, and cefuroxime. Thereafter, these NPs were used as nanopreservatives in white emulsion paint [109].

In another investigation, Sunkar and Nachiyar et al. synthesized AgNPs using endophytic bacterium *Bacillus cereus* isolated from the plant *Garcinia xanthochymus*, which is also known as false mangosteen or Himalayan *Garcinia*. The obtained nanoparticles were spherical AgNPs with their size in the range of 20–40 nm. The studies also demonstrated that the synthesized NPs possessed augmented activity against pathogenic bacterial species such as *Klebsiella pneumoniae*, *Salmonella typhi*, *Escherichia coli*, *Staphylococcus aureus*, and *Pseudomonas aeruginosa* [110].

Various other demonstrations have been reported, wherein the bacteria were employed to synthesize AgNPs; for example, [106] utilized *Pseudomonas stutzeri* AG259 obtained from a silver mine for the fabrication of silver NPs with a well-defined size and sharp morphology, including shapes such as equilateral triangles and hexagons. The study also revealed that the characteristics and morphology of nanoparticles can be mediated by several factors including cultivation conditions such as the time of incubation, composition and pH of growth media, and exposure to light [111].

Karthik et al. demonstrated the extracellular synthesis of AgNPs using bacterial species *Streptomyces* sp. LK3. The hypothesized mechanism involves a reduction of nitrate to nitrite by nitrate reductase, and this mechanism is widely accepted [112]. Some studies have also suggested that the NADH-dependent nitrate reductase-mediated reduction is the key factor in the green synthesis of AgNPs (shown in Figure 5). In the process of reduction, the electron is transferred to the Ag^+, leading to its reduction to metallic Ag [113]. The green synthesis of AgNPs was investigated using nitrate reductase (NR) from *Fusarium oxysporum* [113,114]. In another study, it was also observed that the chemical functionalities of the bacterial cell wall reduced silver ions to metallic silver in the absence of NR enzyme [103].

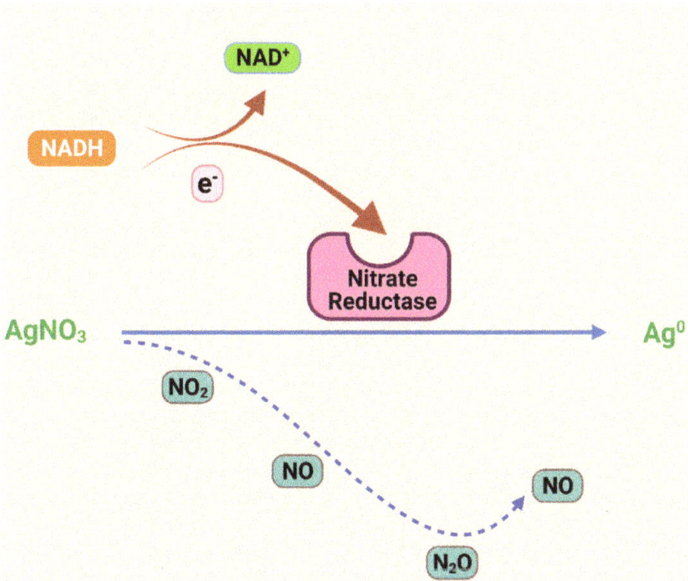

Figure 5. Nitrate reductase-mediated biogenic synthesis of AgNPs.

The synthesis of AgNPs using *Plectonema boryanum* UTEX 485, a filamentous cyanobacterium, revealed the participation of proteins during synthesis [115]. The culture supernatants of *Escherichia coli*, *Enterobacter cloacae*, and *Klebsiella pneumoniae* have also resulted in speedy formation of AgNPs [116]. Shahverdi et al. studied the biosynthesis of AgNPs using culture supernatants of different strains of enterobacteria including *Bacillus cereus*, *Bacillus subtilis*, *Escherichia coli*, *Enterobacter cloacae*, *Klebsiella pneumoniae*, *Lactobacillus acidophilus*, *Staphylococcus aureus*, *Pseudomonas aeruginosa*, *Candida albicans*, *Aspergillus Niger*, and *Enterobacter cloacae* [109,116]. In this approach, culture flasks of enterobacteria were incubated at 35 °C for about 24 h, and this rapid process led to the formation of AgNPs in just 5 min of contact with the culture supernatants. Extensive scientific research has been conducted to study the extracellular green synthesis of AgNPs using *P. aeruginosa* [117] and *E. coli* [118,119]. Studies suggest that silver-resistant bacterial cell walls can accumulate a maximum amount of silver at 25% of their dry weight biomass [107,120]. The flexibility and cost of bacteria-mediated biosynthesis are dependent on the selection. This was also reported as a suitable method for large-scale production [121]. However, a major setback in the biosynthesis of NPs using bacteria is the time-consuming synthesis procedure and the limited availability of sizes and shapes compared to other methods of synthesis. Due to this reason, fungi-based nanofactories or plant-based materials were further examined in [122].

2.2.2. Fungi-Mediated Green Synthesis of Silver NPs

Considering the limitations of the bacteria-mediated synthesis [106], a plethora of different proteins, the cell mass, the enzymes, or the extracellular components from fungi, such as *Aspergillus flavus*, *Fusarium oxysporum*, *Penicillium brevicompactum*, and *Aspergillus clavatus* [123,124], have been investigated to reduce silver ions in order to synthesize AgNPs. Around the beginning of the 20th century, the first fungi-mediated metal nanoparticle generation was reported as *Verticillium*-mediated AgNP synthesis with an average diameter of about 25 ± 12 nm [125,126], as reviewed in [91]. The suggested pathway for the synthesis of NPs includes a few steps. The first step involves the electrostatic interaction between COO^- groups present in the enzymes of the mycelial cell wall and Ag^+ ions, leading to the absorption of positively charged silver ions on the fungal cell surface. The second step is the reduction of Ag$^+$ ions by the enzymes present in the cell wall, followed by the generation of silver nuclei, which eventually form the AgNPs [127]. In another example [106], the superiority of filamentous fungi over bacteria for the synthesis of NPs was attributed to its innate characteristics such as high metal tolerance, intracellular metal uptake capability, and cell-wall-binding capacity [128]. Previously, [129] incubated fungus *Aspergillus flavus* with AgNO$_3$ solution for 72 h, and it was demonstrated that the AgNPs accumulated on the surface of the cell wall of the fungus. In another study, culture supernatant (CS) [128] obtained from the fungus *Cunninghamella phaeospora* was employed for the creation of AgNPs. The resultant NPs were mostly spherical in shape, having a size of around 12.2 nm. The other characteristics shown by AgNPs included their monodisperse nature and, most importantly, their broad-spectrum antibacterial activity. The fungal species which can be utilized for the biosynthesis of both gold (Au) and silver (Ag) NPs include [108] *Fusarium* sp., *Penicillium* sp., and *Aspergillus* sp. Silver nanoparticles can be synthesized using *Trichoderma viride* fungus [130] via the extracellular biosynthesis method [4]. Stable AgNPs having a size range of 5 to 15 nm can be synthesized using *Fusarium oxysporum*, wherein the NADH-dependent reductase enzyme reduces the silver ions [131]. A recent study presented by [132] explored an in vitro application for the generation of AgNPs using the fungus *Penicillium citrinum*. Similarly to bacteria [91], fungi possess metal tolerance, metal bioaccumulation ability, uptake via an intracellular route, and high binding capacity [91]. In comparison with other microorganisms, fungi are highly beneficial for the synthesis of NPs on an appreciably large scale. For example, as compared to bacteria and plant extracts, the fungal mycelial mesh has a better capacity to tolerate conditions such as pressure, flow, and agitation present in bioreactors and chambers. Fungi secrete various enzymes or several proteins per unit of biomass, which leads to higher yields than the nanoparticles formed in a bacteria-mediated process. Fungi are quick to reproduce and grow. More reductive proteins are secreted via the extracellular route by fungi than bacteria. The extracellular synthesis of NPs is beneficial as resultant NPs are free from redundant cellular components and they do not bind to the biomass; hence, they can be used without further treatment for various purposes [91]. Furthermore, the increased application of fungi in the biosynthesis of NPs in comparison to other microorganisms is attributed to their ease of handling and ecofriendly nature; for example, the nonpathogenic white rot fungus could be utilized for the synthesis of AgNPs on a large scale [133].

2.2.3. Yeast-Mediated Green Synthesis of Silver NPs

Yeasts represent another class of microorganisms explored for the green synthesis of AgNPs [133]. Yeast is a eukaryotic, unicellular organism which has evolved from multicellular antecedents. Yeasts, being chemoorganotrophs, utilize organic compounds such as carbon obtained from sugars as their primary source of energy. They are known to grow well in neutral or slightly acidic environments. Newer methods of cultivation of yeast are devoid of all the exasperating steps, thus resulting in a simpler and easier process [133]. Currently, the focus has shifted from prokaryote-based biosynthesis to eukaryote-mediated green synthesis, which has expanded the scope and broadened the future possibilities for the ecofriendly generation of NPs. In another instance, Korbekandi et al. showed the utilization of AgNPs during biotransformation using *Saccharomyces cerevisiae*, commonly known as brewer's yeast, wherein the synthesis of NPs took place in various parts of the cells, such as in the core of the cells and the cell membrane, adhering to the cell membrane and possibly the exterior of the yeast cells. Thus, they are preferred to bacteria for NP synthesis because of the potential scale-up, easy handling, and regulation at lab scale using only simple nutrients [134].

2.2.4. Algae-Mediated Synthesis of AgNPs

Algae [133] are a diverse group of aquatic organisms. They are eukaryotic species capable of conducting photosynthesis. Thus, in aquatic atmospheres, they are prevalent primary producers, including cyanobacteria. Algae have a wide distribution of sizes [133], varying from microscopic picoplankton to Rhodophyta, which has a size in the macroscopic range. Algal *Chlorella* sp. was reported to possess the ability to accumulate heavy metals such as uranium [135], cadmium, zinc, nickel, and copper [136]. Algae-mediated biosynthesis [133] has been found to be rapid and inexpensive [108]. The selection of algae as a biological source is attributed to its negatively charged cell surface, which has the capacity to accumulate and grow the crystals rapidly. Its associated low cost also makes mass production of nanoparticles a realizable possibility. In the case of *Chaetomorpha linum* algae, the algal metabolites assist in the reduction of $AgNO_3$. The metabolites, essentially terpenoids and flavonoids, aid in the capping and stabilization of nanoparticles. Other metabolites such as polysaccharides are found to be effective in controlling the shape and size of AgNPs [137]. AgNPs can be synthesized using marine algae *Cystophora moniliformis* for the reduction of ions and stabilization of NPs. Patel et al. also stated that AgNPs synthesized by strains of *Coelastrum* sp. and *Botryococcus braunii* exhibited antibacterial activity against tested pathogens such as *B. megaterium, E. coli, B. subtilis, M. luteus, P. aeruginosa*, and *S. aureus*. [138]. Other algal extracts used to produce AgNPs include those obtained from red algal seaweeds of *Laurenciella* sp. and *Laurencia aldingensis* [133]. Significant cytotoxicity was observed against uterine sarcoma MESSA/Dx5, as well as its parental MESSA cell line, when subjected to synthesized AgNP. In contrast, no toxicity was observed against P4 (human foreskin fibroblasts) health cells. The study indicated that the AgNPs can be used as potentially novel *Candida*tes for chemotherapy [139].

2.2.5. Actinomycetes-Mediated Green Synthesis of Silver NPs

Actinomycetes belong to the order of Actinobacteria, a phylum of bacteria called Actinomycetales. Actinobacteria are mostly Gram-positive. They may or may not be anaerobic. As discussed earlier, two mechanisms elaborated in microbe-mediated synthesis are intracellular synthesis and extracellular synthesis, and actinomycetes are good *Candida*tes for both types of synthesis [113]. It was found that the cell wall and cell membrane of actinomycetes contain the NADH-dependent reductase enzyme, which is useful to reduce gold and silver ions, and the secreted proteins such as cytochrome C act as capping and stabilizing agents for the synthesized NPs [140]. In one such demonstration [133], *Streptomyces* species were utilized to generate spherical AgNPs of 20–70 nm size range. The actinomycetes-mediated synthesis of NPs yields stable and polydisperse NPs which usually have remarkable antimicrobial capacity [113]. AgNPs synthesized using a culture supernatant of *Streptomyces* sp. JAR1 showed significant inhibitory activity toward pathogens such as *Candida tropicalis, Salmonella* spp., *E. coli, Scedosporium* spp., *Pseudomonas aeruginosa, Ganoderma* spp., *Fusarium* spp., and *Staphylococcus* aureus [141].

2.3. Enzyme-Based Synthesis of Silver NPs

The purity of available enzymes [4] and their structure make this synthetic route a prospective method to create silver nanoparticles of the desired form. The extracellular synthesis of AgNPs is attributed to the enzymes released by the cells [91]. The enzymes present in the extracts of plants could also act as a reducing agent in this green AgNP synthesis. During the enzyme-based synthesis of AgNPs, a specific enzyme is obtained from the cultural supernatant of lifeforms such as bacteria. Kumar et al. (2007) first demonstrated the in vitro fabrication of AgNPs using the α-NADPH-dependent nitrate reductase enzyme obtained from the cultural supernatant of fungus *Fusarium oxysporum* and phytochelatin [142].

Biomolecule-Mediated Synthesis of Silver NPs

The biosynthesis of AgNPs [15] has been explored using monosaccharides such as glucose, fructose, and galactose several disaccharides such as lactose and maltose, and numerous polysaccharide molecules such as starch, heparin, dextran, chitosan, and pectin. There are convincing studies available on utilizing starch as a stabilizing and reducing agent in this process of synthesizing nanoparticles. Hence, polysaccharides serve as a reducing and a capping agent. In one of the experiments, the in situ method of preparation was utilized for the preparation of green silver nanocomposites. In this experiment, a colloidal silver dispersion was prepared by employing glucose as a reducing agent, water as a solvent, and soluble starch as a stabilizing agent. When the produced green AgNPs were dispersed in a potato starch/glycerol matrix, the process yielded silver nanocomposites with potential application in antimicrobial packaging [143]. According to an experiment performed by Zain et al. (2014), when ascorbic acid was utilized as a reducing agent in chitosan solution for the creation of AgNPs and CuNPs (copper nanoparticles) using microwave heating, the resultant AgNPs were found to possess greater bactericidal activity against bacterial species such as *Bacillus subtilis* and *E. coli* as compared to CuNPs of the same mean size [144]. Another example is represented by levan, which is a polysaccharide mainly derived from plants but is also obtained from microorganisms and curdlan (a bacterial exopolysaccharide), which have all been employed to fabricate appreciably stable AgNPs. The particles synthesized using curdlan as a reducing and stabilizing agent were observed to be mostly spherical in shape with a mean diameter around 15 nm [145]. Furthermore, vitamins and amino acids have proven their suitability, especially for the synthesis of therapeutic nanoparticles [15]. The literature has also mentioned the capping and reducing properties of vitamin B2 for the synthesis of silver and palladium (Pd) nanoparticles. The NPs were found to self-assemble into different shapes depending on the solvent used for their preparation. The AgNPs and PdNPs were found to self-assemble into structures of nanorods and nanowires when solvents such as water and isopropanol were used, respectively. These self-assemblies of AgNPs and PdNPs further catalyzed the reactions of pyrrole and aniline to form polypyrrole and polyaniline nanocomposites [146]. Furthermore, glycerol has recently been considered as a promising green solvent in the synthesis of various metallic NPs. Because of its low toxicity, glycerol is preferred as a cheaper and better alternative to usually employed polyols, including propylene glycol and ethylene glycol [147]. For example, the process of synthesizing AgNPs was accomplished using starch as a capping agent and β-D-glucose as a reducing agent. The resultant nanoparticles were found to be stable and similar in properties (e.g., polydispersity and shape) to AgNPs obtained using conventional techniques of synthesis [148]. AgNP [24] synthesis utilizes water (solvent) and polysaccharides (capping agent) such as starch and heparin in this process. This method is reported to be advantageous as the binding interactions between AgNPs and starch are quite weak and are noted to reverse at higher temperatures, thus assisting the segregation of the obtained NPs. Additionally, surface passivation and a significant prevention of particle aggregation occur due to the extensive hydrogen bonding network [148].

Furthermore, the literature highlighted a synthesis pathway that employs negatively charged heparin as a reducing and stabilizing agent in a mixture of $AgNO_3$ and heparin heated to 70 °C for approximately 8 h [120]. The anionic nature of the sulfonate groups present in heparin facilitates the formation of silver nanoparticles. There is another strategy known as the Tollens method that employs a single-step process to form AgNPs of dictated size [149,150]. In an altered Tollens procedure, positively charged silver ions are reduced by sugars such as fructose, glucose, xylose, and maltose in the presence of ammonia. This procedure yields AgNPs of various shapes with a particle size ranging from 50–200 nm. Furthermore, investigations also revealed that the smallest particle size was obtained at the lowest ammonia concentration [151]. Some scientists also mentioned the autoclave technique for producing AgNPs using starch, whereby, similarly to previous examples, starch acts as a reducing and stabilizing agent [152,153].

2.4. Green Synthesis from Vitamins

The synthesis of gold and palladium nanospheres, nanowires, or nanorods has been reported, wherein vitamin B2 was employed as a reducing and capping agent. Vitamin B2 employed here acts as a reducing agent in the process of synthesis. Such a synthesis approach can be further extended to form silver nanostructures. Ascorbic acid or vitamin C has also been employed as a capping and reducing agent, whereas chitosan is used as a stabilizing agent because it binds to charged metal species [154]. An interesting process for the synthesis of NPs possessing uniform size was developed using ascorbic acid [155]. Even during glycolysis, plants produce H^+ ions along with NAD, which play the role of strong reducing agents in the synthesis of AgNPs [156].

2.5. Ionic Liquid-Mediated Synthesis

Ionic liquids (ILs) have recently emerged as a new option of reaction medium [157] due to their appreciable properties such as low volatility, nonflammability, high chemical and thermal stabilities, designable structures, high ionic conductivity, and broad electrochemical windows. Hence, ionic liquid-mediated [158] synthesis is gaining preference in the green chemistry world through their use as green electrolytes. Chemical and electrochemical AgNP synthesis utilizing a similar method was developed for the synthesis of AgX (X = Cl, I) NPs by employing ionic liquids [159]. In another study, ionic liquids of bis(alkylethylenediamine) silver(I)) salts such as bis(*N*-2-ethylhexylethylene diamine) silver (I) nitrate and bis(*N*-hexylethylenediamine) silver (I) hexafluorophosphate were used for the synthesis of AgNPs. It was found that uniform AgNPs were formed successfully via the reduction of bis(N-2-ethylhexylethylenediamine) silver (I) nitrate solution with aqueous $NaBH_4$ but not by the reduction of bis(*N*-hexylethylenediamine) silver (I) hexafluorophosphate [160]. Pringle et al. (2008) demonstrated a single-step process of conducting polymer–AgNP composite synthesis using an ionic liquid solution of silver nitrate. The ionic liquid used for the experiment was 1-ethyl-3-methylimidazolium bis(trifluoromethanesulfonyl)amide [161]. In addition, synthesis of an Ag–carbon hybrid with controlled structure and morphology via a hydrothermal treatment of silver nitrate and glucose was achieved using ionic liquid tetradecyl-3-methylimidazolium tetrafluoroborate as a soft template [162], whereas the synthesis of partially positively charged AgNPs was also achieved using ionic liquid 1-butyl-3-methylimidazolium tetrafluoroborate [163]. In one of the experiments, the electrodeposition of silver from the distillable ionic liquid DIMCARB, formed by mixing CO_2 and Me_2NH in a specific proportion, was carried out [164]. Furthermore, the fabrication of spherical and polygonal AgNPs in ionic liquid $bmimBF_4$ using an electrochemical method was reported. In this case, polyvinylchloride (PVP) was used as the stabilizer. The study revealed that, in comparison with water which is a traditional solvent, ionic liquids offer many advantages including a shortened electrolytic time, reduced PVP dosage, and reduced energy consumption, as the process allows electrolysis without the requirement of mechanical stirring [165].

2.6. Irradiation-Assisted Synthesis

According to the sixth principle of green chemistry, i.e., design for energy efficiency, irradiation-assisted synthesis is an excellent green method [166]. Irradiation-assisted synthesis does not require the use of additional reducing agents, thus avoiding the associated side reactions and toxicity (if any). Accordingly, this method of synthesis also obeys other principles of green chemistry. A number of irradiation methods including laser irradiation, radiolysis, and pulse radiolysis can be employed for the synthesis of AgNPs [1]. AgNPs can be fabricated with a distinct size and shape via laser irradiation [1] of an aqueous solution of silver salt and a surfactant solution of sodium dodecyl sulfate (SDS) [167]. In another instance, a laser was utilized in a photo-sensitization method for AgNP formation by employing benzophenone [168]. AgNPs of around 20 nm were obtained by employing low-power lasers for short irradiation times. However, NPs 5 nm in size were formed when a greater irradiation power was supplied. They were also successfully synthesized via this concept using a mercury lamp [168]. The photosensitized growth of AgNPs was also demonstrated in visible-light irradiation studies, wherein thiophene was employed as a sensitizing dye [169], and AgNP synthesis was completed via the illumination of $Ag(NH_3)^+$ in ethanol [170]. Huang et al. (2009) synthesized green AgNPs using a chitosan aqueous solution as the stabilizing agent under γ-irradiation without isopropanol and an N_2 atmosphere. The minimum inhibitory concentration (MIC) value of the synthesized AgNPs was found to be around 100 ppm when tested against methicillin-resistant *Staphylococcus aureus* and *Aeromonas hydrophila* for their antibacterial efficacy. The resultant AgNPs were found to exhibit good antibacterial activity [171].

2.7. Microwave-Assisted Synthesis

Uniform AgNPs were also fabricated by microwave radiation [1,24] of a carboxymethyl cellulose sodium (CMS) and silver nitrate solution. The NPs generated were uniform and appreciably stable at room temperature for around 2 months [172]. CMS can be employed as both a reducing and a stabilizing agent in the process of formation of AgNPs [172]. Moreover, only $AgNO_3$ is required in the reaction as a reagent. When compared to the general heating treatment [4], this method provides homogeneous heating and appreciable nucleation of noble-metal NPs in a simple manner [173]. It was employed as a fast process for NP formation (a matter of seconds) using irradiation at 50 W [174], wherein red grape pomace was employed as a reducing agent. Furthermore, microwave radiation was applied to create monodisperse AgNPs by employing L-lysine or L-arginine and soluble starch as reducing and protecting agents, respectively [175]. Noroozi et al. (2012) achieved rapid microwave-mediated green synthesis of AgNPs without using a reducing agent. AgNPs were produced by using polyvinylpyrrolidone (PVP) as the stabilizing agent in a water medium. In comparison to silver nanoparticles produced using the conventional heating method, the AgNPs fabricated using microwave irradiation were found to be denser, more uniform, and smaller in size. These AgNPs can be used to formulate size-dependent nanomedicines [176]. While analyzing the microwave-assisted synthesis of green AgNPs using peel extracts of citrus fruits, Kahrilas et al. realized that silver nanoparticles were successfully produced only when orange peel extract was subjected to synthesis, where it worked as a reducing and capping agent. Thus, the study showed a greener alternative to toxic reducing and capping agents [177]. Later, AgNP creation using biomaterials such as sodium alginate was also reported and, interestingly, the synthesized spherical nanoparticles were noted to be stable for about 6 months or more when stored at room temperature. The AgNPs also showed good antibacterial activity toward *Escherichia coli* and *Staphylococcus aureus* [178]. In another study, Albadran and Kamal et al. worked on the optimization and modeling of the green synthesis of AgNPs by employing a one-pot microwave-mediated method. In this study, a microwave-assisted reaction of $AgNO_3$ and cactus extract was carried out to produce a colloidal suspension of AgNPs. The resultant colloid suspension of AgNPs was further investigated for its absorbance and photocatalytic activity in the removal of organic pollutants from wastewater [179]. Anjana et al. (2021) demonstrated the rapid microwave-assisted synthesis of stable AgNPs by employing leaf extracts of *Cyanthillium cinereum*. The resultant AgNPs were found to demonstrate antibacterial activity toward bacterial species such as *Klebsiella pneumoniae* and *Staphylococcus aureus*. In addition, it was suggested that the resultant nanoparticles can be employed as promising *Candida*tes in applications as biosensors and nonenzymatic electrochemical sensors [180].

3. Characterization of Green Silver Nanoparticles

Silver nanoparticles are typically characterized [128] according to their size, surface charge, distribution, surface morphology (shape), aggregation, etc. The physicochemical properties have a notable impact on the biological properties; thus, physicochemical characterization of nanoparticles is typically conducted prior to in vitro and in vivo studies [181]. It has also been stated that numerous physiochemical properties can particularly impact the interactions between the nanomaterials and specific target sites (proteins, cells) [182]. To evaluate these properties, a number of analytical techniques are utilized such as ultraviolet/visible spectroscopy, X-ray diffractometry (XRD), Fourier-transform infrared spectroscopy (FTIR), scanning electron microscopy (SEM), dynamic light scattering (DLS), atomic force microscopy (AFM), transmission electron microscopy (TEM), and zeta potential analysis [183].

3.1. Surface Morphology

SEM and TEM are the two main types of electron microscopy. TEM is popularly used for the characterization of nanomaterials. TEM is used to determine the size, degree of aggregation, dispersion, etc. in nanomaterials. SEM is employed in addition to TEM for the structural analysis of NPs. Compared to SEM, TEM is considered more advantageous as it provides good-quality spatial resolution along with precise analytical measurements. For instance, SEM and TEM were used to clarify the morphology and size of the resultant NPs in [4]. SEM analysis has mostly revealed that the AgNPs formed in the literature were spherical, although a few authors reported irregular [184], triangular [185], hexagonal [129], isotropic [186], polyhedral [187], flake [188], flower [189], pentagonal [190], anisotropic [191], and rod-like structures [192]. Zeta potential values are used to confer colloidal stability to the integrated NPs, allowing the solution or dispersion to resist aggregation [193]. Scientists have employed macroalga *Spirogyra varians* to synthesize NPs with an average size of 35 nm in uniform and quasi-spherical shapes [37]. Abd-Elnaby et al. characterized AgNPs synthesized extracellularly from marine *Streptomyces rochei* MHM13, and SEM confirmed the spherical shape of AgNPs 22 to 85 nm in size [44]. Singh et al. reported the TEM results of AgNPs synthesized using an endophytic fungal supernatant of *Raphanus sativus*, whereby they observed that AgNPs were almost spherical with a size ranging from 10 nm to 30 nm [49]. In another experiment, *W. oryzae* DC6-mediated AgNPs, when studied by FE-TEM, showed a spherical shape with a 10–30 nm size range [53]. Suresh et al. studied AgNPs synthesized using *Delphinium denudatum* root extract using FE-SEM. The synthesized AgNPs were found to be polydisperse and spherical with a size mostly below 85 nm [56].

Atomic force microscopy (AFM) is a scanning probe microscopy (SPM) method with greater resolution, and it is capable of creating 3D images of surfaces at high magnification [194]. When AFM was used for the determination of size and morphology of green AgNPs biosynthesized using an endophytic fungal supernatant of *Raphanus sativus*, the study showed the presence of monodisperse AgNPs having an average particle size of 4 to 28 nm [49]. Praphulla Rao et al. synthesized AgNPs using lemon extract, and further characterization using AFM suggested that the NPs were nearly 12 nm height and 100 nm in width [62]. Bankura et al. presented an AFM image of dextran-mediated AgNPs, showing the formation of well-dispersed NPs in the 10 nm to 60 nm size range [64]. Minhas et al. evaluated the surface properties of biogenic AgNPs synthesized using extracts of *Ulva compressa* L. Kütz. and *Cladophora glomerata* L. Kütz., wherein characterization of NPs using AFM revealed the roughness parameters average roughness (Sa), root-mean-square roughness (Sq), and ten-point height (Sz). For AgNPs synthesized using an extract of *U. compressa*, the measurements obtained were as follows: Sa, 1.01 nm; Sq, 1.48 nm; Sz, 9.09 nm, whereas, for AgNPs synthesized using an extract of *C. glomerata*, the values obtained were as follows: Sa, 0.471 nm; Sq, 0.848 nm; Sz, 5.90 nm [36]. Gopinath et al. utilized the dried fruit body extract of *Tribulus terrestris* for the synthesis of AgNPs. The AFM study revealed the formation of spherical AgNPs that were mostly homogeneous in size with individual particles in the range of 24.631 nm [61].

3.2. UV/Visible Spectroscopy

UV/visible spectra assist in analyzing the relationship between the metal ion concentration, pH, and extract content and the type of AgNPs formed. The emergence of a yellow or light brown-yellow color in a previously colorless mixture is usually indicative of the formation of AgNPs. The optical properties also depend on the particle size and shape [195].

The characteristic peak of AgNPs in a UV/visible spectrum is usually observed around 430 nm; however, as discussed earlier, many factors affect the position of this peak. In the green synthesis of AgNPs using macroalgae *Spirogyra varians* [37], *Streptomyces rochei* MHM13 [44], the endophytic fungal supernatant of *Raphanus sativus* [49], *Weissella oryzae* DC6 [1], *Delphinium denudatum* root [56], *Piper nigrum* leaf and stem [59], and dextran [64], the absorption peaks in the UV/visible spectrum were observed at 430 nm, 410 nm, 426 nm, 432 nm, 416 nm, 460 nm, and 423 nm, respectively.

3.3. SAED

Selected area (electron) diffraction (SAED) is a diffraction technique used to determine the crystal structure of NPs. Studies are usually conducted in a TEM or SEM using electron backscatter diffraction [194]. AgNPs are generally crystalline in nature having an fcc (face-centered cubic) structure. During the characterization of AgNPs synthesized using an endophytic fungal supernatant of *Raphanus sativus*, Singh et al. found bright rings in the SAED pattern, indicating the crystalline nature of NPs [49]. Similar results were obtained in case of green AgNPs synthesized using marine endophytic fungus *Penicillium polonicum* [51]. In another study, Bankura et al. synthesized AgNPs using dextran; during SAED analysis, circular rings corresponding to the (111), (200), (220), (211), and (222) planes were observed in the SAED pattern, indicating the highly crystalline nature of resultant AgNPs [64]. Additionally, the SAED pattern of AgNPs synthesized using gum kondagogu (*Cochlospermum gossypium*) consisted of concentric rings corresponding to the (111), (200), (220), and (311) planes of face-centered cubic (fcc) silver with intermittent bright dots, revealing the highly crystalline nature of these NPs [63].

3.4. XRD

XRD studies are employed to determine the crystal structure of nanomaterials. They are also used to record the size and shape of the unit cell from the peak points via translational symmetry [196]. This technique is basically employed for the analysis of atomic spacing and crystal structures. The observations are based on the constructive interference of a crystalline sample and the monochromatic X-rays generated by a cathode ray tube, which are then filtered to produce monochromatic radiations. This method aids to understand the crystal structure and atomic properties of the synthesized AgNPs, along with their size measurements [197].

In a synthesis study of AgNPs using macroalgae *Spirogyra varians*, four different and important characteristic peaks were observed after 2 h at 38.1°, 44.3°, 64.5°, and 76.4° corresponding to the (111), (200), (220), and (311) planes, respectively. All peaks in the XRD pattern can be readily indexed to the face-centered cubic structure of silver [37]. A similar trend was observed in case of green AgNPs synthesized using an endophytic fungal supernatant of *Raphanus sativus* [49], *Piper nigrum* leaf and stem [59], and dextran [64]. Singh et al. analyzed *Weissella oryzae* DC6-mediated AgNPs using XRD, showing that the XRD pattern of fabricated NPs showed extreme peaks across the whole spectrum of 2θ values ranging from 20–80, and this pattern was analogous to the Braggs reflection of silver nanocrystals [53]. Suresh et al. observed 13 intense peaks (12.25°, 13.12°, 16.07°, 16.68°, 18.60°, 20.30°, 27.24°, 27.63°, 28.30°, 30.31°, 30.89°, 38.45°, and 44.87°) across the whole spectrum of a *Delphinium denudatum* root extract after 2 h, with values ranging from 10 to 70, indicating the presence of both cubic and fcc structured NPs, i.e., a mixed phase of AgNPs. In another study, Shameli et al. prepared PEG-based AgNPs via a green synthesis route and characterized them using XRD. The results indicates peaks at 38.04°, 44.08°, 64.36°, and 77.22° attributed to the crystallinity of the AgNPs [198].

3.5. FTIR

FTIR is used to analyze the functional groups or metabolites (capping/stabilizing agents) employed on the surface of NPs [128]. Salari et al. reported the presence of amino, carboxylic, hydroxyl, and carbonyl groups following an FTIR study of AgNPs synthesized using macroalgae *Spirogyra varians*. The FTIR spectra displayed several peaks attributed to different functional groups including strong broad O–H stretching carboxylic bands around

3423 cm^{-1}, carboxylic/phenolic stretching bands around 2927 cm^{-1}, and quinine OH bands around 1515 and 1429 cm^{-1}. After the synthesis of AgNPs, the characteristics peaks of proteins appearing around 1645 cm^{-1} were found to be shifted [37]. In the FTIR spectrum of AgNPs synthesized using marine *Streptomyces rochei* MHM13, an array of absorbance bands were observed from 400 to 4000 cm^{-1}. Characteristic intense peaks appeared at 3420.14, 2932.23, 2362.37, 1639.20, 1430.92, 1115.62, and 613.252 cm^{-1}, which denoted the presence of capping and stabilizing agents containing several functional groups [44]. The FTIR spectrum of AgNPs synthesized using an endophytic fungal supernatant of *Raphanus sativus* displayed absorption bands at 3145, 1597, 1402, 1109, 1213, 995, 911, 699, and 504 cm^{-1}. These peaks were attributed to several functional groups such as hydroxyl, amino, amino acid, methyl, alkene, and alkyl halides [49]. Suresh et al. observed FTIR peaks at 3354, 2952, 2063, 1651, 1419, 1383, 1354, 1171, 1093, 780, 672, and 605 cm^{-1} for AgNPs synthesized using *Delphinium denudatum* root extract. Comparing the peaks of the spectrum of synthesized AgNPs and *Delphinium denudatum* root extract, they concluded that the polyols and phenols acted as reducing agents, whereas some proteins and metabolites (e.g., terpenoids) with functional groups such as amine, alcohol, ketone, aldehyde, and carboxylic acid acted as capping and stabilizing agents [56]. During the FTIR analysis of AgNPs synthesized using a leaf extract of *P. nigrum*, absorption bands were observed at 3314, 3197, 2897, 2362, 1763, 1668, 1628, 1532, 1480, 1399, 1383, 1335, 1276, 1191, 1122, 884, 823, 750 cm, 656, and 602 cm^{-1}, indicating that the phytochemicals responsible for the synthesis of NPs contained functional groups such as amino, alkyl, carboxyl, carbonyl, nitro, alkyne, and ester. In the case of stem-derived AgNPs, the bands observed at 3697, 3313, 3195, 2298, 1456, 1336, 1670, 1193, 1118, 811, 750, 651, and 601 cm^{-1} allowed concluding that the phytochemicals contained functional groups including amino, nitrile, amide, ester, and alkynes [59]. Furthermore, Shameli et al. prepared PEG-based AgNPs via a green synthesis route and characterized them using FTIR; the interaction between AgNPs and PEG molecules was established by the peak at 1730 cm^{-1} (−C=O carboxylic acid group in gluconic acid) and a shift in the peak at 1007 cm^{-1} toward a lower frequency (Figure 6A–C) [198].

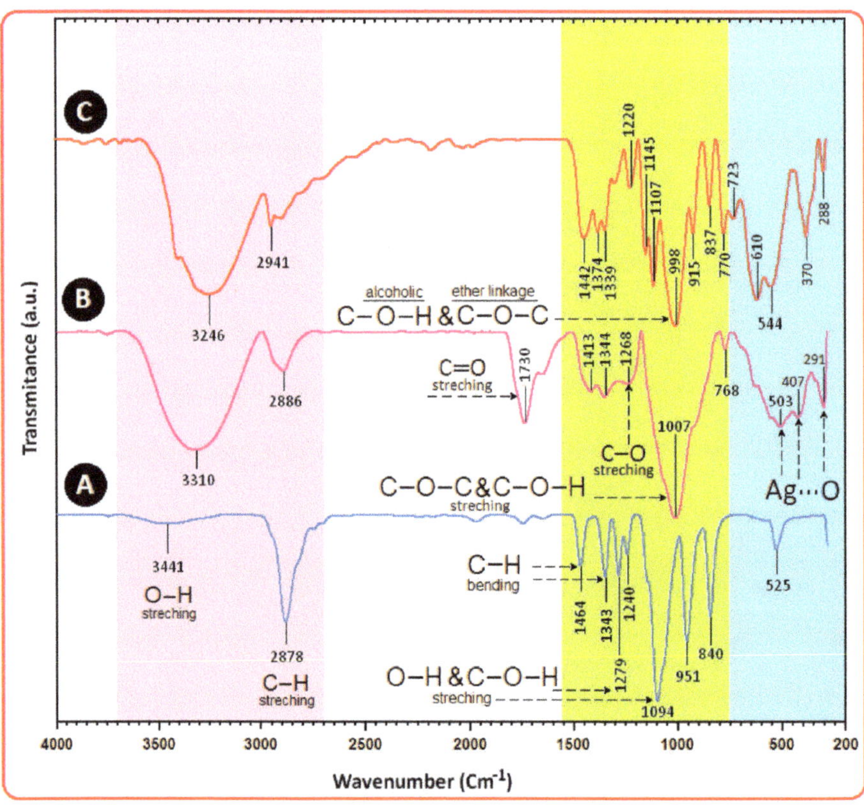

Figure 6. FTIR spectra for PEG (**A**), [Ag(PEG)] for the stirring time of 48 h (**B**) and β-D-glucose (**C**). Figure was adopted from [198]; CC-BY License.

4. Antimicrobial Applications of Green Silver Nanoparticles

4.1. Antibacterial Applications

It is well known that elemental silver and its various compounds [30] have been utilized for decades to preserve water in the form of silver coins/silver vessels. Since the Middle Ages [199], silver has been used as an inhibitory and antibacterial material, highlighting its activity as an antibacterial agent. The antimicrobial properties of AgNPs obtained via green synthesis are summarized in Table 1. In one study, it was proven that AgNPs [199] synthesized by employing black cohosh, geranium, aloe, etc. at a concentration of around 4 ppm exhibited an inhibitory effect on the proliferation of *E. coli*. The NPs synthesized using aloe displayed the highest antibacterial activity, whereas the black cohosh and geranium NPs did not display a notable effect. The aloe-based NPs displayed greater activity toward the growth of *Salmonella typhimurium* than that of *E. coli* [200]. These plant extract-based nanoparticles exhibited greater inhibitory effects against *Salmonella* and *E. coli*, whereas they exerted moderate activity toward *Pseudomonas aeruginosa*, *Kocura rhizophila*, and *Bacillus subtilis*. Some scientists have proposed that AgNPs can show significant activity toward our respiratory system and cell membrane permeability by adhering to their surface [58]. Furthermore, it has been observed that AgNPs can perforate the bacterial cell membrane surface. Moreover, they have also been found to exert greater activity toward Gram-negative bacteria than Gram-positive bacteria [58]. For instance, in one report, AgNPs synthesized using *Aloe vera* via a hydrothermal method were studied for their bactericidal effect against pathogenic Gram-positive *S. epidermidis* and Gram-negative

P. aeruginosa, wherein, at a lower AgNP concentration, there was major lethality exhibited toward Gram-negative bacteria only, whereas higher concentrations also displayed activity toward Gram-positive bacteria [58]. This is possibly because of the beta barrel proteins present, which are commonly known as porins, as well as the thinner peptidoglycan layer in Gram-negative bacteria compared to Gram-positive bacteria. However, it has also been observed that, when the surface area of NPs is increased, the surface energy also increases, thereby increasing their effectiveness [83]. Therefore, even at a lower concentration, smaller NPs possessing a greater surface-area-to-volume ratio exhibit significant antibacterial activity. It was also observed in a report that Gram-negative bacterium *E. coli* displayed a maximum zone of inhibition of 10.75 mm, possibly due to the cell wall of Gram-positive bacteria being composed of a thick peptidoglycan layer forming a rigid structure, leading to difficult penetration of the AgNPs, unlike Gram-negative bacteria, whose cell wall constitutes a thinner peptidoglycan layer. The high bactericidal activity is certainly attributed to the silver cations released from AgNPs acting as reservoirs of Ag^+ (bactericidal agent) ions. These ions from the NPs are speculated to attach to the negatively charged bacterial cell wall and rupture it, leading to protein denaturation and cell death [61].

Several studies support the appreciable antibacterial activity displayed by various green AgNPs. For instance, a study evaluating the activity of Fu-AgNPs toward *K. pneumoniae*, a Gram-negative bacterium which causes *Klebsiella* infections, revealed that the antibacterial activity of these AgNPs was impressive by inhibiting the multiplication of bacterial cells despite their multidrug resistance [37]. In another report, the disc diffusion method was utilized to evaluate the antibacterial activity of AgNPs fabricated from *S. fusiformis* toward human bacterial pathogens such as *E. coli*, *S. aureus*, *K. pneumoniae*, and *P. aeruginosa*, wherein the zone of inhibition was measured to obtain results in line with the proposed characteristic activity of the AgNPs [39]. The results of another experiment indicated that the extracellularly fabricated AgNPs employing *P. deceptionensis* DC5 revealed the highest antimicrobial potential against *V. parahaemolyticus*, followed by *C. albicans*, *S. aureus*, *S. enterica*, and *B. anthracis*.

A well-diffusion method revealed that bio(AgNPs) exhibited the highest antimicrobial activity toward *P. aeruginosa*, *S. aureus*, and *P. mirabilis* (10 mm for all), followed by *E. coli*, *K. pneumoniae*, and *B. subtilis* (6 mm for all). However, no activity was detected toward *S. infantis*. The authors also observed synergistic effects of the green AgNPs with antibiotics against bacterial pathogens. For instance, the antibacterial effect of ampicillin, kanamycin, and tetracycline was enriched in the presence of bio(AgNPs) against *S. aureus*, *K. pneumoniae*, and *P. aeruginosa*, along with improved streptomycin activity toward *P. mirabilis* when combined with these AgNPs [43] (see Table 1).

In line with these results, numerous bacterial pathogens exhibit distinct susceptibilities to AgNPs. A study was conducted on the combined effect of AgNPs with six standard antibiotic discs (ciprofloxacin, streptomycin, ampicillin, tetracycline, gentamicin, and lincomycin) against a few multidrug-resistant pathogenic bacteria [201]. For all tested cases, the resulting inhibition zone diameters were notably increased when the antibiotics were combined with AgNPs. It is, thus, evident that the synergy of AgNPs with antibiotics has appreciable antimicrobial effects, which hinders the development of resistance and improves the antimicrobial properties of the antibiotics, while also decreasing their dosage [44]. In addition to their antibacterial activity, the cytotoxic activity of AgNPs on Hep2 cell lines was also studied using the method of Daikoku et al. (1989); moreover, on similar lines, [40] first reported the cytotoxic activity of AgNPs synthesized using seaweed extracts of *G. corticate*. Moreover, nanoparticles are well known for their target specificity. In one study, silver nanoparticles were synthesized using *B. methylotrophicus* DC3 via an ecofriendly approach, showing antimicrobial activity toward numerous pathogenic microorganisms. Several metals and their respective salts [78] have been reported to possess antibacterial activity toward *H. pylori* [202]. The activity of these metallic agents may be due to the inactivation of *H. pylori* urease [203]. Even particles synthesized via olive leaf extract [78] showed significant antibacterial activities. In another instance, silver nanoparticles synthesized using [204] carob leaf extract displayed significant activity toward *E. coli*. The synthesized nanoparticles also exhibited significant inhibition activity toward *C. albicans*.

AgNPs synthesized [158] using *Spirogyra varians* were found to be 17.6 nm in size, and they acted as an appreciable antibacterial agent. These particles also offered potential effects against [103] multidrug-susceptible and -resistant strains such as ampicillin-resistant *Escherichia coli*, *Pseudomonas aeruginosa*, methicillin-resistant *Staphylococcus aureus* (MRSA), erythromycin-resistant *Streptococcus pyogenes*, and vancomycin-resistant *Staphylococcus aureus* (VRSA). Their antimicrobial activities [92] are also represented by the widespread use of AgNPs in cardiovascular implants. Interestingly, a prosthetic silicone heart valve was the first cardiovascular device to be coated with silver [92]. Here, Ag was employed to prevent bacterial contamination from taking place on the silicone valve. This further reduced the inflammation of the heart due to contamination. However, it was found that silver induces an allergic reaction and inhibits normal fibroblast function, among other effects, in patients during clinical trials. Consequently, the incorporation of AgNPs in medical devices may provide a nontoxic, safe, and antibacterial coating to overcome the stated challenges.

Another very prevalent use of AgNPs is for catheters in hospitals, which show a significant probability of contamination, thus leading to further complications. In one study, the synthesized AgNPs were examined for the inhibition of biofilm formation by *S. aureus* and *P. aeruginosa*. It was observed that a 5–6 µg concentration of AgNPs was sufficient for the inhibition of biofilm formation by these microbes. Therefore, the synthesized NPs clearly presented appreciable antimicrobial and biofilm inhibition potential, suggesting potential application as antimicrobial agents in the future [53]. For instance, polyurethane catheters are coated with AgNPs to create antibacterial catheters. It was observed that such modified catheters can significantly reduce the growth of bacteria up to 72 h in several models involving animals. Silver injury dressings are utilized to treat various injuries, chronic ulcers, burns, toxic epidermal necrolysis, etc. Furthermore, AgNPs utilized in wound dressing reduce the therapeutic time of injury by a standard of 3.35 days and prevent bacterial growth. They do not have any adverse impacts on patients when compared to standard gauze dressing and Ag sulfadiazine. Another study used chitosan AgNPs for dressing wounds, which displayed enhanced therapeutic action compared to 1% Ag sulfadiazine [92].

Many types of orthopedic and orthodontic implants commonly suffer from contamination, leading to further complications in treatment; therefore, AgNPs have also been introduced into plain poly bones to reduce bacterial resistance. They can also aid in decreasing the microbial colonization of coating materials in dentistry and improve antifungal activity. Moreover, AgNPs incorporated into endodontic fillings displayed prolonged activity toward *Staphylococcus aureus*, *Streptococcus milleri*, and *Enterococcus faecalis* [92]. Moreover, the utilization of antimicrobial plant extracts as reducing and capping agents also facilitates the synthesis of nanoparticles possessing greater activity [103]. Thus, if the capping agents themselves have antimicrobial activity, the antimicrobial activity may be further enhanced. The antimicrobial effect of silver nanoparticles is mediated by several factors which should be considered. One important property is the size. In a broader sense, nanoparticles should be smaller than 50 nm to display appreciable antimicrobial activity, while NPs 10–15 nm in size exhibit higher activity. The highest antibacterial activity in the literature was reported [103] in nanoparticles possessing a size of around 5 nm, wherein increased membrane permeability was observed due to the adherence of the NPs onto the cell membrane, further leading to membrane destruction and cell demise [205]. It was stated that AgNPs with an approximate size of 20–80 nm showed toxicity due to Ag^+ ions being released, whereas those 10 nm in size or smaller displayed better cell–particle interactions and displayed greater toxicity, resulting in increased intracellular bioavailability. Another important factor is the shape. The antimicrobial activity of triangular, spherical, and hexagonal AgNPs toward Gram-negative *E. coli* was compared [206]. It was found that hexagonal NPs displayed the greatest activity, whereas triangular AgNPs showed no activity. Thus, their activity may also depend on their shape. For instance, truncated triangular silver nanoplates displayed the greatest antibacterial activity due to their crystallographic surface structures and larger surface-area-to-volume ratios [207]. A few scientists have stated that the antimicrobial activity of NPs does not depend on their shape; thus, the exact mechanism of this dependency is still unknown.

Another considerable factor is the nanoparticle concentration. This can be directly correlated to microbial species [208]. The zeta potential affects the activity of NPs because electrostatic adhesion affects the interplay between particles and the cell membrane. The nanoparticle surface charge displays a direct relationship with the antibacterial activity [209]. For instance, Gram-positive *Bacillus* strains exhibited lower susceptibility to AgNPs compared to Gram-negative ones due to the repulsion between the negatively charged functional groups of biomolecules present on the cell surface and the negatively charged surface of the nanoparticles.

AgNPs [103] have displayed significant antimicrobial activity alone and in combination with antibiotics, mediated by three possible mechanisms: (i) cell wall and membrane damage, (ii) intracellular penetration and damage, and (iii) oxidative stress [83,103,210]. In one study [24], it was shown that AgNPs formed using disaccharides such as maltose and lactose possessed greater activity compared to those synthesized using monosaccharides such as glucose and galactose. This could be due to the smaller size of disaccharides compared to monosaccharides. The antibacterial effects of these nanoparticles may also correspond to [82] a dual mechanism that includes the bactericidal effect of Ag^+ ions and the membrane-disrupting effect of the polymer subunits. Moreover, in other applications [24], silver aerosol NPs were found to be significantly active as antimicrobials toward *B. subtilis* [211]. They have also been utilized in antimicrobial water filters, in activated carbon fiber (ACF) filters [212], and for the removal of bioaerosols. AgNPs and AuNPs synthesized extracellularly using *Fusarium oxysporum* have been employed in sterile clothes for use in hospitals to reduce infections by pathogenic bacteria such as *Staphylococcus aureus*. Thus, it is interesting to that AgNPs may exert better antimicrobial activity than standard antibiotics.

4.2. Antifungal Applications

It is known that there is a high rate of mortality due to fungal infections [102], which continues to increase due to the limited number of new antifungal targets and the development of prophylactic antifungals leading to the emergence of resistant strains. Antifungal drugs can be adsorbed onto the surface of biogenic silver, which is thought to facilitate antifungal drug delivery [213,214]. Green AgNPs [214] also exhibit antifungal activity because of their bio-coating activity complementing their advantageous size. Green synthesized AgNPs were found to display high antifungal activity toward *Cryptococcus* and *Candida* species. AgNPs [215] formed using *Pilimelia columellifera* subsp. *pallida* SL19 showed activity toward fungi responsible for superficial mycoses, i.e., *M. furfur* and *C. albicans*. Silver nanoparticles synthesized using *Mentha pulegium* [216] aqueous extract displayed significant activity toward fluconazole-resistant *Candida albicans*. Moreover, the colonies of plant pathogenic fungi such as *Magnaporthe grisea* and *Bipolaris sorokiniana* were restrained by nano-ionic silver in vitro conditions [217]. AgNPs were also found to be appreciably active as a broad-spectrum fungicide toward *Botrytis cinerea* and *Alternaria alternata*. They also possess significant collaborative activity with several fungicides toward numerous plant pathogenic fungi such as *Fusarium oxysporum* (tomato wilt) and *Penicillium expansum* (apple rot) [218]. Furthermore, disruption of cell membrane formation [83] and further stoppage of fungal reproduction in *C. albanicans* species were reportedly due to the action of AgNPs [219].

One study investigated the antifungal activity of AgNPs toward 10 fungal pathogens, including *Aspergillus* spp., *Candida* spp., and *Fusarium* spp., revealing significant antifungal activity in all cases. In addition, yeast is also one of the main causes of fungal diseases.

In this study, the effect of AgNPs on the growth of yeast was investigated by adding AgNPs to YEPD. The growth curves of *Candida (C. albicans, C. parapsilosis, C. krusei,* and *C. tropicalis)* in the presence of AgNPs revealed that their growth can be completely inhibited by AgNPs acting as fungistatic agents. There is great scope for further research on a profitable alternative to treat various fungal diseases in the future [47].

4.3. Antiparasitic Applications

AgNPs were observed to have larvicidal activity toward *Culex quinquefasciatus* [74], dengue vector *Aedes aegypti* [56], malarial vector *A. subpictus* [190], filariasis vector *C. quinquefasciatus* [220], *Aedes aegypti* [221], *A. subpictu* [220], and other parasites. Although the proper mechanism is not yet known, the denaturation of sulfur-containing proteins and phosphorus-containing DNA by AgNPs further leads to the denaturation of enzymes, organelles, etc., which may be responsible for its activity [222]. Leishmaniasis is a disease caused by parasites of the *Leishmania* genus [223]. The high cost and low availability of antileishmanial drugs, as well as the developing resistance to these drugs, have made the current situation worrisome. However, this parasite is quite sensitive to AgNPs due to the generation of ROS. NPs possess combinatory action against *Leishmania tropica* under UV light [224]. Furthermore, it was observed that miltefosine-doped green AgNPs displayed an increased antileishmanial effect. Another study reported a similar promising activity (IC$_{50}$ value 4.37) of green spherical AgNPs (3–8 nm) by employing *Sargentodoxa cuneata* [225]. The spherical AgNPs (116 nm) synthesized via *Moringa oleifera* extract showed a significant reduction in the average size of leishmaniasis cutaneous lesions in mice [226]. Despite these reports, further studies and trials are necessary to establish a concrete conclusion.

5. Conclusions and Future Perspectives

Scientists around the world usually seem skeptical when a given process employs a significant number of chemicals that may not be as safe as they seem to be, and this uncertainty has given rise to the now widely known concept of "green" processes. Owing to the drawbacks associated with synthetic approaches such as the employment of the reactive and toxic reducing and stabilizing agents that lead to adverse effects, a similar scenario has evolved in nanoparticle synthesis, especially silver nanoparticle fabrication methods. Responding to this challenge, the current review covered various ecofriendly AgNP synthesis methods including phytosynthesis, microbial-mediated synthesis, and enzyme-based synthesis, revealing the potential of various organisms and biomolecules to be employed as bionanofactories for green synthesis. The research has displayed that these bionanofactories are quite cost-effective and environmentally benign, while offering easy scale-up compared to conventional methods. For instance, the phytosynthesis methods revealed that the roots, shoots, bark, leaves, peel, flowers, and fruits of plants can be exploited as bionanofactories that consist of different phytoconstituents such as flavonoids, terpenoids, pectin, sugars, ascorbic acid, and carotenoids which aid in the synthesis by act-ing as reducing and capping agents, in addition to contributing to the therapeutic effect of the formulation. The fact that these plant extracts are easy to handle and readily accessible adds to the advantages of utilizing them in the current scenario. Additionally, microbes such as bacteria, fungi, yeast, algae, and actinomycetes can be employed for AgNP syn-thesis. Studies have demonstrated that the microbial synthesis method includes reactions such as trapping, bioreduction, and capping that can occur intracellularly or extracellularly. However, extracellular synthesis overcomes the limitations of the intracellular method by easing out the difficulties related to large-scale production, cumbersome purification, etc. In contrast to microbial synthesis, the phytosynthesis method seems advantageous as it does not require the complicated process of nurturing cell cultures, which makes it a more economic and industrially viable process.

Furthermore, the enzyme-based synthesis of AgNPs has revealed the potential of vari-ous enzymes such as α-NADPH-dependent nitrate reductase, as well as biomolecules such as glucose, galactose lactose, starch, heparin, and chitosan and vitamins such as vitamin B2 and vitamin C (ascorbic acid), to be employed as bionanofactories. Ionic liquid-mediated synthesis, the irradiation method, and microwave-assisted synthesis have introduced modernization to the green synthesis concept. In addition, the characterization of the synthesized AgNPs plays a crucial role in this process. This involves the determination of size, surface charge, distribution, surface morphology (shape), and aggregation as evaluated by employing a number of analytical techniques such as UV/visible spectroscopy, XRD, FTIR, SEM, AFM, TEM, DLS, and zeta potential analysis. Characterization studies, including in vitro and in vivo studies, have displayed the huge impact of the physicochemical properties of AgNPs on their therapeutic and biological applications.

Hence, we believe that the world is laden with an infinite number of biological species and compounds, some of which have been discovered while others are yet to be explored by humans. This contributes to the positive outlook of this emerging field and its immense potential to add value to the green methods currently available for the fabrication of AgNPs, which have already led to breakthroughs in a variety of fields ranging from therapeutics to diagnostics.

Author Contributions: All authors participated in article preparation. All authors have read and agreed to the published version of the manuscript.

Funding: This research received no external funding.

Acknowledgments: The authors are thankful to the Institute of Chemical Technology, Mumbai, India for providing the facilities to carry out this work. V.J. is thankful to the Finnish Cultural Foundation (Ingrid, Toini and Olavi Martelius foundation). The figures were created using biorender.com. We are thankful for the all the reviewers.

Conflicts of Interest: The authors declare no conflict of interest.

References

1. Sharma, V.K.; Yngard, R.A.; Lin, Y. Silver nanoparticles: Green synthesis and their antimicrobial activities. *Adv. Colloid Interface Sci.* **2009**, *145*, 83–96. [CrossRef]
2. Rana, A.; Yadav, K.; Jagadevan, S. A comprehensive review on green synthesis of nature-inspired metal nanoparticles: Mechanism, application and toxicity. *J. Clean. Prod.* **2020**, *272*, 122880. [CrossRef]
3. Parveen, K.; Banse, V.; Ledwani, L. Green synthesis of nanoparticles: Their advantages and disadvantages. *AIP Conf. Proc.* **2016**, *1724*, 020048.
4. Gour, A.; Jain, N.K. Advances in green synthesis of nanoparticles. *Artif. Cells Nanomed. Biotechnol.* **2019**, *47*, 844–851. [CrossRef]
5. Raveendran, P.; Fu, J.; Wallen, S.L. Completely "Green" Synthesis and Stabilization of Metal Nanoparticles. *J. Am. Chem. Soc.* **2003**, *125*, 13940–13941. [CrossRef]
6. Dyawanapelly, S.; Junnuthula, V.R.; Singh, A. The Holy Grail of Polymer Therapeutics for Cancer Therapy: An Overview on the Pharmacokinetics and Bio Distribution. *Curr. Drug Metab.* **2015**, *16*, 522–537. [PubMed]
7. Junnuthula, V.; Boroujeni, A.S.; Cao, S.; Tavakoli, S.; Ridolfo, R.; Toropainen, E.; Ruponen, M.; van Hest, J.; Urtti, A. Intravitreal Polymeric Nanocarriers with Long Ocular Retention and Targeted Delivery to the Retina and Optic Nerve Head Region. *Pharmaceutics* **2021**, *13*, 445. [CrossRef]
8. Ridolfo, R.; Tavakoli, S.; Junnuthula, V.; Williams, D.S.; Urtti, A.; van Hest, J.C.M. Exploring the Impact of Morphology on the Properties of Biodegradable Nanoparticles and Their Diffusion in Complex Biological Medium. *Biomacromolecules* **2021**, *22*, 126–133. [CrossRef] [PubMed]
9. Devassy, G.; Ramachandran, R.; Jeena, K.; Junnuthula, V.R.; Gopinatha, V.K.; Manju, C.A.; Manohar, M.; Nair, S.V.; Raghavan, S.C.; Koyakutty, M.; et al. Simultaneous release of two drugs from polymer nano-implant inhibits recurrence in glioblastoma spheroids. *Precis. Nanomed.* **2018**, *2*, 218–229. [CrossRef]
10. Albrecht, M.A.; Evans, C.W.; Raston, C.L. Green chemistry and the health implications of nanoparticles. *Green Chem.* **2006**, *8*, 417–432.
11. Hussain, I.; Singh, N.B.; Singh, A.; Singh, H.; Singh, S. Green synthesis of nanoparticles and its potential application. *Biotechnol. Lett.* **2015**, *38*, 545–560.
12. Barbalinardo, M.; Bertacchini, J.; Bergamini, L.; Magarò, M.S.; Ortolani, L.; Sanson, A.; Palumbo, C.; Cavallini, M.; Gentili, D. Surface properties modulate protein corona formation and determine cellular uptake and cytotoxicity of silver nanoparticles. *Nanoscale* **2021**, *13*, 14119–14129. [CrossRef]
13. Barbalinardo, M.; Caicci, F.; Cavallini, M.; Gentili, D. Protein Corona Mediated Uptake and Cytotoxicity of Silver Nanoparticles in Mouse Embryonic Fibroblast. *Small* **2018**, *14*, e1801219. [CrossRef]
14. Ksv, G. Green Synthesis of Iron Nanoparticles Using Green Tea leaves Extract. *J. Nanomed. Biother. Discov.* **2017**, *7*, 1. [CrossRef]

15. Kumar, S.; Lather, V.; Pandita, D. Green synthesis of therapeutic nanoparticles: An expanding horizon. *Nanomedicine* **2015**, *10*, 2451–2471.
16. Husen, A.; Siddiqi, K.S. Phytosynthesis of nanoparticles: Concept, controversy and application. *Nanoscale Res. Lett.* **2014**, *9*, 229. [CrossRef]
17. Narayanan, K.B.; Sakthivel, N. *Green Synthesis of Biogenic Metal Nanoparticles by Terrestrial and Aquatic Phototrophic and Heterotrophic Eukaryotes and Biocompatible Agents; Advances in Colloid and Interface Science*; Elsevier B.V.: Amsterdam, The Netherlands, 2011; Volume 169, pp. 59–79.
18. Hulkoti, N.I.; Taranath, T. Biosynthesis of nanoparticles using microbes—A review. *Colloids Surf. B Biointerfaces* **2014**, *121*, 474–483. [CrossRef] [PubMed]
19. Asmathunisha, N.; Kathiresan, K. A review on biosynthesis of nanoparticles by marine organisms. *Colloids Surf. B Biointerfaces* **2013**, *103*, 283–287. [CrossRef] [PubMed]
20. Schröfel, A.; Kratošová, G.; Šafařík, I.; Šafaříková, M.; Raška, I.; Shor, L. Applications of biosynthesized metallic nanoparticles—A review. *Acta Biomater.* **2014**, *10*, 4023–4042. [CrossRef]
21. Philip, D. Mangifera Indica leaf-assisted biosynthesis of well-dispersed silver nanoparticles. *Spectrochim. Acta Part A Mol. Biomol. Spectrosc.* **2011**, *78*, 327–331. [CrossRef]
22. Philip, D.; Unni, C.; Aromal, S.A.; Vidhu, V. Murraya Koenigii leaf-assisted rapid green synthesis of silver and gold nanoparticles. *Spectrochim. Acta Part A Mol. Biomol. Spectrosc.* **2011**, *78*, 899–904. [CrossRef]
23. Chhabra, R.; Peshattiwar, V.; Pant, T.; Deshpande, A.; Modi, D.; Sathaye, S.; Tibrewala, A.; Dyawanapelly, S.; Jain, R.D.; Dandekar, P. In Vivo Studies of 3D Starch–Gelatin Scaffolds for Full-Thickness Wound Healing. *ACS Appl. Bio Mater.* **2020**, *3*, 2920–2929. [CrossRef]
24. Sarkar, A.; Junnuthula, V.; Dyawanapelly, S. Ocular Therapeutics and Molecular Delivery Strategies for Neovascular Age-Related Macular Degeneration (nAMD). *Int. J. Mol. Sci.* **2021**, *22*, 10594. [CrossRef]
25. Vasileva, P.; Donkova, B.; Karadjova, I.; Dushkin, C. Synthesis of starch-stabilized silver nanoparticles and their application as a surface plasmon resonance-based sensor of hydrogen peroxide. *Colloids Surf. A Physicochem. Eng. Asp.* **2011**, *382*, 203–210. [CrossRef]
26. Filippo, E.; Serra, A.; Buccolieri, A.; Manno, D. Green synthesis of silver nanoparticles with sucrose and maltose: Morphological and structural characterization. *J. Non-Cryst. Solids* **2010**, *356*, 344–350. [CrossRef]
27. Sulaiman, G.M.; Mohammed, W.H.; Marzoog, T.R.; Al-Amiery, A.A.A.; Kadhum, A.A.H.; Mohamad, A.B. Green synthesis, antimicrobial and cytotoxic effects of silver nanoparticles using *Eucalyptus chapmaniana* leaves extract. *Asian Pac. J. Trop. Biomed.* **2013**, *3*, 58–63. [CrossRef]
28. Wang, C.; Kim, Y.J.; Singh, P.; Mathiyalagan, R.; Jin, Y.; Yang, D.C. Green synthesis of silver nanoparticles by *Bacillus methylotrophicus*, and their antimicrobial activity. *Artif. Cells Nanomed. Biotechnol.* **2015**, *44*, 1127–1132.
29. Dhand, V.; Soumya, L.; Bharadwaj, S.; Chakra, S.; Bhatt, D.; Sreedhar, B. Green synthesis of silver nanoparticles using *Coffea arabica* seed extract and its antibacterial activity. *Mater. Sci. Eng. C* **2016**, *58*, 36–43. [CrossRef]
30. Ahmed, S.; Saifullah; Ahmad, M.; Swami, B.L.; Ikram, S. Green synthesis of silver nanoparticles using *Azadirachta indica* aqueous leaf extract. *J. Radiat. Res. Appl. Sci.* **2016**, *9*, 1–7. [CrossRef]
31. Velmurugan, P.; Sivakumar, S.; Young-Chae, S.; Seong-Ho, J.; Pyoung-In, Y.; Jeong-Min, S.; Sung-Chul, H. Synthesis and characterization comparison of peanut shell extract silver nanoparticles with commercial silver nanoparticles and their antifungal activity. *J. Ind. Eng. Chem.* **2015**, *31*, 51–54.
32. Ahmad, S.; Munir, S.; Zeb, N.; Ullah, A.; Khan, B.; Ali, J.; Bilal, M.; Omer, M.; Alamzeb, M.; Salman, S.M.; et al. Green nanotechnology: A review on green synthesis of silver nanoparticles—An ecofriendly approach. *Int. J. Nanomed.* **2019**, *14*, 5087–5107. [CrossRef]
33. Moulton, M.C.; Braydich-Stolle, L.K.; Nadagouda, M.N.; Kunzelman, S.; Hussain, S.M.; Varma, R.S. Synthesis, characterization and biocompatibility of "green" synthesized silver nanoparticles using tea polyphenols. *Nanoscale* **2010**, *2*, 763–770. [CrossRef]
34. Elumalai, E.K.; Kayalvizhi, K.; Silvan, S. Coconut water assisted green synthesis of silver nanoparticles. *J. Pharm. Bioallied Sci.* **2014**, *6*, 241–245. [CrossRef]
35. Shankar, S.S.; Ahmad, A.; Sastry, M. Geranium Leaf Assisted Biosynthesis of Silver Nanoparticles. *Biotechnol. Prog.* **2003**, *19*, 1627–1631. [CrossRef] [PubMed]
36. Minhas, F.T.; Arslan, G.; Gubbuk, I.H.; Akkoz, C.; Ozturk, B.Y.; Asıkkutlu, B.; Arslan, U.; Ersoz, M. Evaluation of antibacterial properties on polysulfone composite membranes using synthesized biogenic silver nanoparticles with *Ulva compressa* (L.) Kütz. and *Cladophora glomerata* (L.) Kütz. extracts. *Int. J. Biol. Macromol.* **2018**, *107*, 157–165. [PubMed]
37. Salari, Z.; Danafar, F.; Dabaghi, S.; Ataei, S.A. Sustainable synthesis of silver nanoparticles using macroalgae *Spirogyra varians* and analysis of their antibacterial activity. *J. Saudi Chem. Soc.* **2016**, *20*, 459–464. [CrossRef]
38. Ravichandran, A.; Subramanian, P.; Manoharan, V.; Muthu, T.; Periyannan, R.; Thangapandi, M.; Ponnuchamy, K.; Pandi, B.; Marimuthu, P.N. Phyto-mediated synthesis of silver nanoparticles using fucoidan isolated from *Spatoglossum asperum* and assessment of antibacterial activities. *J. Photochem. Photobiol. B Biol.* **2018**, *185*, 117–125. [CrossRef]
39. Murugesan, S.; Bhuvaneswari, S.; Sivamurugan, V. Green synthesis, characterization of silver nanoparticles of a marine red alga *Spyridia fusiformis* and their antibacterial activity. *Int. J. Pharm. Pharm. Sci.* **2017**, *9*, 192–197. [CrossRef]

40. Saraniya Devi, J.; Valentin Bhimba, B. Biogenic synthesis by *Gracilaria corticata* for efficient production of biocompatible silver nanoparticles and its applications. *Int. J. Nanopart.* **2013**, *6*, 312–323. [CrossRef]

41. Sonker, A.S.; Pathak, J.; Kannaujiya, V.K.; Sinha, R.P. Characterization and in vitro antitumor, antibacterial and antifungal activities of green synthesized silver nanoparticles using cell extract of *Nostoc* sp. strain HKAR-2. *Can. J. Biotechnol.* **2017**, *1*, 26–37.

42. Singh, G.; Babele, P.K.; Shahi, S.K.; Sinha, R.P.; Tyagi, M.B.; Kumar, A. Green Synthesis of Silver Nanoparticles Using Cell Extracts of *Anabaena doliolum* and Screening of Its Antibacterial and Antitumor Activity. *J. Microbiol. Biotechnol.* **2014**, *24*, 1354–1367. [CrossRef] [PubMed]

43. Buszewski, B.; Railean-Plugaru, V.; Pomastowski, P.; Rafińska, K.; Szultka-Młyńska, M.; Golinska, P.; Wypij, M.; Laskowski, D.; Dahm, H. Antimicrobial activity of biosilver nanoparticles produced by a novel *Streptacidiphilus durhamensis* strain. *J. Microbiol. Immunol. Infect.* **2018**, *51*, 45–54.

44. Abdelnaby, H.; Abo-Elala, G.M.; Abdel-Raouf, U.M.; Hamed, M. Antibacterial and anticancer activity of extracellular synthesized silver nanoparticles from marine *Streptomyces rochei* MHM13. *Egypt. J. Aquat. Res.* **2016**, *42*, 301–312. [CrossRef]

45. Sanjenbam, P.; Gopal, J.; Kannabiran, K. Anticandidal activity of silver nanoparticles synthesized using *Streptomyces* sp. VITPK1. *J. Mycol. Médicale* **2014**, *24*, 211–219. [CrossRef] [PubMed]

46. Manivasagan, P.; Venkatesan, J.; Senthilkumar, K.; Sivakumar, K.; Kim, S.K. Biosynthesis, antimicrobial and cytotoxic effect of silver nanoparticles using a novel *Nocardiopsis* sp. MBRC-1. *BioMed Res. Int.* **2013**, *2013*, 287638. [CrossRef] [PubMed]

47. Xue, B.; He, D.; Gao, S.; Wang, D.; Yokoyama, K.; Wang, L. Biosynthesis of silver nanoparticles by the fungus *Arthroderma fulvum* and its antifungal activity against genera of *Candida*, *Aspergillus* and *Fusarium*. *Int. J. Nanomed.* **2016**, *11*, 1899–1906.

48. Al-Bahrani, R.; Raman, J.; Lakshmanan, H.; Hassan, A.A.; Sabaratnam, V. Green synthesis of silver nanoparticles using tree oyster mushroom *Pleurotus ostreatus* and its inhibitory activity against pathogenic bacteria. *Mater. Lett.* **2017**, *186*, 21–25. [CrossRef]

49. Singh, T.; Jyoti, K.; Patnaik, A.; Singh, A.; Chauhan, R.; Chandel, S. Biosynthesis, characterization and antibacterial activity of silver nanoparticles using an endophytic fungal supernatant of *Raphanus sativus*. *J. Genet. Eng. Biotechnol.* **2017**, *15*, 31–39.

50. Biosynthesis of Silver Nanoparticles Using an Endophytic Fungus, Curvularialunata and Its Antimicrobial Potential. Available online: http://webcache.googleusercontent.com/search?q=cache:x28VJtAJul8J:files.aiscience.org/journal/article/pdf/702700 29.pdf+&cd=2&hl=en&ct=clnk&gl=fi (accessed on 8 September 2021).

51. Neethu, S.; Midhun, S.J.; Radhakrishnan, E.; Jyothis, M. Green synthesized silver nanoparticles by marine endophytic fungus *Penicillium polonicum* and its antibacterial efficacy against biofilm forming, multidrug-resistant *Acinetobacter baumanii*. *Microb. Pathog.* **2018**, *116*, 263–272. [CrossRef]

52. Jo, J.H.; Singh, P.; Kim, Y.J.; Wang, C.; Mathiyalagan, R.; Jin, C.-G.; Yang, D.C. *Pseudomonas deceptionensis* DC5-mediated synthesis of extracellular silver nanoparticles. *Artif. Cells Nanomed. Biotechnol.* **2015**, *44*, 1576–1581. [CrossRef]

53. Singh, P.; Kim, Y.J.; Wang, C.; Mathiyalagan, R.; Yang, D.C. *Weissella oryzae* DC6-facilitated green synthesis of silver nanoparticles and their antimicrobial potential. *Artif. Cells Nanomed. Biotechnol.* **2015**, *44*, 1569–1575. [CrossRef]

54. Nayak, P.S.; Arakha, M.; Kumar, A.; Asthana, S.; Mallick, B.C.; Jha, S. An approach towards continuous production of silver nanoparticles using *Bacillus thuringiensis*. *RSC Adv.* **2016**, *6*, 8232–8242.

55. Gan, L.; Zhang, S.; Zhang, Y.; He, S.; Tian, Y. Biosynthesis, characterization and antimicrobial activity of silver nanoparticles by a halotolerant *Bacillus endophyticus* SCU-L. *Prep. Biochem. Biotechnol.* **2018**, *48*, 582–588. [CrossRef]

56. Suresh, G.; Gunasekar, P.H.; Kokila, D.; Prabhu, D.; Dinesh, D.; Ravichandran, N.; Ramesh, B.; Koodalingam, A.; Siva, G.V. Green synthesis of silver nanoparticles using *Delphinium denudatum* root extract exhibits antibacterial and mosquito larvicidal activities. *Spectrochim. Acta Part A Mol. Biomol. Spectrosc.* **2014**, *127*, 61–66. [CrossRef]

57. He, Y.; Wei, F.; Ma, Z.; Zhang, H.; Yang, Q.; Yao, B.; Huang, Z.; Li, J.; Zeng, C.; Zhang, Q. Green synthesis of silver nanoparticles using seed extract of *Alpinia katsumadai*, and their antioxidant, cytotoxicity, and antibacterial activities. *RSC Adv.* **2017**, *7*, 39842–39851. [CrossRef]

58. Tippayawat, P.; Phromviyo, N.; Boueroy, P.; Chompoosor, A. Green synthesis of silver nanoparticles in aloe vera plant extract prepared by a hydrothermal method and their synergistic antibacterial activity. *PeerJ* **2016**, *4*, e2589. [CrossRef]

59. Paulkumar, K.; Gnanajobitha, G.; Vanaja, M.; RajeshKumar, S.; Malarkodi, C.; Pandian, K.; Annadurai, G. *Piper nigrum* Leaf and Stem Assisted Green Synthesis of Silver Nanoparticles and Evaluation of Its Antibacterial Activity against Agricultural Plant Pathogens. *Sci. World J.* **2014**, *2014*, 829894. [CrossRef]

60. Bankar, A.; Joshi, B.; Kumar, A.R.; Zinjarde, S. Banana peel extract mediated novel route for the synthesis of silver nanoparticles. *Colloids Surf. A Physicochem. Eng. Asp.* **2010**, *368*, 58–63. [CrossRef]

61. Gopinath, V.; MubarakAli, D.; Priyadarshini, S.; Priyadharsshini, N.M.; Thajuddin, N.; Velusamy, P. Biosynthesis of silver nanoparticles from *Tribulus terrestris* and its antimicrobial activity: A novel biological approach. *Colloids Surf. B Biointerfaces* **2012**, *96*, 69–74. [CrossRef] [PubMed]

62. Awwad, A.M.; Salem, N.M.; Abdeen, A.O. Advanced Materials Letters Biosynthesis of silver nanoparticles using Loquat leaf extract and its antibacterial activity. *Adv. Mater. Lett.* **2013**, *2013*, 338–342.

63. Kora, A.J.; Sashidhar, R.; Arunachalam, J. Gum kondagogu (*Cochlospermum gossypium*): A template for the green synthesis and stabilization of silver nanoparticles with antibacterial application. *Carbohydr. Polym.* **2010**, *82*, 670–679. [CrossRef]

64. Bankura, K.; Maity, D.; Mollick, M.; Mondal, D.; Bhowmick, B.; Bain, M.; Chakraborty, A.; Sarkar, J.; Acharya, K.; Chattopadhyay, D. Synthesis, characterization and antimicrobial activity of dextran stabilized silver nanoparticles in aqueous medium. *Carbohydr. Polym.* **2012**, *89*, 1159–1165. [CrossRef] [PubMed]

65. Ashraf, S.; Abbasi, A.Z.; Pfeiffer, C.; Hussain, S.Z.; Khalid, Z.M.; Gil, P.R.; Parak, W.J.; Hussain, I. Protein-mediated synthesis, pH-induced reversible agglomeration, toxicity and cellular interaction of silver nanoparticles. *Colloids Surf. B Biointerfaces* **2013**, *102*, 511–518. [CrossRef] [PubMed]
66. El-Seedi, H.R.; El-Shabasy, R.M.; Khalifa, S.A.M.; Saeed, A.; Shah, A.; Shah, R.; Iftikhar, F.J.; Abdel-Daim, M.M.; Omri, A.; Hajrahand, N.H.; et al. Metal nanoparticles fabricated by green chemistry using natural extracts: Biosynthesis, mechanisms, and applications. *RSC Adv.* **2019**, *9*, 24539–24559. [CrossRef]
67. Liu, J.; Hurt, R.H. Ion Release Kinetics and Particle Persistence in Aqueous Nano-Silver Colloids. *Environ. Sci. Technol.* **2010**, *44*, 2169–2175. [CrossRef] [PubMed]
68. Lee, Y.-J.; Kim, J.; Oh, J.; Bae, S.; Lee, S.; Hong, I.S.; Kim, S.-H. Ion-release kinetics and ecotoxicity effects of silver nanoparticles. *Environ. Toxicol. Chem.* **2011**, *31*, 155–159. [CrossRef]
69. Zhang, W.; Yao, Y.; Sullivan, N.; Chen, Y. Modeling the Primary Size Effects of Citrate-Coated Silver Nanoparticles on Their Ion Release Kinetics. *Environ. Sci. Technol.* **2011**, *45*, 4422–4428. [CrossRef]
70. Liu, J.; Sonshine, D.A.; Shervani, S.; Hurt, R.H. Controlled Release of Biologically Active Silver from Nanosilver Surfaces. *ACS Nano* **2010**, *4*, 6903–6913. [CrossRef]
71. Yuan, Z.; Li, J.; Cui, L.; Xu, B.; Zhang, H.; Yu, C.-P. Interaction of silver nanoparticles with pure nitrifying bacteria. *Chemosphere* **2013**, *90*, 1404–1411. [CrossRef]
72. Forough, M.; Khalil, F. Biological and green synthesis of silver nanoparticles. *Turk. J. Eng. Environ. Sci.* **2010**, *34*, 281–287.
73. Gardea-Torresdey, J.L.; Gomez, E.; Peralta-Videa, J.R.; Parsons, J.G.; Troiani, H.; Jose-Yacaman, M. Alfalfa sprouts: A natural source for the synthesis of silver nanoparticles. *Langmuir* **2003**, *19*, 1357–1361. [CrossRef]
74. Mondal, N.K.; Chowdhury, A.; Dey, U.; Mukhopadhya, P.; Chatterjee, S.; DAS, K.; Datta, J.K. Green synthesis of silver nanoparticles and its application for mosquito control. *Asian Pac. J. Trop. Dis.* **2014**, *4*, S204–S210. [CrossRef]
75. Bar, H.; Bhui, D.K.; Sahoo, G.P.; Sarkar, P.; De, S.P.; Misra, A. Green synthesis of silver nanoparticles using latex of *Jatropha curcas*. *Colloids Surf. A Physicochem. Eng. Asp.* **2009**, *339*, 134–139. [CrossRef]
76. Lin, Z.; Wu, J.; Xue, R.; Yang, Y. Spectroscopic characterization of Au3+ biosorption by waste biomass of *Saccharomyces cerevisiae*. *Spectrochim. Acta Part A Mol. Biomol. Spectrosc.* **2005**, *61*, 761–765. [CrossRef]
77. Vidhu, V.; Aromal, S.A.; Philip, D. Green synthesis of silver nanoparticles using *Macrotyloma uniflorum*. *Spectrochim. Acta Part A Mol. Biomol. Spectrosc.* **2011**, *83*, 392–397. [CrossRef]
78. Amin, M.; Anwar, F.; Janjua, M.R.S.A.; Iqbal, M.A.; Rashid, U. Green Synthesis of Silver Nanoparticles through Reduction with *Solanum xanthocarpum* L. Berry Extract: Characterization, Antimicrobial and Urease Inhibitory Activities against *Helicobacter pylori*. *Int. J. Mol. Sci.* **2012**, *13*, 9923–9941. [CrossRef]
79. Li, S.; Shen, Y.; Xie, A.; Yu, X.; Qiu, L.; Zhang, L.; Zhang, Q. Green synthesis of silver nanoparticles using *Capsicum annuum* L. extract. *Green Chem.* **2007**, *9*, 852. [CrossRef]
80. Ahmad, N.; Sharma, S. Green Synthesis of Silver Nanoparticles Using Extracts of *Ananas comosus*. *Green Sustain. Chem.* **2012**, *2*, 141–147. [CrossRef]
81. Yang, J.; Liu, R.H.; Halim, L. Antioxidant and antiproliferative activities of common edible nut seeds. *LWT Food Sci. Technol.* **2009**, *42*, 1–8. [CrossRef]
82. Singh, A.; Jain, D.; Upadhyay, M.K.; Khandelwal, N.; Verma, H.N. Green synthesis of silver nanoparticles using *Argemone mexicana* leaf extract and evaluation of their antimicrobial activities. *Dig. J. Nanomater. Biostruct.* **2010**, *5*, 483–489.
83. Srikar, S.K.; Giri, D.D.; Pal, D.B.; Mishra, P.K.; Upadhyay, S.N. Green Synthesis of Silver Nanoparticles: A Review. *Green Sustain. Chem.* **2016**, *6*, 34–56. [CrossRef]
84. Jain, S.; Mehata, M.S. Medicinal Plant Leaf Extract and Pure Flavonoid Mediated Green Synthesis of Silver Nanoparticles and their Enhanced Antibacterial Property. *Sci. Rep.* **2017**, *7*, 15867. [CrossRef]
85. Mallikarjuna, K.; Narasimha, G.; Dillip, G.R.; Praveen, B.; Shreedhar, B.; Lakshmi, C.S.; Reddy, B.V.S.; Raju, B.D.P. Green synthesis of silver nanoparticles using *Ocimum* leaf extract and their characterization. *Dig. J. Nanomater. Biostruct.* **2011**, *6*, 181–186.
86. Ahsan, A.; Farooq, M.A.; Bajwa, A.A.; Parveen, A. Green Synthesis of Silver Nanoparticles Using Parthenium Hysterophorus: Optimization, Characterization and In Vitro Therapeutic Evaluation. *Molecules* **2020**, *25*, 3324. [CrossRef]
87. Jha, A.K.; Prasad, K. Green Synthesis of Silver Nanoparticles Using Cycas Leaf. *Int. J. Green Nanotechnol. Phys. Chem.* **2010**, *1*, P110–P117. [CrossRef]
88. Awwad, A.M.; Salem, N.M. Green Synthesis of Silver Nanoparticles by Mulberry Leaves Extract. *Nanosci. Nanotechnol.* **2012**, *2*, 125–128. [CrossRef]
89. Khalil, M.M.; Ismail, E.H.; El-Baghdady, K.Z.; Mohamed, D. Green synthesis of silver nanoparticles using olive leaf extract and its antibacterial activity. *Arab. J. Chem.* **2014**, *7*, 1131–1139.
90. Elavazhagan, T.; Arunachalam, K.D. Memecylon edule leaf extract mediated green synthesis of silver and gold nanoparticles. *Int. J. Nanomed.* **2011**, *6*, 1265–1278. [CrossRef]
91. Rauwel, P.; Küünal, S.; Ferdov, S.; Rauwel, E. A Review on the Green Synthesis of Silver Nanoparticles and Their Morphologies Studied via TEM. *Adv. Mater. Sci. Eng.* **2015**, *2015*, 682749. [CrossRef]
92. Rafique, M.; Sadaf, I.; Tahir, M.B. A review on green synthesis of silver nanoparticles and their applications. *Artif. Cells Nanomed. Biotechnol.* **2016**, *45*, 1272–1291. [CrossRef]

93. Sadeghi, B.; Gholamhoseinpoor, F. A study on the stability and green synthesis of silver nanoparticles using *Ziziphora tenuior* (Zt) extract at room temperature. *Spectrochim. Acta Part A Mol. Biomol. Spectrosc.* **2015**, *134*, 310–315. [CrossRef]

94. Vilchis-Nestor, A.R.; Sánchez-Mendieta, V.; Camacho-López, M.A.; Espinosa, R.M.G.; Camacho-López, M.A.; Arenas-Alatorre, J.A. Solventless synthesis and optical properties of Au and Ag nanoparticles using *Camellia sinensis* extract. *Mater. Lett.* **2008**, *62*, 3103–3105. [CrossRef]

95. Song, J.Y.; Jang, H.-K.; Kim, B.S. Biological synthesis of gold nanoparticles using *Magnolia kobus* and *Diopyros kaki* leaf extracts. *Process Biochem.* **2009**, *44*, 1133–1138.

96. Krishnaraj, C.; Jagan, E.; Rajasekar, S.; Selvakumar, P.; Kalaichelvan, P.; Mohan, N. Synthesis of silver nanoparticles using *Acalypha indica* leaf extracts and its antibacterial activity against water borne pathogens. *Colloids Surf. B Biointerfaces* **2010**, *76*, 50–56. [CrossRef]

97. Sathyavathi, R.; Krishna, M.B.; Rao, S.V.; Saritha, R.; Rao, D.N. Biosynthesis of Silver Nanoparticles Using Coriandrum Sativum Leaf Extract and Their Application in Nonlinear Optics. *Adv. Sci. Lett.* **2010**, *3*, 138–143. [CrossRef]

98. Dubey, S.P.; Lahtinen, M.; Särkkä, H.; Sillanpää, M. Bioprospective of *Sorbus aucuparia* leaf extract in development of silver and gold nanocolloids. *Colloids Surf. B Biointerfaces* **2010**, *80*, 26–33. [CrossRef]

99. Dubey, S.P.; Lahtinen, M.; Sillanpää, M. Green synthesis and characterizations of silver and gold nanoparticles using leaf extract of *Rosa rugosa*. *Colloids Surf. A Physicochem. Eng. Asp.* **2010**, *364*, 34–41. [CrossRef]

100. Das, S.K.; Liang, J.; Schmidt, M.; Laffir, F.; Marsili, E. Biomineralization Mechanism of Gold by Zygomycete Fungi *Rhizopous oryzae*. *ACS Nano* **2012**, *6*, 6165–6173. [PubMed]

101. Ahluwalia, V.; Kumar, J.; Sisodia, R.; Shakil, N.A.; Walia, S. Green synthesis of silver nanoparticles by *Trichoderma harzianum* and their bio-efficacy evaluation against *Staphylococcus aureus* and *Klebsiella pneumonia*. *Ind. Crop. Prod.* **2014**, *55*, 202–206. [CrossRef]

102. Liu, L.; Liu, T.; Tade, M.; Wang, S.; Li, X.; Liu, S. Less is more, greener microbial synthesis of silver nanoparticles. *Enzym. Microb. Technol.* **2014**, *67*, 53–58. [CrossRef] [PubMed]

103. Roy, A.; Bulut, O.; Some, S.; Mandal, A.K.; Yilmaz, M.D. Green synthesis of silver nanoparticles: Biomolecule-nanoparticle organizations targeting antimicrobial activity. *RSC Adv.* **2019**, *9*, 2673–2702. [CrossRef]

104. Otari, S.V.; Patil, R.M.; Ghosh, S.J.; Thorat, N.D.; Pawar, S.H. Intracellular synthesis of silver nanoparticle by actinobacteria and its antimicrobial activity. *Spectrochim. Acta Part A Mol. Biomol. Spectrosc.* **2015**, *136*, 1175–1180.

105. Singh, D.; Rathod, V.; Ninganagouda, S.; Hiremath, J.; Singh, A.K.; Mathew, J. Optimization and characterization of silver nanoparticle by endophytic fungi *Penicillium* sp. isolated from *Curcuma longa* (Turmeric) and application studies against MDR *E. coli* and *S. aureus*. *Bioinorg. Chem. Appl.* **2014**, *2014*, 408021. [CrossRef] [PubMed]

106. Velusamy, P.; Kumar, G.V.; Jeyanthi, V.; Das, J.; Pachaiappan, R. Bio-Inspired Green Nanoparticles: Synthesis, Mechanism, and Antibacterial Application. *Toxicol. Res.* **2016**, *32*, 95–102. [PubMed]

107. Mondal, A.H.; Yadav, D.; Mitra, S.; Mukhopadhyay, K. Biosynthesis of silver nanoparticles using culture supernatant of *Shewanella* sp. ARY1 and their antibacterial activity. *Int. J. Nanomed.* **2020**, *15*, 8295–8310.

108. Chandra, H.; Kumari, P.; Bontempi, E.; Yadav, S. Medicinal plants: Treasure trove for green synthesis of metallic nanoparticles and their biomedical applications. *Biocatal. Agric. Biotechnol.* **2020**, *24*, 101518. [CrossRef]

109. Oladipo, I.C.; Lateef, A.; Azeez, M.A.; Asafa, T.B.; Yekeen, T.A.; Akinboro, A.; Akinwale, A.S.; Gueguim-Kana, E.B.; Beukes, L.S. Green Synthesis and Antimicrobial Activities of Silver Nanoparticles using Cell Free-Extracts of *Enterococcus species*. *Not. Sci. Biol.* **2017**, *9*, 196–203. [CrossRef]

110. Sunkar, S.; Nachiyar, C.V. Biogenesis of antibacterial silver nanoparticles using the endophytic bacterium *Bacillus cereus* isolated from *Garcinia xanthochymus*. *Asian Pac. J. Trop. Biomed.* **2012**, *2*, 953–959. [CrossRef]

111. Klaus, T.; Joerger, R.; Olsson, E.; Granqvist, C.-G. Silver-based crystalline nanoparticles, microbially fabricated. *Proc. Natl. Acad. Sci. USA* **1999**, *96*, 13611–13614. [CrossRef]

112. Karthik, L.; Kumar, G.; Kirthi, A.V.; Rahuman, A.A.; Rao, K.V.B. *Streptomyces* sp. LK3 mediated synthesis of silver nanoparticles and its biomedical application. *Bioprocess Biosyst. Eng.* **2013**, *37*, 261–267. [CrossRef]

113. Golinska, P.; Wypij, M.; Ingle, A.P.; Gupta, I.; Dahm, H.; Rai, M. Biogenic synthesis of metal nanoparticles from actinomycetes: Biomedical applications and cytotoxicity. *Appl. Microbiol. Biotechnol.* **2014**, *98*, 8083–8097. [CrossRef]

114. Gudikandula, K.; Vadapally, P.; Charya, M.S. Biogenic synthesis of silver nanoparticles from white rot fungi: Their characterization and antibacterial studies. *OpenNano* **2017**, *2*, 64–78. [CrossRef]

115. Lengke, M.F.; Fleet, M.E.; Southam, G. Biosynthesis of Silver Nanoparticles by Filamentous Cyanobacteria from a Silver(I) Nitrate Complex. *Langmuir* **2007**, *23*, 2694–2699. [CrossRef] [PubMed]

116. Shahverdi, A.R.; Minaeian, S.; Shahverdi, H.R.; Jamalifar, H.; Nohi, A.-A. Rapid synthesis of silver nanoparticles using culture supernatants of Enterobacteria: A novel biological approach. *Process Biochem.* **2007**, *42*, 919–923. [CrossRef]

117. Quinteros, M.A.; Aiassa Martínez, I.M.; Dalmasso, P.R.; Páez, P.L. Silver Nanoparticles: Biosynthesis Using an ATCC Reference Strain of *Pseudomonas aeruginosa* and Activity as Broad Spectrum Clinical Antibacterial Agents. *Int. J. Biomater.* **2016**, *2016*, 5971047. [CrossRef]

118. Koilparambil, D.; Kurian, L.C.; Vijayan, S.; Manakulam Shaikmoideen, J. Green synthesis of silver nanoparticles by *Escherichia coli*: Analysis of antibacterial activity. *J. Water Environ. Nanotechnol.* **2016**, *1*, 63–74.

119. Gandhi, H.; Khan, S. Biological Synthesis of Silver Nanoparticles and Its Antibacterial Activity. *J. Nanomed. Nanotechnol.* **2016**, *7*, 2–4.

120. Huang, H.; Yang, X. Synthesis of polysaccharide-stabilized gold and silver nanoparticles: A green method. *Carbohydr. Res.* **2004**, *339*, 2627–2631. [CrossRef]
121. Samadi, N.; Golkaran, D.; Eslamifar, A.; Jamalifar, H.; Fazeli, M.R.; Mohseni, F.A. Intra/extracellular biosynthesis of silver nanoparticles by an autochthonous strain of *Proteus mirabilis* isolated from photographic waste. *J. Biomed. Nanotechnol.* **2009**, *5*, 247–253. [CrossRef]
122. Kharissova, O.V.; Dias, H.V.R.; Kharisov, B.I.; Pérez, B.O.; Pérez, V.M.J. The greener synthesis of nanoparticles. *Trends Biotechnol.* **2013**, *31*, 240–248. [PubMed]
123. Marshall, M.J.; Beliaev, A.S.; Dohnalkova, A.C.; Kennedy, D.; Shi, L.; Wang, Z.; Boyanov, M.I.; Lai, B.; Kemner, K.M.; McLean, J.; et al. c-Type Cytochrome-Dependent Formation of U(IV) Nanoparticles by *Shewanella oneidensis*. *PLoS Biol.* **2006**, *4*, 1324–1333. [CrossRef] [PubMed]
124. Ober, C.K. Self-assembly: Persistence pays off. *Science* **2002**, *296*, 859–861. [CrossRef]
125. Wei, X.; Luo, M.; Li, W.; Yang, L.; Liang, X.; Xu, L.; Kong, P.; Liu, H. Synthesis of silver nanoparticles by solar irradiation of cell-free *Bacillus amyloliquefaciens* extracts and AgNO3. *Bioresour. Technol.* **2012**, *103*, 273–278. [CrossRef]
126. Liu, L.; Cañizares, M.; Monger, W.; Perrin, Y.; Tsakiris, E.; Porta, C.; Shariat, N.; Nicholson, L.; Lomonossoff, G.P. Cowpea mosaic virus-based systems for the production of antigens and antibodies in plants. *Vaccine* **2005**, *23*, 1788–1792. [CrossRef]
127. Mukherjee, P.; Ahmad, A.; Mandal, D.; Senapati, S.; Sainkar, S.R.; Khan, M.I.; Parishcha, R.; Ajaykumar, P.V.; Alam, M.; Kumar, R.; et al. Fungus-Mediated Synthesis of Silver Nanoparticles and Their Immobilization in the Mycelial Matrix: A Novel Biological Approach to Nanoparticle Synthesis. *Nano Lett.* **2001**, *1*, 515–519. [CrossRef]
128. Abdelghany, T.M.; Al-Rajhi, A.M.H.; Al Abboud, M.A.; AlAwlaqi, M.M.; Magdah, A.G.; Helmy, E.A.M.; Mabrouk, A.S. Recent Advances in Green Synthesis of Silver Nanoparticles and Their Applications: About Future Directions. A Review. *BioNanoScience* **2017**, *8*, 5–16. [CrossRef]
129. Vigneshwaran, N.; Ashtaputre, N.; Varadarajan, P.; Nachane, R.; Paralikar, K.; Balasubramanya, R. Biological synthesis of silver nanoparticles using the fungus *Aspergillus flavus*. *Mater. Lett.* **2007**, *61*, 1413–1418.
130. Fayaz, M.; Tiwary, C.S.; Kalaichelvan, P.; Venkatesan, R. Blue orange light emission from biogenic synthesized silver nanoparticles using *Trichoderma viride*. *Colloids Surf. B Biointerfaces* **2010**, *75*, 175–178. [PubMed]
131. Ahmad, A.; Mukherjee, P.; Senapati, S.; Mandal, D.; Khan, M.; Kumar, R.; Sastry, M. Extracellular biosynthesis of silver nanoparticles using the fungus *Fusarium oxysporum*. *Colloids Surf. B Biointerfaces* **2003**, *28*, 313–318. [CrossRef]
132. Honary, S.; Barabadi, H.; Gharaei-Fathabad, E.; Naghibi, F. Green Synthesis of Silver Nanoparticles Induced by the Fungus *Penicillium citrinum*. *Trop. J. Pharm. Res.* **2013**, *12*, 7–11.
133. Balaji, D.; Basavaraja, S.; Deshpande, R.; Mahesh, D.B.; Prabhakar, B.; Venkataraman, A. Extracellular biosynthesis of functionalized silver nanoparticles by strains of *Cladosporium cladosporioides* fungus. *Colloids Surf. B Biointerfaces* **2009**, *68*, 88–92. [PubMed]
134. Korbekandi, H.; Mohseni, S.; Jouneghani, R.M.; Pourhossein, M.; Iravani, S. Biosynthesis of silver nanoparticles using *Saccharomyces cerevisiae*. *Artif. Cells Nanomed. Biotechnol.* **2014**, *44*, 235–239.
135. Kumar, K.S.K.; Dahms, H.-U.; Won, E.-J.; Lee, J.-S.; Shin, K.-H. Microalgae—A promising tool for heavy metal remediation. *Ecotoxicol. Environ. Saf.* **2014**, *113*, 329–352. [CrossRef] [PubMed]
136. Fraile, A.; Penche, S.; González, F.; Blázquez, M.L.; Muñoz, J.A.; Ballester, A. Biosorption of copper, zinc, cadmium and nickel by *Chlorella vulgaris*. *Chem. Ecol.* **2005**, *21*, 61–75.
137. Chaudhary, R.; Nawaz, K.; Khan, A.K.; Hano, C.; Abbasi, B.H.; Anjum, S. An Overview of the Algae-Mediated Biosynthesis of Nanoparticles and Their Biomedical Applications. *Biomolecules* **2020**, *10*, 1498. [CrossRef] [PubMed]
138. Patel, V.; Berthold, D.; Puranik, P.; Gantar, M. Screening of cyanobacteria and microalgae for their ability to synthesize silver nanoparticles with antibacterial activity. *Biotechnol. Rep.* **2014**, *5*, 112–119. [CrossRef]
139. Vieira, A.P.; Stein, E.M.; Andreguetti, D.X.; Colepicolo, P.; Ferreira, A.M.D.C. Preparation of silver nanoparticles using aqueous extracts of the red algae *Laurencia aldingensis* and *Laurenciella* sp. And their cytotoxic activities. *J. Appl. Phycol.* **2016**, *28*, 2615–2622. [CrossRef]
140. Abdeen, S.; Geo, S.; Praseetha, P.K.; Dhanya, R.P. Biosynthesis of silver nanoparticles from Actinomycetes for therapeutic applications. *Int. J. Nano Dimens.* **2014**, *5*, 155–162.
141. Chauhan, R.; Kumar, A.; Abraham, J. A Biological Approach to Synthesis of Silver Nanoparticles with Streptomyces sp JAR1 and its Antimicrobial Activity. *Sci. Pharm.* **2013**, *81*, 607–621.
142. Kumar, S.A.; Abyaneh, M.K.; Gosavi, S.W.; Kulkarni, S.K.; Pasricha, R.; Ahmad, A.; Khan, M.I. Nitrate reductase-mediated synthesis of silver nanoparticles from AgNO3. *Biotechnol. Lett.* **2007**, *29*, 439–445. [CrossRef]
143. Cheviron, P.; Gouanvé, F.; Espuche, E. Green synthesis of colloid silver nanoparticles and resulting biodegradable starch/silver nanocomposites. *Carbohydr. Polym.* **2014**, *108*, 291–298. [CrossRef]
144. Zain, N.M.; Stapley, A.; Shama, G. Green synthesis of silver and copper nanoparticles using ascorbic acid and chitosan for antimicrobial applications. *Carbohydr. Polym.* **2014**, *112*, 195–202. [CrossRef] [PubMed]
145. Yan, J.-K.; Cai, P.-F.; Cao, X.-Q.; Ma, H.-L.; Zhang, Q.; Hu, N.-Z.; Zhao, Y.-Z. Green synthesis of silver nanoparticles using 4-acetamido-TEMPO-oxidized curdlan. *Carbohydr. Polym.* **2013**, *97*, 391–397. [CrossRef] [PubMed]
146. Nadagouda, M.N.; Varma, R.S. Green synthesis of Ag and Pd nanospheres, nanowires, and nanorods using vitamin B2: Catalytic polymerisation of aniline and pyrrole. *J. Nanomater.* **2008**, *2008*, 782358. [CrossRef]

147. Díaz-Álvarez, A.E.; Cadierno, V. Glycerol: A promising Green Solvent and Reducing Agent for Metal-Catalyzed Transfer Hydrogenation Reactions and Nanoparticles Formation. *Appl. Sci.* **2013**, *3*, 55–69.

148. Raveendran, P.; Fu, J.; Wallen, S.L. A simple and "green" method for the synthesis of Au, Ag, and Au–Ag alloy nanoparticles. *Green Chem.* **2006**, *8*, 34–38.

149. Hoekstra, J.; Beale, A.M.; Soulimani, F.; Versluijs-Helder, M.; Geus, J.W.; Jenneskens, L.W. Shell decoration of hydrothermally obtained colloidal carbon spheres with base metal nanoparticles. *N. J. Chem.* **2015**, *39*, 6593–6601. [CrossRef]

150. He, Y.; Wu, X.; Lu, G.; Shi, G. A facile route to silver nanosheets. *Mater. Chem. Phys.* **2006**, *98*, 178–182. [CrossRef]

151. Kvítek, L.; Prucek, R.; Panáček, A.; Novotný, R.; Hrbáč, J.; Zbořil, R. The influence of complexing agent concentration on particle size in the process of SERS active silver colloid synthesis. *J. Mater. Chem.* **2005**, *15*, 1099–1105.

152. Christy, A.J.; Umadevi, M. Synthesis and characterization of monodispersed silver nanoparticles. *Adv. Nat. Sci. Nanosci. Nanotechnol.* **2012**, *3*, 035013. [CrossRef]

153. Batabyal, S.K.; Basu, C.; Das, A.R.; Sanyal, G.S. Green Chemical Synthesis of Silver Nanowires and Microfibers Using Starch. *J. Biobased Mater. Bioenergy* **2007**, *1*, 143–147. [CrossRef]

154. Shao, Y.; Wu, C.; Wu, T.; Yuan, C.; Chen, S.; Ding, T.; Ye, X.; Hu, Y. Green synthesis of sodium alginate-silver nanoparticles and their antibacterial activity. *Int. J. Biol. Macromol.* **2018**, *111*, 1281–1292. [CrossRef] [PubMed]

155. Malassis, L.; Dreyfus, R.; Murphy, R.J.; Hough, L.A.; Donnio, B.; Murray, C.B. One-step green synthesis of gold and silver nanoparticles with ascorbic acid and their versatile surface post-functionalization. *RSC Adv.* **2016**, *6*, 33092–33100.

156. Ahmad, N.; Sharma, S.; Singh, V.N.; Shamsi, S.F.; Fatma, A.; Mehta, B.R. Biosynthesis of Silver Nanoparticles from *Desmodium triflorum*: A Novel Approach Towards Weed Utilization. *Biotechnol. Res. Int.* **2011**, *2011*, 454090. [CrossRef] [PubMed]

157. Goharshadi, E.K.; Ding, Y.; Jorabchi, M.N.; Nancarrow, P. Ultrasound-assisted green synthesis of nanocrystalline ZnO in the ionic liquid [hmim][NTf2]. *Ultrason. Sonochem.* **2009**, *16*, 120–123. [CrossRef] [PubMed]

158. Tsai, T.-H.; Thiagarajan, S.; Chen, S.-M. Green Synthesis of Silver Nanoparticles Using Ionic Liquid and Application for the Detection of Dissolved Oxygen. *Electroanalysis* **2010**, *22*, 680–687. [CrossRef]

159. Rodil, E.; Aldous, L.; Hardacre, C.; Lagunas, M.C. Preparation of AgX (X = Cl, I) nanoparticles using ionic liquids. *Nanotechnology* **2008**, *19*, 105603. [CrossRef]

160. Iida, M.; Baba, C.; Inoue, M.; Yoshida, H.; Taguchi, E.; Furusho, H. Ionic Liquids of Bis(alkylethylenediamine)silver(I) Salts and the Formation of Silver(0) Nanoparticles from the Ionic Liquid System. *Chem. Eur. J.* **2008**, *14*, 5047–5056. [CrossRef]

161. Pringle, J.M.; Ngamna, O.W.; Lynam, C.; Wallace, G.G.; Forsyth, M.; Macfarlane, D.R. One-Step Synthesis of Conducting Polymer-Noble Metal Nanoparticle Composites using an Ionic Liquid. *Adv. Funct. Mater.* **2008**, *18*, 2031–2040. [CrossRef]

162. Wu, S.; Ding, Y.S.; Zhang, X.M.; Tang, H.O.; Chen, L.; Li, B. Structure and morphology controllable synthesis of Ag/carbon hybrid with ionic liquid as soft-template and their catalytic properties. *J. Solid State Chem.* **2008**, *181*, 2171–2177. [CrossRef]

163. Kang, S.W.; Char, K.; Kang, Y.S. Novel Application of Partially Positively Charged Silver Nanoparticles for Facilitated Transport in Olefin/Paraffin Separation Membranes. *Chem. Mater.* **2008**, *20*, 1308–1311. [CrossRef]

164. Bhatt, A.I.; Bond, A.M. Electrodeposition of silver from the 'distillable' ionic liquid, DIMCARB in the absence and presence of chemically induced nanoparticle formation. *J. Electroanal. Chem.* **2008**, *619–620*, 1–10. [CrossRef]

165. Li, N.; Bai, X.; Zhang, S.; Gao, Y.; Zheng, L.; Zhang, J.; Ma, H. Synthesis of Silver Nanoparticles in Ionic Liquid by a Simple Effective Electrochemical Method. *J. Dispers. Sci. Technol.* **2008**, *29*, 1059–1061. [CrossRef]

166. Erythropel, H.C.; Zimmerman, J.B.; de Winter, T.M.; Petitjean, L.; Melnikov, F.; Lam, C.H.; Lounsbury, A.W.; Mellor, K.E.; Janković, N.Z.; Tu, Q.; et al. The Green ChemisTREE: 20 years after taking root with the 12 principles. *Green Chem.* **2018**, *20*, 1929–1961.

167. Abid, J.P.; Wark, A.W.; Brevet, P.F.; Girault, H.H. Preparation of silver nanoparticles in solution from a silver salt by laser irradiation. *Chem. Commun.* **2002**, *7*, 792–793. [CrossRef]

168. Eustis, S.; Krylova, G.; Eremenko, A.; Smirnova, N.; Schill, A.W.; El-Sayed, M. Growth and fragmentation of silver nanoparticles in their synthesis with a fs laser and CW light by photo-sensitization with benzophenone. *Photochem. Photobiol. Sci.* **2004**, *4*, 154–159. [CrossRef] [PubMed]

169. Sudeep, P.K.; Kamat, P.V. Photosensitized growth of silver nanoparticles under visible light irradiation: A mechanistic investigation. *Chem. Mater.* **2005**, *17*, 5404–5410. [CrossRef]

170. Zhang, L.; Yu, J.C.; Yip, H.Y.; Li, Q.; Kwong, K.W.; Xu, A.W.; Wong, P.K. Ambient Light Reduction Strategy to Synthesize Silver Nanoparticles and Silver-Coated TiO_2 with Enhanced Photocatalytic and Bactericidal Activities. *Langmuir* **2003**, *19*, 10372–10380. [CrossRef]

171. Huang, N.; Radiman, S.; Lim, H.N.; Khiew, P.; Chiu, W.S.; Lee, K.; Syahida, A.; Hashim, R.; Chia, C.H. γ-Ray assisted synthesis of silver nanoparticles in chitosan solution and the antibacterial properties. *Chem. Eng. J.* **2009**, *155*, 499–507. [CrossRef]

172. Chen, J.; Wang, J.; Zhang, X.; Jin, Y. Microwave-assisted green synthesis of silver nanoparticles by carboxymethyl cellulose sodium and silver nitrate. *Mater. Chem. Phys.* **2008**, *108*, 421–424.

173. Li, A.; Kaushik, M.; Li, C.-J.; Moores, A. Microwave-Assisted Synthesis of Magnetic Carboxymethyl Cellulose-Embedded Ag–Fe3O4 Nanocatalysts for Selective Carbonyl Hydrogenation. *ACS Sustain. Chem. Eng.* **2015**, *4*, 965–973. [CrossRef]

174. Baruwati, B.; Varma, R.S. High value products from waste: Grape pomace extract-a three-in-one package for the synthesis of metal nanoparticles. *ChemSusChem* **2009**, *2*, 1041–1044. [PubMed]

175. Hu, B.; Wang, S.-B.; Wang, K.; Zhang, M.; Yu, S.-H. Microwave-Assisted Rapid Facile "Green" Synthesis of Uniform Silver Nanoparticles: Self-Assembly into Multilayered Films and Their Optical Properties. *J. Phys. Chem. C* **2008**, *112*, 11169–11174. [CrossRef]

176. Noroozi, M.; Zakaria, A.; Moksin, M.M.; Wahab, Z.A.; Abedini, A. Green Formation of Spherical and Dendritic Silver Nanostructures under Microwave Irradiation without Reducing Agent. *Int. J. Mol. Sci.* **2012**, *13*, 8086–8096.

177. Kahrilas, G.A.; Wally, L.M.; Fredrick, S.J.; Hiskey, M.; Prieto, A.L.; Owens, J.E. Microwave-Assisted Green Synthesis of Silver Nanoparticles Using Orange Peel Extract. *ACS Sustain. Chem. Eng.* **2013**, *2*, 367–376. [CrossRef]

178. Zhao, X.; Xia, Y.; Li, Q.; Ma, X.; Quan, F.; Geng, C.; Han, Z. Microwave-assisted synthesis of silver nanoparticles using sodium alginate and their antibacterial activity. *Colloids Surf. A Physicochem. Eng. Asp.* **2014**, *444*, 180–188. [CrossRef]

179. Albadran, F.H.; Kamal, I.M. Synthesize of green silver nanoparticles by one pot microwave-assisted technique: Modeling and optimization. *Period. Eng. Nat. Sci.* **2020**, *8*, 1591–1599.

180. Anjana, V.N.; Joseph, M.; Francis, S.; Joseph, A.; Koshy, E.P.; Mathew, B. Microwave assisted green synthesis of silver nanoparticles for optical, catalytic, biological and electrochemical applications. *Artif. Cells Nanomed. Biotechnol.* **2021**, *49*, 438–449. [CrossRef]

181. Shin, S.W.; Song, I.H.; Um, S.H. Role of Physicochemical Properties in Nanoparticle Toxicity. *Nanomaterials* **2015**, *5*, 1351–1365. [CrossRef]

182. Jo, D.H.; Kim, J.H.; Lee, T.G.; Kim, J.H. Size, surface charge, and shape determine therapeutic effects of nanoparticles on brain and retinal diseases. *Nanomed. Nanotechnol. Biol. Med.* **2015**, *11*, 1603–1611. [CrossRef]

183. Gurunathan, S.; Han, J.W.; Kim, E.S.; Park, J.H.; Kim, J.-H. Reduction of graphene oxide by resveratrol: A novel and simple biological method for the synthesis of an effective anticancer nanotherapeutic molecule. *Int. J. Nanomed.* **2015**, *10*, 2951–2969.

184. Jagtap, U.B.; Bapat, V.A. Green synthesis of silver nanoparticles using *Artocarpus heterophyllus* Lam. seed extract and its antibacterial activity. *Ind. Crop. Prod.* **2013**, *46*, 132–137. [CrossRef]

185. Vijayaraghavan, K.; Nalini, S.K.; Prakash, N.U.; Nk, U.P. One step green synthesis of silver nano/microparticles using extracts of *Trachyspermum ammi* and *Papaver somniferum*. *Colloids Surf. B Biointerfaces* **2012**, *94*, 114–117. [CrossRef] [PubMed]

186. Vigneshwaran, N.; Kathe, A.A.; Varadarajan, P.; Nachane, R.P.; Balasubramanya, R. Biomimetics of silver nanoparticles by white rot fungus, *Phaenerochaete chrysosporium*. *Colloids Surf. B Biointerfaces* **2006**, *53*, 55–59.

187. Ortega-Arroyo, L.; Martin-Martinez, E.S.; Aguilar-Mendez, M.A.; Cruz-Orea, A.; Hernandez-Pérez, I.; Glorieux, C. Green synthesis method of silver nanoparticles using starch as capping agent applied the methodology of surface response. *Starch-Stärke* **2013**, *65*, 814–821. [CrossRef]

188. Kaviya, S.; Santhanalakshmi, J.; Viswanathan, B. Biosynthesis of silver nano-flakes by *Crossandra infundibuliformis* leaf extract. *Mater. Lett.* **2012**, *67*, 64–66. [CrossRef]

189. Sreekanth, T.V.M.; Nagajyothi, P.C.; Lee, K.D. Dioscorea batatas Rhizome-Assisted Rapid Biogenic Synthesis of Silver and Gold Nanoparticles. *Synth. React. Inorg. Met. Chem.* **2012**, *42*, 567–572. [CrossRef]

190. Rajakumar, G.; Rahuman, A.A. Larvicidal activity of synthesized silver nanoparticles using *Eclipta prostrata* leaf extract against filariasis and malaria vectors. *Acta Trop.* **2011**, *118*, 196–203. [CrossRef]

191. Sant, D.G.; Gujarathi, T.R.; Harne, S.R.; Ghosh, S.; Kitture, R.; Kale, S.; Chopade, B.A.; Pardesi, K.R. *Adiantum philippense* L. Frond Assisted Rapid Green Synthesis of Gold and Silver Nanoparticles. *J. Nanopart.* **2013**, *2013*, 182320. [CrossRef]

192. Sathishkumar, G.; Gobinath, C.; Karpagam, K.; Hemamalini, V.; Premkumar, K.; Sivaramakrishnan, S. Phyto-synthesis of silver nanoscale particles using *Morinda citrifolia* L. and its inhibitory activity against human pathogens. *Colloids Surf. B Biointerfaces* **2012**, *95*, 235–240. [CrossRef] [PubMed]

193. Ramaye, Y.; Dabrio, M.; Roebben, G.; Kestens, V. Development and Validation of Optical Methods for Zeta Potential Determination of Silica and Polystyrene Particles in Aqueous Suspensions. *Materials* **2021**, *14*, 290. [CrossRef]

194. Mourdikoudis, S.; Pallares, R.M.; Thanh, N.T. Characterization techniques for nanoparticles: Comparison and complementarity upon studying nanoparticle properties. *Nanoscale* **2018**, *10*, 12871–12934. [CrossRef] [PubMed]

195. Liz-Marzán, L.M. Nanometals: Formation and Color*. In *Colloidal Synthesis of Plasmonic Nanometals*; Jenny Stanford Publishing: Singapore, 2020; Volume 7, pp. 26–31. [CrossRef]

196. Giannini, C.; Ladisa, M.; Altamura, D.; Siliqi, D.; Sibillano, T.; De Caro, L. X-ray Diffraction: A Powerful Technique for the Multiple-Length-Scale Structural Analysis of Nanomaterials. *Crystals* **2016**, *6*, 87. [CrossRef]

197. Smiechowicz, E.; Niekraszewicz, B.; Kulpinski, P. Optimisation of AgNP Synthesis in the Production and Modification of Antibacterial Cellulose Fibres. *Materials* **2021**, *14*, 4126. [CrossRef] [PubMed]

198. Shameli, K.; Bin Ahmad, M.; Jazayeri, S.D.; Sedaghat, S.; Shabanzadeh, P.; Jahangirian, H.; Mahdavi, M.; Abdollahi, Y. Synthesis and Characterization of Polyethylene Glycol Mediated Silver Nanoparticles by the Green Method. *Int. J. Mol. Sci.* **2012**, *13*, 6639–6650. [CrossRef]

199. Okafor, F.; Janen, A.; Kukhtareva, T.; Edwards, V.; Curley, M. Green synthesis of silver nanoparticles, their characterization, application and antibacterial activity. *Int. J. Environ. Res. Public Health* **2013**, *10*, 5221–5238.

200. Loo, Y.Y.; Rukayadi, Y.; Nor-Khaizura, M.-A.; Kuan, C.H.; Chieng, B.W.; Nishibuchi, M.; Radu, S. In Vitro Antimicrobial Activity of Green Synthesized Silver Nanoparticles against Selected Gram-negative Foodborne Pathogens. *Front. Microbiol.* **2018**, *9*, 1555. [PubMed]

201. Yuan, Y.-G.; Peng, Q.-L.; Gurunathan, S. Effects of Silver Nanoparticles on Multiple Drug-Resistant Strains of *Staphylococcus aureus* and *Pseudomonas aeruginosa* from Mastitis-Infected Goats: An Alternative Approach for Antimicrobial Therapy. *Int. J. Mol. Sci.* **2017**, *18*, 569. [CrossRef]

202. Amin, M.; Iqbal, M.S.; Hughes, R.W.; Khan, S.A.; Reynolds, P.A.; Enne, V.I.; Sajjad-ur-Rahman; Mirza, A.S. Mechanochemical synthesis and in vitro anti-*Helicobacter pylori* and uresase inhibitory activities of novel zinc(II)famotidine complex. *J. Enzym. Inhib. Med. Chem.* **2010**, *25*, 383–390.

203. Zaborska, W.; Krajewska, B.; Olech, Z. Heavy Metal Ions Inhibition of Jack Bean Urease: Potential for Rapid Contaminant Probing. *J. Enzym. Inhib. Med. Chem.* **2004**, *19*, 65–69.

204. Awwad, A.M.; Salem, N.M.; Abdeen, A.O. Green synthesis of silver nanoparticles using carob leaf extract and its antibacterial activity. *Int. J. Ind. Chem.* **2013**, *4*, 29.

205. Ivask, A.; Kurvet, I.; Kasemets, K.; Blinova, I.; Aruoja, V.; Suppi, S.; Vija, H.; Kakinen, A.; Titma, T.; Heinlaan, M.; et al. Size-Dependent Toxicity of Silver Nanoparticles to Bacteria, Yeast, Algae, Crustaceans and Mammalian Cells In Vitro. *PLoS ONE* **2014**, *9*, e102108. [CrossRef]

206. El-Zahry, M.R.; Refaat, I.H.; Mohamed, H.A.; Rosenberg, E.; Lendl, B. Utility of surface enhanced Raman spectroscopy (SERS) for elucidation and simultaneous determination of some penicillins and penicilloic acid using hydroxylamine silver nanoparticles. *Talanta* **2015**, *144*, 710–716. [CrossRef] [PubMed]

207. Kulkarni, A.P.; Srivastava, A.A.; Harpale, P.M.; Zunjarrao, R.S. Plant mediated synthesis of silver nanoparticles—Tapping the unexploited sources. *J. Nat. Prod. Plant Resour.* **2011**, *1*, 100–107.

208. Kim, J.S.; Kuk, E.; Yu, K.N.; Kim, J.-H.; Park, S.; Lee, H.J.; Kim, S.H.; Park, Y.K.; Park, Y.H.; Hwang, C.-Y.; et al. Antimicrobial effects of silver nanoparticles. *Nanomed. Nanotechnol. Biol. Med.* **2007**, *3*, 95–101. [CrossRef]

209. El Badawy, A.M.; Silva, R.G.; Morris, B.; Scheckel, K.G.; Suidan, M.T.; Tolaymat, T.M. Surface charge-dependent toxicity of silver nanoparticles. *Environ. Sci. Technol.* **2011**, *45*, 283–287. [CrossRef]

210. Chopade, B.A.; Ghosh, S.; Patil, S.; Ahire, M.; Kitture, R.; Jabgunde, A.; Kale, S.; Pardesi, K.; Cameotra, S.S.; Bellare, J.; et al. Synthesis of silver nanoparticles using *Dioscorea bulbifera* tuber extract and evaluation of its synergistic potential in combination with antimicrobial agents. *Int. J. Nanomed.* **2012**, *7*, 483–496.

211. Yoon, K.-Y.; Byeon, J.H.; Park, J.-H.; Ji, J.H.; Bae, G.N.; Hwang, J. Antimicrobial Characteristics of Silver Aerosol Nanoparticles against *Bacillus subtilis* Bioaerosols. *Environ. Eng. Sci.* **2008**, *25*, 289–294. [CrossRef]

212. Ki, Y.Y.; Jeong, H.B.; Chul, W.P.; Hwang, J. Antimicrobial effect of silver particles on bacterial contamination of activated carbon fibers. *Environ. Sci. Technol.* **2008**, *42*, 1251–1255.

213. Scorzoni, L.; Silva, A.C.A.D.P.E.; Marcos, C.M.; Assato, P.; De Melo, W.C.M.A.; Oliveira, H.; Costa-Orlandi, C.B.; Giannini, M.J.M.; Fusco-Almeida, A.M. Antifungal Therapy: New Advances in the Understanding and Treatment of Mycosis. *Front. Microbiol.* **2017**, *8*, 36. [CrossRef] [PubMed]

214. Ahmad, A.; Wei, Y.; Syed, F.; Tahir, K.; Taj, R.; Khan, A.U.; Hameed, M.U.; Yuan, Q. Amphotericin B-conjugated biogenic silver nanoparticles as an innovative strategy for fungal infections. *Microb. Pathog.* **2016**, *99*, 271–281. [CrossRef]

215. Khan, N.T.; Mushtaq, M. Determination of Antifungal Activity of Silver Nanoparticles Produced from *Aspergillus niger*. *Biol. Med.* **2017**, *9*, 1–4. [CrossRef]

216. Wypij, M.; Czarnecka, J.; Dahm, H.; Rai, M.; Golinska, P. Silver nanoparticles from *Pilimelia columellifera* subsp. pallida SL19 strain demonstrated antifungal activity against fungi causing superficial mycoses. *J. Basic Microbiol.* **2017**, *57*, 793–800. [CrossRef]

217. Jo, Y.-K.; Kim, B.H.; Jung, G. Antifungal Activity of Silver Ions and Nanoparticles on Phytopathogenic Fungi. *Plant Dis.* **2009**, *93*, 1037–1043.

218. Ali, S.M.; Yousef, N.M.H.; Nafady, N.A. Application of biosynthesized silver nanoparticles for the control of land snail *Eobania vermiculata* and some plant pathogenic fungi. *J. Nanomater.* **2015**, *2015*, 218904. [CrossRef]

219. Kim, K.-J.; Sung, W.S.; Suh, B.K.; Moon, S.-K.; Choi, J.-S.; Kim, J.G.; Lee, D.G. Antifungal activity and mode of action of silver nano-particles on *Candida albicans*. *BioMetals* **2008**, *22*, 235–242. [CrossRef] [PubMed]

220. Santhoshkumar, T.; Rahuman, A.A.; Rajakumar, G.; Marimuthu, S.; Bagavan, A.; Jayaseelan, C.; Zahir, A.A.; Elango, G.; Kamaraj, C. Synthesis of silver nanoparticles using *Nelumbo nucifera* leaf extract and its larvicidal activity against malaria and filariasis vectors. *Parasitol. Res.* **2010**, *108*, 693–702. [CrossRef] [PubMed]

221. Kumar, V.; Yadav, S.K. Synthesis of stable, polyshaped silver, and gold nanoparticles using leaf extract of *Lonicera japonica* L. *Int. J. Green Nanotechnol. Biomed.* **2011**, *3*, 281–291. [CrossRef]

222. Gnanadesigan, M.; Anand, M.; Ravikumar, S.; Maruthupandy, M.; Vijayakumar, V.; Selvam, S.; Dhineshkumar, M.; Kumaraguru, A. Biosynthesis of silver nanoparticles by using mangrove plant extract and their potential mosquito larvicidal property. *Asian Pac. J. Trop. Med.* **2011**, *4*, 799–803. [CrossRef]

223. de Vries, H.J.C.; Reedijk, S.H.; Schallig, H.D.F.H. Cutaneous Leishmaniasis: Recent Developments in Diagnosis and Management. *Am. J. Clin. Dermatol.* **2015**, *16*, 99–109. [CrossRef]

224. Allahverdiyev, A.M.; Abamor, E.; Bagirova, M.; Ustundag, C.B.; Kaya, C.; Rafailovich, M. Antileishmanial effect of silver nanoparticles and their enhanced antiparasitic activity under ultraviolet light. *Int. J. Nanomed.* **2011**, *6*, 2705–2714. [CrossRef] [PubMed]

225. Ahmad, A.; Syed, F.; Shah, A.; Khan, Z.; Tahir, K.; Khan, A.U.; Yuan, Q. Silver and gold nanoparticles from *Sargentodoxa cuneata*: Synthesis, characterization and antileishmanial activity. *RSC Adv.* **2015**, *5*, 73793–73806. [CrossRef]
226. El-Khadragy, M.; AlOlayan, E.M.; Metwally, D.M.; El-Din, M.F.S.; Alobud, S.S.; Alsultan, N.I.; AlSaif, S.S.; Awad, M.A.; Moneim, A.E.A. Clinical Efficacy Associated with Enhanced Antioxidant Enzyme Activities of Silver Nanoparticles Biosynthesized Using Moringa oleifera Leaf Extract, Against Cutaneous Leishmaniasis in a Murine Model of Leishmania major. *Int. J. Environ. Res. Public Health* **2018**, *15*, 1037.

International Journal of
Molecular Sciences

Review

Effect of Cavity Disinfectants on Adhesion to Primary Teeth—A Systematic Review

Ana Coelho [1,2,3,*,†], Inês Amaro [1,†], Ana Apolónio [1], Anabela Paula [1,2,3], José Saraiva [1], Manuel Marques Ferreira [2,3,4], Carlos Miguel Marto [2,3,5,6] and Eunice Carrilho [1,2,3]

[1] Faculty of Medicine, Institute of Integrated Clinical Practice, University of Coimbra, 3000-075 Coimbra, Portugal; ines.amaros@hotmail.com (I.A.); anaclaudiia21@gmail.com (A.A.); anabelabppaula@sapo.pt (A.P.); ze-93@hotmail.com (J.S.); eunicecarrilho@gmail.com (E.C.)

[2] Area of Environment Genetics and Oncobiology (CIMAGO), Faculty of Medicine, Coimbra Institute for Clinical and Biomedical Research (iCBR), University of Coimbra, 3000-548 Coimbra, Portugal; m.mferreira@netcabo.pt (M.M.F.); cmiguel.marto@uc.pt (C.M.M.)

[3] Clinical Academic Center of Coimbra (CACC), 3004-561 Coimbra, Portugal

[4] Faculty of Medicine, Institute of Endodontics, University of Coimbra, 3000-075 Coimbra, Portugal

[5] Faculty of Medicine, Institute of Biophysics, University of Coimbra, 3004-548 Coimbra, Portugal

[6] Faculty of Medicine, Institute of Experimental Pathology, University of Coimbra, 3004-548 Coimbra, Portugal

* Correspondence: anasofiacoelho@gmail.com

† These authors contributed equally to this work.

Abstract: Some authors have been proposing the use of cavity disinfectants in order to reduce, or even eliminate, the effect of the microorganisms present in a dental cavity before a restoration is placed. The aim of this study was to evaluate the effect of different cavity disinfectants on bond strength and clinical success of composite and glass ionomer restorations on primary teeth. The research was conducted using Cochrane Library, PubMed/MEDLINE, SCOPUS, and Web of Science for articles published up to February 2021. The search was performed according to the PICO strategy. The evaluation of the methodological quality of each in vitro study was assessed using the CONSORT checklist for reporting in vitro studies on dental materials. Sixteen in vitro studies and one in situ study fulfilled the inclusion criteria and were analyzed. Chlorhexidine was the most studied cavity disinfectant, and its use does not compromise dentin bonding. Sodium hypochlorite is a promising alternative, but more research on its use is required to clearly state that it can safely be used as a cavity disinfectant for primary teeth. Although other disinfectants were studied, there is a low-level evidence attesting their effects on adhesion, therefore their use should be avoided.

Keywords: cavity disinfectants; primary teeth; adhesion; bond strength

Citation: Coelho, A.; Amaro, I.; Apolónio, A.; Paula, A.; Saraiva, J.; Ferreira, M.M.; Marto, C.M.; Carrilho, E. Effect of Cavity Disinfectants on Adhesion to Primary Teeth—A Systematic Review. *Int. J. Mol. Sci.* **2021**, *22*, 4398. https://doi.org/10.3390/ijms22094398

Academic Editors: Sotiris Hadjikakou, Christina N. Banti and Andreas K. Rossos

Received: 31 March 2021
Accepted: 22 April 2021
Published: 22 April 2021

Publisher's Note: MDPI stays neutral with regard to jurisdictional claims in published maps and institutional affiliations.

1. Introduction

Dental caries has a high prevalence worldwide, affecting more than 2.4 thousand million adults and 620 thousand children with primary teeth [1]. It can be defined as a multifactorial pathology arising from the interaction between dental structure and microbial biofilm, due to an imbalance between remineralization and demineralization, with the last one prevailing [2,3].

Although complete removal of the decayed and necrotic tissue is directly related to restorations' clinical success, cariogenic bacteria can be pushed deep inside the dentinal tubules while removing the carious tissue and remain viable for a long time. In fact, the remaining of cariogenic bacteria in the cavity can be associated with the development of secondary caries [4,5].

According to Dalkilic et al. [6], fermenting microorganisms can remain viable for 139 days in a restored cavity. Moreover, bacteria present in the smear layer may remain viable and proliferate, allowing their metabolism products to reach and to cause inflammatory changes in the dental pulp. Bacteria penetration through restoration and teeth interface (microinfiltration) can also explain restorations' failure [7–9].

As so, some authors have been proposing the use of cavity disinfectants in order to reduce, or even eliminate, the effect of the microorganisms present in a dental cavity before a restoration is placed [8–10].

Among the available disinfectants, chlorhexidine, sodium hypochlorite, and fluoridated solutions are the most used. Despite their benefits, their effect on adhesion to dentin, especially that of primary teeth, is still unknown [7,11,12].

Among the pediatric population, dental caries treatment is the most common procedure to be performed in a dental appointment [12]. Restorations' success rate is associated with dentist's experience and patient's collaboration. However, one of the most common causes of failure is the development of secondary caries [13–15].

Thereby, the aim of this systematic review was to evaluate the effect of the application of different cavity disinfectants on bond strength and clinical success of composite and glass ionomer restorations on primary teeth.

2. Results

Initial research on electronic databases resulted in 585 articles. After evaluating titles and abstracts, 41 articles were selected for full text analysis, and of those, 17 studies fulfilled the inclusion and exclusion criteria. The flowchart of the data selection process is detailed in Figure 1.

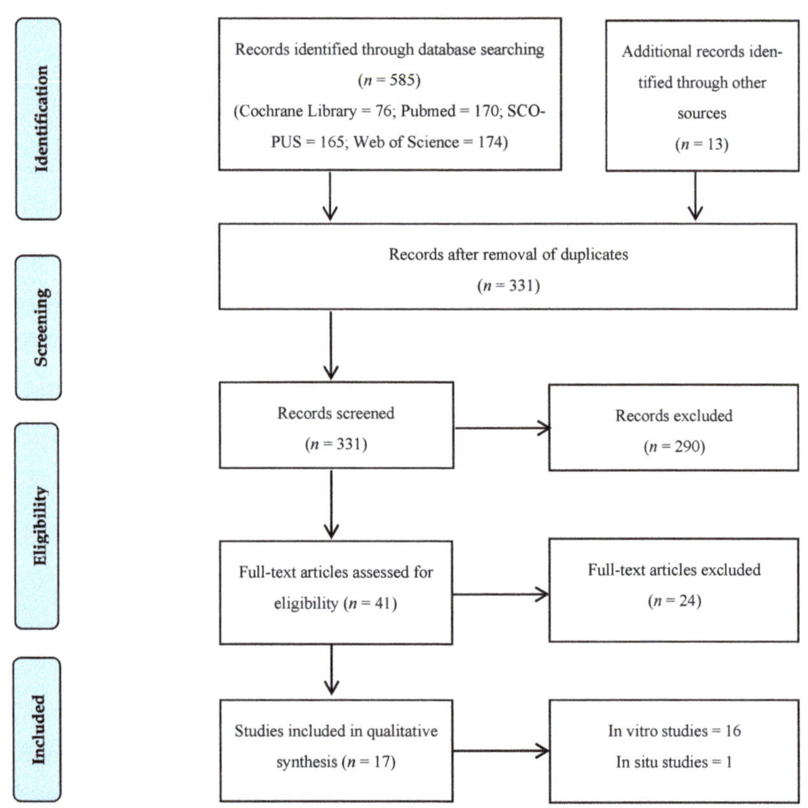

Figure 1. Flowchart of the data selection process.

Sixteen in vitro studies [12,16–30] were included in this systematic review.

The earliest study was published in 2003 [12], and the most recent one in 2020 [29].

Most authors used primary molars [12,16,18–28,30], but Monghini et al. [17] evaluated canines, and Mohammadi et al. [29] used anterior teeth. Sample size varied from 2 [25] to 20 [28,31] teeth per group.

Even though all authors studied healthy dentin, Ersin et al. [18] additionally evaluated carious dentin, and Lenzi et al. [22,23] also evaluated demineralized dentin (artificially induced lesions).

After extraction, teeth were stored in thymol [12,18,28], chloramine [16,22,23,27,30], distilled water [16,20,22–25,27,30,32], saline solution [17,26], or sodium azide [17,21]. Ricci et al. [19] and Mohammadi [29] did not report data on the storage medium used after teeth extraction.

All authors used water to store the specimens after adhesive experiments and before bond strength evaluation.

All authors reported results on adhesion to composite resin. Only Ersin et al. [18] also reported results on adhesion to glass ionomer materials.

Most of the authors reported the use of 2% chlorhexidine [12,18–20,22,23,25,28,29] as a cavity disinfectant. A few studies reported results on the application of sodium hypochlorite [16,24,27], Er:YAG laser [17,21,26], KTP laser [25], ozone [25], doxycycline [29], ethylenediaminetetraacetic acid (EDTA) [29], propolis [25], and Aqua-prep™ (Bisco, Schaumburg, IL, USA) [26].

Except for Vieira et al. [12], all of the authors studying the effect of 2% chlorhexidine as a cavity disinfectant [18–20,22,23,25,28,29] reported positive results, allowing for maintenance or a statistically significant increase in bond strength values. These values ranged from 7.58 $\pm$ 3.18 MPa [25] to 66.45 $\pm$ 8.3 MPa [28] in resin specimens and from 7.1 $\pm$ 5.2 MPa to 14.4 $\pm$ 6.6 MPa [18] in caries-affected dentin in glass ionomer specimens. Vieira et al. [12] were the only authors applying chlorhexidine before etching the specimens with phosphoric acid.

The authors evaluating the effect of the application of sodium hypochlorite tested different concentrations, ranging from 2.5% [27] to 10% [16]. Regardless of the concentration, all authors [16,24,27] reported positive results, allowing for maintenance or even a statistically significant increase in bond strength values. These values ranged from 9.9 $\pm$ 0.2 MPa [16] to 18.45 $\pm$ 2.30 MPa [24].

The Er:YAG laser was evaluated by three studies [17,21,26]. Monghini et al. [17] reported statistically significant negative results when testing the laser with three different working parameters. However, Scatena et al. [21] did not find statistically significant differences regarding bond strength results for different focal distances (mm), and Yildiz et al. [26] even reported a statistically significant increase in bond strength values. The bond strength values of these studies ranged from 5.07 $\pm$ 2.62 MPa [21] to 20.57 $\pm$ 9.02 MPa [26].

Oznurhan et al. [25] assessed the use of a KTP laser as a cavity disinfectant and found no statistically significant differences when comparing its results to the ones of the control group (9.58 $\pm$ 2.92 and 6.38 $\pm$ 2.47 MPa, respectively).

Gaseous ozone and ozonated water [25] were also tested as cavity disinfectants. The authors reported a maintenance of the bond strength values when using gaseous ozone (5.84 $\pm$ 2.62 MPa vs. 6.38 $\pm$ 2.47 MPa for the control group) and a statistically significant increase of the bond strength values when using ozonated water (11.12 $\pm$ 2.41 MPa vs. 6.38 $\pm$ 2.47 MPa for the control group).

Aqua-prep™ [26], an aqueous solution of fluoride and hydroxyethyl methacrylate (HEMA), 2% Doxycycline [29], 17% EDTA [29], and 30% propolis [25] were all evaluated in only one study each, and no statistically significant differences were found between test and control groups.

Relevant information on each in vitro study is summarized in Table 1.

Table 1. Characteristics of the in vitro studies included in the systematic review.

Authors, Year	Groups (n)	Teeth	Storage	Materials	Results (MPa)
Vieira et al., 2003 [12]	G_1—37% phosphoric acid + adhesive (10) + resin G_2—2% CHX + 37% phosphoric acid + adhesive (10) + resin	Molars	0.1% Thymol	Adhesive: 3M Single Bond (3M, USA) Resin: Filtek^{TM} Z250 (3M, USA)	G_1: 19.88 ± 1.04 G_2: 17.99 ± 1.15 G_1*/G_2
Correr et al., 2004 [16]	G_1—35% phosphoric acid + adhesive 1 (15) G_2—35% phosphoric acid + 10% NaOCl + adhesive 1 (15) G_3—37% phosphoric acid + adhesive 2 (15) G_4—37% phosphoric acid + 10% NaOCl + adhesive 2 (15) G_5–Adhesive 3 (15) G_6—10% NaOCl + adhesive 3 (15) + resin	Molars	0.5% Chloramine	Adhesive: 1–3M Single Bond; 2–Prime & Bond 2.1$^{®}$ (Dentsply, Brazil); 3–Clearfill^{TM} SE Bond (Kuraray, Houston, TX, USA) Resin: Filtek^{TM} Z250 (3M, USA)	G_1: 15.8 ± 1.9 G_2: 14.6 ± 1.3 G_3: 10.2 ± 0.7 G_4: 9.9 ± 0.2 G_5: 13.3 ± 1.2 G_6: 10.7 ± 1.0 G_1*/G_3
Monghini et al., 2004 [17]	G_1—None (12) G_2—Laser Er;YAG 60 mJ/2 Hz (12) G_3—Laser Er;YAG 80 mJ/2 Hz (12) G_4—Laser Er;YAG 100 mJ/2 Hz (12) + 35% phosphoric acid + adhesive + resin	Canines	0.9% Saline solution with 0.4% sodium azide	Adhesive: 3M Single Bond Laser: Kavo Key Laser 2 (Kavo Dental, Germany) Resin: Filtek^{TM} Z250	G_1:17.89 ± 4.75 G_2:12.34 ± 4.85 G_3:10.30 ± 3.67 G_4:10.41 ± 4.20 G_1*/G_2;G_3;G_4
Ersin et al., 2009 [18]	G_1—25% polyacrlylic acid + 2% CHX + GIC 1 (sound dentin) (3) G_2—25% polyacrlylic acid + 2% CHX + GIC 1 (carious dentin) (3) G_3—25% polyacrlylic acid + GIC 1 (sound dentin) (3) G_4—25% polyacrlylic acid + GIC 1 (carious dentin) (3) G_5—2% CHX + GIC 2 (sound dentin) (3) G_6—2% CHX + GIC 2 (carious dentin) (3) G_7—GIC 2 (sound dentin) (3) G_8—GIC 2 (carious dentin) (3) G_9—37% phosphoric acid + 2% CHX + adhesive + resin (sound dentin) (3) G_{10}-37% phosphoric acid + 2% CHX + adhesive + resin (carious dentin) (3) G_{11}—37% phosphoric acid + adhesive + resin (sound dentin) (3) G_{12}—37% phosphoric acid + adhesive + resin (carious dentin) (3)	Molars	0.1% Thymol	Adhesive: Prime & Bond$^{®}$; GIC: 1–Ketac^{TM} Molar (3M, Germany); 2–Vitremer^{TM} (3M, USA) Resin–Surefil^{TM} (Dentsply, USA)	G_1: 8.7 ± 4.3 G_2: 7.1 ± 5.2 G_3: 9.2 ± 5.2 G_4: 10.3 ± 6.6 G_5: 12.4 ± 5.7 G_6: 14.4 ± 6.6 G_7: 11.2 ± 4.8 G_8: 13.8 ± 4.9 G_9: 22.9 ± 6.9 G_{10}: 23.2 ± 6.2 G_{11}: 20.2 ± 5.8 G_{12}: 22.1 ± 6.2 G_9*/G_1;G_2;G_3;G_4;G_5;G_6;G_7;G_8 G_{10}*/G_1;G_2;G_3;G_4;G_5;G_6;G_7;G_8 G_{11}*/G_1;G_2;G_3;G_4;G_5;G_6;G_7;G_8 G_{12}*/G_1;G_2;G_3;G_4;G_5;G_6;G_7;G_8

Table 1. *Cont.*

Authors, Year	Groups (n)	Teeth	Storage	Materials	Results (MPa)
Ricci et al., 2010 [19]	35% phosphoric acid + G_1—2% CHX + adhesive 1 (4) G_2—deionized water + adhesive 1 (4) G_3—2% CHX + adhesive 2 (4) G_4—deionized water + adhesive 2 (4) G_5—2% CHX + adhesive 3 (4) G_6—deionized water + adhesive 3 (4) + resin	Molars	NA	Adhesive: 1–Adper^{TM} Single Bond (3M, USA); 2–Prime & Bond NT$^®$ (Dentsply, USA); 3–Excite$^®$ DSC (Ivoclar, Liechtenstein) Resin: Filtek^{TM} Z250	G_1: 47.4 ± 9.5 G_2: 41.4 ± 11.9 G_3: 48.0 ± 9.8 G_4: 40.8 ± 13.4 G_5: 45.2 ± 9.2 G_6: 43.4 ± 12.0 G_1*/G_2; G_3*/G_4
Leitune et al., 2011 [20]	37% phosphoric acid + G_1—Adhesive (24 h) (10) G_2—Adhesive (6 months) (10) G_3—2% CHX + Adhesive (24 h) (10) G_4—2% CHX + Adhesive (6 months) (10)	Molars	Distilled water	Adhesive: Adper^{TM} Scotchbond^{TM} Multi Purpose (3M, USA) Resin: Filtek^{TM} Z250	G_1: 22.37 ± 3.69 G_2: 19.93 ± 2.05 G_3: 22.30 ± 3.66 G_4: 24.48 ± 2.24 G_2*/G_4
Scatena et al., 2011 [21]	G_1–None (10) G_2—Laser Er:YAG (80 mJ, 11 mm) (10) G_3—Laser Er:YAG (80 mJ, 12 mm) (10) G_4—Laser Er:YAG (80 mJ, 16 mm) (10) G_5—Laser Er:YAG (80 mJ, 17 mm) (10) G_6—Laser Er:YAG (80 mJ, 20 mm) (10) + 37% phosphoric acid + adhesive + resin	Molars	0.4% Sodium azide	Laser: Kavo Key Laser 2 Adhesive: 3M Single Bond Resin: Filtek^{TM} Z250	G_1: 7.32 ± 3.83 G_2: 5.07 ± 2.62 G_3: 6.49 ± 1.64 G_4: 7.71 ± 0.66 G_5: 7.33 ± 0.02 G_6: 9.65 ± 2.41 G_2*/G_4;G_6
Manfro et al., 2012 [30]	37% phosphoric acid + G_1—water + adhesive (7) G_2—water + adhesive (12 months) (7) G_3—0.5% CHX + adhesive (7) G_4—0.5% CHX + adhesive (12 months) (7) G_5—2% CHX + adhesive (7) G_6—2% CHX + adhesive (12 months) (7) + resin	Molars	0.5% Chloramine	Adhesive: 3M Single Bond Resin: Filtek^{TM} Z250	G_1: 50.8 ± 12.8 G_2: 20.4 ± 3.7 G_3: 49.3 ± 2.6 G_4: 32.3 ± 7.9 G_5: 44.0 ± 8.7 G_6: 34.6 ± 5.1 G_1*/G_2; G_2*/G_4;G_6; G_3*/G_4; G_5*/G_6
Lenzi et al., 2012 [22]	35% phosphoric acid + G_1—distilled water + adhesive (sound dentin) (5) G_2—2% CHX + adhesive (sound dentin) (5) G_3—distilled water + adhesive (artificial caries) (5) G_4—2% CHX + adhesive (artificial caries) (5)	Molars	0.5% Chloramine	Adhesive: Adper^{TM} Single Bond 2 Resin: Filtek^{TM} Z250	G_1: 30.8 ± 2.2 G_2: 32.8 ± 3.8 G_3: 24.5 ± 3.8 G_4: 25.6 ± 3.6 G_1*/G_3;G_4; G_2*/G_3;G_4

Table 1. *Cont.*

Authors, Year	Groups (n)	Teeth	Storage	Materials	Results (MPa)
Aras et al., 2013 [24]	G_1—37% phosphoric acid (10) G_2—37% phosphoric acid + 5% NaOCl (10) G_3—5% NaOCl + 37% phosphoric acid (10) + adhesive + resin	Molars	Distilled water	Adhesive: Gluma® Confort Bond (Herause-Kulzer, Germany) Resin: Charisma® (Herause-Kulzer, Germany)	G_1: 14.51 ± 2.89 G_2: 18.45 ± 2.30 G_3: 17.06 ± 2.99 G_1*/G_2
Lenzi et al., 2014 [23]	35% phosphoric acid + G_1—distilled water + adhesive (sound dentin) (5) G_2—distilled water + adhesive (sound dentin) (6 months) (5) G_3—2% CHX (without rinsing) + adhesive (sound dentin) (5) G_4—2% CHX (without rinsing) + adhesive (sound dentin) (6 months) (5) G_5—distilled water + adhesive (artificial lesion) (5) G_6—distilled water + adhesive (artificial lesion) (6 months) (5) G_7—2% CHX (without rinsing) + adhesive (artificial lesion) (5) G_8—2% CHX (without rinsing) + adhesive (artificial lesion) (6 months) (5)	Molars	Distilled water	Adhesive: Adper™ Single Bond Resin: Filtek™ Z250	G_1: 30.7 ± 2.2 G_2: 25.9 ± 5.7 G_3: 32.8 ± 3.8 G_4: 31.3 ± 2.6 G_5: 26.2 ± 5.4 G_6: 20.0 ± 3.9 G_7: 28.3 ± 3.4 G_8: 26.9 ± 5.9 G_1*/G_5;G_7; G_2*/G_6;G_8 G_3*/G_5;G_7 G_4*/G_6;G_8
Oznurhan et al., 2015 [25]	G_1—2% CHX (2) G_2—30% propolis (2) G_3—Gaseous ozone (2) G_4—Ozonated water (2) G_5—Laser KTP (2) G_6—None (2) + adhesive + resin	Molars	Distilled water	Adhesive: Adper™ Prime & Bond NT® Resin: Tetric® N-Ceram (Ivoclar Vivadent, Liechenstein) Laser: Smartlite D (Deka, Italy)	G_1: 7.58 ± 3.18 G_2: 7.42 ± 2.28 G_3: 5.84 ± 2.62 G_4: 11.12 ± 2.41 G_5: 9.58 ± 2.92 G_6: 6.38 ± 2.47 G_3*/G_5; G_4*/G_1/G_2/G_3/G_6
Yildiz et al., 2015 [26]	G_1—37% phosphoric acid (3) G_2—37% phosphoric acid + Aqua-Prep™ (without rinsing) (3) G_3—Laser Er:YAG (10 Hz, 8 mm) (3) + adhesive + resin	Molars	Saline solution	Adhesive: Adper™ Single Bond 2 Resin: Filtek™ Z250 Laser: Fidelis Plus III (Fotona, Slovenia) Aqua-Prep™ (Bisco, USA)	G_1: 14.28 ± 5.22 G_2: 18.35 ± 7.94 G_3: 20.57 ± 9.02 G_1*/G_3
Bahrololoomi et al., 2017 [27]	35% phosphoric acid + G_1—none (14) G_2—2.5% NaOCl (14) G_3—5.25% NaOCl (14) + adhesive + resin	Molars	0.5% Chloramine	Adhesive: One-Step® Plus (Bisco, USA) Resin: AELITE (Bisco, USA)	G_1: 13.56 ± 3.36 G_2: 13.53 ± 3.64 G_3: 14.36 ± 3.64

Table 1. *Cont.*

Authors, Year	Groups (n)	Teeth	Storage	Materials	Results (MPa)
Ebrahimi et al., 2018 [28]	G_1—37% phosphoric acid + adhesive 1 (20) G_2—37% phosphoric acid + adhesive 1 (3 months) (20) G_3—37% phosphoric acid + adhesive 1 + 2% CHX (without rinsing) (20) G_4—37% phosphoric acid + adhesive 1 + 2% CHX (without rinsing) (3 months) (20) G_5—Adhesive 2 (20) G_6—Adhesive 2 (3 months) (20) G_7—Adhesive 2 (Primer) + 2% CHX (without rinsing) + adhesive 2 (bond) (20) G_8—Adhesive 2 (primer) + 2% CHX (without rinsing) + adhesive 2 (bond) (3months) (20)	Molars	0.1% Thymol + water	Adhesive: 1–Adper™ Single Bond 2–Clearfil™ SE Bond Resin: Filtek™ Z250	G_1: 25.43 ± 12.94 G_2: 39.96 ± 21.75 G_3: 66.45 ± 8.3 G_4: 39.02 ± 23.29 G_5: 47.83 ± 19.83 G_6: 53.36 ± 18.05 G_7: 46.25 ± 9.34 G_8: 56.4 ± 22.18 G_1*/G_3
Mohammadi et al., 2020 [29]	37% phosphoric acid + G_1–PBS (15) G_2—2% CHX (without rinsing) (15) G_3—2% Doxycycline (without rinsing) (15) G_4—17% EDTA (15) + adhesive	Anterior teeth	-	Adhesive: Adper™ Single Bond 2 Resin: Filtek™ Z250	G_1: 6.20 ± 2.11 G_2: 5.60 ± 2.69 G_3: 8.82 ± 3.29 G_4: 7.50 ± 3.94 G_2*/G_3

CHX–Chlorhexidine; EDTA–Ethylenediaminetetraacetic Acid; GIC–Glass Ionomer Cement; NaOCl–Sodium hypochlorite; *–Statistically significant difference ($p < 0.05$).

No clinical studies were identified, and only one in situ study regarding the use of a cavity disinfectant in primary teeth was evaluated. Ricci et al. [31] developed a split-mouth experimental protocol that included children aged between 8 and 11 years with at least two contralateral primary molars with small carious lesions. Chlorhexidine was used as a cavity disinfectant after enamel and dentin were etched with 35% phosphoric acid. The solution was removed with absorbent papers, and the cavities were restored with Prime & Bond NT® (Dentsply, York, PA, USA) and Filtek™ Z250 (3M, Saint Paul, MN, USA). All the procedures were done under rubber dam, and the teeth were collected later, after exfoliation. The teeth were grouped according to the time of oral function after restoration: up to 30 days, 1 to 5 months, 10 to 12 months, and 18 to 20 months. A progressive decrease in bond strength values was reported for control and experimental groups as the time in oral function increased. However, a statistically significant decrease was reported sooner for the control group (it started after 1–5 months, while for the experimental group it started after 10–12 months). Also, significantly higher bond strength values were reported for the experimental group after 1–5 and 18–20 months.

Quality Assessment

Methodological quality assessment outcomes are presented in Table 2. All studies presented accurate information regarding each item from 1 to 10. However, none of them provided results with confidence intervals. In addition, only two studies [26,29] reported study limitations and sources of potential bias (item 12).

Table 2. Modified CONSORT checklist for reporting in vitro studies of dental materials.

Studies	Item									
	1 Abstract	2a Introduction (Background)	2b Introduction (Objectives)	3 Methods (Intervention)	4 Methods (Outcomes)	10 Methods (Statistical Methods)	11 Results (Outcomes and Estimation)	12 Discussion (Limitations)	13 Other Information (Funding)	14 Other Information (Protocol)
Vieira et al., 2003 [12]	Yes	Yes	Yes	Yes	Yes	Yes	Yes [a]	No	No	No
Correr et al., 2004 [16]	Yes	Yes	Yes	Yes	Yes	Yes	Yes [a]	No	No	No
Monghini et al., 2004 [17]	Yes	Yes	Yes	Yes	Yes	Yes	Yes [a]	No	No	No
Ersin et al., 2009 [18]	Yes	Yes	Yes	Yes	Yes	Yes	Yes [a]	No	No	No
Ricci et al., 2010 [19]	Yes	Yes	Yes	Yes	Yes	Yes	Yes [a]	No	Yes	No
Leitune et al., 2011 [20]	Yes	Yes	Yes	Yes	Yes	Yes	Yes [a]	No	No	No
Scatena et al., 2011 [21]	Yes	Yes	Yes	Yes	Yes	Yes	Yes [a]	No	No	No
Manfro et al., 2012 [30]	Yes	Yes	Yes	Yes	Yes	Yes	Yes [a]	No	No	No
Lenzi et al., 2012 [22]	Yes	Yes	Yes	Yes	Yes	Yes	Yes [a]	No	Yes	No

Table 2. *Cont.*

Studies	1 Abstract	2a Introduction (Background)	2b Introduction (Objectives)	3 Methods (Intervention)	4 Methods (Outcomes)	10 Methods (Statistical Methods)	11 Results (Outcomes and Estimation)	12 Discussion (Limitations)	13 Other Information (Funding)	14 Other Information (Protocol)
						Item				
Aras et al., 2013 [24]	Yes	Yes	Yes	Yes	Yes	Yes	Yes [a]	No	Yes	No
Lenzi et al., 2014 [23]	Yes	Yes	Yes	Yes	Yes	Yes	Yes [a]	No	Yes	No
Oznurhan et al., 2015 [25]	Yes	Yes	Yes	Yes	Yes	Yes	Yes [a]	No	Yes	No
Yildiz et al., 2015 [26]	Yes	Yes	Yes	Yes	Yes	Yes	Yes [a]	Yes	Yes	No
Bahrololoomi et al., 2017 [27]	Yes	Yes	Yes	Yes	Yes	Yes	Yes [a]	No	Yes	No
Ebrahimi et al., 2018 [28]	Yes	Yes	Yes	Yes	Yes	Yes	Yes [a]	No	Yes	No
Mohammadi et al., 2020 [29]	Yes	Yes	Yes	Yes	Yes	Yes	Yes [a]	Yes	Yes	No

[ε] No confidence interval.

3. Discussion

A cavity disinfectant must not only have a strong antimicrobial effect but also not compromise the adhesion of the restorative material to the dental substrates [7,32]. The majority of the studies on this topic reports results on permanent teeth, but the structural and mechanical properties of the primary teeth make it necessary to carry out experimental protocols testing this type of teeth [33,34]. Compared to permanent teeth, primary teeth have thinner enamel and dentin, are less mineralized due to their lower concentration of calcium and potassium ions, have a hybrid layer more prone to be degraded [35], and their dentin has a lower tubule density [18,36,37]. This may explain why bond strength values of composite materials in primary teeth are lower than those of permanent teeth [38].

Dental adhesion may be affected not only by the cavity disinfectant used but also by the dental substrate. In order to minimize its effect, it is recommended to perform adhesion tests in the superficial dentin of healthy teeth, ideally without restorations [39]. Deep dentin is mainly composed of dentinal tubules and a small percentage of intertubular dentin. Superficial dentin has a higher percentage of organic components (collagen) and of intertubular dentin and a lower number of dentinal tubules [40–42].

The differences between healthy and caries-affected dentin should also be underlined. The caries-affected dentin is more porous and softer due to its partial demineralization, which leads to a less effective adhesion [43–45]. In fact, some of the articles included in this systematic review evaluated the effect of a cavity disinfectant in healthy and affected dentin [18,22,23], and Lenzi et al. [22,23] reported significant lower bond strength values for the affected-dentin groups.

Besides dentin's quality (superficial/deep dentin, permanent/primary teeth, healthy/carious dentin, amount of collagen and number, diameter, orientation, and size of dentinal tubules), moisture, contaminants, adhesive systems, solvents, and phosphoric acid/acidic primers are all factors affecting bond strength to dentin [46–49]. As so, the inclusion of at least one control group per study was mandatory for a study to be included in this systematic review.

All of the studies reported the use of a storage medium before the samples were submitted to the experimental protocol. The ISO/TS 11405/2015 (Dentistry–Testing of adhesion to tooth structure) [39] provides guidance for testing adhesion between dental substrates and restorative materials. This ISO/TS recommends the use of a 0.5% chloramine solution or of distilled water as a storage medium for the extracted teeth. If chloramine is chosen, it should be replaced by distilled water after one week. Despite these recommendations, some authors used other solutions, such as thymol [12,18,28]. The use of other solutions is not recommended by the ISO/TS 11405/2015, since it may affect dentin's mechanical properties. In fact, Santana et al. [50] reported that the use of thymol as a storage medium led to impaired adhesion.

After the restorations were made, all authors stated that the samples were kept in water, which is exactly the recommendation of the ISO/TS 11405/2015 (ISO 3696:1987, grade 3) [51].

Almost all authors reported results on adhesion to molars, which is also in line with the recommendations of the ISO/TS 11405/201545. However, Monghini et al. [17] and Mohammadi et al. [29] used anterior teeth.

Most authors [12,18–20,22,23,25,28,29] evaluated the effect of chlorhexidine as a cavity disinfectant. Chlorhexidine has been widely used in dentistry, mainly because of its antimicrobial properties, including against *Streptococcus mutans*, and of its antiplaque effect [52–55]. Chlorhexidine is also well known for its ability to inhibit matrix metalloproteinases due to its strong collagenolytic activity, reducing the degradation of the hybrid layer [56,57], which may justify the positive results reported by almost all authors. Although only Ersin et al. [18] evaluated the effect of chlorhexidine on the adhesion to a glass ionomer material, the authors also reported positive results.

Similar results were previously reported for permanent teeth [58], which makes chlorhexidine the most consensual cavity disinfectant to be used in clinical practice.

Not only adhesion to dentin is not only adequate after its use but, as stated by some authors [59,60], it can even be enhanced. As so, chlorhexidine presents as a safe and effective product to be used as a cavity disinfectant.

Sodium hypochlorite is commonly used as a cavity disinfectant due to its favorable properties: antibacterial action against aerobic bacteria, such as *S. mutans*, wettability, and deproteinization [61–65]. Although all authors studying the effect of the use of sodium hypochlorite as a cavity disinfectant in primary teeth reported positive results, only three articles [16,24,27] were identified. Since there are just a few studies reporting results on primary teeth and that the use of sodium hypochlorite as a cavity disinfectant in permanent teeth is still a matter of discussion [58], caution is required when choosing this product as a cavity disinfectant.

Initially presented as an alternative to the use of burs for cavity preparation, the Erbium:Ytrium (Er:YAG) laser was first introduced in 1989 by Hibst and Keller [66]. From then on, lasers have been used in numerous dentistry fields such as oral surgery, periodontics, endodontics, and prosthodontics [67]. However, similarly to what was reported for permanent teeth [58], there is no consensus regarding the use of lasers as cavity disinfectants, with some authors reporting an impairment of adhesion [17], and others reporting maintenance or even an enhancement of the bond strength values [21,26]. Moreover, even though some authors did not report secondary side effects [66,68,69], lasers may lead to overheating of the dental structures, which may induce pulp injuries, hydroxyapatite changes, and excessive dentin dehydration [17,70–76]. Given the results, the use of lasers as a cavity disinfection method should be avoided.

Both gaseous ozone and ozonated water have been recently introduced as alternatives to cavity disinfection due to their known antimicrobial and strong antioxidant properties. Polydorou et al. [77] reported that gaseous ozone eliminated 99.9% of the microorganisms in carious lesions in 20 s. In addition to its great antimicrobial activity (including against *S. mutans*), ozone also has antifungal and antiviral properties [78]. Authors analyzing the effect of either ozonated water or gaseous ozone on adhesion reported positive results [25], which may be justified by the opening of the dentinal tubules caused by oxygen [79–83]. Although there is limited information about the use of ozone as a cavity disinfectant in primary teeth, it looks like a promising alternative.

EDTA is an organic compound responsible for chelating calcium and potassium ions and for selective removal of hydroxyapatite crystals, which allows for the maintenance of the collagen matrix [84,85]. It is widely used in endodontics to improve shaping of the entire root canal system and to dissolve the inorganic components of the smear layer [86]. Although the reported results were positive (no differences on bond strength values after using it as a cavity disinfectant), only one study [29] evaluated it. A few articles on permanent teeth [58] also showed that EDTA presents as a promising alternative, but there is a clear need for further research.

Aqua-prep™ [26], 2% doxycycline [29], and 30% propolis [25] were all evaluated by studies included in this review, and the reported results were positive, but only one article was included for each product. Given the limited scientific evidence associated with these products (even in permanent teeth [58]), their use as cavity disinfectants should be avoided.

The limitations of this systematic review mainly reflect the shortcomings of the included articles. No clinical studies on the topic were identified, and such studies are essential to analyze the effects of the different cavity disinfectants when applied in the oral cavity. In addition, there is no information on the best application time and on the durability of bond interfaces over time. Also, there are several studies reporting results on different adhesive systems (total etch, self-etch, universal) but given the different methods applied, it is impossible to draw conclusions regarding this matter.

Further studies with standardized protocols should be developed to allow solid conclusions and recommendations concerning this issue. The effect of the incorporation of cavity disinfectants into adhesive systems must also be evaluated, since it may reduce clinical steps, which is of great importance in pediatric dentistry.

4. Materials and Methods

The present systematic review was registered on the International Prospective Register of Systematic Reviews (PROSPERO) platform (ID CRD42020199614) and followed the PRISMA protocol (Preferred Reporting Items for Systematic Reviews and Meta-Analyses Protocols) [87].

The research questions were developed according to the PICO (Population, Intervention, Comparison, Outcome) methodology, as described in Table 3.

Table 3. Problem, Intervention, Comparison, Outcome (PICO) strategy.

Parameter	In Vitro Studies	Clinical/In Situ Studies
P (Population)	Primary teeth / dentin discs	Children in need of a restoration
I (Intervention)	Restoration with prior application of a cavity disinfectant	
C (Comparison)	Conventional restoration	
O (Outcome)	Effect of cavity disinfection on dentin bond strength	Effect of cavity disinfection on clinical success

The inclusion and exclusion criteria are presented in Table 4.

Table 4. Inclusion and Exclusion Criteria.

Inclusion Criteria	Primary teeth evaluation
	Bond strength/clinical success evaluation
	Existence of a control group
	Evaluation of commercially available adhesive systems and composite resins or glass ionomer
	Application of only one cavity disinfectant per experimental group
	Report of results as mean and standard deviation
Exclusion Criteria	Permanent teeth evaluation
	Evaluation of teeth with endodontic treatment
	Evaluation of adhesion of cements, posts, sealants, or brackets
	Use of experimental adhesive systems or of mixtures of adhesives with disinfectants
	Revisions, animal or cell studies, letters, abstracts, comments, and clinical cases

An electronic research was conducted in Cochrane Library (www.cochranelibrary.com), PubMed/MEDLINE (pubmed.ncbi.nlm.nih.gov), SCOPUS (www.scopus.com), and Web of Science (webofknowledge.com). The research keys used in each database can be found in Table 5.

The search was limited to articles published until 14 February 2021, with no restrictions on region, language, or year of publication. A manual search for other references in reviews and in the included articles was performed.

Duplicate articles were removed with Endnote 20 (Clarivate™, Boston, MA, USA). Two independent reviewers analyzed titles, abstracts, and full texts, and a third one's opinion was obtained when necessary.

Selected articles were read by the same two independent authors, who collected the following data on the in vitro studies: authors and year of publication, number of elements per group (n), materials used (cavity disinfectant, type of adhesive system, and restorative material), storage, and bond strength results.

Regarding the clinical/in situ studies, the following data were acquired: authors and year of publication, type of teeth, number and ages of children per group (n), materials used (cavity disinfectant, type of adhesive system, and restorative material), and results (pigmentation, marginal gaps, or existence of carious lesions).

Table 5. Search keys used in the different databases.

Database	Search keys
Cochrane Library	#1 MeSH descriptor: [Dentin] explode all trees #2 dentin #3 cavity #4 MeSH descriptor: [Disinfection] explode all trees #5 disinfect* #6 antibacteria* #7 MeSH descriptor: [Anti-Bacterial Agents] explode all trees #8 chlorhexidine #9 MeSH descriptor: [Chlorhexidine] explode all trees #10 "sodium hypochlorite" #11 MeSH descriptor: [Sodium Hypochlorite] explode all trees #12 laser #13 MeSH descriptor: [Lasers] explode all trees #14 ozone #15 MeSH descriptor: [Ozone] explode all trees #16 "aloe vera" #17 MeSH descriptor: [Aloe] explode all trees #18 ethanol #19 MeSH descriptor: [Ethanol] explode all trees #20 EDTA #21 MeSH descriptor: [Edetic Acid] explode all trees #22 "green tea" #23 EGCG #24 "bond strength" #25 adhesion #26 adhesive #27 MeSH descriptor: [Dental Cements] explode all trees #28 primary #29 deciduous #30 MeSH descriptor: [Tooth, Deciduous] explode all trees #31 temporary #32 #1 OR #2 OR #3 #33 #4 OR #5 OR #6 OR #7 OR #8 OR #9 OR #10 OR #11 OR #12 OR #13 OR #14 OR #15 OR #16 OR #17 OR #18 OR #19 OR #20 OR #21 OR #22 OR #23 #34 #24 OR #25 OR #26 OR #27 #35 #28 OR #29 OR #30 OR #31 #36 #32 AND #33 AND #34 AND #35
PubMed	(dentin[MeSH Terms] OR dentin OR cavity) AND (disinfection[MeSH Terms] OR disinfect* OR antibacteria* OR agents, antibacterial[MeSH Terms] OR chlorhexidine[MeSH Terms] OR chlorhexidine OR "sodium hypochlorite" OR sodium hypochlorite[MeSH Terms] OR laser OR lasers[MeSH Terms] OR ozone OR ozone[MeSH Terms] OR "aloe vera" OR aloe[MeSH Terms] OR ethanol OR ethanol[MeSH Terms] OR EDTA OR Edetic acid[MeSH Terms] OR "green tea" OR EGCG) AND ("bond strength" OR adhesion OR adhesive OR adhesives[MeSH Terms]) AND (deciduous tooth[MeSH Terms] OR deciduous OR primary OR temporary)
SCOPUS	TITLE-ABS-KEY (dentin OR cavity) AND TITLE-ABS-KEY (disinfect* OR antibacterial* OR chlorhexidine OR "sodium hypochlorite" OR laser OR ozone OR "aloe vera" OR ethanol OR EDTA OR "green tea" OR EGCG) AND TITLE-ABS-KEY ("bond strength" OR adhesion OR adhesive) AND TITLE-ABS-KEY (primary OR deciduous OR temporary)
Web of Science	TS= ((dentin[MeSH Terms] OR dentin OR cavity) AND (disinfect* OR antibacteria* OR chlorhexidine OR "sodium hypochlorite" OR laser OR ozone OR "aloe vera" OR ethanol OR EDTA OR "green tea" OR EGCG) AND ("bond strength" OR adhesion or adhesive) AND (primary OR deciduous OR temporary))

Quality Assessment

The evaluation of the methodological quality of each in vitro study was assessed using the modified Consolidated Standards of Reporting Trials (CONSORT) checklist [88] for reporting in vitro studies on dental materials. When applying this checklist, items 5 to 9 could not be evaluated, since these are designed to evaluate sample standardization.

Two authors assessed the risk of bias independently, and any disagreement was solved by consensus.

5. Conclusions

Chlorhexidine is the most studied cavity disinfectant, and according to the results, its use does not compromise adhesion to primary dentin. Sodium hypochlorite is a promising alternative, but more research on its effects on adhesion is required to clearly state that it can be safely used as a cavity disinfectant for primary teeth. Although other disinfectants were studied, there is a low-level evidence attesting their effects on adhesion; therefore, their use should be avoided.

There is a clear need for researchers to conduct well-designed in vitro and clinical studies so more options can be identified, and the long-term effect on adhesion can be evaluated.

Author Contributions: Conceptualization, A.C., I.A., and E.C.; methodology, A.C., I.A., A.P., C.M.M.; software, A.C., I.A.; validation, A.C., I.A., A.P., C.M.M., M.M.F., E.C.; data extraction and analysis, A.C., I.A., A.A., J.S., E.C.; writing-original draft preparation, A.C., I.A., A.A., J.S.; writing-review and editing, A.C., I.A., A.P., C.M.M., M.M.F., E.C.; supervision, A.C., I.A., E.C. All authors have read and agreed to the published version of the manuscript.

Funding: This research received no external funding.

Institutional Review Board Statement: Not applicable.

Informed Consent Statement: Not applicable.

Conflicts of Interest: The authors declare no conflict of interest.

References

1. Kassebaum, N.J.; Bernabé, E.; Dahiya, M.; Bhandari, B.; Murray, C.J.L.; Marcenes, W. Global burden of untreated caries: A systematic review and metaregression. *J. Dent. Res.* **2015**, *94*, 650–658. [CrossRef] [PubMed]
2. Pitts, N.B.; Zero, D.T.; Marsh, P.D.; Ekstrand, K.; Weintraub, J.A.; Ramos-Gomez, F.; Tagami, J.; Twetman, S.; Tsakos, G.; Ismail, A. Dental caries. *Nat. Rev. Dis. Primers* **2017**, *3*, 17030. [CrossRef] [PubMed]
3. Askar, H.; Krois, J.; Göstemeyer, G.; Bottenberg, P.; Zero, D.; Banerjee, A.; Schwendicke, F. Secondary caries: What is it, and how it can be controlled, detected, and managed? *Clin. Oral Investig.* **2020**, *24*, 1869–1876. [CrossRef] [PubMed]
4. Singhal, D.K.; Acharya, S.; Thakur, A.S. Microbiological analysis after complete or partial removal of carious dentin using two different techniques in primary teeth: A randomized clinical trial. *Dent. Res. J.* **2016**, *13*, 30–37. [CrossRef] [PubMed]
5. Anderson, M.H.; Loesche, W.J.; Charbeneau, G.T. Bacteriologic study of a basic fuchsin caries-disclosing dye. *J. Prosthet. Dent.* **1985**, *54*, 51–55. [CrossRef]
6. Dalkilic, E.E.; Arisu, H.D.; Kivanc, B.H.; Uctasli, M.B.; Omurlu, H. Effect of different disinfectant methods on the initial microtensile bond strength of a self-etch adhesive to dentin. *Lasers Med. Sci.* **2012**, *27*, 819–825. [CrossRef]
7. Elkassas, D.W.; Fawzi, E.M.; Zohairy, A. The effect of cavity disinfectants on the micro-shear bond strength of dentin adhesives. *Eur. J. Dent.* **2014**, *8*, 184–190. [CrossRef] [PubMed]
8. Say, E.C.; Koray, F.; Tarim, B.; Soyman, M.; Gülmez, T. In vitro effect of cavity disinfectants on the bond strength of dentin bonding systems. *Quintessence Int.* **2004**, *35*, 56–60.
9. Hiraishi, N.; Yiu, C.K.Y.; King, N.M.; Tay, F.R. Effect of 2% chlorhexidine on dentin microtensile bond strengths and nanoleakage of luting cements. *J. Dent.* **2009**, *37*, 440–448. [CrossRef]
10. Colares, V.; Franca, C.; Filho, H.A.A. O tratamento restaurador atraumático nas dentições decídua e permanente. *Rev. Port. Estomatol. Med. Dentária Cir. Maxilofac.* **2009**, *50*, 35–41. [CrossRef]
11. Suma, N.K.; Shashibhushan, K.K. Effect of Dentin Disinfection with 2% Chlorhexidine Gluconate and 0.3% Iodine on Dentin Bond Strength: An in vitro Study. *Int. J. Clin. Pediatric Dent.* **2017**, *10*, 223–228. [CrossRef] [PubMed]
12. Vieira, R.S.; Silva, I.A. Bond strength to primary tooth dentin following disinfection with a chlorhexidine solution: An in vitro study. *Pediatric Dent.* **2003**, *25*, 49–52.
13. Franzon, R.; Opdam, N.J.; Guimarães, L.F.; Demarco, F.F.; Casagrande, L.; Haas, A.N.; Araújo, F.B. Randomized controlled clinical trial of the 24-months survival of composite resin restorations after one-step incomplete and complete excavation on primary teeth. *J. Dent.* **2015**, *43*, 235–1241. [CrossRef] [PubMed]
14. Demarco, F.F.; Corrêa, M.B.; Cenci, M.S.; Moraes, R.R.; Opdam, N.J.M. Longevity of posterior composite restorations: Not only a matter of materials. *Dent. Mater.* **2012**, *28*, 87–101. [CrossRef] [PubMed]
15. Sande, F.H.; Collares, K.; Correa, M.B.; Cenci, M.S.; Demarco, F.F.; Opdam, N. Restoration Survival: Revisiting Patients' Risk Factors through a Systematic Literature Review. *Oper. Dent.* **2016**, *41*, S7–S26. [CrossRef] [PubMed]

16. Correr, G.M.; Puppin-Rontani, R.M.; Correr-Sobrinho, L.; Sinhoret, M.A.C.; Consani, S. Effect of sodium hypochlorite on dentin bonding in primary teeth. *J. Adhes. Dent.* **2004**, *6*, 307–312. [PubMed]
17. Monghini, E.M.; Wanderley, R.L.; Pécora, J.D.; Dibb, P.R.G.; Corona, S.A.M.; Borsatto, M.C. Bond Strength to Dentin of Primary Teeth Irradiated with Varying Er: YAG Laser Energies and SEM Examination of the Surface Morphology. *Lasers Surg. Med.* **2004**, *34*, 254–259. [CrossRef] [PubMed]
18. Ersin, N.K.; Candan, U.; Aykut, A.; Eronat, C.; Belli, S. No Adverse Effect to Bonding Following Caries. *J. Dent. Child.* **2009**, *76*, 20–27.
19. Ricci, H.A.; Sanabe, M.E.; Costa, C.A.S.; Hebling, J. Effect of chlorhexidine on bond strength of two-step etch-and-rinse adhesive systems to dentin of primary and permanent teeth. *Am. J. Dent.* **2010**, *23*, 128–132.
20. Leitune, V.C.B.; Portella, F.F.; Bohn, P.V.; Collares, F.M.; Samuel, S.M.W. Influence of chlorhexidine application on longitudinal adhesive bond strength in deciduous teeth. *Braz. Oral Res.* **2011**, *25*, 388–392. [CrossRef]
21. Scatena, C.; Torres, C.P.; Gomes-Silva, J.M.; Contente, M.; Pécora, J.D.; Palma-Dibb, R.G.; Borsatto, M.C. Shear strength of the bond to primary dentin: Influence of Er: YAG laser irradiation distance. *Lasers Med. Sci.* **2011**, *26*, 293–297. [CrossRef]
22. Lenzi, T.L.; Tedesco, T.K.; Soares, F.Z.M.; Loguercio, A.D.; Rocha, R.O. Chlorhexidine does not increase immediate bond strength of etch-and-rinse adhesive to caries-affected dentin of primary and permanent teeth. *Braz. Dent. J.* **2012**, *23*, 438–442. [CrossRef]
23. Lenzi, T.L.; Tedesco, T.K.; Soares, F.Z.M.; Loguercio, A.D.; Rocha, R.O. Chlorhexidine application for bond strength preservation in artificially-created caries-affected primary dentin. *Int. J. Adhes. Adhes.* **2014**, *54*, 51–56. [CrossRef]
24. Aras, S.; Küçükeçmen, H.C.; Öaroğlu, S.I. Deproteinization treatment on bond strengths of primary, mature and immature permanent tooth enamel. *J. Clin. Pediatric Dent.* **2013**, *37*, 275–280. [CrossRef]
25. Oznurhan, F.; Ozturk, C.; Ekci, E.S. Effects of different cavity-disinfectants and potassium titanyl phosphate laser on microtensile bond strength to primary dentin. *Niger. J. Clin. Pract.* **2015**, *18*, 400–404. [CrossRef] [PubMed]
26. Yildiz, E.; Karaarslan, E.S.; Simsek, M.; Cebe, F.; Ozsevik, A.S.; Ozturk, B. Effect of a re-wetting agent on bond strength of an adhesive to primary and permanent teeth dentin after different etching techniques. *Niger. J. Clin. Pract.* **2015**, *18*, 364–370. [CrossRef] [PubMed]
27. Bahrololoomi, Z.; Dadkhah, A.; Alemrajabi, M. The Effect of Er: YAG laser irradiation and different concentrations of sodium hypochlorite on shear bond strength of composite to primary teeth's dentin. *J. Lasers Med. Sci.* **2017**, *8*, 29–35. [CrossRef] [PubMed]
28. Ebrahimi, M.; Naseh, A.; Abdollahi, M.; Shirazi, A.S. Can chlorhexidine enhance the bond strength of self-etch and etch-and-rinse systems to primary teeth dentin? *J. Contemp. Dent. Pract.* **2018**, *19*, 404–408.
29. Mohammadi, N.; Parsaie, Z.; Jafarpour, D.; Bizolm, F. Effect of different matrix metalloproteinase inhibitors on shear bond strength of composite attached to primary teeth dentin. *Eur. J. Gen. Dent.* **2020**, *9*, 147–151. [CrossRef]
30. Manfro, A.R.G.; Reis, A.; Loguercio, A.D.; Imparato, J.C.P.; Raggio, D.P. Effect of different concentrations of chlorhexidine on bond strength of primary dentin. *Pediatric Dent.* **2012**, *34*, 11E–15E.
31. Ricci, H.A.; Sanabe, M.E.; Costa, C.A.S.; Pashley, D.H.; Hebling, J. Chlorhexidine increases the longevity of in vivo resin-dentin bonds. *Eur. J. Oral Sci.* **2010**, *118*, 411–416. [CrossRef] [PubMed]
32. Jowkar, Z.; Farpour, N.; Koohpeima, F.; Mokhtari, M.J.; Shafiei, F. Effect of silver nanoparticles, zinc oxide nanoparticles and titanium dioxide nanoparticles on microshear bond strength to enamel and dentin. *J. Contemp. Dent. Pract.* **2018**, *19*, 1405–1412.
33. Koutsi, V.; Noonan, R.G.; Horner, J.A.; Simpson, M.D.; Matthews, W.G.; Pashley, D.H. The effect of dentin depth on the permeability and ultrastructure of primary molars. *Pediatric Dent.* **1994**, *16*, 29–35.
34. Dourda, A.O.; Moule, A.J.; Young, W.G. A morphometric analysis of the cross-sectional area of dentine occupied by dentinal tubules in human third molar teeth. *Int. Endod. J.* **1994**, *27*, 184–189. [CrossRef] [PubMed]
35. Hashimoto, M.; Ohno, H.; Endo, K.; Kaga, M.; Sano, H.; Oguchi, H. The effect of hybrid layer thickness on bond strength: Demineralized dentin zone of the hybrid layer. *Dent. Mater.* **2000**, *16*, 406–411. [CrossRef]
36. Angker, L.; Nockolds, C.; Swain, M.V.; Kilpatrick, N. Quantitative analysis of the mineral content of sound and carious primary dentine using BSE imaging. *Arch. Oral Biol.* **2004**, *49*, 99–107. [CrossRef]
37. Nor, J.; Dennison, J.; Edwardsa, C.; Feigal, R. Dentin bonding: SEM comparison teeth. *Am. Acad. Pediatric Dent.* **1997**, *19*, 246–252.
38. Sung, E.C.; Chenard, T.; Caputo, A.A.; Amodeo, M.; Chung, E.M.; Rizoiu, I.M. Composite resin bond strength to primary dentin prepared with ER, CR: YSSG laser. *J. Clin. Pediatric Dent.* **2005**, *30*, 45–50. [CrossRef]
39. *ISO/TS 11405:2015 Dental Materials—Testing of Adhesion to Tooth Structure*; International Organisation for Standardization: Geneva, Switzerland, 2015.
40. Uceda-Gómez, N.; Reis, A.; Carrilho, M.R.O.; Loguercio, A.D.; Filho, L.E.R. Effect of sodium hypochlorite on the bond strength of an adhesive system to superficial and deep dentin. *J. Appl. Oral Sci.* **2003**, *11*, 223–228. [CrossRef]
41. Ramos, R.P.; Chimello, D.T.; Chinelatti, M.A.; Nonaka, T.; Pécora, J.D.; Dibb, R.G.P. Effect of Er: YAG laser on bond strength to dentin of a self-etching primer and two single-bottle adhesive systems. *Lasers Surg. Med.* **2002**, *31*, 164–170. [CrossRef] [PubMed]
42. Yu, H.H.; Zhang, L.; Yu, F.; Li, F.; Liu, Z.Y.; Chen, J.H. Epigallocatechin-3-gallate and Epigallocatechin-3-O-(3-O-methyl)-gallate Enhance the Bonding Stability of an Etch-and-Rinse Adhesive to Dentin. *Materials* **2017**, *10*, 183. [CrossRef] [PubMed]
43. Pashley, D.H.; Carvalho, R.M. Dentine permeability and dentine adhesion. *J. Dent.* **1997**, *25*, 355–372. [CrossRef]
44. Pashley, E.L.; Talman, R.; Horner, J.A.; Pashley, D.H. Permeability of normal versus carious dentin. *Dent. Traumatol.* **1991**, *7*, 207–211. [CrossRef] [PubMed]

45. Swift, E.J.J. Dentin/enamel adhesives: Review of the literature. *Pediatric Dent.* **2002**, *24*, 456–461.
46. Powers, J.M.; O'Keefe, K.L.; Pinzon, L.M. Factores affecting in vitro bond strength of bonding agents to human dentin. *Odontology* **2003**, *91*, 1–6. [CrossRef] [PubMed]
47. Carvalho, R.M.; Mendonça, J.S.; Santiago, S.L.; Silveira, R.R.; Garcia, F.C.P.; Tay, F.R.; Pashley, D.H. Effects of HEMA/Solvent combinations on bond strength to dentin. *J. Dent. Res.* **2003**, *82*, 597–601. [CrossRef] [PubMed]
48. Guo, J.; Wang, L.; Zhu, J.; Yang, J.; Zhu, H. Impact of dentinal tubule orientation on dentin bond strength. *Curr. Med. Sci.* **2018**, *38*, 721–726. [CrossRef]
49. Lima, D.M.; Candido, M.S.M. Effect of dentin on the shear bond strength of different adhesive systems. *Rev. Gaúcha Odontol.* **2012**, *60*, 149–161.
50. Santana, F.R.; Pereira, J.C.; Pereira, C.A.; Neto, F.A.J.; Soares, C.J. Influence of method and period of storage on the microtensile bond strength of indirect composite resin restorations to dentine. *Braz. Oral Res.* **2008**, *22*, 352–357. [CrossRef]
51. *ISO 3696:1987—Water for Analytical Laboratory Use—Specification and Test Methods*; International Organisation for Standardization: Geneva, Switzerland, 1987.
52. Kang, H.J.; Moon, H.J.; Shin, D.H. Effect of different chlorhexidine application times on microtensile bond strength to dentin in Class I cavities. *Restor. Dent. Endod.* **2012**, *37*, 9. [CrossRef]
53. Coelho, A.; Paula, A.; Carrilho, T.; Silva, M.J.; Botelho, M.F.; Carrilho, E. Chlorhexidine mouthwash as an anticaries agent: A systematic review. *Quintessence Int.* **2017**, *48*, 585–591.
54. Haydari, M.; Bardakci, A.G.; Koldsland, O.C.; Aass, A.M.; Sandvik, L.; Preus, H.R. Comparing the effect of 0.06%, 0.12% and 0.2% Chlorhexidine on plaque, bleeding and side effects in an experimental gingivitis model: A parallel group, double masked randomized clinical trial. *BMC Oral Health* **2017**, *17*, 1–8. [CrossRef]
55. Kandaswamy, S.K.; Sharath, A.; Priya, P.G. Comparison of the Effectiveness of Probiotic, Chlorhexidine-based Mouthwashes, and Oil Pulling Therapy on Plaque Accumulation and Gingival Inflammation in 10- to 12-year-old Schoolchildren: A Randomized Controlled Trial. *Int. J. Clin. Pediatric Dent.* **2018**, *11*, 66–70. [CrossRef]
56. Hebling, J.; Pashley, D.H.; Tjäderhane, L.; Tay, F.R. Chlorhexidine arrests subclinical degradation of dentin hybrid layers in vivo. *J. Dent. Res.* **2005**, *84*, 741–746. [CrossRef] [PubMed]
57. Pashley, D.H.; Tay, F.R.; Yiu, C.; Hashimoto, M.; Breschi, L.; Carvalho, R.M.; Ito, S. Collagen Degradation by Host-derived Enzymes during Aging. *J. Dent. Res.* **2004**, *83*, 216–221. [CrossRef]
58. Coelho, A.; Amaro, I.; Rascão, B.; Marcelino, I.; Paula, A.; Saraiva, J.; Spagnuolo, G.; Ferreira, M.M.; Marto, C.M.; Carrilho, E. Effect of Cavity Disinfectants on Dentin Bond. Strength and Clinical Success of Composite Restorations—A Systematic Review of In Vitro, In Situ and Clinical Studies. *Int. J. Mol. Sci.* **2020**, *22*, 353. [CrossRef]
59. Pappas, M.; Burns, D.R.; Moon, P.C.; Coffey, J.P. Influence of a 3-step tooth disinfection procedure on dentin bond strength. *J. Prosthet. Dent.* **2005**, *93*, 545–550. [CrossRef] [PubMed]
60. Carrilho, M.R.O.; Carvalho, R.M.; Goes, M.F.; Hipólito, V.; Geraldeli, S.; Tay, F.R.; Pashley, D.H.; Tjäderhane, L. Chlorhexidine preserves dentin bond in vitro. *J. Dent. Res.* **2007**, *86*, 90–94. [CrossRef] [PubMed]
61. Salles, M.M.; Badaró, M.M.; Arruda, C.N.F.; Leite, V.; Silva, C.; Watanabe, E.; Oliveira, V.; Paranhos, H. Antimicrobial activity of complete denture cleanser solutions based on sodium hypochlorite and Ricinus communis—A randomized clinical study. *J. Appl. Oral Sci.* **2015**, *23*, 637–6342. [CrossRef] [PubMed]
62. Estrela, C.; Estrela, C.R.A.; Decurcio, D.A.; Hollanda, A.C.B.; Silva, J.A. Antimicrobial efficacy of ozonated water, gaseous ozone, sodium hypochlorite and chlorhexidine in infected human root canals. *Int. Endod. J.* **2007**, *40*, 85–93. [CrossRef] [PubMed]
63. Arslan, S.; Ozbilge, H.; Kaya, E.G.; Er, O. In vitro antimicrobial activity of propolis, BioPure MTAD, sodium hypochlorite, and chlorhexidine on Enterococcus faecalis and Candida albicans. *Saudi Med. J.* **2011**, *32*, 479–483.
64. Cha, H.S.; Shin, D.H. Antibacterial capacity of cavity disinfectants against Streptococcus mutans and their effects on shear bond strength of a self-etch adhesive. *Dent. Mater. J.* **2016**, *35*, 147–152. [CrossRef] [PubMed]
65. Ahuja, B.; Yeluri, R.; Baliga, S.; Munshi, A.K. Enamel deproteinization before acid etching–a scanning electron microscopic observation. *J. Clin. Pediatric Dent.* **2010**, *35*, 169–172. [CrossRef] [PubMed]
66. Hibst, R.; Keller, U. Experimental studies of the application of the Er: YAG laser on dental hard substances: I. Measurement of the ablation rate. *Lasers Surg. Med.* **1989**, *9*, 338–344. [CrossRef] [PubMed]
67. Franke, M.; Taylor, A.W.; Lago, A.; Fredel, M.C. Influence of Nd: YAG laser irradiation on an adhesive restorative procedure. *Oper. Dent.* **2006**, *31*, 604–609. [CrossRef] [PubMed]
68. Keller, U.; Hibst, R. Experimental studies of the application of the Er: YAG laser on dental hard substances: II. Light microscopic and SEM investigations. *Lasers Surg. Med.* **1989**, *9*, 345–351. [CrossRef]
69. Tokonabe, H.; Kouji, R.; Watanabe, H.; Nakamura, Y.; Matsumoto, K. Morphological changes of human teeth with Er: YAG laser irradiation. *J. Clin. Laser Med. Surg.* **1999**, *17*, 7–12. [CrossRef]
70. Nelson, D.G.; Jongebloed, W.L.; Featherstone, J.D. Laser irradiation of human dental enamel and dentine. *N. Z. Dent. J.* **1986**, *82*, 74–77.
71. Hossain, M.; Nakamura, Y.; Yamada, Y.; Kimura, Y.; Nakamura, G.; Matsumoto, K. Ablation depths and morphological changes in human enamel and dentin after Er: YAG laser irradiation with or without water mist. *J. Clin. Laser Med. Surg.* **1999**, *17*, 105–109. [CrossRef]

72. Armengol, V.; Jean, A.; Rohanizadeh, R.; Hamel, H. Scanning electron microscopic analysis of diseased and healthy dental hard tissues after Er: YAG laser irradiation: In vitro study. *J. Endod.* **1999**, *25*, 543–546. [CrossRef]
73. Gonçalves, M.; Corona, S.A.M.; Palma-Dibb, R.G.; Pécora, J.D. Influence of pulse repetition rate of Er: YAG laser and dentin depth on tensile bond strength of dentin-resin interface. *J. Biomed. Mater. Res.* **2008**, *86*, 477–482. [CrossRef]
74. Ferreira, L.S.; Apel, C.; Francci, C.; Simoes, A.; Eduardo, C.P.; Gutknecht, N. Influence of etching time on bond strength in dentin irradiated with erbium lasers. *Lasers Med. Sci.* **2010**, *25*, 849–854. [CrossRef]
75. Martínez-Insua, A.; Dominguez, L.S.; Rivera, F.G.; Santana-Penín, U.A. Differences in bonding to acid-etched or Er: YAG-laser-treated enamel and dentin surfaces. *J. Prosthet. Dent.* **2000**, *84*, 280–288. [CrossRef] [PubMed]
76. Munck, J.; Meerbeek, B.; Yudhira, R.; Lambrechts, P.; Vanherle, G. Micro-tensile bond strength of two adhesives to Erbium:YAG-lased vs. bur-cut enamel and dentin. *Eur. J. Oral Sci.* **2002**, *110*, 322–329. [CrossRef] [PubMed]
77. Polydorou, O.; Pelz, K.; Hahn, P. Antibacterial effect of an ozone device and its comparison with two dentin-bonding systems. *Eur. J. Oral Sci.* **2006**, *114*, 349–353. [CrossRef] [PubMed]
78. Bocci, V.A. Scientific and medical aspects of ozone therapy. State of the art. *Arch. Med. Res.* **2006**, *37*, 425–435. [CrossRef]
79. Nagayoshi, M.; Kitamura, C.; Fukuizumi, T.; Nishihara, T.; Terashita, M. Antimicrobial effect of ozonated water on bacteria invading dentinal tubules. *J. Endod.* **2004**, *30*, 778–781. [CrossRef]
80. Baysan, A.; Lynch, E. Effect of ozone on the oral microbiota and clinical severity of primary root caries. *Am. J. Dent.* **2004**, *17*, 56–60. [PubMed]
81. Baysan, A.; Whiley, R.A.; Lynch, E. Antimicrobial effect of a novel ozone-generating device on micro-organisms associated with primary root carious lesions in vitro. *Caries Res.* **2000**, *34*, 498–501. [CrossRef]
82. Castillo, A.; Galindo-Moreno, P.; Avila, G.; Valderrama, M.; Liébana, J.; Baca, P. In vitro reduction of mutans streptococci by means of ozone gas application. *Quintessence Int.* **2008**, *39*, 827–831.
83. Fagrell, T.G.; Dietz, W.; Lingström, P.; Steiniger, F.; Norén, J.G. Effect of ozone treatment on different cariogenic microorganisms in vitro. *Swed. Dent. J.* **2008**, *32*, 139–147. [PubMed]
84. Wang, J.; Song, W.; Zhu, L.; Wei, X. A comparative study of the microtensile bond strength and microstructural differences between sclerotic and Normal dentine after surface pretreatment. *BMC Oral Health* **2019**, *19*, 1–10. [CrossRef]
85. Thompson, J.M.; Agee, K.; Sidow, S.; McNally, K.; Lindsey, K.; Borke, J.; Elsalanty, M.; Tay, F.R. Inhibition of endogenous dentin matrix metalloproteinases by ethylenediaminetetraacetic acid. *J. Endod.* **2012**, *38*, 62–65. [CrossRef] [PubMed]
86. Youm, S.H.; Jung, K.H.; Son, S.A.; Kwon, Y.H.; Park, J.K. Effect of dentin pretreatment and curing mode on the microtensile bond strength of self-adhesive resin cements. *J. Adv. Prosthodont.* **2015**, *7*, 317–322. [CrossRef] [PubMed]
87. Liberati, A.; Altman, D.G.; Tetzlaff, J.; Mulrow, C.; Gøtzsche, P.C.; Ioannidis, J.P.A.; Clarke, M.; Devereaux, P.J.; Kleijnen, J.; Moher, D. The PRISMA statement for reporting systematic reviews and meta-analyses of studies that evaluate health care interventions: Explanation and elaboration. *PLoS Med.* **2009**, *62*, 1–34.
88. Faggion, C.M. Guidelines for reporting pre-clinical in vitro studies on dental materials. *J. Evid. Based Dent. Pract.* **2012**, *12*, 182–189. [CrossRef] [PubMed]

International Journal of
Molecular Sciences

Review

Antibacterial Titanium Implants Biofunctionalized by Plasma Electrolytic Oxidation with Silver, Zinc, and Copper: A Systematic Review

Ingmar A. J. van Hengel *, Melissa W. A. M. Tierolf, Lidy E. Fratila-Apachitei, Iulian Apachitei and Amir A. Zadpoor

Department of Biomechanical Engineering, Faculty of Mechanical, Maritime, and Materials Engineering, Delft University of Technology, 2628 CD Delft, The Netherlands; melissa_tierolf@hotmail.com (M.W.A.M.T.); e.l.fratila-apachitei@tudelft.nl (L.E.F.-A.); i.apachitei@tudelft.nl (I.A.); a.a.zadpoor@tudelft.nl (A.A.Z.)
* Correspondence: i.a.j.vanhengel@tudelft.nl

Citation: van Hengel, I.A.J.; Tierolf, M.W.A.M.; Fratila-Apachitei, L.E.; Apachitei, I.; Zadpoor, A.A. Antibacterial Titanium Implants Biofunctionalized by Plasma Electrolytic Oxidation with Silver, Zinc, and Copper: A Systematic Review. *Int. J. Mol. Sci.* **2021**, *22*, 3800.

https://doi.org/10.3390/ijms22073800

Academic Editor: Sotiris Hadjikakou

Received: 16 February 2021
Accepted: 5 April 2021
Published: 6 April 2021

Publisher's Note: MDPI stays neutral with regard to jurisdictional claims in published maps and institutional affiliations.

Abstract: Patients receiving orthopedic implants are at risk of implant-associated infections (IAI). A growing number of antibiotic-resistant bacteria threaten to hamper the treatment of IAI. The focus has, therefore, shifted towards the development of implants with intrinsic antibacterial activity to prevent the occurrence of infection. The use of Ag, Cu, and Zn has gained momentum as these elements display strong antibacterial behavior and target a wide spectrum of bacteria. In order to incorporate these elements into the surface of titanium-based bone implants, plasma electrolytic oxidation (PEO) has been widely investigated as a single-step process that can biofunctionalize these (highly porous) implant surfaces. Here, we present a systematic review of the studies published between 2009 until 2020 on the biomaterial properties, antibacterial behavior, and biocompatibility of titanium implants biofunctionalized by PEO using Ag, Cu, and Zn. We observed that 100% of surfaces bearing Ag (Ag-surfaces), 93% of surfaces bearing Cu (Cu-surfaces), 73% of surfaces bearing Zn (Zn-surfaces), and 100% of surfaces combining Ag, Cu, and Zn resulted in a significant (i.e., >50%) reduction of bacterial load, while 13% of Ag-surfaces, 10% of Cu-surfaces, and none of Zn or combined Ag, Cu, and Zn surfaces reported cytotoxicity against osteoblasts, stem cells, and immune cells. A majority of the studies investigated the antibacterial activity against *S. aureus*. Important areas for future research include the biofunctionalization of additively manufactured porous implants and surfaces combining Ag, Cu, and Zn. Furthermore, the antibacterial activity of such implants should be determined in assays focused on prevention, rather than the treatment of IAIs. These implants should be tested using appropriate in vivo bone infection models capable of assessing whether titanium implants biofunctionalized by PEO with Ag, Cu, and Zn can contribute to protect patients against IAI.

Keywords: plasma electrolytic oxidation; additive manufacturing; titanium bone implants; antibacterial biomaterials; surface biofunctionalization; implant-associated infection

1. Introduction

Implant-associated infections (IAI) are a devastating complication for patients receiving bone implants in total joint arthroplasty, trauma surgeries, and malignant bone tumor resections [1–3]. These infections form a tremendous burden for both patients and society. As the number of implantations continues to grow annually [4–6], the need for a cure increases. Given that the treatment of such infections is highly costly from both financial and societal points of view, the focus has shifted towards the prevention of IAI through the development of implants with intrinsic antibacterial activity.

Antibiotics form the primary source of antibacterial agents used to treat bacterial infections. However, a vast number of IAI is caused by *Staphylococci* and multiple strains have developed high levels of antibiotic resistance [7,8], raising concerns for the future treatments of IAI. Infection by methicillin-resistant *Staphylococcus aureus* (MRSA) highly complicates the treatment of IAI and adversely affects the treatment outcomes [9,10]. Other antibacterial agents are, therefore, being investigated. Metallic elements, such as Ag, Cu, and Zn have shown strong antibacterial behavior against a wide microbial spectrum, including resistant bacterial strains [11–14].

Ag has excellent antibacterial properties, but may also induce cytotoxicity [15,16]. Cu and Zn, on the other hand, exhibit lower levels of antibacterial behavior, but are essential trace elements. Furthermore, they have been found to enhance the cytocompatibility of implant surfaces [17,18]. Therefore, combining these elements may result in the right balance between antibacterial behavior, chemical biocompatibility, and osteogenic response [19,20].

The local administration of antibacterial agents at the implant site was shown to greatly complement the systemic administration of antibiotics [21,22]. The side effects of such agents can also be prevented as the required antibacterial dose is generally lower [23]. To deliver antibacterial agents locally, the surface of the implants can be biofunctionalized through surface treatment techniques. Antibacterial agents can be attached to implants either as a coating layer, embedded directly onto the implant surface, or incorporated as part of a converted surface layer [24].

Antibacterial agents can be deposited onto the implant surface by means of polymeric, ceramic or metallic coatings. To produce these coatings, usually low temperatures are used and therefore little interaction occurs with the implant substrate. Coatings have a tendency to be thin and fragile, thereby limiting the availability of the antibacterial agent and hampering their use during surgical implantation. To enhance the diffusion, the antibacterial agent can be incorporated in a biodegradable polymer coating. In this way implants were manufactured that contain Ag [25,26], Cu [27], and Zn [28]. Polymeric coatings can be attached onto an implant by dipping and drying, sol-gel technology, spray drying, layer-by-layer manufacturing, and self-assembly monolayers. Downsides are the limited mechanical and chemical stability, local inflammatory response due to degradation products, and uncontrolled release kinetics.

Another strategy is direct embedding of the antibacterial agent into the implant surface. In this way, no new material is added on top of the substrate, but the composition of the outermost layer of the implant substrate is altered. Examples of such methods are ion implantation, plasma immersion ion implantation [29], and in situ reduction [30]. Advantages are that the implant surface morphology remains intact, and the corrosive and biocompatible properties of the substrate material retained. However, this strategy is difficult to perform on complex geometries and does not allow for optimization of the surface morphology.

A third approach to incorporate Ag, Cu, and Zn in the implant surface is through generation of a converted surface layer. One such technique is plasma electrolytic oxidation (PEO), which was investigated to biofunctionalize the surface of highly porous implants made of specific metallic biomaterials [31]. During PEO, the native titanium oxide layer is transformed into a crystalline and microporous surface in a swift and single-step process.

Through the addition of antibacterial elements into the PEO electrolyte, these elements become part of the converted surface layer and result in a surface exhibiting antibacterial behavior [32,33]. Due to the tight embedding of the antibacterial agents into the surface, the release of these ions can be controlled and the undesired circulation of agents can be prevented, thereby avoiding nanotoxic effects [34]. PEO was applied to generate titanium implants with antibacterial properties using Ag, Cu, and Zn [35–37]. In addition to the antibacterial behavior, PEO biofunctionalized surfaces were shown to enhance osseointegration and stimulate bony ingrowth in vivo [38,39].

Bone implants are increasingly produced through additive manufacturing (AM), as this allows free-form fabrication and customized treatment for patients. AM allows for the fabrication of highly porous implants with vast internal surface areas, which may make the implants more prone to infection, while at the same time providing a challenging surface to modify through surface biofunctionalization techniques. PEO is capable of biofunctionalizing the surface of complex geometries. In addition, the parameters of the PEO process can be controlled, which allows to tailor the chemistry of the surface layer [40,41]. Furthermore, the synthesized surface layer adheres strongly to the implant substrate. Moreover, the method is easily scalable towards clinically sized implants. Limitations of PEO are that the surface morphology and chemistry of the surface are modified simultaneously and this makes the individual tuning of these properties difficult. Furthermore, the exact mechanism of plasma discharging is still unknown, and thereby the fine-tuning of the PEO processing parameters difficult to predict [42].

In order to develop clinically relevant antibacterial implants, it is important to assess the progress made in this area and compare the outcomes of different studies. As most implants available for current clinical use are made of titanium, we performed a systematic review on titanium implants biofunctionalized by PEO using Ag, Cu, and Zn. In order to illustrate the progress made in this area, we screened the studies published between 2009 and December 2020. This area of research involved several scientific disciplines, including engineering, material sciences, microbiology, and orthopedics. We, therefore, analyzed a broad spectrum of aspects including the implant substrate, PEO parameters, surface characteristics, antibacterial assays, and cytocompatibility testing.

2. Methods

2.1. Literature Search

A comprehensive electronic search was performed using Scopus and Google Scholar search engines up until December 2020. In addition, a global screening was performed using PubMed. The article search was conducted using different combinations of the following keywords: plasma electrolytic oxidation, micro-arc oxidation, antibacterial activity, Ag, Cu, and Zn. To ensure that relevant publications were not excluded, combinations of subject headings, text-word terms, and the Boolean operators AND and OR were used. The searches were limited to those studies published in English between 2009 and 2020. The reference lists of the included eligible studies were scanned to ensure no eligible studies were omitted. The last search date was 24 December, 2020. This systematic review was written according to the PRISMA (Preferred Reporting Items for Systematic Review and Meta-Analyses) statement [43].

2.2. Inclusion and Exclusion Criteria

The inclusion criteria were—(1) the surface modification technique: plasma electrolytic oxidation (PEO), micro-arc oxidation (MAO), or anodic spark deposition (ASD); (2) implant substrate: titanium and its alloys; (3) antibacterial agents: Ag, Cu and Zn; (4) metallic-based antibacterial agents should have been incorporated in PEO-modified Ti-based surfaces; and (5) assessment of the antibacterial behavior should have been performed. A study was excluded if it did not report any outcome variable. Furthermore, studies were not eligible for inclusion when—(1) articles were not published in English; (2) no surface modification technique was utilized; (3) PEO was performed in combination with other surface modification techniques or treatments; (4) no antibacterial testing was performed; and (5) study was of one of the following document types: reviews, patents, conference abstracts/papers, and case reports.

2.3. Study Selection

The titles and abstracts were screened to assess the suitability of the search results. Subsequently, the full-text of the studies selected in the first stage of screening were analyzed to assess whether or not they satisfied the inclusion criteria.

2.4. Risk of Bias

The methodological details of the included studies were analyzed to minimize the risks of biases in the individual studies. Furthermore, excluding grey literature in Google Scholar decreased the risk of biases in the evaluation.

2.5. Data Extraction

Extracted information included the type of the titanium substrate, electrolyte composition, PEO processing parameters, surface topography, XRD phase composition, surface content of the incorporated elements, the release profile of the metallic (i.e., Ag, Cu, and Zn) ions, antibacterial assays, tested pathogens, eukaryotic cell types, and the outcomes (i.e., antibacterial behavior and cytocompatibility). The results were considered significant when $p < 0.05$.

2.6. Search Results

A total of 1261 studies were identified in the two search engines: 1190 from Google Scholar and 71 from Scopus. After screening the titles and abstracts, 1158 studies were excluded. The primary reasons for exclusion were no antibacterial or biocompatibility tests, PEO performed in combination with other surface modification techniques, and document types: reviews, patents, conference abstracts/papers, citations and case reports. As a result, 103 studies were selected for full-text analysis. The analysis led to the exclusion of 59 studies, as they failed to meet the inclusion criteria. Finally, 49 studies were included in this systematic review and were used for a qualitative analysis of their data and for comparison with each other. A flow diagram was created to represent the entire systematic search of the relevant studies (Supplementary Figure S1). The outline of the review is presented in Figure 1.

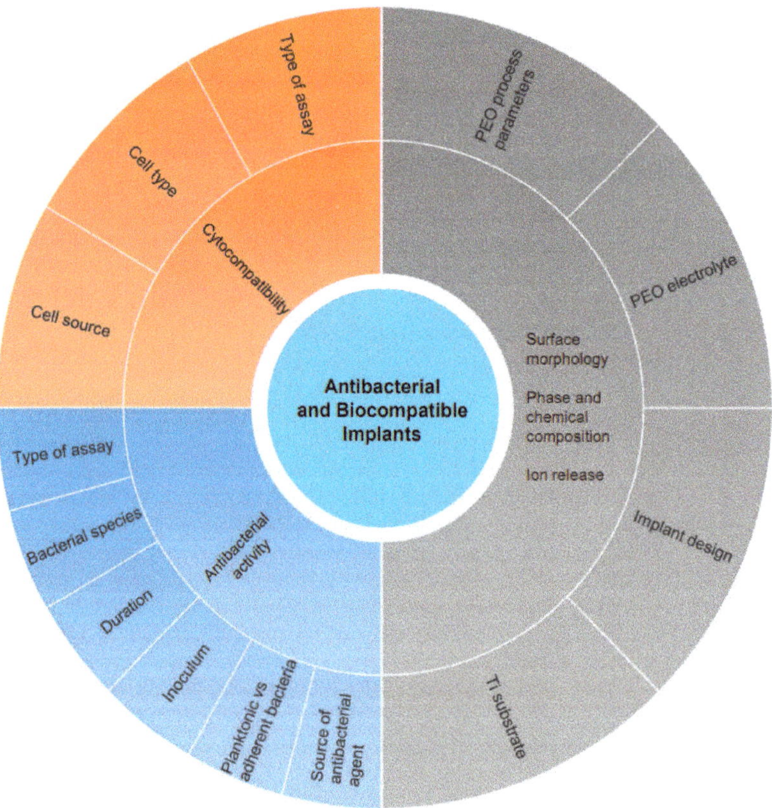

Figure 1. A graphical presentation of the outline of this systematic review.

3. Summary of Study Characteristics

A summary of the study characteristics is presented in Figure 2. Of the analyzed studies, 43% used Ag, 26% used Cu, and 21% worked with Zn, while 9% investigated a combination of Ag, Cu, and Zn (i.e., using two or more metallic agents. Various types of parameters were reported in the studies (Figure 2A), including the PEO processing parameters (98%), phase composition (87%), surface content of the incorporated elements (80%), and ion release kinetics (48%). Furthermore, 92% of the studies quantified the antibacterial activity, which was reported to be >50% for 100% of the studies using Ag, 93% of the studies using Cu, and 73% of those employing Zn, as well as 100% of the studies combining multiple metallic agents (Figure 2B). Of those studies, 57% tested the efficacy of the surfaces against *S. aureus*, 31% of the studies tested their specimens against *E. coli*, while 12% of the studies chose other bacterial species. Furthermore, the antibacterial activity was determined against adherent bacteria in 42% of the studies, while 35% of the studies assessed the antibacterial activity of their specimens against planktonic bacteria, and 23% assessed both.

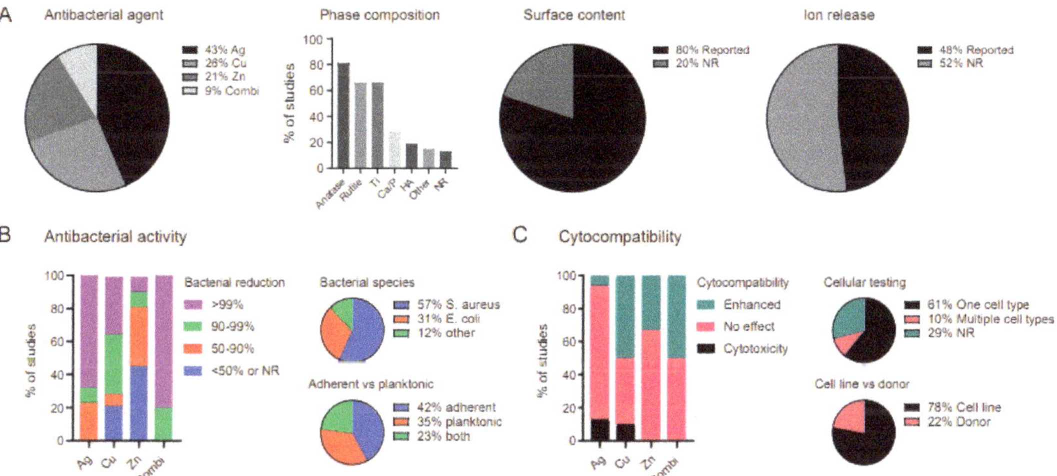

Figure 2. An overview of the (**A**) biomaterial, (**B**) antibacterial, and (**C**) cytocompatibility specifications of the studies included in this systematic review of the literature. Combi: combination of Ag, Cu, and/or Zn, HA: hydroxyapatite, NR: not reported.

Cytocompatibility was tested in 71% of all studies, of which 10% tested against multiple cell types (Figure 2C). Of the studies assessing the cytocompatibility of their specimens, 78% used a cell line while 22% used cells obtained from a donor. The addition of the metallic antibacterial agent resulted in cytotoxicity for 13% of the Ag studies, 10% of the Cu studies, 0% of the Zn studies, and 0% of the studies combining two or more metals. Meanwhile, improved cell response (i.e., enhanced cell viability and/or osteogenic differentiation) was observed for 7% of the Ag surfaces, 50% of the Cu surfaces, and 33% of the Zn surfaces, as well as for 50% of the surfaces combining Ag, Cu, and Zn.

4. Synthesis and Characterization of PEO Biofunctionalized Surfaces

PEO is an electrochemical process that converts the outer oxide layer of valve metals into a ceramic surface layer and is applied to enhance corrosion resistance [44], dielectric properties [45], and biocompatibility [46] of the substrates. A PEO setup has two electrodes: the cathode and anode (Figure 3A). Usually, either a constant current or voltage is applied, leading to the formation of an oxide layer on the anode (i.e., the specimen to be treated). After dielectric breakdown, the oxide layer is thickened by spark discharges that lead to pore formation [47] (Figure 3B). As the process continues, the sparks become more intense, resulting in the formation of larger pores.

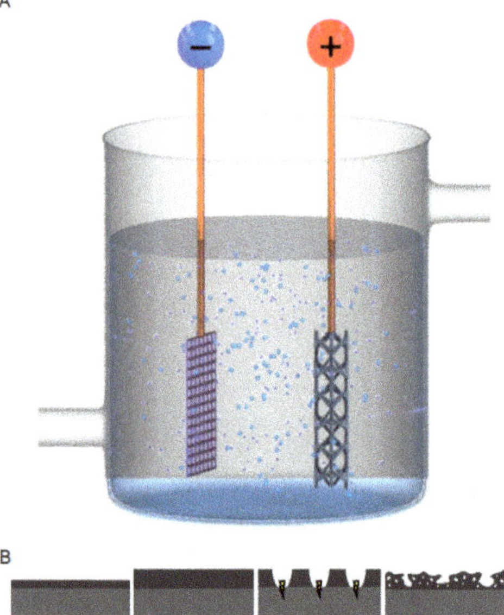

Figure 3. (**A**) A schematic drawing of the plasma electrolytic oxidation (PEO) setup with a cathode and an anode (implant). (**B**) During PEO processing, initially the titanium oxide layer grew outwards. After dielectric breakdown, plasma discharge occurred at the surface, resulting in a highly porous structure.

PEO biofunctionalization results in an altered surface morphology and chemical composition. In order to relate the antibacterial activity to certain surface characteristics, the surface of the biofunctionalized specimens is usually characterized (Tables 1–4). The important surface parameters in this regard are the surface topography, chemical composition, phase composition, and ion release profile. In the following sections, we will discuss the results regarding each of these parameters in more detail.

Table 1. The methodological details of the included studies in which Ag was used as the antibacterial agent.

	PEO Processing Parameters									
Substrate	# of Exp Groups with Ag	Electrolyte	Voltage (V)	Current Density (A/dm^2)	Time (min)	Surface Topography	Phase Composition	Surface Content of Ag	Cumulative Ag Ion Release (ppb)	Ref.
Ti6Al7Nb	2	0.02 M CA, 0.15 M Ca-GP, and (0.3 and 3.0) g/L Ag NPs	-	20	5	Porous structures (<5 μm)	-	-	12—day 7 89—day 7	[15]
Ti6Al4V	2	0.15 M CA, 0.02 M Ca-GP, and 3.0 g/L Ag NPs	-	20	5	Micro- and nano-porous structures with Ag NPs of 7–25 nm	Ti, anatase, rutile, HA, CaTiO$_3$, and Ca$_3$(PO$_4$)$_2$	-	138—day 28 600—day 28	[31]
Ti6Al7Nb	1	0.15 M CA, 0.02 M Ca-GP, and 3.0 g/L Ag NPs	-	20	5	Porous structures (<3 μm) with Ag NPs of 37 nm	Ti, anatase, and rutile	0.03 wt%	-	[32]
CP-Ti	3	0.4 M CA, 0.04 M β-GP, and (0.00003, 0.00006 and 0.004 M) AgNO$_3$	380–420	-	180	Irregular and rough morphology with spherical particles and flakes	Rutile, α-TCP, β-Ca$_2$P$_2$O$_7$, and HA	<0.1 wt% <0.1 wt% 0.21–0.45 wt%	-	[35]
CP-Ti	1	0.15 M CA, 0.05 M NaH$_2$PO$_4$, 0.25 mM AgNO$_3$	280–320	-	6	Porous surface with 1.5 μm pore size and 8.5% pore density	Anatase, rutile	0.13 at%	48—day 18	[48]
CP-Ti	3	0.4 g/L NaOH, 4.0 g/L NaH$_2$PO$_4$, and 0.1–1.0 g/L Ag NPs	400	-	5	Homogenous porous surface layer	Ti, anatase, rutile	1.5 at% 3.5 at% 5.8 at%	40—day 7 200—day 7 240—day 7	[49]
Ti6Al4V	2	0.15 M CA, 0.02 M Ca-GP, 0.3 M SrA, and 3.0 g/L Ag NPs	-	20	5	Uniform coverage with a micro-/nanopores. Addition of SrA resulted in smaller pore size.	Ti, anatase, rutile, HA, SrTiO$_3$, Sr$_2$Ca(PO$_4$)$_2$	-	1500—day 28 1800—day 28	[50]
CP-Ti	3	100 mM Ca-GP, 150 mM CA, 0,5, and 10 mM AgNO$_3$	-	2.51	10	Porous oxide layer for 0 and 5 mM Ag, non-porous surface for 10 mM Ag	Anatase, α-Ti	0.5 at% 1.5 at% 3.0 at%	300—day 28 3000—day 28 10^4—day 28	[51]

Table 1. *Cont.*

	PEO Processing Parameters									
Substrate	# of Exp Groups with Ag	Electrolyte	Voltage (V)	Current Density (A/dm^2)	Time (min)	Surface Topography	Phase Composition	Surface Content of Ag	Cumulative Ag Ion Release (ppb)	Ref.
CP-Ti, Ti-40Nb	2	Na$_2$HPO$_4$, NaOH, β-Ca$_3$(PO$_4$)$_2$, and 0.3—1 g/L AgNO$_3$	200–450	-	5–10	Uniformly distributed β-TCP particles over a porous surface with 0–8 μm pore sizes	Anatase, α-TCP, β-TCP	0.2 at% 0.8 at%	-	[52]
CP-Ti	4	Na$_2$HPO$_4$, NaOH, β-Ca$_3$(PO$_4$)$_2$, and 1 g/L AgNO$_3$	200–450	-	5–10	Uniformly distributed β-TCP particles over a porous surface with 0–8 μm pore sizes	Anatase, α-TCP, β-TCP	0.3 at% 0.5 at% 0.8 at%	-	[53]
CP-Ti	3	0.1 M CA, 0.06 M NaH$_2$P, and 0.01—0.05 M Ag$_2$O NPs	-	10	10	Porous structure with typical micro-sized pores	Anatase, rutile	1.6 wt% 3.1 wt% 5.8 wt%	2000—day 28 4000—day 28 10^4—day 28	[54]
CP-Ti	1	CA, Na$_2$HPO$_4$, and 0.0025 M Ag-A	380	-	5	Flake-like morphology with regional Ag particles of <200 nm	Ti, anatase, rutile, HA, and CaTiO3	4.6 wt%	-	[55]
CP-Ti	3	20.5 g/L CA, 7.2 G/L Na$_2$HPO$_4$, and (0.0005, 0.001, and 0.002) M Ag-A	400	-	5	Micro-porous structures with Ag NPs surrounding micro-pores	Ti, anatase, rutile, HA, and CaTiO3	1.14 wt%	-	[56]
Ti6Al4V	1	20.5 g/L CA, 7.2 g/L Na$_2$HPO$_4$, and 0.001 M Ag-A	400	-	5	Micro-porous structures with Ag NPs of <100 nm surrounding micro-pores	Ti, anatase, rutile, HA, and CaTiO3	0.7 wt%	1500—day 14	[57]
Ti6Al4V	2	CA, β-GP and (0.1 and 0.4) g/L AgNO$_3$	400	-	5	Granular and needle-like morphology with Ag NPs of 20–30 nm	Ti, anatase, rutile, HA, and CaTiO$_3$	0.6 wt% 2.1 wt%	2500—day 14 8000—day 14	[58]
Ti-29Nb-13Ta-4.6Zr	2	0.15 M CA, 0.1 M Ca-GP, and (0.0005 and 0.0025) M AgNO$_3$	-	2.51	10	Porous structures (<10 μm)	-	0.01 wt% 0.01 wt%	-	[59]

Table 1. *Cont.*

| | | | | PEO Processing Parameters | | | | | |
Substrate	# of Exp Groups with Ag	Electrolyte	Voltage (V)	Current Density (A/dm^2)	Time (min)	Surface Topography	Phase Composition	Surface Content of Ag	Cumulative Ag Ion Release (ppb)	Ref.
CP-Ti	3	0.1 M KOH, 0.015 M $K_4P_2O_7$, and (0.1, 0.3 and 0.5) g/L Ag NPs	-	10	5	Micro-porous structures with Ag NPs of <20 nm (3–7.5 μm)	-	0.53 at% 0.69 at% 0.80 at%	12.2—day 1 22.7—day 1 28.8—day 1	[60]
CP-Ti	1	0.3 M CA, 0.02 M GP, and 0.62 g/L Ag NPs	290	-	10	Porous structures with volcano top-like micro-pores	Ti, anatase, and rutile	1.07 at%	-	[61]
CP-Ti	1	0.3 M CA, 0.02 M GP, and 0.62 g/L Ag NPs	290	-	10	Porous structures with Ag NPs of <100 nm	Ti, anatase, and rutile	-	-	[62]
Ti6Al4V	1	Pure water and AgPURE^{TM} W10 nanosilver suspension	-	20	0.5	Flake-like morphology with Ag particles of <200 nm	-	3.6 at%	-	[63]
Ti6Al4V	2	0.2 M CA, 0.02 M β-GP, and (0.005 and 0.05) g/L Ag NPs	387 ± 3 385 ± 2	8	3	Porous structures with volcano top-like micro-pores (<3 μm)	Ti, rutile, and HA	<0.1 wt% <0.1 wt%	-	[64]
CP-Ti	3	2.0 g/L $NaH_2PO_4 \cdot 2H_2O$, 5.0 g/L CA, and 0.1, 0.5, and 0.8 g/L Ag-A	500	-	5	Porous structures uniformly covering surface	Ti, anatase, rutile, HA, $CaTiO_3$	0.8 at% 1.5 at% 2.2 at%	264—day 7 813—day 7 1110—day 7	[65]
CP-Ti	2	NTA, $Ca(OH)_2$, and 180 mg/L Ag NPs	250–300	-	5	Rough, thick oxide layer with a highly porous structure	-	0.3 wt% 0.7 wt%	-	[66]

Ag-A: silver acetate, CA: calcium acetate, Ca-GP: calcium glycerophosphate, GP: glycerophosphate, HA: hydroxyapatite, KOH: potassium hydroxide, NPs: nanoparticles, NTA: nitrilotriacetic acid, SrA: strontium acetate, TCP: tricalcium phosphate.

Table 2. The methodological details of the included studies in which Cu was used as the antibacterial agent.

	PEO Processing Parameters									
Substrate	# of Exp Groups with Cu	Electrolyte	Voltage (V)	Current Density (A/dm^2)	Time (min)	Surface Topography	Phase Composition	Surface Content of Cu	Cumulative Cu Ion Release (ppb)	Ref.
CP-Ti	1	0.1 M CA, 0.05 M GP, and 0.05 M Cu(OAc)$_2$	-	16.5	4	Micro-porous or crater structures (3–5 μm) with nano-grains of 30–50 nm	Ti and anatase	1.4 ± 0.08 wt%	-	[36]
CP-Ti, Ti-40Nb	2	H$_3$PO$_4$, 50–75 g/L CaCO$_3$, 40–60 g/L Cu-substituted HA	200–450	-	5–10	Uniformly distributed β-TCP particles over a porous coating surface with 0–8 μm pore sizes.	Anatase, β-TCP, α-TCP, Ca$_2$P$_2$O$_7$	0.1 at% 0.2 at%	-	[52]
CP-Ti	1	0.02 M C$_{12}$H$_{22}$CaO$_{14}$, 0.01 M (NaPO$_3$)$_6$, 0.02 M C$_{12}$H$_{22}$CuO$_{14}$	NR	NR	6	Porous surface with irregularly shaped and sized pores	-	-	-	[67]
CP-Ti	2	0.1 M CA, 0.06 M NaH$_2$P, 5–10 g/L Na$_2$Cu-EDTA	-	10	10	Highly porous area with micro-sized pores and a rough less porous area	-	2.3 wt% 4.2 wt%	3.3/cm^2—day 8 8.1/cm^2—day 8	[68]
CP-Ti	3	H$_3$PO$_4$, 300–600 g/L Cu(NO$_3$)$_2$·H$_2$O	450	-	5	With increasing Cu-salt levels sharpening of pores	Ti, anatase	0.54 at% 0.55 at% 0.72 at%	-	[69]
Ti6Al4V	2	11 g/L KOH, 10 g/L EDTA-CuNa$_2$, 5 or 15 g/L phytic acid	-	10	3	Uniformly distributed three-dimensional porous structure	Anatase, rutile, and TiP$_2$O$_7$	1.01 wt% 1.92 wt%	192—day 8 197—day 8	[70]
CP-Ti	1	0.2 M CA monohydrate, 0.02 M NaH$_2$PO$_4$, 0.01 M CuA monohydrate	-	3.25	5	Volcanic uniform porous morphology with 1–5 μm pores	Ti, rutile, anatase, Ca$_3$(PO$_4$)$_2$	5.05 at%	32.8—day 14	[71]
CP-Ti	4	0.2 M CA, 0.02 M β-GP, and (0.00125, 0.0025, 0.00375, and 0.005) M Cu(OAc)$_2$	450	-	1.5	Micro-porous structures (1–4 μm)	Ti, anatase, and rutile	0.67 wt% 1.17 wt% 1.51 wt% 1.93 wt%	6.75—day 21 - - 60.2—day 21	[72]
CP-Ti	2	0.1 M Na$_2$, 0.25 M NaOH, 0.1 M CA, 0.02 M Na$_2$SiO$_3$, and (0.0002 and 0.002) M CuSO$_4$	250	-	5	Macro-pores or crater structures (>100 μm) with nano-grains	-	-	411.3—day 2 27.0—day 2	[73]

Table 2. *Cont.*

PEO Processing Parameters										
Substrate	# of Exp Groups with Cu	Electrolyte	Voltage (V)	Current Density (A/dm^2)	Time (min)	Surface Topography	Phase Composition	Surface Content of Cu	Cumulative Cu Ion Release (ppb)	Ref.
CP-Ti	1	15 g/L NaH$_2$PO$_4$, 2 g/L NaOH, and 3.0 g/L Cu NPs	-	20	5	Porous structures (<5 µm) with Cu NPs of <60 nm	Ti, anatase, and rutile	-	-	[74]
CP-Ti	2	15 g·L-1 NaH$_2$PO$_4$, 2 g/L NaOH, and (0.3 and 3.0) g/L Cu NPs	470 ± 3 465 ± 3	20	5	Micro-porous structures (1–5 µm)	Ti, anatase	1.30 at% 2.76 at%	0.117—day 1 0.135—day 1	[75]
Ti6Al4V	3	Phosphate electrolyte with (2,6 and 10) g/L Cu$_2$O NPs	450	-	15	Micro-porous structures (<30 µm) with Cu$_2$O NPs of 20–30 nm	Ti, anatase, rutile, Cu, Cu$_2$O, and CuO	16.0 wt% 23.2 wt% 24.5 wt%	-	[76]
CP-Ti	1	0.002 M CA, 0.02 M β-GP, and 0.0013 M Cu(OAc)$_2$	480	-	2	Micro-porous structures (1–4 µm)	Ti, anatase, and rutile	0.77 wt%	4.5—day 7	[77]
Ti6Al4V	1	50 g/L Na$_2$SiO$_3$ and 4 g/L Cu$_2$O NPs	350	-	15	Porous structures (<3 µm) with Cu$_2$O NPs of 20–50 nm	Ti, anatase, rutile, Cu, Cu$_2$O, and CuO	27.27 wt%	-	[78]

CA: calcium acetate, Ca-GP: calcium glycerophosphate, CuA: copper acetate, GP: glycerophosphate, HA: hydroxyapatite, KOH: potassium hydroxide, NPs: nanoparticles, NR: not reported, TCP: tricalcium phosphate.

Table 3. The methodological details of the included studies in which Zn was used as the antibacterial agent.

				PEO Processing Parameters						
Substrate	# of Exp Groups with Zn	Electrolyte	Voltage (V)	Current Density (A/dm^2)	Time (min)	Surface Topography	Phase Composition	Surface Content of Zn	Cumulative Zn Ion Release (ppb)	Ref.
CP-Ti	3	20 g/L Na$_3$PO$_4$, 4 g/L NaOH, and (5, 10, and 15) g/L NPs	301 304 310	1000	7	Porous structures with ZnO NPs of 25 nm (<1.51–0.98 μm)	Ti, anatase, and rutile	20 wt% 25 wt% 35 wt%	-	[37]
CP-Ti, Ti-40Nb	2	H3PO4, 50–75 g/L CaCO$_3$, 40–60 g/L Zn-substituted HA	200–450	-	5–10	Uniformly distributed β-TCP particles over a porous coating surface with 0–8 μm pore sizes	Anatase, β-TCP, α-TCP, Ca$_2$P$_2$O$_7$	0.28 at% 0.4 at%	-	[52]
Ti6Al4V	1	50 g/L Na$_2$SiO$_3$ and 4 g/L ZnO NPs	350	-	15	Porous structures (<3 μm) with ZnO NPs of 20–50 nm	Ti, anatase, rutile, and ZnO	35.54 wt%	-	[78]
CP-Ti	2	0.1 M CA, 0.06 M NaH$_2$P, 0.02 M Na$_2$Zn-EDTA, or 0.02 M ZnO NPs	-	10	10	Porous surface at micrometer scale	Anatase, rutile, ZnO	-	-	[79]
CP-Ti	3	0.15 M CA, 0.1 M Ca-GP, 0.5–2.5 mM ZnCl$_2$	-	2.51	10	Continuous porous surface with circular pores of 5.3 μm in size	α-Ti, anatase	3.3 at%	250—day 7	[80]
CP-Ti	1	15 g EDTA-$_2$Na, 8.8 g Ca(CH$_3$COO)$_2$·H$_2$O, 6.3 g Ca(H$_2$PO$_4$)·H$_2$O, 7.1 g Na$_2$SiO$_3$·9H$_2$O, 5 g NaOH, 6 mL H$_2$O$_2$, 8.5 g Zn(CH$_3$COO)$_2$ in 1 L	350–500	-	7	Porous and rough surface with 1–3 μm pore sizes increasing voltages resulting in decreasing pore density and increased pore sizes	Ti, anatase, rutile	2 at%	250—day 15	[81]
CP-Ti	1	0.15 M CA, 0.15 M Ca-GP, and 0.02 M ZnA	350	-	1	Porous structures with volcano-shaped structures	Ti, anatase, and rutile	9.7 at%	300—day 1 <1000—day 28	[82]
CP-Ti	3	0.1 M CA, 0.05 GP, and (0.02, 0.04, and 0.06) M ZnA	-	16.5	4	Porous (<5 μm) with nano-grains of 20–100 nm	Ti, anatase, and rutile	4.6 ± 0.7 wt% 7.1 ± 0.6 wt% 9.3 ± 0.8 wt%	1180—day 14 2235—day 14 3620—day 14	[83]

Table 3. *Cont.*

		PEO Processing Parameters								
Substrate	# of Exp Groups with Zn	Electrolyte	Voltage (V)	Current Density (A/dm^2)	Time (min)	Surface Topography	Phase Composition	Surface Content of Zn	Cumulative Zn Ion Release (ppb)	Ref.
CP-Ti	1	0.02 M CA, 0.15 M Ca-GP, and 0.06 M ZnA	-	30	5	Porous structures (<5 μm)	Ti, anatase, and rutile	8.7 at%	-	[84]
CP-Ti	3	0.1 M CA, 0.025 M $Na_5P_3O_{10}$, and (0.01, 0.03, and 0.05) M ZnA	380	-	20	Micro-porous structures	-	0.199 at% 0.574 at% 1.995 at%	-	[85]
Ti-15Mo	3	0.1 M $Ca(H_2PO_2)_2$, 10 g/L ZnO, or 25 g/L $Zn_3(PO_4)_2$ or 10 g/L $Ca_3(PO_4)_2$ and 10 g/L $Zn_3(PO_4)_2$ particles	300	15	5	Porous oxide layer with micropores	-	1.5 at% 1.1 at% 0.2 at%	115—week 16 64—week 16 60—week 16	[86]

CA: calcium acetate, Ca-GP: calcium glycerophosphate, GP: glycerophosphate, HA: hydroxyapatite, KOH: potassium hydroxide, NPs: nanoparticles, NR: not reported, TCP: tricalcium phosphate, ZnA: zinc acetate.

Table 4. The methodological details of the included studies in which multiple antibacterial agents were used.

				PEO Processing Parameters						
Substrate	# of Exp groups	Electrolyte	Voltage (V)	Current Density (A/dm²)	Time (min)	Surface Topography	Phase Composition	Surface Content of Zn	Cumulative Ion Release (ppb)	Ref.
				Ag and Cu						
Ti6Al4V	6	0.15 M CA, 0.02 M Ca-GP, 0.75–3.0 g/L Ag, and/or Cu NPs in ratios 0–100%	-	20	5	Homogeneous porous surface with circular pores. Ag and/or Cu NPs scattered on surface.	-	-	Day 28: 1491 (Ag)/- 1906 (Ag)/- 1573 (Ag)/1527 (Cu) 1425 (Ag)/1392 (Cu) 1291 (Ag)/1225 (Cu) -/1981 (Cu)	[19]
				Ag and Zn						
Ti6Al4V	6	0.15 M CA, 0.02 M Ca-GP, 0.75–3.0 g/L Ag, and/or Zn NPs in ratios 0–100%	-	20	5	Homogeneous porous surface with circular pores. Ag and/or Zn NPs scattered on surface.	-	-	Day 28: 1491 (Ag)/- 1906 (Ag)/- 1573 (Ag)/1467 (Zn) 1682 (Ag)/1697 (Zn) 1749 (Ag)/1678 (Zn) -/2281 (Zn)	[20]
CP-Ti	3	0.1 M CA, 0.02 M β-GP, 0.25 g·L-1 SDBS, 0.1 M ZnA, and 6 g/L Ag NPs	390	-	0.5 1.5 2	Micro-porous structures with nano-grains of 5–40 nm and Ag NPs of <20 nm (1–4 μm)	Ti, anatase, rutile, and ZnO	1.06 (Ag)/22.19 (Zn) 1.42 (Ag)/26.93 (Zn) 1.56 (Ag)/29.38 (Zn)	Week 36 - 684 (Ag)/6880 (Zn)	[87]
				Cu and Zn						
CP-Ti	5	0.002 M CA, 0.02 M β-GP, (0, 0.005, 0.01, 0.02, and 0.04) M ZnA and 0.0013 M Cu(OAc)₂	480	-	2	Micro-porous structures (1–4 μm)	Ti, anatase, and rutile	0.77 (Cu) 0.62 (Cu)/1.79 (Zn) 0.55 (Cu)/2.53 (Zn) 0.39 (Cu)/6.47 (Zn) 0.33 (Cu)/8.92 (Zn)	Day 20: 4.5 (Cu) 3.2 (Cu)/7.8 (Zn) 2.7 (Cu)/23.2 (Zn) 2.3 (Cu)/64.5 (Zn) 1.9 (Cu)/94.9 (Zn)	[77]
Ti6Al4V	9	3–9 g/L KOH, 5–11 g/L phytic acid, 2–10 g/L EDTA-CuNa₂, 2–10 g/L EDTA-ZnNa₂	-	11	3	Porous surface with increasing pore sizes for increased levels of Cu and/or Zn in surface	Ti, anatase	-/3.47 (Zn) -/9.84 (Zn) -/7.90 (Zn) 0.61 (Cu)/11.41 (Zn) 0.98 (Cu)/4.42 (Zn) 2.15 (Cu)/5.42 (Zn) -/5.64 (Zn) 1.25 (Cu)/6.71 (Zn) 4.18 (Cu)/2.89 (Zn)	-	[88]

CA: calcium acetate, Ca-GP: calcium glycerophosphate, GP: glycerophosphate, HA: hydroxyapatite, KOH: potassium hydroxide, NPs: nanoparticles, NR: not reported, SDBS: sodium dodecyl benzene sulfonate, TCP: tricalcium phosphate, ZnA: zinc acetate.

4.1. Titanium Substrate

Of the reviewed studies, most used commercially pure (CP) titanium (62%), followed by Ti6Al4V (23%), Ti6Al7Nb (4%) [15,32], Ti40Nb [52], Ti29Nb13Ta4.6Zr [59], and Ti15Mo [86]. Titanium is used for bone implants because of its mechanical properties, corrosion resistance and chemical biocompatibility [46,89]. Ti6Al4V has a higher strength to weight ratio than CP titanium and is, therefore, the natural choice for load-bearing applications, such as joint replacing implants, while CP titanium is more frequently applied for non-load bearing applications, such as maxillofacial implants [90]. Clinical studies comparing the long-term outcomes of patients treated with either CP-Ti or Ti-alloys are lacking [91,92].

Ti6Al4V implants may release vanadium and aluminum ions that can induce cytotoxicity [93]. Other alloys employing niobium have, therefore, been developed, including Ti6Al7Nb and Ti40Nb, which have similar mechanical properties, but do not induce cytotoxicity [94]. In addition, the cytotoxic effects of Al and/or V can be mitigated by PEO, since it reduces the ion release of those species [89]. PEO is easily scalable and can be applied to human-sized implants [95]. In order to translate the results from in vitro studies, it is, therefore, interesting to investigate the antibacterial behavior of substrates that are designed and produced like an implant, for instance, through additive manufacturing. This also highlights one of the advantages of PEO, namely that it can be applied on highly porous surfaces [31].

4.2. PEO Electrolyte

The bioactivity of PEO-biofunctionalized implant surfaces is determined for a large part by the composition of the PEO electrolyte, as the elements in the electrolyte eventually make up the chemical composition of the implant surface. More than 50% of the studies included in this systematic review used electrolytes with Ca and P elements. The presence of Ca and P in the electrolyte can result in the formation of hydroxyapatite, which forms more than 60% of bone tissue and is associated with a Ca/P ratio of 1.67 [96,97]. Calcium acetate and calcium glycerophosphate were the primary source of Ca, while $CaCO_3$ [52] and $C_{12}H_{22}CaO_{14}$ [67] were also used in some studies. P is usually added in the form of calcium glycerophosphate, β-glycerophosphate, H_3PO_4 [52,69], $K_4P_2O_7$ [60], NaH_2PO_4 [48,49,65,71,74,75], $NaPO_3$ [67], or $Na_5P_3O_{10}$ [85]. Another element used in about 30% of the included studies is Na in the form of NaOH, NaH_2PO_4 [48,49,65,71,74,75], $NaPO_3$ [67], $Na_5P_3O_{10}$ [85], or Na_2SiO_3 [73,76,78,98]. The addition of Na roughens the surface and enhances the Ca/P ratio [99], which has been shown to enhance the osteogenic cell response [100,101]. In addition, the implantation of Na through plasma immersion has been found to stimulate the osteogenic differentiation of cells [102]. Moreover, KOH [60,70,88] is used as an alternative base for NaOH given its similar effects on osteogenic differentiation [103].

4.3. PEO Processing Parameters

The electrical parameters of the PEO process affect the surface morphology [42], including the porosity [104], pore size [105], pore shape [106], and pore density [107], as well as the surface chemistry [83,84]. Of the included studies, 54% controlled the voltage, 31% controlled the current density, and 13% controlled both, while 1 study did not report the PEO processing parameters. The oxidation times ranged between 0 and 180 min, with 21% between 0–4 min, 50% between 5–9 min, 19% between 10–14 min, 6% between 15–19 min, and 4% ≥ 20 min. As the current density, voltage, or oxidation time increases, the spark discharge energy amplifies, affecting the mass of the oxide layer formed by a single pulse and resulting in enhanced growth of the oxide layer [40,108]. Furthermore, as temperature of the local discharge area increases, the plasma effect is enhanced, resulting in larger pore sizes and the transformation of amorphous TiO_2 to anatase and rutile phases. Meanwhile, the intensity of the spark discharge enhances with time, meaning that prolonged oxidation times result in the formation of hydroxyapatite on the implant surface [109,110]. As such,

PEO processing parameters largely affect the chemical and phase composition as well as the surface topography of the implant surface.

4.4. Surface Morphology

As PEO greatly affects the surface topography of titanium surfaces, all studies investigated the surface topography by scanning electron microscopy (SEM) and most studies reported a porous surface topography with rounded pores (Figure 4A). PEO transforms the native titanium oxide layer into a highly porous surface with interconnected porous networks, which is frequently described as a volcanic landscape with micropores that are <10 µm in diameter. In addition, flake-like morphologies [35,55,63] and needle-like structures [58] are often observed. Furthermore, the thickness and porosity of the oxide layer were shown to depend on the composition of the PEO electrolyte and PEO processing parameters [54,111]. The specifications of the surface morphology in turn were shown to greatly influence the antibacterial behavior [112] and osteogenic properties [113,114] of the implant surfaces.

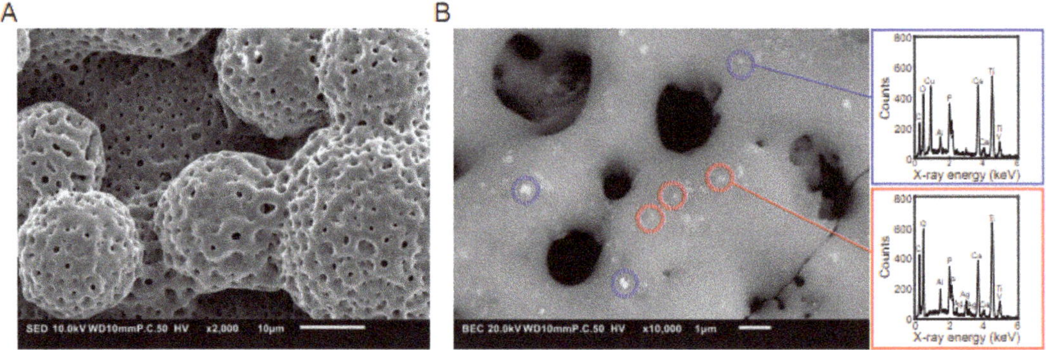

Figure 4. (**A**) SEM images of the typical surface morphology of titanium implants after PEO processing. (**B**) EDS analysis of the implant surface to characterize its chemical composition with spectrum of Cu (blue) and Ag (red) nanoparticles.

4.5. Phase Composition by XRD

One component of the surface that plays a major role in the biological behavior is the phase composition of the implants [115]. These phases can be analyzed with X-ray diffraction (XRD). Among the included studies, 87% analyzed the phase composition. Of those, all studies analyzed Ti phases and observed bare Ti (66%), anatase (81%), and/or rutile (66%). Some studies observed both Ti and anatase, but no studies reported solely Ti and rutile. This is in line with the observation that during PEO processing, first the metastable anatase is formed, which then turns into the stable rutile [116]. While all studies that performed XRD analysis identified the TiO_2 phases, not all studies analyzed the other phases formed by the elements incorporated from the electrolyte. Since many PEO electrolytes contain both Ca and P, 19% of the studies observed hydroxyapatite [31,35,50,55–58,64,65] and 28% contain other Ca/P phases including α-TCP [35,52,53], β-TCP [52,53], TiP_2O_7 [70], $CaTiO_3$ [31,55–58,65], $Ca_2P_2O_7$ [35,52], and $Ca_3(PO_4)_2$ [31,71]. In addition, phases with Cu, Cu_2O, and CuO [76,78], as well as ZnO [78,79,87] were observed.

These phases were shown to affect the biological response. For instance, TiO_2 is transformed from an amorphous phase into crystalline anatase and rutile phases that were shown to produce reactive oxygen species (ROS) [117], which in turn contribute to the desired antibacterial behavior [118].

4.6. Content of the Antibacterial Elements Incorporated in the PEO Layers

The antibacterial activity of Ag, Cu, and Zn may be present on the implant surface depending on the dose [119–121]. Therefore, it is important to quantify the content of these elements on the implant surface after PEO biofunctionalization. This analysis is usually done either by energy-dispersive X-ray spectroscopy (EDS; Figure 4B) or X-ray photoelectron spectroscopy (XPS). Among the included studies, 80% reported the elemental composition of the surface, while 20% did not. The studies generally reported the elemental composition either in terms of atomic% or weight% and found them to correlate with the amount of Ag, Cu, and Zn dispersed in the PEO electrolyte. The amount of Ag incorporated in the implant surfaces tended to be lower (1.35 ± 1.82 wt%) than Cu (7.70 ± 10.17 wt%) and Zn (18.79 ± 12.06 wt%), reflecting the lower minimal inhibitory concentration (MIC) of Ag (0.03–8 µg/mL) as compared to Cu (256–448 µg/mL) and Zn (765 µg/mL) [122]. However, EDS does not exclusively measure the elemental composition of the surface but may penetrate deeper into the oxide layer. This is an important point, because it is not clear to what extent the species present deeper inside the oxide layer, which can be up to 10 µm in thickness, and contribute to the antibacterial properties of biofunctionalized implants [15]. The amount of active agents present on the implant surface may not be directly related to the antibacterial activity, since the form in which the element is present on the surface (i.e., ionic species, nanocrystals, or nanoparticles) affects the antibacterial properties as well [123,124].

4.7. Ion Release

An important antibacterial mechanism is through the release of metallic ions from the implant surface. These released ions do not only play a role in contact-killing, but also target planktonic bacteria in the implant surrounding, as this area could form a niche for bacteria [125]. Ion release was studied in 48% of the included studies and was measured from 12 h up to 56 days. Overall, the release of Ag, Cu, and Zn ions was found to be higher for the implant surfaces with a higher elemental content and a higher concentration of the active agents in the PEO electrolyte. The combination of Ag with Cu or Zn NPs on the implant surface resulted in enhanced Cu or Zn release while the Ag release was reduced in the first 24 h [19,20]. Similarly, higher concentrations of zinc acetate added to copper acetate resulted in enhanced Zn ion release while Cu ion release was reduced with higher concentrations of zinc acetate [77]. This may stem from galvanic coupling favoring the oxidation and release of one element over the other [126,127]. When studied in detail, this may allow for controlled release profiles and accompanying antibacterial effects.

Ion release results depend on the liquid in which these measurements are performed. Frequently used liquids are phosphate-buffered saline (PBS) and simulated body fluid (SBF) [128]. Ion release does not only depend on the surface content, but also on the form in which the antibacterial agent is present on the surface (i.e., as ionic species, nanoparticles, or other forms) [124]. Ideally, one could control the release of ions to not only prevent infection immediately after surgery, but also ward off late implant-associated infections [129]. However, comparing the reported ions release kinetics is difficult due to the different units, specimen designs, and measurement setups being used. In addition to the previously mentioned parameters, the surface area plays an important role in determining the concentration of the released ions, as a larger area allows for more agents to be incorporated on the surface, in turn leading to a higher release rate [31]. The reported concentrations of release ions should, therefore, be normalized with respect to the surface area of the specimens to enable direct comparison between different studies. The information regarding the surface area is generally not reported in the studies, rendering a direct comparison impossible.

5. Antibacterial Properties

Surface biofunctionalization by PEO with Ag, Cu, and Zn results in antibacterial surfaces. In the following section, we will first compare the antibacterial activity of PEO biofunctionalized titanium implants bearing Ag, Cu, and Zn found by in vitro and ex vivo studies (Table 5). Then, we will discuss the factors that determine the antibacterial activity. First of all, the types of the bacterial species and strains were shown to affect the susceptibility and resistance of bacteria to antibacterial agents [130], their ability to infect host cells [131], and their pathogenicity [132]. Moreover, the type of assay, the inoculation dose, and the culture time used in the studies may affect the observed antibacterial activity. Finally, the activity against adherent and/or planktonic bacteria is discussed, as the adherence of bacteria may initiate biofilm formation, while planktonic bacteria form a source for reinfection and host cell invasion [133].

Table 5. Antibacterial tests and results on PEO-modified Ti-surfaces bearing single or multiple elements.

Bacterial Species	Bacterial Strain	Source	Analysis Method	Duration (h)	Test Inoculum	Planktonic/ Adherent	Main Outcomes	Ref.
					Ag			
MRSA	AMC201	Ag NPs	Modified version of JIS Z 2801:2000	24	10^7 CFU/mL	Adherent	After 24 h: 98 and 99.75% reduction by incorporation of 0.3 and 3 g/L Ag NPs	[15]
MRSA	AMC201	Ag NPs	Petrifilm^{TM} assay Zone of inhibition CFU count SEM Ex vivo	48	10^3–10^8 CFU/mL	Adherent	Significantly reduced numbers of viable bacterial colonies by incorporation of Ag NPs in the surface after 15 min. Four-logs reduction in the numbers of viable bacterial colonies in the ex vivo infection model by incorporation of Ag, compared with a 2-logs reduction in absence of Ag after 24 h. Prevention biofilm formation for at least 48 h	[31]
MRSA	AMC201	Ag NPs	Modified version of JIS Z 2801:2000	24	10^7 CFU/mL	Adherent	100% killed by incorporation of 0.03wt% Ag at 24 h	[32]
S. aureus *E. coli*	ATCC6538 ATCC25922	AgNO$_3$	Spread plate analysis	24	$1.6 \cdot 10^5$ CFU/mL	Planktonic	After 24 h: >99.8 reduction by incorporation of >0.1 wt% Ag, compared with a reduction of 20% in absence of Ag	[35]
E. coli	ATCC25933	AgNO$_3$	Spread plate analysis	12	10^6 CFU/mL	Adherent	After 12 h: >99.9% eradication of *E. coli*	[48]
S. aureus *E. coli*	ATCC6538 ATCC25922	Ag NPs	CFU count Fluorescence measurement	24	0.0001 OD$_{590}$	Adherent	After 24 h: complete eradication for *E. coli* and 6-log reduction for *S.aureus* with 5.8 at% Ag Stronger antibacterial effect on *E. coli* compared to *S. aureus*	[49]
MRSA	USA300	Ag NPs	Zone of inhibition CFU count SEM Ex vivo	48	10^4–10^7 CFU/mL	Adherent Planktonic	After 24 h: enhanced zone of inhibition for PT-AgSr samples compared to PT-Ag samples. Complete eradiation of adherent and planktonic bacteria in vitro and ex vivo. After 48 h: prevention of biofilm formation in Ag-containing surfaces.	[50]
S.aureus *E.coli*	NBRC122135 NBRC3972	AgNO$_3$	ISO 22196:2007	24	0.4–$3.0 \cdot 10^6$ CFU/mL	Adherent	After 24 h: >0.05 mM Ag in PEO electrolyte reduced bacteria >90%. Inhibitory effect was stronger for *E. coli* compared to *S. aureus*	[51]

Table 5. *Cont.*

Bacterial Species	Bacterial Strain	Source	Analysis Method	Duration (h)	Test Inoculum	Planktonic/ Adherent	Main Outcomes	Ref.
S. aureus	209P	$AgNO_3$	Spread plate analysis	2	500 CFU/mL	Planktonic	After 2 h: 53% reduction in CFU after incubation in supernatant	[52]
S. aureus	ATCC6538-P	$AgNO_3$	Spread plate analysis	2	250 CFU/mL	Planktonic	After 2 h: 70% reduction in CFU and 45% antibacterial rate for >0.3%at Ag	[53]
S. aureus	NR	Ag_2O	Spread plate analysis	24	10^5 CFU/mL	Adherent	After 24 h: antibacterial rate >1 with 5.8wt% Ag	[54]
S. aureus *E. coli*	ATCC6538 ATCC25822	Ag-A	Spread plate analysis	24	$2.5 \cdot 10^5$ CFU/mL	Planktonic	At 24 h: 99.9 and 58.3% reduction of E. coli for 4.6 wt% Ag and Ag-free. At 24 h: 99.8 and 47.8% reduction of S. aureus for 4.6 wt% Ag and Ag-free	[55]
S. aureus	ATCC6538	Ag-A	Modified version of JIS Z 2801:2000	24	$2.5 \cdot 10^5$ CFU/mL	Planktonic	After 24 h: 99.98% reduction by incorporation of 1.14 wt% Ag	[56]
S. mutans	ATCC25175	Ag-A	Spread plate analysis SEM	16.5	$1.5 \cdot 10^8$ CFU/mL	Adherent	After 16.5 h: 67% reduction by incorporation of 0.7 wt% Ag	[57]
E. coli	ATCC25822	$AgNO_3$	Spread plate analysis	24	10^9 CFU/mL	Planktonic	After 24 h: 97.4 and 99.2% reduction by incorporation of 0.6 and 2.1 wt% Ag, compared with a reduction of 22.7% in absence of Ag: Ag-free PEO-modified surface	[58]
E. coli	NBRC3972	$AgNO_3$	ISO 22196:2011	24	$5 \cdot 10^6$ CFU/mL	Planktonic	100% killed in presence of 0.01 wt% Ag at 24 h	[59]
E. coli	ATCC25922	Ag NPs	Spread plate analysis	24	10^6 CFU/mL	Planktonic	100% killed by incorporation of 0.53 wt% Ag within 12 h	[60]
S. sanguinis	IAL1832	Ag NPs	Spread plate analysis	24	10^7 CFU/mL	Planktonic	At 24 h: 62 and 53% reduction by incorporation of 1.9wt% Ag, compared to pure Ti and the Ag-free PEO-modified surface, respectively	[61]
S. epidermidis	ATCC35984	Ag NPs	Spread plate analysis SEM	18	10^6 CFU/mL	Adherent Planktonic	100% killed by incorporation of 3.6at% Ag within 12 h	[63]
P. gingivalis	NR	Ag NPs	Microbial Viability Assay SEM	24	10^7 CFU/mL	Adherent	Reduction of the bacterial viability to 21–31% by incorporation of <0.1wt% Ag at 8 h, compared with a mean viability of 96.6% in absence of Ag in the PEO-modified surface	[64]

Table 5. *Cont.*

Bacterial Species	Bacterial Strain	Source	Analysis Method	Duration (h)	Test Inoculum	Planktonic/ Adherent	Main Outcomes	Ref.
E. coli *S. aureus* MRSA	ATCC25922 ATCC6538 Mu50	Ag-A	CFU count SEM	24	0.0005 OD_{590}	Adherent	4–6 log inhibition of *E. coli*, 3–5 log inhibition of *S. aureus*, and 2–5 log inhibition of MRSA after 24 h for 0.1 and 0.5 and 0.8 g/L Ag respectively	[65]
S. aureus	B 918	Ag NPs	Spread plate analysis	24	10^6 CFU/mL	Adherent	Lower amounts of adherent bacteria after 2 h. No inhibition at later time points	[66]
					Cu			
S. aureus	NR	$Cu(OAc)_2$	Spread plate analysis	4	10^6 CFU/mL	Planktonic	Significantly reduced numbers of bacterial colonies by incorporation of 1.4 wt% Cu in the surface after 4 h	[36]
S. aureus	209P	Cu-substituted HA	Spread plate analysis	2	500 CFU/mL	Planktonic	After 2 h: 27% reduction in optical density after incubation in supernatant	[52]
S. aureus	NR	$C_{12}H_{22}$-CuO_{14}	Spread plate analysis SEM	24	10^4 CFU/mL	Adherent	After 24 h: 100% antibacterial rate on Cu surfaces Morphological changes and disrupted membrane of bacterial cells.	[67]
S. aureus	ATCC6538	EDTA-$CuNa_2$	Live/dead staining SEM	24	10^5 CFU/mL	Adherent	After 24 h: more dead bacteria on Cu surface compared to Ti control. Shape changes and membrane disruption of bacteria under SEM	[68]
E. coli	ATCC25922	$Cu(NO_3)_2 \cdot H_2O$	Zone of inhibition Adhesion test	24	10^8 CFU/mL	Adherent Planktonic	After 24 h: zone of inhibition around 0.54–0.72 wt% Cu. No bacterial cells adhering after 24 h	[69]
S. aureus *E. Coli*	ATCC43300 ATCC25922	EDTA-$CuNA_2$	Spread plate analysis	24	$5 \cdot 10^5$ CFU/mL	Adherent	After 24 h: complete eradication of S. aureus and E. coli for 1.92 wt% Cu. After 14 days no antibacterial activity.	[70]
S. aureus	ATCC6538	CuA monohydrate	Spread plate analysis	24	10^5 CFU/mL	Adherent	After 24 h: >99% growth reduction with 5.05 at% Cu in the surface.	[71]

Table 5. *Cont.*

Bacterial Species	Bacterial Strain	Source	Analysis Method	Duration (h)	Test Inoculum	Planktonic/ Adherent	Main Outcomes	Ref.
S. aureus	ATCC25923	$Cu(OAc)_2$	Spread plate analysis Live/dead staining SEM	96	10^5 CFU/mL	Adherent Planktonic	At 6 h: 0.6×10^5 CFU/cm^2 on 1.93 wt% Cu-PEO and 1.5×10^5 CFU/cm^2 on Cu-free. At 24 h: 0.6×10^5 CFU/cm^2 on 1.93 wt% Cu-PEO and 9.7×10^5 CFU/cm^2 on Cu-free. At 6 h: 1.0×10^5 CFU/mL for 1.93 wt% Cu- PEO and 3.8×10^5 CFU/mL on Cu-free. At 24 h: 5.2×10^5 CFU/mL for 1.93 wt% Cu-PEO and 200×10^5 CFU/mL on Cu-free.	[72]
S. aureus	NR	$CuSO_4$	Macrophage bactericidal assay SEM	2	10^7 CFU/mL	Planktonic	Significantly enhanced macrophage-bactericidal capacity on 2 mM Cu-incorporated PEO-modified surface	[73]
S. aureus *E. coli*	NR	Cu NPs	Live/dead staining	24	10^5 CFU/mL	Adherent	Majority of bacteria killed after 24 h	[74]
S. aureus	NR	Cu NPs	Spread plate analysis Live/dead staining SEM	24	10^7 CFU/mL	Adherent Planktonic	100% killed by incorporation of 2.76 at% Cu at 24 h	[75]
E. coli	CMCC44102	Cu_2O NPs	ASTM G21-13	24	NR	Adherent	At 24 h: 99.74% killed by incorporation of 10 g·L-1 Cu_2O NPs, compared to 95.25% killed in absence of Cu in the PEO-modified surface	[76]
Zn								
S. aureus *E. coli*	ATCC25923 ATCC25922	ZnO NPs	ASTM G21-1996	24	10^6 CFU/mL	Planktonic	After 24 h: reduced numbers of viable colonies by incorporation of Zn compared with Zn-free surfaces	[37]
S. aureus	209P	Zn-substituted HA	Spread plate analysis	2	500 CFU/mL	Planktonic	After 2 h: 40% reduction in optical density after incubation in supernatant	[52]
E. coli	NR	ZnO NPs Zn-EDTA	Measurement of OD_{600}	24	NR	Planktonic	After 24 h: 50% reduction in OD_{600} values of culture medium	[79]
E. coli	NBRC3972	$ZnCl_2$	Spread plate analysis	24	$4.9 \cdot 10^6$ CFU/mL	Adherent	After 24 h: less than 1 log reduction	[80]

Table 5. *Cont.*

Bacterial Species	Bacterial Strain	Source	Analysis Method	Duration (h)	Test Inoculum	Planktonic/ Adherent	Main Outcomes	Ref.
S. aureus *E. coli*	ATCC25923 ATCC25922	ZnA	Spread plate analysis SEM	24	10^7 CFU/mL	Planktonic	After 24 h: 40% enhanced antibacterial rate on *E.coli*. No effect on *S. aureus*	[81]
S. aureus *P. aeruginosa*	NR	ZnA	Live/dead staining SEM	24	OD_{600}~0.35	Adherent	Significantly reduced numbers of viable colonies by incorporation of 9.7at% Zn at 6 and 24 h	[82]
S. aureus *E. coli*	ATCC25923 ATCC25922	ZnA	Spread plate analysis SEM	24	10^7 CFU/mL	Adherent	At 24 h: 40.2, 99.2 and 100% reduction of *E. coli* for 4.6, 7.1, and 9.3 wt% Zn. At 24 h: 96.3, 99.5, and 99.8% reduction of *S. aureus* for 4.6, 7.1, and 9.3 wt% Zn	[83]
S. aureus *E. coli*	NR	ZnA	Spread plate analysis Live/dead staining SEM	24	10^5 CFU/mL	Adherent Planktonic	>90% killed at 24 h	[84]
S. mutans	ATCC 25175	ZnA	Spread plate analysis SEM	48	10^9 CFU/mL	Adherent	At 24 h: 62.54, 69.84 and 79.19% reduction for 0.199, 0.574 and 1.995at% Zn	[85]
S. aureus MRSA *S. epidermidis*	ATCC25923 MRSA1030 ATCC700296 *S. epidermidis* 15560	ZnO and $Zn_3(PO_4)_2$ particles	Spread plate analysis	4	10^6 CFU/mL	Adherent	After 4 h: no growth inhibition for *S. aureus* and MRSA, and 90% eradication of *S. epidermidis* on Zn-bearing surfaces.	[86]
Ag and Cu								
MRSA	USA300	Ag and Cu NPs	Zone of inhibition CFU count SEM Ex vivo	24	10^4–10^7 CFU/mL	Adherent Planktonic	After 24 h: zone of inhibition and eradication of adhering and planktonic bacteria in vitro and ex vivo for surface containing >50% Ag and Cu NPs. No antibacterial properties for solely Cu NP-bearing surfaces and controls.	[19]
Ag and Zn								
MRSA	USA300	Ag and Zn NPs	Zone of inhibition CFU count SEM Ex vivo	24	10^4–10^7 CFU/mL	Adherent Planktonic	After 24 h: zone of inhibition and eradication of adhering and planktonic bacteria in vitro and ex vivo for surface containing >50% Ag and Zn NPs. No antibacterial properties for solely Zn NP bearing surfaces and controls.	[20]

Table 5. *Cont.*

Bacterial Species	Bacterial Strain	Source	Analysis Method	Duration (h)	Test Inoculum	Planktonic/ Adherent	Main Outcomes	Ref.
S. aureus	ATCC25923	Ag NPs and ZnA	Spread plate analysis SEM	24	10^5 CFU/mL	Adherent Planktonic	At 24 h: 4.1, 2.5, and $2.4 \cdot 10^3$ CFU/cm^2 on Ag and Zn co-doped surfaces compared with $2.3 \cdot 10^6$ CFU/cm^2 on polished Ti, respectively. Significantly reduced numbers of viable colonies by incorporation of Ag NPs and Zn compared to polished Ti.	[87]
					Cu and Zn			
S. aureus	ATCC25923	Cu(OAc)$_2$ ZnA	Spread plate analysis Live/dead staining SEM	24	10^5 CFU/mL	Adherent Planktonic	At 6 h: 2.63, 1.47, and $0.84 \cdot 10^5$ CFU/cm^2 on Cu and Zn co-doped surfaces compared with 1.8, and $8.5 \cdot 10^5$ CFU/cm-2 on Cu-single doped and Cu-free surfaces, respectively. At 24 h: 3.72, 2.89, and $1.32 \cdot 10^5$ CFU/cm^2 on Cu and Zn co-doped surfaces compared to 2.89 and $16 \cdot 10^5$ CFU/cm^2 on Cu-single doped and Cu-free surfaces, respectively. Significantly reduced number of viable colonies by incorporation of >2.53 wt% Zn and <0.55 wt% Cu, compared to 0.77 wt% Cu	[77]
E. coli	CMCC44102	Cu$_2$O and ZnO NPs	ASTM G21-13	24	10^6 CFU/mL	Planktonic	PEO-modified surfaces bearing Cu$_2$O NPs demonstrated a superior antibacterial activity~100% killed, compared with PEO-modified surfaces bearing ZnO NPs	[78]
MRSA *S. aureus* *E. coli*	ATCC43300 CGMCC12465 CGMCC13373	EDTA-CuNa$_2$ EDTA-ZnNa$_2$	Spread plate analysis	24	10^6 CFU/mL	Adherent	After 24 h: complete prevention of growth with >6 g/L Cu or Zn in PEO electrolyte against MRSA, *S. aureus* and *E. coli*.	[88]

Ag-A: silver acetate, ASTM: American Society for Testing and Materials, CFU: colony forming unit, CuA: copper acetate HA; hydroxyapatite, JIS: Japanese Industrial Standards, NPs: nanoparticles, NR: not reported, SEM: scanning electron microscopy, ZnA: zinc acetate.

5.1. Comparing Antibacterial Activities of Ag, Cu, and Zn

All the included studies reported antibacterial activity. Guidelines designate a material as antibacterial when it induces a >99.9% (i.e., 3-log) reduction in the number of viable bacteria [134]. However, this is a guideline for treatment, while the required reduction in the bacterial load for the prevention of IAI is not known. In fact, 48% of the studies using Ag, 14% of the studies with Cu, 10% of the studies with Zn, and 80% of the studies that combined these metallic agents reduced the bacterial load by >99.9%. This indicates that surfaces biofunctionalized with Ag demonstrate the highest degree of antibacterial activity, while Cu and Zn were less effective, which is not surprising given the much lower MIC for Ag as compared to Cu and Zn [122]. Interestingly, combining Ag, Cu, and Zn resulted in much higher levels of antibacterial activity, while the doses of single elements can be reduced [19,20,87,88].

Studies that focused on the antibacterial mechanisms of Ag, Cu, and Zn NPs suggest that two antibacterial mechanisms play a role: ion release killing [135] and the generation of reactive oxygen species (ROS) [136]. Ions released from the implant diffuse across the bacterial cell wall and penetrate into bacteria where vital bacterial structures are targeted. Meanwhile, ROS are highly reactive and cause lysis of the bacterial cell wall. It was found that Cu showed the best antibacterial activity as a result of contact killing [137], while Ag exhibited most of its antibacterial activity through both ion release and contact killing [138]. Furthermore, the synergistic antibacterial properties of AgNPs and Zn ions were observed to stem from long-range Zn ion release and contact-killing effects from Ag through microgalvanic coupling [29,139].

We plotted a 3D graph showing the correlation between antibacterial activity, cytocompatibility, and surface content of the antibacterial agent for the titanium substrates biofunctionalized by PEO with Ag, Cu, or Zn (Figure 5). Very few studies reported all of these 3 parameters. This analysis shows that Ag indeed resulted in the highest levels of antibacterial activity at lower doses compared to Cu and Zn, yet also induced cytotoxicity more frequently. However, a direct comparison between the included studies, and thereby of Ag, Cu, and Zn bearing surfaces, was hampered by a large number of variables that differ in the various studies and are addressed in the next paragraphs of this section.

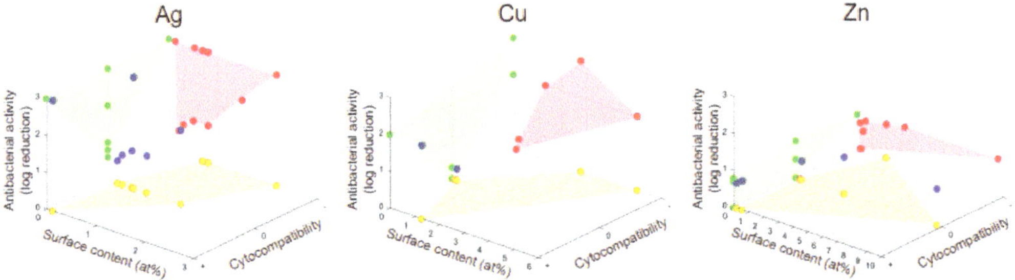

Figure 5. The relation between the antibacterial activity, cytocompatibility, and surface content for titanium surfaces biofunctionalized by PEO with Ag, Cu, or Zn. The reported antibacterial activity as a function of surface content and cytocompatibility is depicted by the blue dots. The green, red, and yellow projections enable a comparison between the parameters. Cytocompatibility is depicted as cytotoxicity (−), no effect (0), or enhanced cytocompatibility (+).

5.2. Bacterial Species and Strains

Antibacterial results are affected by the tested bacterial species. Of the reviewed studies, 57% used *S. aureus*, 31% *E. coli*, and 12% other bacterial species, including *S. epidermidis* [63,86], *S. sanguinis* [61], *S. mutans* [57,85], *P. aeruginosa* [82], and *P. gingivalis* [64]. Given that Ag, Cu, and Zn form an alternative to antibiotics, it is important to analyze the results on antibiotic resistant bacteria, such as MRSA, which are involved in up to 32% of fracture-related infections [140,141]. MRSA was investigated in 9 studies and found to be strongly inhibited by Ag [15,31,32,50,65], Ag and Cu [19], Ag and Zn [20], Cu and Zn [88] bearing surfaces, while one study that included Zn surfaces did not observe any inhibition [86]. Thukkaram et al. observed that the antibacterial effect of Ag containing surfaces against MRSA was lower compared to *S. aureus* and *E. coli*, although with increasing doses of Ag, all bacterial species were targeted equally [65]. Furthermore, testing on multiple species was performed in 19% of the included studies. No studies tested multiple species in a single experiment (i.e., co-culture of multiple species), which would be of interest given that 10–20% of IAI are induced by polymicrobial infections [142,143].

We can, thus, conclude that most studies investigated antibacterial behavior against *S. aureus*. This bacterial species causes 20–46% of IAI [144–146]. Other gram-positive species, such as *Streptococci* caused up to 10% and *Enterococci* 3–7% of cases [147]. *Enterococci* have not been tested in studies with PEO-treated surfaces bearing Ag, Cu, or Zn. Gram-negative bacteria, such as *Pseudomonas aeruginosa* and *Enterobacteriaceae* induce 6–17% of IAI [143,148]. Given the relatively low rate of IAI induced by *Enterobacteriaceae*, it is surprising that 31% of the studies investigated the effects of the implant surfaces on *E. coli*. While some studies that analyzed both *S. aureus* and *E. coli* reported a stronger antibacterial effect against *E. coli* as compared to *S. aureus* [49,51,55,65,83,98], others reported a similar antibacterial effect for both species [35,37,70,74,84,88]. Interestingly, up to 42% of IAI in patients were caused by culture-negative (i.e., undefined) bacteria [149,150] and therefore warrant an antibacterial agent effective against a wide antimicrobial spectrum.

Among bacterial species, different levels of sensitivity to antibacterial agents have been reported [151], including against Ag and Cu [152]. To what extent the differences between strains plays a role depends on the bacterial species. The differences between strains in terms of their MIC/MBC values was found to be negligible for *S. aureus,* but were quite large in the case of *E. coli* strains [153]. It is, therefore, important that the bacterial strain is properly reported, which was done only in 79% of the included studies. Only one study, conducted by Leśniak-Ziółkowska et al., compared different strains within a bacterial species, namely *S. aureus* (ATCC 25,923 and clinical MRSA 1030) and *S. epidermidis* (ATCC 700,296 and clinical 15560) [86]. No strain-dependent differences were observed after 4 h using a bacterial adhesion test.

5.3. Source of Antibacterial Agent

Antibacterial behavior depends not exclusively on the antibacterial agent, but also on the form in which Ag, Cu, and Zn are added to the PEO electrolyte and are subsequently incorporated onto the titanium implant surface [124]. Ag, Cu, and Zn elements are either completely dissolved in the electrolyte or are added in the form of NPs that form a suspension. The former will end-up in the form of chemical compounds present all over the surface, while the latter (NPs) are spread over the surface. NPs may form a reservoir from which ions are released, thereby ensuring prolonged antibacterial activity [154]. In addition, the shape of the NPs determines the antibacterial activity as the surface-to-volume ratio affects the ion release and, thus, the efficacy of the surface biofunctionalization process [155]. Ionic forms only induce antibacterial activity through the action of ions, while NPs also produce reactive oxygen species and induce contact-killing [156]. Among the included studies, 33% used NPs, 64% employed ionic species, and only a study by Zhang et al., combined ions and NPs [87]. This study combined Ag NPs with Zn acetate, which resulted in much higher release of Zn ions compared to Ag ions. Furthermore, the antibacterial activity was assessed against both adherent and planktonic *S. aureus* after 24 h. The developed surface demonstrated significant antibacterial behavior with increasing concentrations of Ag and Zn leading to further reduction of viable bacteria. The authors reasoned that the antibacterial activity stems from ROS generation by both Ag and Zn as well as Ag$^+$ release. Moreover, both Ag and Zn ion concentrations remained below cytotoxicity levels and thus stressed the utility of combining these elements. Studies that investigate the differences in the antibacterial properties induced by NP and ionic forms are lacking.

5.4. Analysis Method

Antibacterial properties can be investigated by different assays. Properties often investigated are the antibacterial leaching activity, the killing of adherent bacteria, and the prevention of biofilm formation. Although most of the included studies used only one antibacterial assay (53%), the use of several assays is required for the assessment of the various types of antibacterial properties [157]. Therefore, 32, 8, and 8% of the included studies used 2, 3, and 4 assays, respectively. To determine the leaching effects of the antibacterial ions released from the PEO surfaces, a zone of inhibition assay or a Kirby-Bauer assay is often used. The number of bacteria can be quantified either through a direct CFU count, by spread plate analysis, or by staining the live cells using a fluorescent dye. A few studies referred to ISO [51,59] and ASTM [37,76,78] standards. With SEM, adherent bacteria and/or biofilm formation can be visualized in a non-quantitative manner. A wide variety in the type of assays used in the studies was found, with spread plate analysis (33%), SEM (24%), and viability fluorescence imaging (12%) being the most frequently applied assays.

In addition to in vitro assays, ex vivo models were explored, in which infected implants biofunctionalized with Ag and Cu, Zn, or Sr are inserted into a murine femur [19,20,31,50]. Subsequently, the number of CFU present are quantified (e.g., after 24 h). Although this ex vivo model does not allow to assess the effects of the implants on the immune system or bony ingrowth, some of the other in vivo effects such as those of the extracellular matrix and bone tissue [158] can be captured to some extent. Indeed, the gene expression profile of osteocytes was found to be similar between an ex vivo bone infection model and tissue samples from IAI patients [159]. Thus far, no study has tested the antibacterial activity of titanium implants biofunctionalized by PEO with Ag, Cu, and Zn in vivo.

5.5. Duration and Inoculum of Antibacterial Assay

Over two thirds of IAIs are initiated during surgery [160]. A rapid antibacterial response to prevent the adherence of the bacteria that enter the human body peri-operatively is, therefore, desired. Almost all of the included studies (94%) tested the antibacterial properties within 24 h and 10% even within 2 h. However, IAI can also be initiated long after surgery, stemming from hematogenous origins. Prolonged antibacterial activity is, thus, desirable too [72,85,161]. Zhang et al., reported on the antibacterial activity of Cu-containing surfaces for longer periods of time [72]. It was observed that the number of viable adherent bacteria was significantly reduced on surfaces containing 0.67–1.98 wt% Cu up to 96 h. However, this was one of the few studies aiming to assess long-term antibacterial behavior, since prolonged in vitro culture of bacteria is challenging. Research into late IAI is, therefore, primarily performed in vivo [162,163].

The inoculum used in the antibacterial assays is another factor determining the antibacterial behavior of PEO-biofunctionalized implants. The exact number of bacteria required for IAI is unknown, but it was shown that the presence of a foreign body can reduce the infection dose by 6 orders of magnitude [164] due to a hampered immune response [165]. The inoculum used in the included studies varied widely between 250 [53] and 10^9 CFU/mL [58,85], and was not reported in two studies. Currently, most inocula are presented per volume or as a measure of optical density. However, the surface area of the implant is also of importance, as more area with more incorporated antibacterial agent is likely to have a greater antibacterial effect. Therefore, presenting the inoculum per volume per surface area would support comparative analyses of different studies.

5.6. Planktonic vs. Adherent Bacteria

As both planktonic and adherent bacteria play an important role in IAIs, antibacterial implants should target both types of bacteria. Planktonic bacteria are present in the fluid and tissue surrounding the implant and have shown to be a reservoir for late-stage reinfections [125]. Once the bacteria adhere to the implant, bacteria should be targeted in order to prevent biofilm formation as this would induce bacterial resistance to antibiotic treatment [166]. In this respect, 42% of the included studies investigated antibacterial activity against adherent, 35% against planktonic, and 23% against both planktonic and adherent bacteria. Targeting both planktonic and adherent bacteria should, therefore, be emphasized more in future studies.

6. Biocompatibility

In addition to antibacterial properties, PEO-biofunctionalized implant surfaces should not induce cytotoxicity, and ideally even enhance cell response and bony ingrowth. The compatibility of the implants with mammalian cells is, therefore, an important topic that needs to be thoroughly investigated for any such implant. Several of the included studies report the results of such in vitro cytocompatibility experiments, which are affected by the type of the assay, cell type, and cell source (Supplementary Table S1).

6.1. Cytocompatibility of Ag, Cu, and Zn Surfaces

Cytocompatibility was investigated in 71% of studies. In those studies, Ag induced cytotoxicity in 13% of the studies, while 10% of the studies investigating Cu and 0% of those employing Zn reported cytotoxic effects. None of the studies combining Ag, Cu, and Zn reported cytotoxicity. Cell response of the implants was improved in 7% of the studies using Ag, 50% of the studies focused on Cu, and 33% of the studies with Zn, as well as for 50% of the studies in which two or more antibacterial agents were combined. The control group often consists of PEO biofunctionalized surfaces without antibacterial elements. Cytotoxicity is, therefore, not considered a major concern by the vast majority of the included studies. Indeed, Cu and to somewhat lesser extent Zn were shown to improve the cytocompatibility of PEO-treated implants.

6.2. Type of Assay

Several processes that occur in bone regeneration were investigated in vitro. Cells need to attach to the implant surface [167], spread [168], stay viable [169], proliferate, differentiate towards the osteogenic lineage [170], and eventually form an extracellular matrix [171]. Indicators for the bone regeneration process include cell morphology [172], expression of osteogenic markers [173], metabolic activity [174], and the production of specific proteins [175]. The parameters studied the most in the included studies were viability and proliferation (analyzed in 56% of the included studies), followed by adhesion and attachment (36%), differentiation (25%), cell spreading (22%), matrix calcification and mineralization (11%), metabolic activity (8%), gene expression (8%), morphology (3%), cell seeding (3%), and other assays (6%) including protein production, mitochondrial functioning, and cytokine production.

6.3. Cell Type

The cellular response was shown to differ in in vitro experiments between different cell types [176,177]. In the reviewed studies, pre-osteoblasts (32%), osteosarcoma cells (22%), fibroblasts (20%), MSCs (17%) and SV-HFO, macrophages, adipose stem cells, and endothelial cells (each in 1 study) were used. Pre-osteoblasts and MSCs are the main cells responsible for bone formation [178,179]. Osteosarcoma and SV-HFO cells [180] are immortalized cells stemming from the osteogenic lineage. However, osteosarcoma was shown to stem from defective differentiation [181]. Since these titanium implants will be used in bone tissue, it was surprising that 29% of the studies did not analyze the effects of the implants on bone-forming cells. Other cell types may support bone formation through indirect pathways. Endothelial cells play a role in angiogenesis, which plays a major role in bone regeneration as blood vessels carry nutrients and oxygen and facilitate the transport of immune cells to the regenerating bone tissue [182]. Meanwhile, macrophages form an important part of the immune response against IAI. Any potential toxicity of the synthesized implants against this cell type is of concern, as it may hamper the clearance of infections [16,183]. Finally, fibroblasts were shown to regulate osteoblast activity through tight junction interactions [184].

6.4. Cell Source

About 22% of the included studies used primary cells, whereas 78% utilized cell lines. Primary cells are more representative of the clinical situation, as they have been isolated from donors. However, their variability is high. Cells from multiple donors, therefore, need to be tested [185]. Cell lines, on the other hand, are homogenous and stable, while exhibiting little variability. However, their immortalized nature makes them differ from the clinical situation [186]. Furthermore, the source of animal species from which the cells were derived differed greatly between the included studies, with 56% using murine cells, 34% human cells, and 10% rat cells. The osteogenic differentiation capacity of stem cells is known to differ between human, mice, and rat MSCs [187,188]. These differences in animal species make it difficult (if not impossible) to directly compare the cytocompatibility results reported in the different studies.

7. Discussion

In order to prevent IAI, the biofunctionalization of titanium implants by PEO using Ag, Cu, and Zn as the active agents has gained significant momentum in the last decade. Therefore, we systematically reviewed the progress made on those implants and summarized the various types of properties measured for such types of PEO-biofunctionalized implants.

7.1. Antibacterial Results

From the results of this study, it can be concluded that Ag is the most potent antibacterial agent followed by Cu and Zn. It is important to stress that different studies utilized different experimental protocols to determine the antibacterial properties of PEO-biofunctionalized implants. It was shown that titanium surfaces bearing Ag, Cu, and Zn can kill bacteria through antibacterial leaching activity, contact killing, and the formation of ROS [156,189]. These properties cannot be assessed in a single assay. The use of multiple assays is, therefore, warranted to support the claim of antibacterial activity [157]. Finally, it is important to make sure that the assays assess infection prevention rather than infection treatment.

Furthermore, the bacterial species and strains used were found to affect the level of antibacterial activity. For instance, surfaces demonstrating antibacterial activity against *E. coli* may not do the same against *S. aureus* [51,55,98]. Most studies investigated the antibacterial activity of the implants against *S. aureus* or *E. coli*. While a large proportion of IAI was induced by *S. aureus*, only a small proportion of infections were caused by *E. coli* [144,147]. The rationale for choosing *E. coli* was, thus, primarily methodological convenience rather than clinical prevalence. Meanwhile, *S. epidermidis* or polymicrobial infections were rarely studied, even though they cause a significant proportion of IAI [142–144]. Moreover, the antibacterial behavior of PEO-biofunctionalized implants should be assessed in environments co-habited by multiple bacterial species, as this was shown to influence the resistance profiles of bacteria [130].

The antibacterial experiments aimed to mimic the clinical situation as closely as possible. In this respect, both adherent and planktonic bacteria should be warded off, as adherent bacteria can form biofilms [133], while planktonic bacteria may infect the peri-implant tissue and form a reservoir for late-stage reinfection [125]. Furthermore, an antibacterial implant should prevent infections that occur immediately after surgery, as that is the point where most IAI occur [160], as well as late-stage infections from hematogenous origins [161]. At the moment, the focus primarily lies on preventing early-stage infections. Ultimately, Ag, Cu, and Zn may form an alternative to antibiotics, as bacteria are developing ever-growing degrees of antibiotics resistance [144,190]. As such, the development of resistance against

Ag, Cu, and Zn and combination thereof is worthwhile to investigate given that resistance against Ag, Cu, and Zn was reported in vitro [191–193] and in patients [194].

The observed antibacterial activity depends on a wide variety of factors described in this review, including the titanium substrate, composition of the PEO electrolyte, and PEO processing parameters that in turn affect the surface morphology, phase composition, surface content of the incorporated antibacterial agent, and ion release profile. These parameters determine the antibacterial properties and biocompatibility of the implants. The measured antibacterial properties are highly dependent on the bacterial species and strains used, experimental techniques, the duration of the assays, bacterial inoculum, and the type of bacteria against which the implant performance is measured (i.e., planktonic and/or adherent). As for biocompatibility, the type of the assays, cell type, and cell source could all influence the final read-outs. These factors varied between the studies included in this review and make a one-to-one comparison between the different studies challenging.

The antibacterial activity is dependent on the dose of Ag, Cu, and Zn present on the surface of the titanium implants [119–121]. It is, therefore, essential to determine the amount of these elements present on the surface. In addition, the Ag, Cu, and Zn ions released from the implant surface are responsible for a significant part of the antibacterial activity, which is why it is important to measure the concentration of the ions released from the implant surface. From the results, it is clear that the surfaces bearing Ag had much lower elemental content and ion release as compared to those bearing Cu and Zn, which was expected due to the lower MIC of Ag as compared to Cu and Zn [122]. Both the surface content and ion release were also dependent on the surface area, as a larger surface area allows for the incorporation of a greater amount of elements and, thus, increased ion release [31]. Therefore, describing these properties relative to the surface area may aid in a comparison between the results of different studies.

7.2. Biocompatibility

Most of the included studied found cytotoxicity to be a minor concern, with Ag inducing cytotoxicity in 13% of the studies. It was striking that 29% of the included studies did not investigate the effects of the implants on bone-forming cells, even though the implants are intended for bone tissue. In addition, cytotoxicity against other cell types, such as endothelial cells and immune cells is of interest, as these cells contribute to bone regeneration as well [195,196]. Furthermore, the use of cell lines vs. donor cells and different mammalian species complicates the comparisons between different studies [197]. Moreover, biocompatibility needs to be investigated both in vitro and in vivo, as the results of in vitro and in vivo experiments are known to differ, for instance, in the case of Ag-bearing surfaces [16].

Another way to enhance the cytocompatibility of PEO-biofunctionalized implants is by combining two or more antibacterial metals (i.e., Ag, Cu, and Zn), as synergic effects between various such agents are reported to exist [19,20] and could be used to reduce the concentration of Ag [126,198]. In addition, combining these elements with other osteogenic elements, such as Sr [50] may enhance their antibacterial and biocompatible properties. Finally, the combination of multiple antibacterial elements significantly reduces the risk of the development of bacterial resistance, thereby ensuring that the prolonged use of these elements will remain possible [199].

PEO is frequently applied in combination with other surface treatments, such as hydrothermal treatment [200] and physical vapor deposition [201] to alter the chemical and phase composition of the surface. This may result in improved antibacterial behavior [98]. Furthermore, hydrothermal treatment has resulted in the enhanced formation of hydroxyapatite crystals, yet may reduce corrosion resistance too [202]. A major disadvantage of these additional surface treatments is that they make the entire process lengthier and more complex, thus making it more difficult to upscale the production of clinically sized implants.

7.3. Towards Clinically Relevant Implants

A decade of PEO biofunctionalization of titanium implants with Ag, Cu, and Zn confirmed the great potential of this method as an effective, fast, and scalable process. At the moment, however, the research on antibacterial PEO-biofunctionalized titanium implants is still far away from clinical application, as the research was primarily conducted in vitro with few studies also exploring ex vivo models [20,50]. Furthermore, PEO was shown to enhance the osteogenic capacity of titanium implants in vivo [38,39,203], including surfaces bearing Zn [204]. However, these studies did not analyze the antibacterial properties of such implants, which should be evaluated using bone infection models [205]. In this respect, a major limitation of the state-of-the-art techniques is their limited relevance for the assessment of the preventive potential of antibacterial implants (as opposed to their treatment potential). However, studying prevention requires a much larger sample size, as it is associated with lower bacterial loads, meaning that infections are less likely to occur. This lower risk of infection has major ethical and financial implications. In addition, future implants will most likely be fabricated by AM and as such be highly porous. Not only is the risk of infection of such volume-porous implants higher, their IAI treatment is also highly challenging due to their usually high degree of bony ingrowth that may cause significant bone loss during their removal. The development of antibacterial surface treatments for such types of implants is, thus, highly relevant. In fact, the additional surface area of such implants may be exploited to enhance the bioactivity of PEO biofunctionalized implants [31].

8. Conclusions

In order to combat IAI, the biofunctionalization of titanium implants by Ag, Cu, and Zn has gained significant momentum in recent years and resulted in the synthesis of potent antibacterial and biocompatible surfaces. Implant biofunctionalized with Ag, Cu, and Zn demonstrated significant antibacterial behavior against a wide bacterial spectrum, including antibiotic-resistant bacterial strains. However, the antibacterial properties of these implants were primarily investigated in vitro and occasionally ex vivo. Furthermore, many studies do not reach sufficiently high antibacterial levels, as indicated by international guidelines. Moreover, the biofunctionalization of volume-porous AM implants has not been investigated extensively. Finally, combining Ag, Cu, and Zn on the surface of titanium implants was shown to result in potent antibacterial surfaces with reduced cytotoxicity. In order to take the PEO biofunctionalization of titanium implants by Ag, Cu, and Zn to clinical settings, in vivo studies should be conducted using relevant infection models for both solid and volume-porous bone implants.

Supplementary Materials: The following are available online at https://www.mdpi.com/article/10.3390/ijms22073800/s1, Figure S1: Flow diagram of the systematic literature search., Table S1: The biocompatibility of PEO-modified Ti-based surfaces bearing single or multiple elements.

Funding: The research for this paper was financially supported by the Prosperos project, funded by the Interreg VA Flanders—The Netherlands program, CCI grant no. 2014TC16RFCB046.

Conflicts of Interest: The authors declare no conflict of interest.

References

1. Zimmerli, W. Clinical presentation and treatment of orthopaedic implant-associated infection. *J. Intern. Med.* **2014**, *276*, 111–119. [CrossRef]
2. Govaert, G.A.M.; Kuehl, R.; Atkins, B.L.; Trampuz, A.; Morgenstern, M.; Obremskey, W.T.; Verhofstad, M.H.J.; McNally, M.A.; Metsemakers, W.J.; Fracture-Related Infection (FRI) Consensus Group. Fracture-related infection consensus, diagnosing fracture-related infection: Current concepts and recommendations. *J. Orthop. Trauma* **2020**, *34*, 8–17. [CrossRef]
3. Kapoor, S.K.; Thiyam, R. Management of infection following reconstruction in bone tumors. *J. Clin. Orthop. Trauma* **2015**, *6*, 244–251. [CrossRef]
4. Singh, J.A. Epidemiology of knee and hip arthroplasty: A systematic review. *Open Orthop. J.* **2011**, *5*, 80–85. [CrossRef]

5.   Kremers, H.M.; Larson, D.R.; Crowson, C.S.; Kremers, W.K.; Washington, R.E.; Steiner, C.A.; Jiranek, W.A.; Berry, D.J. Prevalence of total hip and knee replacement in the United States. *J. Bone Jt. Surg. Am.* **2015**, *97*, 1386–1397. [CrossRef]
6.   Singh, J.A.; Yu, S.; Chen, L.; Cleveland, J.D. Rates of total joint replacement in the United States: Future projections to 2020–2040 using the national inpatient sample. *J. Rheumatol.* **2019**, *46*, 1134–1140. [CrossRef]
7.   Chambers, H.F.; Deleo, F.R. Waves of resistance: Staphylococcus aureus in the antibiotic era. *Nat. Rev. Microbiol.* **2009**, *7*, 629–641. [CrossRef]
8.   Guo, Y.; Song, G.; Sun, M.; Wang, J.; Wang, Y. Prevalence and therapies of antibiotic-resistance in Staphylococcus aureus. *Front. Cell. Infect. Microbiol.* **2020**, *10*, 107. [CrossRef]
9.   Teterycz, D.; Ferry, T.; Lew, D.; Stern, R.; Assal, M.; Hoffmeyer, P.; Bernard, L.; Uckay, I. Outcome of orthopedic implant infections due to different staphylococci. *Int. J. Infect. Dis.* **2010**, *14*, e913–e918. [CrossRef]
10. Cho, O.H.; Bae, I.G.; Moon, S.M.; Park, S.Y.; Kwak, Y.G.; Kim, B.N.; Yu, S.N.; Jeon, M.H.; Kim, T.; Choo, E.J.; et al. Therapeutic outcome of spinal implant infections caused by Staphylococcus aureus: A retrospective observational study. *Medicine* **2018**, *97*, e12629. [CrossRef]
11. Gross, T.M.; Lahiri, J.; Golas, A.; Luo, J.; Verrier, F.; Kurzejewski, J.L.; Baker, D.E.; Wang, J.; Novak, P.F.; Snyder, M.J. Copper-containing glass ceramic with high antimicrobial efficacy. *Nat. Commun.* **2019**, *10*, 1979. [CrossRef]
12. Lara, H.H.; Ayala-Núñez, N.V.; Ixtepan Turrent, L.d.C.; Rodríguez Padilla, C. Bactericidal effect of silver nanoparticles against multidrug-resistant bacteria. *World J. Microbiol. Biotechnol.* **2009**, *26*, 615–621. [CrossRef]
13. Nanda, A.; Saravanan, M. Biosynthesis of silver nanoparticles from Staphylococcus aureus and its antimicrobial activity against MRSA and MRSE. *Nanomedicine* **2009**, *5*, 452–456. [CrossRef] [PubMed]
14. Siddiqi, K.S.; Rahman, A.U.; Tajuddin; Husen, A. Properties of Zinc Oxide nanoparticles and their activity against microbes. *Nanoscale Res. Lett.* **2018**, *13*, 141. [CrossRef]
15. Necula, B.S.; Van Leeuwen, J.P.; Fratila-Apachitei, L.E.; Zaat, S.A.; Apachitei, I.; Duszczyk, J. In vitro cytotoxicity evaluation of porous TiO(2)-Ag antibacterial coatings for human fetal osteoblasts. *Acta Biomater.* **2012**, *8*, 4191–4197. [CrossRef]
16. Croes, M.; Bakhshandeh, S.; Van Hengel, I.A.J.; Lietaert, K.; Van Kessel, K.P.M.; Pouran, B.; Van der Wal, B.C.H.; Vogely, H.C.; Van Hecke, W.; Fluit, A.C.; et al. Antibacterial and immunogenic behavior of silver coatings on additively manufactured porous titanium. *Acta Biomater.* **2018**, *81*, 315–327. [CrossRef]
17. Bergemann, C.; Zaatreh, S.; Wegner, K.; Arndt, K.; Podbielski, A.; Bader, R.; Prinz, C.; Lembke, U.; Nebe, J.B. Copper as an alternative antimicrobial coating for implants–An in vitro study. *World J. Transpl.* **2017**, *7*, 193–202. [CrossRef] [PubMed]
18. Ding, Q.; Zhang, X.; Huang, Y.; Yan, Y.; Pang, X. In vitro cytocompatibility and corrosion resistance of zinc-doped hydroxyapatite coatings on a titanium substrate. *J. Mater. Sci.* **2014**, *50*, 189–202. [CrossRef]
19. Van Hengel, I.A.J.; Tierolf, M.; Valerio, V.P.M.; Minneboo, M.; Fluit, A.C.; Fratila-Apachitei, L.E.; Apachitei, I.; Zadpoor, A.A. Self-defending additively manufactured bone implants bearing silver and copper nanoparticles. *J. Mater. Chem. B* **2020**, *8*, 1589–1602. [CrossRef]
20. Van Hengel, I.A.J.; Putra, N.E.; Tierolf, M.; Minneboo, M.; Fluit, A.C.; Fratila-Apachitei, L.E.; Apachitei, I.; Zadpoor, A.A. Biofunctionalization of selective laser melted porous titanium using silver and zinc nanoparticles to prevent infections by antibiotic-resistant bacteria. *Acta Biomater.* **2020**, *107*, 325–337. [CrossRef]
21. Yoshitani, J.; Kabata, T.; Arakawa, H.; Kato, Y.; Nojima, T.; Hayashi, K.; Tokoro, M.; Sugimoto, N.; Kajino, Y.; Inoue, D.; et al. Combinational therapy with antibiotics and antibiotic-loaded adipose-derived stem cells reduce abscess formation in implant-related infection in rats. *Sci. Rep.* **2020**, *10*, 11182. [CrossRef] [PubMed]
22. Cavanaugh, D.L.; Berry, J.; Yarboro, S.R.; Dahners, L.E. Better prophylaxis against surgical site infection with local as well as systemic antibiotics. An in vivo study. *J. Bone Jt. Surg. Am.* **2009**, *91*, 1907–1912. [CrossRef]
23. Metsemakers, W.J.; Fragomen, A.T.; Moriarty, T.F.; Morgenstern, M.; Egol, K.A.; Zalavras, C.; Obremskey, W.T.; Raschke, M.; McNally, M.A.; Fracture-Related Infection (FRI) Consensus Group. Evidence-based recommendations for local antimicrobial strategies and dead space management in fracture-related infection. *J. Orthop. Trauma* **2020**, *34*, 18–29. [CrossRef]
24. Celis, J.P.; Drees, D.; Huq, M.Z.; Wu, P.Q.; De Bonte, M. Hybrid processes—A versatile technique to match process requirements and coating needs. *Surf. Coat. Technol.* **1999**, *113*, 165–181. [CrossRef]
25. Kumar, R.; Munstedt, H. Silver ion release from antimicrobial polyamide/silver composites. *Biomaterials* **2004**, *26*, 2081–2088. [CrossRef]
26. Zaporojtchenko, V.; Podschun, R.; Schurmann, U.; Kulkarni, A.; Faupel, F. Physico-chemical and antimicrobial properties of co-sputtered Ag–Au/PTFE nanocomposite coatings. *Nanotechnology* **2006**, *17*, 4904–4908. [CrossRef]
27. Gollwitzer, H.; Haenle, M.; Mittelmeier, W.; Heidenau, F.; Harasser, N. A biocompatible sol–gel derived titania coating for medical implants with antibacterial modification by copper integration. *AMB Express* **2018**, *8*, 1–9. [CrossRef] [PubMed]
28. Su, Y.; Wang, K.; Gao, J.; Yang, Y.; Qin, Y.; Zheng, Y.; Zhu, D. Enhanced cytocompatibility and antibacterial property of zinc phosphate coating on biodegradable zinc materials. *Acta Biomater.* **2019**, *98*, 174–185. [CrossRef]
29. Cao, H.; Liu, X.; Meng, F.; Chu, P.K. Biological actions of silver nanoparticles embedded in titanium controlled by micro-galvanic effects. *Biomaterials* **2011**, *32*, 693–705. [CrossRef]
30. Piwonski, I.; Kadziola, K.; Kisielewska, A.; Soliwoda, K.; Wolszczak, M.; Lisowska, K.; Wronska, N.; Felczak, A. The effect of the deposition parameters on size, distribution and antimicrobial properties of photoinduced silver nanoparticles on titania coatings. *Appl. Surf. Sci.* **2011**, *257*, 7076–7082. [CrossRef]

31. Van Hengel, I.A.J.; Riool, M.; Fratila-Apachitei, L.E.; Witte-Bouma, J.; Farrell, E.; Zadpoor, A.A.; Zaat, S.A.J.; Apachitei, I. Selective laser melting porous metallic implants with immobilized silver nanoparticles kill and prevent biofilm formation by methicillin-resistant Staphylococcus aureus. *Biomaterials* **2017**, *140*, 1–15. [CrossRef] [PubMed]

32. Necula, B.S.; Fratila-Apachitei, L.E.; Zaat, S.A.; Apachitei, I.; Duszczyk, J. In vitro antibacterial activity of porous TiO2-Ag composite layers against methicillin-resistant Staphylococcus aureus. *Acta Biomater.* **2009**, *5*, 3573–3580. [CrossRef]

33. Rizwan, M.; Alias, R.; Zaidi, U.Z.; Mahmoodian, R.; Hamdi, M. Surface modification of valve metals using plasma electrolytic oxidation for antibacterial applications: A review. *J. Biomed. Mater. Res. A* **2018**, *106*, 590–605. [CrossRef]

34. Asharani, P.V.; Low Kah Mun, G.; Prakash Hande, M.; Valiyaveettil, S. Cytotoxicity and genotoxicity of silver nanoparticles in human cells. *ACS Nano* **2009**, *3*, 279–290. [CrossRef]

35. Song, W.H.; Ryu, H.S.; Hong, S.H. Antibacterial properties of Ag (or Pt)-containing calcium phosphate coatings formed by micro-arc oxidation. *J. Biomed. Mater. Res. A* **2009**, *88*, 246–254. [CrossRef] [PubMed]

36. Zhu, W.; Zhang, Z.; Gu, B.; Sun, J.; Zhu, L. Biological activity and antibacterial property of nano-structured TiO2 coating incorporated with Cu prepared by micro-arc oxidation. *J. Mater. Sci. Technol.* **2013**, *29*, 237–244. [CrossRef]

37. Roknian, M.; Fattah-Alhosseini, A.; Gashti, S.O.; Keshavarz, M.K. Study of the effect of ZnO nanoparticles addition to PEO coatings on pure titanium substrate: Microstructural analysis, antibacterial effect and corrosion behavior of coatings in Ringer's physiological solution. *J. Alloy. Compd.* **2018**, *740*, 330–345. [CrossRef]

38. Santos-Coquillat, A.; Martínez-Campos, E.; Mohedano, M.; Martínez-Corriá, R.; Ramos, V.; Arrabal, R.; Matykina, E. In vitro and in vivo evaluation of PEO-modified titanium for bone implant applications. *Surf. Coat. Technol.* **2018**, *347*, 358–368. [CrossRef]

39. Chung, C.J.; Su, R.T.; Chu, H.J.; Chen, H.T.; Tsou, H.K.; He, J.L. Plasma electrolytic oxidation of titanium and improvement in osseointegration. *J. Biomed. Mater. Res. B Appl. Biomater.* **2013**, *101*, 1023–1030. [CrossRef]

40. Martin, J.; Melhem, A.; Shchedrina, I.; Duchanoy, T.; Nominé, A.; Henrion, G.; Czerwiec, T.; Belmonte, T. Effects of electrical parameters on plasma electrolytic oxidation of aluminium. *Surf. Coat. Technol.* **2013**, *221*, 70–76. [CrossRef]

41. Van Hengel, I.A.J.; Laçin, M.; Minneboo, M.; Fratila-Apachitei, L.E.; Apachitei, I.; Zadpoor, A.A. The effects of plasma electrolytically oxidized layers containing Sr and Ca on the osteogenic behavior of selective laser melted Ti6Al4V porous implants. *J. Mater. Sci. Eng. C* **2021**, *124*. [CrossRef]

42. Clyne, T.W.; Troughton, S.C. A review of recent work on discharge characteristics during plasma electrolytic oxidation of various metals. *Int. Mater. Rev.* **2018**, *64*, 127–162. [CrossRef]

43. Moher, D.; Liberati, A.; Tetzlaff, J.; Altman, D.G. Preferred reporting items for systematic reviews and meta-analyses: The PRISMA statement. *PLoS Med.* **2009**, *6*, e1000097. [CrossRef] [PubMed]

44. Yerokhin, A.L.; Nie, X.; Leyland, A.; Matthews, A.; Dowey, S.J. Plasma electrolysis for surface engineering. *Surf. Coat. Technol.* **1999**, *122*, 73–93. [CrossRef]

45. Curran, J.A.; Clyne, T.W. Thermo-physical properties of plasma electrolytic oxide coatings on aluminium. *Surf. Coat. Technol.* **2005**, *199*, 168–176. [CrossRef]

46. Wang, Y.; Yu, H.; Chen, C.; Zhao, Z. Review of the biocompatibility of micro-arc oxidation coated titanium alloys. *Mater. Des.* **2015**, *85*, 640–652. [CrossRef]

47. Matykina, E.; Skeldon, P.; Thompson, G.E. Fundamental and practical evaluations of PEO coatings of titanium. *Int. Heat Treat. Surf. Eng.* **2013**, *3*, 45–51. [CrossRef]

48. Zhang, L.; Li, B.; Zhang, X.; Wang, D.; Zhou, L.; Li, H.; Liang, C.; Liu, S.; Wang, H. Biological and antibacterial properties of TiO2 coatings containing Ca/P/Ag by one-step and two-step methods. *Biomed. Microdevices* **2020**, *22*, 24. [CrossRef]

49. Thukkaram, M.; Cools, P.; Nikiforov, A.; Rigole, P.; Coenye, T.; Van Der Voort, P.; Du Laing, G.; Vercruysse, C.; Declercq, H.; Morent, R.; et al. Antibacterial activity of a porous silver doped TiO2 coating on titanium substrates synthesized by plasma electrolytic oxidation. *Appl. Surf. Sci.* **2020**, *500*. [CrossRef]

50. Van Hengel, I.A.J.; Gelderman, F.S.A.; Athanasiadis, S.; Minneboo, M.; Weinans, H.; Fluit, A.C.; Van der Eerden, B.C.J.; Fratila-Apachitei, L.E.; Apachitei, I.; Zadpoor, A.A. Functionality-packed additively manufactured porous titanium implants. *Materials Today Bio* **2020**, *7*. [CrossRef]

51. Shimabukuro, M.; Tsutsumi, Y.; Yamada, R.; Ashida, M.; Chen, P.; Doi, H.; Nozaki, K.; Nagai, A.; Hanawa, T. Investigation of realizing both antibacterial property and osteogenic cell compatibility on titanium surface by simple electrochemical treatment. *ACS Biomater. Sci. Eng.* **2019**, *5*, 5623–5630. [CrossRef]

52. Sedelnikova, M.B.; Komarova, E.G.; Sharkeev, Y.P.; Ugodchikova, A.V.; Mushtovatova, L.S.; Karpova, M.R.; Sheikin, V.V.; Litvinova, L.S.; Khlusov, I.A. Zn-, Cu- or Ag-incorporated micro-arc coatings on titanium alloys: Properties and behavior in synthetic biological media. *Surf. Coat. Technol.* **2019**, *369*, 52–68. [CrossRef]

53. Sedelnikova, M.B.; Komarova, E.G.; Sharkeev, Y.P.; Ugodchikova, A.V.; Tolkacheva, T.V.; Rau, J.V.; Buyko, E.E.; Ivanov, V.V.; Sheikin, V.V. Modification of titanium surface via Ag-, Sr- and Si-containing micro-arc calcium phosphate coating. *Bioact. Mater.* **2019**, *4*, 224–235. [CrossRef]

54. Lv, Y.; Wu, Y.; Lu, X.; Yu, Y.; Fu, S.; Yang, L.; Dong, Z.; Zhang, X. Microstructure, bio-corrosion and biological property of Ag-incorporated TiO2 coatings: Influence of Ag2O contents. *Ceram. Int.* **2019**, *45*, 22357–22367. [CrossRef]

55. Teker, D.; Muhaffel, F.; Menekse, M.; Karaguler, N.G.; Baydogan, M.; Cimenoglu, H. Characteristics of multi-layer coating formed on commercially pure titanium for biomedical applications. *Mater. Sci. Eng. C Mater. Biol. Appl.* **2015**, *48*, 579–585. [CrossRef] [PubMed]

56. Teker Aydogan, D.; Muhaffel, F.; Menekse Kilic, M.; Karabiyik Acar, O.; Cempura, G.; Baydogan, M.; Karaguler, N.G.; Torun Kose, G.; Czyrska-Filemonowicz, A.; Cimenoglu, H. Optimisation of micro-arc oxidation electrolyte for fabrication of antibacterial coating on titanium. *Mater. Technol.* **2017**, *33*, 119–126. [CrossRef]
57. Dilek Aydogan, T.; Muhaffel, F.; Acar, O.K.; Topcuoglu, E.N.; Kulekci, H.G.; Kose, G.T.; Baydogan, M.; Cimenoglu, H. Surface modification of Ti6Al4V by micro-arc oxidation in AgC2H3O2-containing electrolyte. *Surf. Innov.* **2018**, *6*, 277–285. [CrossRef]
58. Muhaffel, F.; Cempura, G.; Menekse, M.; Czyrska-Filemonowicz, A.; Karaguler, N.; Cimenoglu, H. Characteristics of multi-layer coatings synthesized on Ti6Al4V alloy by micro-arc oxidation in silver nitrate added electrolytes. *Surf. Coat. Technol.* **2016**, *307*, 308–315. [CrossRef]
59. Tsutsumi, Y.; Niinomi, M.; Nakai, M.; Shimabukuro, M.; Ashida, M.; Chen, P.; Doi, H.; Hanawa, T. Electrochemical surface treatment of a β-titanium alloy to realize an antibacterial property and bioactivity. *Metals* **2016**, *6*, 76. [CrossRef]
60. Shin, K.R.; Kim, Y.S.; Kim, G.W.; Yang, H.W.; Ko, Y.G.; Shin, D.H. Effects of concentration of Ag nanoparticles on surface structure and in vitro biological responses of oxide layer on pure titanium via plasma electrolytic oxidation. *Appl. Surf. Sci.* **2015**, *347*, 574–582. [CrossRef]
61. Marques, I.D.; Barao, V.A.; Da Cruz, N.C.; Yuan, J.C.; Mesquita, M.F.; Ricomini-Filho, A.P.; Sukotjo, C.; Mathew, M.T. Electrochemical behavior of bioactive coatings on cp-Ti surface for dental application. *Corros. Sci.* **2015**, *100*, 133–146. [CrossRef]
62. Da Silva Viera Marques, I.; Alfaro, M.F.; Saito, M.T.; Da Cruz, N.C.; Takoudis, C.; Landers, R.; Mesquita, M.F.; Nociti, F.H., Jr.; Mathew, M.T.; Sukotjo, C.; et al. Biomimetic coatings enhance tribocorrosion behavior and cell responses of commercially pure titanium surfaces. *Biointerphases* **2016**, *11*, 031008. [CrossRef]
63. Gasquères, C.; Schneider, G.; Nusko, R.; Maier, G.; Dingeldein, E.; Eliezer, A. Innovative antibacterial coating by anodic spark deposition. *Surf. Coat. Technol.* **2012**, *206*, 3410–3414. [CrossRef]
64. Zhou, L.; Lü, G.-H.; Mao, F.-F.; Yang, S.-Z. Preparation of biomedical Ag incorporated hydroxyapatite/titania coatings on Ti6Al4V alloy by plasma electrolytic oxidation. *Chin. Phys. B* **2014**, *23*. [CrossRef]
65. Thukkaram, M.; Coryn, R.; Asadian, M.; Esbah Tabaei, P.S.; Rigole, P.; Rajendhran, N.; Nikiforov, A.; Sukumaran, J.; Coenye, T.; Van Der Voort, P.; et al. Fabrication of microporous coatings on titanium implants with improved mechanical, antibacterial, and cell-interactive properties. *ACS Appl. Mater. Interfaces* **2020**, *12*, 30155–30169. [CrossRef]
66. Oleshko, O.; Liubchak, I.; Husak, Y.; Korniienko, V.; Yusupova, A.; Oleshko, T.; Banasiuk, R.; Szkodo, M.; Matros-Taranets, I.; Kazek-Kesik, A.; et al. In vitro biological characterization of silver-doped anodic oxide coating on titanium. *Materials* **2020**, *13*, 4359. [CrossRef] [PubMed]
67. Zhao, Q.; Yi, L.; Hu, A.; Jiang, L.; Hong, L.; Dong, J. Antibacterial and osteogenic activity of a multifunctional microporous coating codoped with Mg, Cu and F on titanium. *J. Mater. Chem. B* **2019**, *7*, 2284–2299. [CrossRef] [PubMed]
68. Zhang, X.; Peng, Z.; Lu, X.; Lv, Y.; Cai, G.; Yang, L.; Dong, Z. Microstructural evolution and biological performance of Cu-incorporated TiO2 coating fabricated through one-step micro-arc oxidation. *Appl. Surf. Sci.* **2020**, *508*. [CrossRef]
69. Rokosz, K.; Hryniewicz, T.; Kacalak, W.; Tandecka, K.; Raaen, S.; Gaiaschi, S.; Chapon, P.; Malorny, W.; Matysek, D.; Pietrzak, K.; et al. Porous coatings containing copper and phosphorus obtained by plasma electrolytic oxidation of titanium. *Materials* **2020**, *13*, 828. [CrossRef] [PubMed]
70. Liang, T.; Wang, Y.; Zeng, L.; Liu, Y.; Qiao, L.; Zhang, S.; Zhao, R.; Li, G.; Zhang, R.; Xiang, J.; et al. Copper-doped 3D porous coating developed on Ti-6Al-4V alloys and its in vitro long-term antibacterial ability. *Appl. Surf. Sci.* **2020**, *509*. [CrossRef]
71. He, X.; Zhang, G.; Zhang, H.; Hang, R.; Huang, X.; Yao, X.; Zhang, X. Cu and Si co-doped microporous TiO2 coating for osseointegration by the coordinated stimulus action. *Appl. Surf. Sci.* **2020**, *503*. [CrossRef]
72. Zhang, L.; Guo, J.; Huang, X.; Zhang, Y.; Han, Y. The dual function of Cu-doped TiO2 coatings on titanium for application in percutaneous implants. *J. Mater. Chem. B* **2016**, *4*, 3788–3800. [CrossRef] [PubMed]
73. Huang, Q.; Li, X.; Elkhooly, T.A.; Liu, X.; Zhang, R.; Wu, H.; Feng, Q.; Liu, Y. The Cu-containing TiO2 coatings with modulatory effects on macrophage polarization and bactericidal capacity prepared by micro-arc oxidation on titanium substrates. *Colloids Surf B Biointerfaces* **2018**, *170*, 242–250. [CrossRef]
74. Yao, X.; Zhang, X.; Wu, H.; Tian, L.; Ma, Y.; Tang, B. Microstructure and antibacterial properties of Cu-doped TiO2 coating on titanium by micro-arc oxidation. *Appl. Surf. Sci.* **2014**, *292*, 944–947. [CrossRef]
75. Zhang, X.; Li, J.; Wang, X.; Wang, Y.; Hang, R.; Huang, X.; Tang, B.; Chu, P.K. Effects of copper nanoparticles in porous TiO2 coatings on bacterial resistance and cytocompatibility of osteoblasts and endothelial cells. *Mater. Sci. Eng. C Mater. Biol. Appl.* **2018**, *82*, 110–120. [CrossRef]
76. Zhao, D.; Lu, Y.; Zeng, X.; Wang, Z.; Liu, S.; Wang, T. Antifouling property of micro-arc oxidation coating incorporating Cu2O nanoparticles on Ti6Al4V. *Surf. Eng.* **2017**, *33*, 796–802. [CrossRef]
77. Zhang, L.; Guo, J.; Yan, T.; Han, Y. Fibroblast responses and antibacterial activity of Cu and Zn co-doped TiO2 for percutaneous implants. *Appl. Surf. Sci.* **2018**, *434*, 633–642. [CrossRef]
78. Zhao, D.; Lu, Y.; Wang, Z.; Zeng, X.; Liu, S.; Wang, T. Antifouling properties of micro arc oxidation coatings containing Cu2O/ZnO nanoparticles on Ti6Al4V. *Int. J. Refract. Met. Hard Mater.* **2016**, *54*, 417–421. [CrossRef]
79. Zhang, X.; Li, C.; Yu, Y.; Lu, X.; Lv, Y.; Jiang, D.; Peng, Z.; Zhou, J.; Zhang, X.; Sun, S.; et al. Characterization and property of bifunctional Zn-incorporated TiO2 micro-arc oxidation coatings: The influence of different Zn sources. *Ceram. Int.* **2019**, *45*, 19747–19756. [CrossRef]

80. Shimabukuro, M.; Tsutsumi, Y.; Nozaki, K.; Chen, P.; Yamada, R.; Ashida, M.; Doi, H.; Nagai, A.; Hanawa, T. Chemical and biological roles of Zinc in a porous titanium dioxide layer formed by micro-arc oxidation. *Coatings* **2019**, *9*, 705. [CrossRef]

81. Du, Q.; Wei, D.; Wang, Y.; Cheng, S.; Liu, S.; Zhou, Y.; Jia, D. The effect of applied voltages on the structure, apatite-inducing ability and antibacterial ability of micro arc oxidation coating formed on titanium surface. *Bioact. Mater.* **2018**, *3*, 426–433. [CrossRef] [PubMed]

82. Sopchenski, L.; Popat, K.; Soares, P. Bactericidal activity and cytotoxicity of a zinc doped PEO titanium coating. *Thin Solid Film.* **2018**, *660*, 477–483. [CrossRef]

83. Hu, H.; Zhang, W.; Qiao, Y.; Jiang, X.; Liu, X.; Ding, C. Antibacterial activity and increased bone marrow stem cell functions of Zn-incorporated TiO2 coatings on titanium. *Acta Biomater.* **2012**, *8*, 904–915. [CrossRef]

84. Zhang, X.; Wang, H.; Li, J.; He, X.; Hang, R.; Huang, X.; Tian, L.; Tang, B. Corrosion behavior of Zn-incorporated antibacterial TiO2 porous coating on titanium. *Ceram. Int.* **2016**, *42*, 17095–17100. [CrossRef]

85. Zhao, B.H.; Zhang, W.; Wang, D.N.; Feng, W.; Liu, Y.; Lin, Z.; Du, K.Q.; Deng, C.F. Effect of Zn content on cytoactivity and bacteriostasis of micro-arc oxidation coatings on pure titanium. *Surf. Coat. Technol.* **2013**, *228*, S428–S432. [CrossRef]

86. Leśniak-Ziółkowska, K.; Kazek-Kęsik, A.; Rokosz, K.; Raaen, S.; Stolarczyk, A.; Krok-Borkowicz, M.; Pamuła, E.; Gołda-Cępa, M.; Brzychczy-Włoch, M.; Simka, W. Electrochemical modification of the Ti-15Mo alloy surface in solutions containing ZnO and Zn3(PO4)2 particles. *Mater. Sci. Eng. C* **2020**, *115*. [CrossRef] [PubMed]

87. Zhang, L.; Gao, Q.; Han, Y. Zn and Ag Co-doped anti-microbial TiO2 coatings on Ti by micro-arc oxidation. *J. Mater. Sci. Technol.* **2016**, *32*, 919–924. [CrossRef]

88. Wang, Y.; Zhao, S.; Li, G.; Zhang, S.; Zhao, R.; Dong, A.; Zhang, R. Preparation and in vitro antibacterial properties of anodic coatings co-doped with Cu, Zn, and P on a Ti–6Al–4V alloy. *Mater. Chem. Phys.* **2020**, *241*. [CrossRef]

89. Matykina, E.; Arrabal, R.; Mingo, B.; Mohedano, M.; Pardo, A.; Merino, M.C. In vitro corrosion performance of PEO coated Ti and Ti6Al4V used for dental and orthopaedic implants. *Surf. Coat. Technol.* **2016**, *307*, 1255–1264. [CrossRef]

90. Wauthle, R.; Ahmadi, S.M.; Amin Yavari, S.; Mulier, M.; Zadpoor, A.A.; Weinans, H.; Van Humbeeck, J.; Kruth, J.P.; Schrooten, J. Revival of pure titanium for dynamically loaded porous implants using additive manufacturing. *Mater. Sci. Eng. C Mater. Biol. Appl.* **2015**, *54*, 94–100. [CrossRef]

91. Shah, F.A.; Thomsen, P.; Palmquist, A. Osseointegration and current interpretations of the bone-implant interface. *Acta Biomater.* **2019**, *84*, 1–15. [CrossRef]

92. Shah, F.A.; Trobos, M.; Thomsen, P.; Palmquist, A. Commercially pure titanium (cp-Ti) versus titanium alloy (Ti6Al4V) materials as bone anchored implants–Is one truly better than the other? *Mater. Sci. Eng. C Mater. Biol. Appl.* **2016**, *62*, 960–966. [CrossRef] [PubMed]

93. Elias, C.N.; Fernandes, D.J.; Souza, F.M.D.; Monteiro, E.d.S.; Biasi, R.S.D. Mechanical and clinical properties of titanium and titanium-based alloys (Ti G2, Ti G4 cold worked nanostructured and Ti G5) for biomedical applications. *J. Mater. Res. Technol.* **2019**, *8*, 1060–1069. [CrossRef]

94. Challa, V.S.; Mali, S.; Misra, R.D. Reduced toxicity and superior cellular response of preosteoblasts to Ti-6Al-7Nb alloy and comparison with Ti-6Al-4V. *J. Biomed. Mater. Res. A* **2013**, *101*, 2083–2089. [CrossRef]

95. Krząkała, A.; Kazek-Kęsik, A.; Simka, W. Application of plasma electrolytic oxidation to bioactive surface formation on titanium and its alloys. *RSC Adv.* **2013**, *3*. [CrossRef]

96. Rafieerad, A.R.; Ashra, M.R.; Mahmoodian, R.; Bushroa, A.R. Surface characterization and corrosion behavior of calcium phosphate-base composite layer on titanium and its alloys via plasma electrolytic oxidation: A review paper. *Mater. Sci. Eng. C Mater. Biol. Appl.* **2015**, *57*, 397–413. [CrossRef] [PubMed]

97. Lugovskoy, A.; Lugovskoy, S. Production of hydroxyapatite layers on the plasma electrolytically oxidized surface of titanium alloys. *Mater. Sci. Eng. C Mater. Biol. Appl.* **2014**, *43*, 527–532. [CrossRef] [PubMed]

98. Du, Q.; Wei, D.; Liu, S.; Cheng, S.; Hu, N.; Wang, Y.; Li, B.; Jia, D.; Zhou, Y. The hydrothermal treated Zn-incorporated titania based microarc oxidation coating: Surface characteristics, apatite-inducing ability and antibacterial ability. *Surf. Coat. Technol.* **2018**, *352*, 489–500. [CrossRef]

99. Lee, D.K.; Hwang, D.R.; Sohn, Y.-S. Surface properties of plasma electrolytic oxidation coatings on Ti-alloy in phosphate–based electrolytes with the addition of sodium metasilicate. *Mol. Cryst. Liq. Cryst.* **2019**, *687*, 7–13. [CrossRef]

100. Mohedano, M.; Guzman, R.; Arrabal, R.; Lopez Lacomba, J.L.; Matykina, E. Bioactive plasma electrolytic oxidation coatings–the role of the composition, microstructure, and electrochemical stability. *J. Biomed. Mater. Res. B Appl. Biomater.* **2013**, *101*, 1524–1537. [CrossRef]

101. Lu, X.; Mohedano, M.; Blawert, C.; Matykina, E.; Arrabal, R.; Kainer, K.U.; Zheludkevich, M.L. Plasma electrolytic oxidation coatings with particle additions—A review. *Surf. Coat. Technol.* **2016**, *307*, 1165–1182. [CrossRef]

102. Maitz, M.F.; Poon, R.W.; Liu, X.Y.; Pham, M.T.; Chu, P.K. Bioactivity of titanium following sodium plasma immersion ion implantation and deposition. *Biomaterials* **2005**, *26*, 5465–5473. [CrossRef]

103. Cai, K.; Lai, M.; Yang, W.; Hu, R.; Xin, R.; Liu, Q.; Sung, K.L. Surface engineering of titanium with potassium hydroxide and its effects on the growth behavior of mesenchymal stem cells. *Acta Biomater.* **2010**, *6*, 2314–2321. [CrossRef]

104. Zhang, X.; Yao, Z.; Jiang, Z.; Zhang, Y.; Liu, X. Investigation of the plasma electrolytic oxidation of Ti6Al4V under single-pulse power supply. *Corros. Sci.* **2011**, *53*, 2253–2262. [CrossRef]

105. Shokouhfar, M.; Dehghanian, C.; Montazeri, M.; Baradaran, A. Preparation of ceramic coating on Ti substrate by plasma electrolytic oxidation in different electrolytes and evaluation of its corrosion resistance: Part II. *Appl. Surf. Sci.* **2012**, *258*, 2416–2423. [CrossRef]
106. Galvis, O.A.; Quintero, D.; Castaño, J.G.; Liu, H.; Thompson, G.E.; Skeldon, P.; Echeverría, F. Formation of grooved and porous coatings on titanium by plasma electrolytic oxidation in H2SO4/H3PO4 electrolytes and effects of coating morphology on adhesive bonding. *Surf. Coat. Technol.* **2015**, *269*, 238–249. [CrossRef]
107. Durdu, S.; Bayramoğlu, S.; Demirtaş, A.; Usta, M.; Üçışık, A.H. Characterization of AZ31 Mg Alloy coated by plasma electrolytic oxidation. *Vacuum* **2013**, *88*, 130–133. [CrossRef]
108. Li, Z.; Yuan, Y.; Jing, X. Effect of current density on the structure, composition and corrosion resistance of plasma electrolytic oxidation coatings on Mg–Li alloy. *J. Alloy. Compd.* **2012**, *541*, 380–391. [CrossRef]
109. Montazeri, M.; Dehghanian, C.; Shokouhfar, M.; Baradaran, A. Investigation of the voltage and time effects on the formation of hydroxyapatite-containing titania prepared by plasma electrolytic oxidation on Ti–6Al–4V alloy and its corrosion behavior. *Appl. Surf. Sci.* **2011**, *257*, 7268–7275. [CrossRef]
110. Kossenko, A.; Lugovskoy, S.; Astashina, N.; Lugovskoy, A.; Zinigrad, M. Effect of time on the formation of hydroxyapatite in PEO process with hydrothermal treatment of the Ti-6Al-4V alloy. *Glass Phys. Chem.* **2013**, *39*, 639–642. [CrossRef]
111. Necula, B.S.; Apachitei, I.; Tichelaar, F.D.; Fratila-Apachitei, L.E.; Duszczyk, J. An electron microscopical study on the growth of TiO2-Ag antibacterial coatings on Ti6Al7Nb biomedical alloy. *Acta Biomater* **2011**, *7*, 2751–2757. [CrossRef]
112. Hasan, J.; Jain, S.; Padmarajan, R.; Purighalla, S.; Sambandamurthy, V.K.; Chatterjee, K. Multi-scale surface topography to minimize adherence and viability of nosocomial drug-resistant bacteria. *Mater. Des.* **2018**, *140*, 332–344. [CrossRef]
113. Zhang, Y.; Chen, S.E.; Shao, J.; Van den Beucken, J. Combinatorial surface roughness effects on osteoclastogenesis and osteogenesis. *ACS Appl. Mater. Interfaces* **2018**, *10*, 36652–36663. [CrossRef]
114. Sun, L.; Pereira, D.; Wang, Q.; Barata, D.B.; Truckenmuller, R.; Li, Z.; Xu, X.; Habibovic, P. Controlling growth and osteogenic differentiation of osteoblasts on microgrooved polystyrene surfaces. *PLoS ONE* **2016**, *11*, e0161466.
115. Yamada, M.; Ueno, T.; Tsukimura, N.; Ikeda, T.; Nakagawa, K.; Hori, N.; Suzuki, T.; Ogawa, T. Bone integration capability of nanopolymorphic crystalline hydroxyapatite coated on titanium implants. *Int. J. Nanomed.* **2012**, *7*, 859–873.
116. Lin, G.-W.; Chen, J.-S.; Tseng, W.; Lu, F.-H. Formation of anatase TiO2 coatings by plasma electrolytic oxidation for photocatalytic applications. *Surf. Coat. Technol.* **2019**, *357*, 28–35. [CrossRef]
117. Nosaka, Y.; Nosaka, A.Y. Generation and detection of reactive oxygen species in photocatalysis. *Chem. Rev.* **2017**, *117*, 11302–11336. [CrossRef]
118. Memar, M.Y.; Ghotaslou, R.; Samiei, M.; Adibkia, K. Antimicrobial use of reactive oxygen therapy: Current insights. *Infect. Drug Resist.* **2018**, *11*, 567–576. [CrossRef]
119. Li, K.; Xia, C.; Qiao, Y.; Liu, X. Dose-response relationships between copper and its biocompatibility/antibacterial activities. *J. Trace Elem. Med. Biol.* **2019**, *55*, 127–135. [CrossRef]
120. Pazos-Ortiz, E.; Roque-Ruiz, J.H.; Hinojos-Márquez, E.A.; López-Esparza, J.; Donohué-Cornejo, A.; Cuevas-González, J.C.; Espinosa-Cristóbal, L.F.; Reyes-López, S.Y. Dose-dependent antimicrobial activity of silver nanoparticles on polycaprolactone fibers against gram-positive and gram-negative bacteria. *J. Nanomater.* **2017**, *2017*, 4752314. [CrossRef]
121. Horky, P.; Skalickova, S.; Urbankova, L.; Baholet, D.; Kociova, S.; Bytesnikova, Z.; Kabourkova, E.; Lackova, Z.; Cernei, N.; Gagic, M.; et al. Zinc phosphate-based nanoparticles as a novel antibacterial agent: In vivo study on rats after dietary exposure. *J. Anim. Sci. Biotechnol.* **2019**, *10*, 17. [CrossRef]
122. Ferraris, S.; Spriano, S. Antibacterial titanium surfaces for medical implants. *Mater. Sci. Eng. C Mater. Biol. Appl.* **2016**, *61*, 965–978. [CrossRef]
123. Kedziora, A.; Speruda, M.; Krzyzewska, E.; Rybka, J.; Lukowiak, A.; Bugla-Ploskonska, G. Similarities and differences between silver ions and silver in nanoforms as antibacterial agents. *Int. J. Mol. Sci.* **2018**, *19*, 444. [CrossRef]
124. Stabryla, L.M.; Johnston, K.A.; Millstone, J.E.; Gilbertson, L.M. Emerging investigator series: It's not all about the ion: Support for particle-specific contributions to silver nanoparticle antimicrobial activity. *Environ. Sci. Nano* **2018**, *5*, 2047–2068. [CrossRef]
125. Riool, M.; De Boer, L.; Jaspers, V.; Van der Loos, C.M.; Van Wamel, W.J.B.; Wu, G.; Kwakman, P.H.S.; Zaat, S.A.J. Staphylococcus epidermidis originating from titanium implants infects surrounding tissue and immune cells. *Acta Biomater.* **2014**, *10*, 5202–5212. [CrossRef] [PubMed]
126. Jin, G.; Qin, H.; Cao, H.; Qian, S.; Zhao, Y.; Peng, X.; Zhang, X.; Liu, X.; Chu, P.K. Synergistic effects of dual Zn/Ag ion implantation in osteogenic activity and antibacterial ability of titanium. *Biomaterials* **2014**, *35*, 7699–7713. [CrossRef]
127. Jin, G.; Qin, H.; Cao, H.; Qiao, Y.; Zhao, Y.; Peng, X.; Zhang, X.; Liu, X.; Chu, P.K. Zn/Ag micro-galvanic couples formed on titanium and osseointegration effects in the presence of S. aureus. *Biomaterials* **2015**, *65*, 22–31. [CrossRef] [PubMed]
128. Kokubo, T.; Takadama, H. How useful is SBF in predicting in vivo bone bioactivity? *Biomaterials* **2006**, *27*, 2907–2915. [CrossRef]
129. Uhm, S.H.; Kwon, J.S.; Song, D.H.; Lee, E.J.; Jeong, W.S.; Oh, S.; Kim, K.N.; Choi, E.H.; Kim, K.M. Long-term antibacterial performance and bioactivity of plasma-engineered ag-nps/tio(2). *J. Biomed. Nanotechnol.* **2016**, *12*, 1890–1906. [CrossRef]
130. Otarigho, B.; Falade, B.O. Analysis of antibiotics resistant genes in different strains of Staphylococcus aureus. *Bioinformation* **2018**, *14*, 113–122. [CrossRef] [PubMed]

131. Cremet, L.; Broquet, A.; Brulin, B.; Jacqueline, C.; Dauvergne, S.; Brion, R.; Asehnoune, K.; Corvec, S.; Heymann, D.; Caroff, N. Pathogenic potential of Escherichia coli clinical strains from orthopedic implant infections towards human osteoblastic cells. *Pathog. Dis.* **2015**, *73*, ftv065. [CrossRef]

132. Zimmerli, W.; Moser, C. Pathogenesis and treatment concepts of orthopaedic biofilm infections. *FEMS Immunol. Med. Microbiol.* **2012**, *65*, 158–168. [CrossRef]

133. McConoughey, S.J.; Howlin, R.; Granger, J.F.; Manring, M.M.; Calhoun, J.H.; Shirtliff, M.; Kathju, S.; Stoodley, P. Biofilms in periprosthetic orthopedic infections. *Future Microbiol.* **2014**, *9*, 987–1007. [CrossRef]

134. Barry, A.L.; National Committee for Clinical Laboratory Standards. *Methods for Determining Bactericidal Activity of Antimicrobial Agents; Approved Guideline*; CLSI: Wayne, PA, USA, 1999.

135. Jia, Z.; Xiu, P.; Li, M.; Xu, X.; Shi, Y.; Cheng, Y.; Wei, S.; Zheng, Y.; Xi, T.; Cai, H.; et al. Bioinspired anchoring AgNPs onto micro-nanoporous TiO2 orthopedic coatings: Trap-killing of bacteria, surface-regulated osteoblast functions and host responses. *Biomaterials* **2016**, *75*, 203–222. [CrossRef] [PubMed]

136. Applerot, G.; Lellouche, J.; Lipovsky, A.; Nitzan, Y.; Lubart, R.; Gedanken, A.; Banin, E. Understanding the antibacterial mechanism of CuO nanoparticles: Revealing the route of induced oxidative stress. *Small* **2012**, *8*, 3326–3337. [CrossRef]

137. Hans, M.; Erbe, A.; Mathews, S.; Chen, Y.; Solioz, M.; Mücklich, F. Role of copper oxides in contact killing of bacteria. *Langmuir* **2013**, *29*, 16160–16166. [CrossRef]

138. Afzal, M.A.F.; Kalmodia, S.; Kesarwani, P.; Basu, B.; Balani, K. Bactericidal effect of silver-reinforced carbon nanotube and hydroxyapatite composites. *J. Biomater. Appl.* **2012**, *27*, 967–978. [CrossRef]

139. Cao, H.; Qiao, Y.; Liu, X.; Lu, T.; Cui, T.; Meng, F.; Paul, C.K. Electron storage mediated dark antibacterial action of bound silver nanoparticles: Smaller is not always better. *Acta Biomater.* **2013**, *9*, 5100–5110. [CrossRef] [PubMed]

140. Torbert, J.T.; Joshi, M.; Moraff, A.; Matuszewski, P.E.; Holmes, A.; Pollak, A.N.; O'Toole, R.V. Current bacterial speciation and antibiotic resistance in deep infections after operative fixation of fractures. *J. Orthop. Trauma* **2015**, *29*, 7–17. [CrossRef] [PubMed]

141. Chen, A.F.; Schreiber, V.M.; Washington, W.; Rao, N.; Evans, A.R. What is the rate of methicillin-resistant Staphylococcus aureus and Gram-negative infections in open fractures? *Clin. Orthop. Relat. Res.* **2013**, *471*, 3135–3140. [CrossRef] [PubMed]

142. Tande, A.J.; Patel, R. Prosthetic joint infection. *Clin. Microbiol. Rev.* **2014**, *27*, 302–345. [CrossRef] [PubMed]

143. Corvec, S.; Portillo, M.E.; Pasticci, B.M.; Borens, O.; Trampuz, A. Epidemiology and new developments in the diagnosis of prosthetic joint infection. *Int. J. Artif. Organs* **2012**, *35*, 923–934. [CrossRef]

144. Campoccia, D.; Montanaro, L.; Arciola, C.R. The significance of infection related to orthopedic devices and issues of antibiotic resistance. *Biomaterials* **2006**, *27*, 2331–2339. [CrossRef]

145. Montanaro, L.; Speziale, P.; Campoccia, D.; Ravaioli, S.; Cangini, I.; Pietrocola, G.; Giannini, S.; Arciola, C.R. Scenery of Staphylococcus implant infections in orthopedics. *Future Microbiol.* **2011**, *6*, 1329–1349. [CrossRef]

146. Arciola, C.R.; Campoccia, D.; An, Y.H.; Baldassarri, L.; Pirini, V.; Donati, M.E.; Pegreffi, F.; Montanaro, L. Prevalence and antibiotic resistance of 15 minor staphylococcal species colonizing orthopedic implants. *Int. J. Artif. Organs* **2006**, *29*, 395–401. [CrossRef] [PubMed]

147. Moriarty, T.F.; Kuehl, R.; Coenye, T.; Metsemakers, W.J.; Morgenstern, M.; Schwarz, E.M.; Riool, M.; Zaat, S.A.J.; Khana, N.; Kates, S.L.; et al. Orthopaedic device-related infection: Current and future interventions for improved prevention and treatment. *EFORT Open Rev.* **2016**, *1*, 89–99. [CrossRef]

148. Del Pozo, J.L.; Patel, R. Infection associated with prosthetic joints. *N. Engl. J. Med.* **2009**, *361*, 787–794. [CrossRef]

149. Tan, T.L.; Kheir, M.M.; Shohat, N.; Tan, D.D.; Kheir, M.; Chen, C.; Parvizi, J. Culture-negative periprosthetic joint infection: An update on what to expect. *J. Bone Jt. Surg.* **2018**, *3*, e0060. [CrossRef] [PubMed]

150. Yoon, H.K.; Cho, S.H.; Lee, D.Y.; Kang, B.H.; Lee, S.H.; Moon, D.G.; Kim, D.H.; Nam, D.C.; Hwang, S.C. A review of the literature on culture-negative periprosthetic joint infection: Epidemiology. *Diagn. Treat. Knee Surg. Relat. Res.* **2017**, *29*, 155–164. [CrossRef]

151. Van Opijnen, T.; Dedrick, S.; Bento, J. Strain dependent genetic networks for antibiotic-sensitivity in a bacterial pathogen with a large pan-genome. *PLoS Pathog.* **2016**, *12*, e1005869. [CrossRef] [PubMed]

152. Losasso, C.; Belluco, S.; Cibin, V.; Zavagnin, P.; Micetic, I.; Gallocchio, F.; Zanella, M.; Bregoli, L.; Biancotto, G.; Ricci, A. Antibacterial activity of silver nanoparticles: Sensitivity of different Salmonella serovars. *Front. Microbiol.* **2014**, *5*, 227. [CrossRef] [PubMed]

153. Ruparelia, J.P.; Chatterjee, A.K.; Duttagupta, S.P.; Mukherji, S. Strain specificity in antimicrobial activity of silver and copper nanoparticles. *Acta Biomater.* **2008**, *4*, 707–716. [CrossRef] [PubMed]

154. Shivaram, A.; Bose, S.; Bandyopadhyay, A. Understanding long-term silver release from surface modified porous titanium implants. *Acta Biomater.* **2017**, *58*, 550–560. [CrossRef] [PubMed]

155. Zhai, Y.; Hunting, E.R.; Wouters, M.; Peijnenburg, W.J.; Vijver, M.G. Silver nanoparticles, ions, and shape governing soil microbial functional diversity: Nano shapes micro. *Front. Microbiol.* **2016**, *7*, 1123. [CrossRef]

156. Reidy, B.; Haase, A.; Luch, A.; Dawson, K.A.; Lynch, I. Mechanisms of silver nanoparticle release, transformation and toxicity: A critical review of current knowledge and recommendations for future studies and applications. *Materials* **2013**, *6*, 2295–2350. [CrossRef] [PubMed]

157. Balouiri, M.; Sadiki, M.; Ibnsouda, S.K. Methods for in vitro evaluating antimicrobial activity: A review. *J. Pharm. Anal.* **2016**, *6*, 71–79. [CrossRef] [PubMed]

158. Hudson, M.C.; Ramp, W.K.; Frankenburg, K.P. Staphylococcus aureus adhesion to bone matrix and bone-associated biomaterials. *FEMS Microbiol. Lett.* **1999**, *173*, 279–284. [CrossRef] [PubMed]

159. Yang, D.; Wijenayaka, A.R.; Solomon, L.B.; Pederson, S.M.; Findlay, D.M.; Kidd, S.P.; Atkins, G.J. Novel insights into staphylococcus aureus deep bone infections: The involvement of osteocytes. *mBio* **2018**, *9*. [CrossRef]

160. Zimmerli, W.; Trampuz, A.; Ochsner, P.E. Prosthetic-joint infections. *N. Engl. J. Med.* **2004**, *351*, 1645–1654. [CrossRef]

161. Barrett, L.; Atkins, B. The clinical presentation of prosthetic joint infection. *J. Antimicrob. Chemother.* **2014**, *69*, i25–i27. [CrossRef]

162. Jorgensen, N.P.; Meyer, R.L.; Dagnaes-Hansen, F.; Fuursted, K.; Petersen, E. A modified chronic infection model for testing treatment of Staphylococcus aureus biofilms on implants. *PLoS ONE* **2014**, *9*, e103688.

163. Lebeaux, D.; Chauhan, A.; Rendueles, O.; Beloin, C. From in vitro to in vivo models of bacterial biofilm-related infections. *Pathogens* **2013**, *2*, 288–356. [CrossRef] [PubMed]

164. Zimmerli, W.; Waldvogel, F.A.; Vaudaux, P.; Nydegger, U.E. Pathogenesis of foreign body infection: Description and characteristics of an animal model. *J. Infect. Dis.* **1982**, *146*, 487–497. [CrossRef] [PubMed]

165. Christo, S.N.; Diener, K.R.; Bachhuka, A.; Vasilev, K.; Hayball, J.D. Innate immunity and biomaterials at the nexus: Friends or foes. *Biomed. Res. Int.* **2015**, *2015*, 342304. [CrossRef] [PubMed]

166. Smith, A.W. Biofilms and antibiotic therapy: Is there a role for combating bacterial resistance by the use of novel drug delivery systems? *Adv. Drug Deliv. Rev.* **2005**, *57*, 1539–1550. [CrossRef]

167. Chan, K.H.; Zhuo, S.; Ni, M. Priming the surface of orthopedic implants for osteoblast attachment in bone tissue engineering. *Int. J. Med. Sci.* **2015**, *12*, 701–707. [CrossRef] [PubMed]

168. Jayaraman, M.; Meyer, U.; Bühner, M.; Joos, U.; Wiesmann, H.-P. Influence of titanium surfaces on attachment of osteoblast-like cells in vitro. *Biomaterials* **2004**, *25*, 625–631. [CrossRef]

169. Yao, J.J.; Lewallen, E.A.; Trousdale, W.H.; Xu, W.; Thaler, R.; Salib, C.G.; Reina, N.; Abdel, M.P.; Lewallen, D.G.; Van Wijnen, A.J. Local cellular responses to titanium dioxide from orthopedic implants. *Biores Open Access* **2017**, *6*, 94–103. [CrossRef]

170. Stewart, C.; Akhavan, B.; Wise, S.G.; Bilek, M.M.M. A review of biomimetic surface functionalization for bone-integrating orthopedic implants: Mechanisms, current approaches, and future directions. *Prog. Mater. Sci.* **2019**, *106*. [CrossRef]

171. Shekaran, A.; Garcia, A.J. Extracellular matrix-mimetic adhesive biomaterials for bone repair. *J. Biomed. Mater. Res. A* **2011**, *96*, 261–272. [CrossRef]

172. Long, E.G.; Buluk, M.; Gallagher, M.B.; Schneider, J.M.; Brown, J.L. Human mesenchymal stem cell morphology, migration, and differentiation on micro and nano-textured titanium. *Bioact. Mater.* **2019**, *4*, 249–255. [CrossRef]

173. Jiang, H.; Hong, T.; Wang, T.; Wang, X.; Cao, L.; Xu, X.; Zheng, M. Gene expression profiling of human bone marrow mesenchymal stem cells during osteogenic differentiation. *J. Cell. Physiol.* **2019**, *234*, 7070–7077. [CrossRef] [PubMed]

174. Klontzas, M.E.; Vernardis, S.I.; Heliotis, M.; Tsiridis, E.; Mantalaris, A. Metabolomics analysis of the osteogenic differentiation of umbilical cord blood mesenchymal stem cells reveals differential sensitivity to osteogenic agents. *Stem Cells Dev.* **2017**, *26*, 723–733. [CrossRef] [PubMed]

175. Bennett, K.P.; Bergeron, C.; Acar, E.; Klees, R.F.; Vandenberg, S.L.; Yener, B.; Plopper, G.E. Proteomics reveals multiple routes to the osteogenic phenotype in mesenchymal stem cells. *BMC Genom.* **2007**, *8*, 380. [CrossRef] [PubMed]

176. Hosseini, F.S.; Soleimanifar, F.; Ardeshirylajimi, A.; Vakilian, S.; Mossahebi-Mohammadi, M.; Enderami, S.E.; Khojasteh, A.; Zare Karizi, S. In vitro osteogenic differentiation of stem cells with different sources on composite scaffold containing natural bioceramic and polycaprolactone. *Artif. Cells Nanomed. Biotechnol.* **2019**, *47*, 300–307. [CrossRef]

177. Pineiro-Ramil, M.; Sanjurjo-Rodriguez, C.; Castro-Vinuelas, R.; Rodriguez-Fernandez, S.; Fuentes-Boquete, I.M.; Blanco, F.J.; Diaz-Prado, S.M. Usefulness of Mesenchymal cell lines for bone and cartilage regeneration research. *Int. J. Mol. Sci.* **2019**, *20*, 6286. [CrossRef]

178. Florencio-Silva, R.; Sasso, G.R.; Sasso-Cerri, E.; Simoes, M.J.; Cerri, P.S. Biology of bone tissue: Structure, function, and factors that influence bone cells. *Biomed. Res. Int.* **2015**, *2015*, 421746. [CrossRef]

179. Qin, Y.; Guan, J.; Zhang, C. Mesenchymal stem cells: Mechanisms and role in bone regeneration. *Postgrad. Med. J.* **2014**, *90*, 643–647. [CrossRef]

180. Chiba, H.; Sawada, N.; Ono, T.; Ishii, S.; Mori, M. Establishment and characterization of Simian virus-40 immortalized osteoblastic cell line from normal human bone. *Jpn. J. Cancer Res.* **1993**, *84*, 290–297. [CrossRef] [PubMed]

181. Wagner, E.R.; Luther, G.; Zhu, G.; Luo, Q.; Shi, Q.; Kim, S.H.; Gao, J.L.; Huang, E.; Gao, Y.; Yang, K.; et al. Defective osteogenic differentiation in the development of osteosarcoma. *Sarcoma* **2011**, *2011*, 325238. [CrossRef]

182. Hankenson, K.D.; Dishowitz, M.; Gray, C.; Schenker, M. Angiogenesis in bone regeneration. *Injury* **2011**, *42*, 556–561. [CrossRef] [PubMed]

183. Razzi, F.; Fratila-Apachitei, L.E.; Fahy, N.; Bastiaansen-Jenniskens, Y.M.; Apachitei, I.; Farrell, E.; Zadpoor, A.A. Immunomodulation of surface biofunctionalized 3D printed porous titanium implants. *Biomed. Mater.* **2020**, *15*, 035017. [CrossRef] [PubMed]

184. Pirraco, R.P.; Cerqueira, M.T.; Reis, R.L.; Marques, A.P. Fibroblasts regulate osteoblasts through gap junctional communication. *Cytotherapy* **2012**, *14*, 1276–1287. [CrossRef] [PubMed]

185. Pennings, I.; Van Dijk, L.A.; Van Huuksloot, J.; Fledderus, J.O.; Schepers, K.; Braat, A.K.; Hsiao, E.C.; Barruet, E.; Morales, B.M.; Verhaar, M.C.; et al. Effect of donor variation on osteogenesis and vasculogenesis in hydrogel cocultures. *J. Tissue Eng. Regen. Med.* **2019**, *13*, 433–445. [CrossRef] [PubMed]

186. Zhao, L.; Li, G.; Chan, K.M.; Wang, Y.; Tang, P.F. Comparison of multipotent differentiation potentials of murine primary bone marrow stromal cells and mesenchymal stem cell line C3H10T1/2. *Calcif. Tissue Int.* **2009**, *84*, 56–64. [CrossRef] [PubMed]

187. Scuteri, A.; Donzelli, E.; Foudah, D.; Caldara, C.; Redondo, J.; D'Amico, G.; Tredici, G.; Miloso, M. Mesengenic differentiation: Comparison of human and rat bone marrow mesenchymal stem cells. *Int. J. Stem Cells* **2014**, *7*, 127–134. [CrossRef] [PubMed]

188. Levi, B.; Nelson, E.R.; Brown, K.; James, A.W.; Xu, D.; Dunlevie, R.; Wu, J.C.; Lee, M.; Wu, B.; Commons, G.W.; et al. Longaker, Differences in osteogenic differentiation of adipose-derived stromal cells from murine, canine, and human sources in vitro and in vivo. *Plast. Reconstr. Surg.* **2011**, *128*, 373–386. [CrossRef]

189. Zhao, L.; Wang, H.; Huo, K.; Cui, L.; Zhang, W.; Ni, H.; Zhang, Y.; Wu, Z.; Chu, P.K. Antibacterial nano-structured titania coating incorporated with silver nanoparticles. *Biomaterials* **2011**, *32*, 5706–5716. [CrossRef]

190. Drago, L.; De Vecchi, E.; Bortolin, M.; Zagra, L.; Romano, C.L.; Cappelletti, L. Epidemiology and antibiotic resistance of late prosthetic knee and hip infections. *J. Arthroplast.* **2017**, *32*, 2496–2500. [CrossRef]

191. Panacek, A.; Kvitek, L.; Smekalova, M.; Vecerova, R.; Kolar, M.; Roderova, M.; Dycka, F.; Sebela, M.; Prucek, R.; Tomanec, O.; et al. Bacterial resistance to silver nanoparticles and how to overcome it. *Nat. Nanotechnol.* **2018**, *13*, 65–71. [CrossRef]

192. Richard, D.; Ravigne, V.; Rieux, A.; Facon, B.; Boyer, C.; Boyer, K.; Grygiel, P.; Javegny, S.; Terville, M.; Canteros, B.I.; et al. Adaptation of genetically monomorphic bacteria: Evolution of copper resistance through multiple horizontal gene transfers of complex and versatile mobile genetic elements. *Mol. Ecol.* **2017**, *26*, 2131–2149. [CrossRef]

193. Cavaco, L.M.; Hasman, H.; Aarestrup, F.M. Zinc resistance of Staphylococcus aureus of animal origin is strongly associated with methicillin resistance. *Vet. Microbiol.* **2011**, *150*, 344–348. [CrossRef]

194. Percival, S.L.; Bowler, P.G.; Russell, D. Bacterial resistance to silver in wound care. *J. Hosp. Infect.* **2005**, *60*, 1–7. [CrossRef] [PubMed]

195. Saran, U.; Gemini Piperni, S.; Chatterjee, S. Role of angiogenesis in bone repair. *Arch. Biochem. Biophys.* **2014**, *561*, 109–117. [CrossRef]

196. Lee, J.; Byun, H.; Madhurakkat Perikamana, S.K.; Lee, S.; Shin, H. Current advances in immunomodulatory biomaterials for bone regeneration. *Adv. Healthc. Mater.* **2019**, *8*, e1801106. [CrossRef] [PubMed]

197. Geraghty, R.J.; Capes-Davis, A.; Davis, J.M.; Downward, J.; Freshney, R.I.; Knezevic, I.; Lovell-Badge, R.; Masters, J.R.; Meredith, J.; Stacey, G.N.; et al. Guidelines for the use of cell lines in biomedical research. *Br. J. Cancer* **2014**, *111*, 1021–1046. [CrossRef] [PubMed]

198. Bankier, C.; Matharu, R.K.; Cheong, Y.K.; Ren, G.G.; Cloutman-Green, E.; Ciric, L. Synergistic antibacterial effects of metallic nanoparticle combinations. *Sci. Rep.* **2019**, *9*, 16074. [CrossRef] [PubMed]

199. Worthington, R.J.; Melander, C. Combination approaches to combat multidrug-resistant bacteria. *Trends Biotechnol.* **2013**, *31*, 177–184. [CrossRef]

200. Sobolev, A.; Valkov, A.; Kossenko, A.; Wolicki, I.; Zinigrad, M.; Borodianskiy, K. Bioactive coating on Ti alloy with high osseointegration and antibacterial Ag nanoparticles. *ACS Appl. Mater. Interfaces* **2019**, *11*, 39534–39544. [CrossRef]

201. Durdu, S. Characterization, bioactivity and antibacterial properties of copper-based TiO2 bioceramic coatings fabricated on Titanium. *Coatings* **2018**, *9*, 1. [CrossRef]

202. Fazel, M.; Salimijazi, H.R.; Shamanian, M.; Apachitei, I.; Zadpoor, A.A. Influence of hydrothermal treatment on the surface characteristics and electrochemical behavior of Ti-6Al-4V bio-functionalized through plasma electrolytic oxidation. *Surf. Coat. Technol.* **2019**, *374*, 222–231. [CrossRef]

203. Park, T.-E.; Choe, H.-C.; Brantley, W.A. Bioactivity evaluation of porous TiO2 surface formed on titanium in mixed electrolyte by spark anodization. *Surf. Coat. Technol.* **2013**, *235*, 706–713. [CrossRef]

204. He, J.; Feng, W.; Zhao, B.H.; Zhang, W.; Lin, Z. In vivo effect of Titanium implants with porous Zinc-containing coatings prepared by plasma electrolytic oxidation method on osseointegration in rabbits. *Int. J. Oral Maxillofac. Implant.* **2018**, *33*, 298–310. [CrossRef] [PubMed]

205. Croes, M.; De Visser, H.; Meij, B.P.; Lietart, K.; Van der Wal, B.C.H.; Vogely, H.C.; Fluit, A.C.; Boel, C.H.E.; Alblas, J.; Weinans, H.; et al. Data on a rat infection model to assess porous titanium implant coatings. *Data Brief* **2018**, *21*, 1642–1648. [CrossRef] [PubMed]

International Journal of
Molecular Sciences

MDPI

Review

Effect of Cavity Disinfectants on Dentin Bond Strength and Clinical Success of Composite Restorations—A Systematic Review of In Vitro, In Situ and Clinical Studies

Ana Coelho [1,2,3,*,†], Inês Amaro [1,†], Beatriz Rascão [1], Inês Marcelino [1], Anabela Paula [1,2,3], José Saraiva [1], Gianrico Spagnuolo [4,5], Manuel Marques Ferreira [2,3,6], Carlos Miguel Marto [1,2,3,7,8] and Eunice Carrilho [1,2,3]

[1]   Institute of Integrated Clinical Practice, Faculty of Medicine, University of Coimbra, 3000-075 Coimbra, Portugal; ines.amaros@hotmail.com (I.A.); beatriznmfrascao@gmail.com (B.R.); inescarvalhomarcelino@gmail.com (I.M.); anabelabppaula@sapo.pt (A.P.); ze-93@hotmail.com (J.S.); cmiguel.marto@uc.pt (C.M.M.); eunicecarrilho@gmail.com (E.C.)
[2]   Area of Environment Genetics and Oncobiology (CIMAGO), Coimbra Institute for Clinical and Biomedical Research (iCBR), Faculty of Medicine, University of Coimbra, 3000-548 Coimbra, Portugal; m.mferreira@netcabo.pt
[3]   Clinical Academic Center of Coimbra (CACC), 3004-561 Coimbra, Portugal
[4]   Department of Neurosciences, Reproductive and Odontostomatological Sciences, University of Naples "Federico II", 80131 Napoli, Italy; gspagnuo@unina.it
[5]   Institute of Dentistry, I. M. Sechenov First Moscow State Medical University, 119435 Moscow, Russia
[6]   Institute of Endodontics, Faculty of Medicine, University of Coimbra, 3000-075 Coimbra, Portugal
[7]   Institute of Biophysics, Faculty of Medicine, University of Coimbra, 3004-548 Coimbra, Portugal
[8]   Institute of Experimental Pathology, Faculty of Medicine, University of Coimbra, 3004-548 Coimbra, Portugal
*   Correspondence: anasofiacoelho@gmail.com
†   These authors contributed equally to this work.

Citation: Coelho, A.; Amaro, I.; Rascão, B.; Marcelino, I.; Paula, A.; Saraiva, J.; Spagnuolo, G.; Marques Ferreira, M.; Miguel Marto, C.; Carrilho, E.; et al. Effect of Cavity Disinfectants on Dentin Bond Strength and Clinical Success of Composite Restorations—A Systematic Review of In Vitro, In Situ and Clinical Studies. *Int. J. Mol. Sci.* **2021**, *22*, 353. https://doi.org/10.3390/ijms22010353

Received: 13 December 2020
Accepted: 27 December 2020
Published: 31 December 2020

Publisher's Note: MDPI stays neutral with regard to jurisdictional claims in published maps and institutional affiliations.

Abstract: Cavity disinfection becomes an important step before a dental restorative procedure. The disinfection can be obtained cleaning the dental cavity with antimicrobial agents before the use of adhesive systems. The aim of this study was to conduct a systematic review on the effect of different cavity disinfectants on restorations' adhesion and clinical success. A search was carried out through the Cochrane Library, PubMed, and Web of Science. In vitro and in situ studies reporting results on dentin bond strength tests, and clinical studies published until August 2020, in English, Spanish and Portuguese were included. The methodological quality assessment of the clinical studies was carried out using the Revised Cochrane risk-of-bias tool. Chlorhexidine could preserve adhesion to dentin. EDTA and ethanol had positive results that should be further confirmed. Given the significant lack of scientific evidence, the use of lasers, fluoridated agents, sodium hypochlorite, or other products as cavity disinfectants should be avoided. Chlorhexidine is a safe option for cavity disinfection with adequate preservation of adhesion to dentin. Moreover, future researches should be focused on the efficacy of these disinfectants against cariogenic bacteria and their best application methods.

Keywords: cavity disinfection; antimicrobial substances; chlorhexidine; adhesion; bonding; dental caries

1. Introduction

Dental caries is the most prevalent pathology in the oral cavity, affecting most of the world population. Caries results from the interaction between dental structure and microbial biofilm, highly organized and formed on its surface, being characterized by the alternating phenomena of demineralization and remineralization [1–3]. Under pathological conditions, demineralization overcomes remineralization, leading to the dissolution of hard tissues of the tooth, degradation of collagen fibers and impairment of the mechanical properties of dentin, resulting in caries [1,2,4].

In situations where remineralization is insufficient to resolve the pathology, the treatment of dental caries consists in the removal of infected tissue and subsequent rehabilitation. However, during the removal of decayed tissue, there is the possibility of remaining viable bacteria in the cavity, which can compromise the success of rehabilitation, causing the appearance of a recurrence. On the other hand, rehabilitation failure may be related to tooth and/or restoration fracture and secondary caries, which often occurs at the interface between restorative material and dentin [5–8].

Dentin is considered an intrinsically moist and heterogeneous tissue, which makes adhesion to this tissue a more sensitive adhesive technique when compared to enamel [2,9].

Despite the evolution of adhesive systems, it is known that, over time, the hybrid layer suffers degradation, causing loss of adhesive resistance, which influences the longevity of restorations. The degradation of the adhesive interface is related to several factors, such as oral fluids and bacteria present in situ, leading to degradation of polymers and other organic components. Thus, cavity disinfection becomes an important step prior to the restorative procedure. This is described as cleaning the dental cavity with antimicrobial agents before the use of adhesive systems, making it as innocuous as possible [10].

Among the available disinfectants, chlorhexidine is the most used one. However, despite its beneficial effects, its impact on adhesion is still unclear [3,10,11].

The aim of this study was to conduct a systematic literature review, through the analysis of articles on the effect of different methods of cavity disinfection on adhesion and clinical success of restorations. As so, the research questions were developed according to the PICO (Population, Intervention, Comparison, Outcome) framework—Table 1.

Table 1. Population, Intervention, Comparison and Outcome (PICO) Strategy.

Parameter	In Vitro/In Situ	Clinical
Population	Teeth or dentin discs	Patients in need of a restoration
Intervention	Restoration with prior application of cavity disinfection methods	
Comparison	Conventional restoration	
Outcome	Effect of cavity disinfection on dentin adhesion (bond strength)	Effect of cavity disinfection on the clinical success of restoration

2. Results

2.1. Study Selection

Initial screening of electronic databases yielded a total of 5645 articles. After removal of duplicated studies, a total of 3967 titles and abstracts were evaluated. Overall, a total of 638 potentially relevant articles were selected after an evaluation of titles and abstracts. Full text of these articles was obtained and thoroughly evaluated, and of these, 154 unique articles filled the inclusion criteria and were subsequently included in the systematic review. Two studies reported results for both in situ and in vitro experiments. The flowchart of the data selection process is presented in Figure 1.

2.2. In Vitro Studies

2.2.1. Study Characteristics

One hundred and forty-seven in vitro studies were included in this review. Relevant information regarding each one of the studies was collected and can be found in Supplementary Materials S1.

The earliest study was published in 1992 [12] and the most recent ones were published in 2020 [13–18].

Almost all authors stated the use of premolars and/or permanent molars in all experimental procedures except for Yazici et al. [19] that only indicated the use of human teeth without specification on tooth type. The sample size ranged from 2 [20,21] to 20 [22] teeth per group.

From all of the selected teeth, only dentin substrates were put to use in the experimental protocols and the majority of the authors used healthy dentin as a test substrate. However, some of the authors also used infected dentin [14,23–26] or dentin presenting with artificial caries [27–29] made possible by a laboratory caries-induction procedure. In addition, some studies referred the use of deep and/or superficial dentin [30–38], however in most of the included studies this differentiation was not pointed out.

As regards to the storage medium to preserve teeth after extraction until its use in the experimental procedures, most of the authors reported the use of thymol [15,17,18,25,28,35,36,39–79], chloramine [13,14,23,26,27,29,30,32–34,38,80–111] or water [19–21,31,112–131]. However some other authors chose other storage solutions such as saline solution [22,24,37,132–144], sodium azide [22,135–137,145,146], alcohol [12], sodium hypochlorite [147] or formaldehyde [148]. In nine of the included studies [149–157] there was no available information regarding this issue.

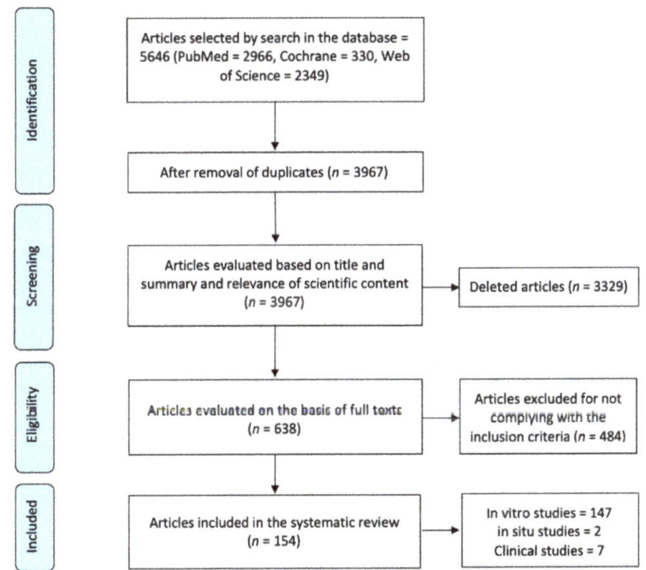

Figure 1. Flowchart of the selection process of the studies.

The majority of the authors [12–21,23,25–28,30–36,38–41,43–48,50–56,58–61,63–65,67–71,73–97,99–110,112–124,126–137,139–149,152,154,156–158] reported the use of water as a storage medium or the incubation of teeth with a 100% humidity after experimental bonding procedures and before bond strength tests. In some of the studies other options for storage were considered such as artificial saliva [24,29,49,57,62,111,150], saline solution [37,125] and sodium hypochlorite [22]. Six authors [66,72,124,138,151,155] did not provide information on this matter. One of the authors, Silva et al. [98], conducted a study on the effect of chlorhexidine and storage media on bond strength values. The authors used distilled water, mineral oil and sodium hypochlorite as storage media to create three independent groups for each experimental procedure.

2.2.2. Interventions—Cavity Disinfection Methods

Most of the studies reported results on the application of cavity disinfectant agents such as chlorhexidine [12–17,23,24,26,27,36,39,42,53,56,57,61,62,65–70,72,73,75,77,79,81,86–88,90, 91,98,100,102–105,111,113,122–124,126,128,132–135,137,142,143,147,148,152–155,157], laser systems [14,19–22,28,29,31,33,37,38,40,41,43,44,46,49–52,54,58,60,63,71,76,78,80,82,83,88,89, 94–96,101,107,112,114–117,119,120,125,131,139–141,143–145,151,156] and sodium hypochlorite (NaOCl) [18,22,25,30,32,34,35,45,47,64,66,70,74,85,90,97,98,106,110,115,118,121,129,130, 132,138,149,150,152,154,158]. Further disinfectant solutions were also evaluated in some of the studies: EDTA based solutions [12,23,39,84,100,110,118,130,132,147], fluoridated agents [28,55,69,74,99,106], green tea extract/epigallocatechin gallate (EGCG)/catechin [56, 57,67,136,137], ozone [26,48,93,133,143] and ozonated water [93], ethanol [26,59,61,75], tetracyclines [103,104,132,153], hydrogen peroxide [152,154], hypochlorous acid (HOCl) [25,47], boric acid [133,157], silver/zinc/titanium nanoparticles [73,108], aloe vera [65,134], urushiol [66,70], iodine based solutions [127,147,155], proanthocyanidin [53], ferrous sulfate (FeSO$_4$) [57], benzalkonic chloride [100], grape seed extract [109] and glutaraldehyde based solutions (Gluma®, Kulzer, Hanau, Germany) [13,146].

2.2.3. Effects of Interventions—Outcomes

Chlorhexidine

The studies evaluating the effects of chlorhexidine as a cavity disinfectant method tested different concentrations, ranging from 0.002% [86] to 5% [23,24]. However, most of the authors evaluated a 2% concentration [13–16,24,26,27,36,39,42,53,56,57,61,62,65–70,73,75,77,79, 81,87,88,90,91,98,100,102–105,111,122–124,126,128,132–135,137,142,143,147,148,152–155,157].

Regarding its performance as a cavity disinfectant, chlorhexidine presented with positive results in the majority of the studies [16,17,24,36,39,42,53,56,57,61,62,65–69,72,73,75,77, 79,81,86,87,90,91,98,100,102–105,111,113,122–124,128,132,134,135,137,142,143,148,153,157], allowing for maintenance or even an increase in bond strength values. On the other hand, a few authors [14,15,26,27,133,154,155] reported undesirable results when using chlorhexidine translating in a decrease in bond strength.

Some studies also evaluated the effects of chlorhexidine when combining it with different adhesive systems. Three of the authors [88,147,152] evaluated the effect of pretreatment with chlorhexidine before the application of two different adhesive systems and all of them reported a decrease in bond strength values when using a self-etch system and an increase or maintenance of values when using an etch-and-rinse system when comparing to control groups. On the contrary, Elkassas et al. [132] stated a decrease in bond strength values when using the etch-and-rinse system and an increase when using the self-etch system. However, five studies [79,113,122,128,137] did not find differences in bond strength when comparing the application of self-etch and etch-and-rinse adhesive systems. Sharma et al. [147] still reported that the use of 2% chlorhexidine before the application of a self-etch adhesive decreased bond strength values but when reducing the concentration to 1% the authors reported an increase in adhesive forces. Universal adhesive systems were also tested in some of the studies, presenting with positive results regarding bond strength. Say et al. [39] and Campos et al. [42] did not find differences between the use of an etch-and-rinse and a universal system as well as Akturk et al. [157] who also did not find differences between a self-etch and a universal system. Bravo et al. [123,124] also conducted studies testing the three different adhesive systems—etch-and-rinse, self-etch and universal—but find no differences between them. One other author tested the same universal system either as a self-etch or as an etch-and-rinse and concluded that bond strength values increased when using the universal system in etch-and-rinse mode and decreased when using it as a self-etch [102].

Laser Systems

Twenty-nine studies [19,28,29,33,37,38,41,43,44,54,58,78,80,82,83,89,94,95,107,112,115–117, 119,125,140,141,144,145] reported the use of the ER:YAG laser, and six [14,50,88,89,96,131] the use of the Er,Cr:YSGG laser. Other laser systems such as the CO$_2$ [20,49,71,151], diode [76,156], Nd:YAP [139], Excimer: KrF [46], ArF [114], Femtosecond [52] and the Ti:sapphire [119] lasers were also reported by some authors although to a lesser extent.

-ER:YAG LASER

Ten of the included studies [19,28,29,38,41,80,94,112,116,125] assessing the use of the ER:YAG laser reported a maintenance or increase in bond strength values, even when testing the laser with different working parameters [19,38,94,112]. However, Baraba et al. [95] tested different pulse durations—50, 100 and 300 µs—and reported worse adhesion results for all except for the medium short pulse (100 µs). Corona et al. [145] also compared different focal distances (mm) and concluded that adhesion values decreased for all except when the laser was used in a defocused mode (17 mm). Shirani et al. [58] tested the same variable—0.5, 2, 4, 11 mm—reporting positive results regarding bond strength but only for the 0.5 and 2 mm focal distances. Oliveira et al. [54] tested four laser settings (energy/repetition rate) but found no differences between them regarding bond strength alterations.

Nine studies [33,44,82,83,89,115,117,140,141] provided with negative results regarding adhesive forces when comparing to control groups and Gonçalves et al. [33] also reported a decrease in bond strength results when testing either on deep or superficial dentin.

Some of the studies also evaluated the effects of the ER:YAG laser combined with different adhesive procedures. Ramos et al. [78] and OLIVEIRA et al. [54] tested three different systems (self-etch, etch-and-rinse and universal) and reported that bond strength results were favourable when using the self-etch and universal adhesives but decreased when using the etch-and-rinse. Two studies [43,119] evaluated self-etch and etch-and-rinse adhesives being that in the study by Sierpinsky et al. [43], bond strength values decreased for all tested adhesives and in the one by Portillo et al. [119] there was only a decrease when using the etch-and-rinse system.

Two studies [94,144] tested the replacement of phosphoric acid with the laser and also tested its combined action and concluded that when the replacement was performed there was a decrease in adhesion but when used together bond strength values underwent no alterations. Davari et al. [144] also tested the combination of laser + phosphoric acid and reported worse results as well as Alahghemand et al. [37] who described similar results in deep dentin. Kucukilmaz et al. [107] also tested the replacement of phosphoric acid with laser and reported worse results when using the laser.

-Er,Cr:YSGG Laser

Three studies [88,96,131] reported favourable results regarding this intervention, presenting with the maintenance or increase in bond strength. Chou et al. [131] tested some different laser parameters but reported no differences in adhesion forces for none of them. Çelik et al. [88] tested two adhesive systems—self-etch and etch-and-rinse—applying the laser before the phosphoric acid in the etch-and-rinse system, and also reported positive results for both of them.

However, Ferreira et al. [89] reported negative results with a decrease in bond strength values when applying phosphoric acid followed by laser application. A significant decrease in adhesive forces was also reported by Carvalho et al. [50] who tested a self-etch and an etch-and-rinse adhesive making the replacement of phosphoric acid with the laser in the etch-and-rinse system.

-CO$_2$ Laser

All four studies [20,49,71,151] assessing this laser as a cavity disinfectant agent reported a decrease in bond strength values when comparing to control groups.

-Nd:YAP Laser

As for the Nd:YAP laser system, only one study [139] evaluated its effects. The authors did not use any phosphoric acid nor adhesive system together with the laser in none of the experimental groups and reported an immediate decrease in bond strength.

-Diode

Only two studies [76,156] evaluated the diode laser as a disinfectant method. Zabeu et al. [156] evaluated two adhesive systems, concluding that regarding the etch-and-rinse system (where the laser was applied after the adhesive application) the bond strength values decreased immediately after the procedure and after 12 months. However, when testing the self-etch system these differences were not verified in neither one of the tested time periods. Kasraei et al. [76] reported opposite results since regarding the etch-and-rinse system, bond strength values did not decrease whether the laser was applied before or after the adhesive.

-Excimer: KrF, ArF

Regarding the Excimer: KrF laser, Eugénio et al. [46] reported a decrease in bond strength when the laser was used prior to the phosphoric acid and when it was used as a substitute of the phosphoric acid.

Regarding the Excimer: ArF laser, only one study was included [114] and the authors reported no alterations in bond strength values.

-Femtosecond

Regarding this intervention, only the study by Gerhardt et al. [52] was included in which different laser parameters were tested and concluded that there was a decrease in adhesive forces when using it at 80 μm/128 s. However, when using the laser as a substitute of the adhesive primer this decrease was not verified.

-Ti:sapphire

Regarding the Ti:sapphire laser, only one study was included, the one by Portillo et al. [119], in which three adhesive systems were tested—one etch-and-rinse and two self-etch—and the laser was always applied after the adhesive. Favourable results were only achieved when using the one-step self-etch adhesive.

Fluoridated Agents

Regarding the use of fluoridated agents, studies including the use of solutions such as silver diamine fluoride [28], ammonium hexafluorsilicate [28], sodium fluoride [69], Riva Star (SDI, Victoria, Australia) [99] and titanium tetrafluoride (TiF_4) [55,74,106] were included.

-Silver Diamine Fluoride and Ammonium Hexafluorsilicate

As for these disinfection methods, only one study [28] evaluated both of them and concluded that bond strength values decreased when using either of them and when testing in both healthy and artificial caries-affected dentin.

-Sodium Fluoride

Only one study [69] evaluated this solution as a cavity disinfectant, reporting positive results regarding adhesive forces when used in a concentration of 1.23%.

-Titanium Tetrafluoride

Three studies [55,74,106] evaluated the effects of TiF_4 as a disinfection method. Two of them [74,106] revealed positive outcomes, both using a 4% TiF_4 concentration. In the third study, by Bridi et al. [55], two self-etch systems were tested with a 2.5% TiF_4 concentration and with one of them there was a decrease of adhesive forces.

-Riva Star

The study by Koizumi et al. [99] was the only study including Riva Star. The authors reported a decrease in bond strength when using both self-etch and etch-and-rinse adhesive systems.

Sodium Hypochlorite

The studies evaluating the effects of NaOCl as a cavity disinfectant tested different concentrations, ranging from 0.5% [85,130] to 10% [18,32,35,45,110,115,121,150].

Regarding its action as a cavity disinfectant, NaOCl performed well in some of the studies, allowing for an increase and/or preservation of bond strength values [64,74,129,130]. Kunawarote et al. [25] also reported positive outcomes when testing either in healthy or infected dentin and Aguilera et al. [34] also did not find differences when testing deep nor superficial dentin when using a 5% NaOCl solution. However, two studies [32,35] revealed better results when testing a 10% NaOCl solution in superficial dentin while another study [30] reported an increase in bond strength when testing a 5% NaOCl solution in deep dentin.

Negative results regarding bond strength were also reported [18,45,70,106,150] and one of the studies [150] maintained these results even after 6 months. Kunawarote et al. [47] tested the effects of NaOCl in different application times (5, 15 and 30 s) and reported that as the application time increased, the results of adhesive forces worsened, with statistically significant results at 30 s.

Also, some of the studies tested the effect of NaOCl when used in combination with different adhesive systems, either self-etch [35,97,118], total-etch [35,115,121] and/or universal [118] and all reported positive results regarding bond strength. Abo et al. [85] even tested eight adhesive systems and reported an overall positive outcome immediately after the procedure as well as after 1 year.

Three authors [92,132,152] reported the use of two adhesive systems—self-etch and etch-and-rinse—and achieved opposite results since Mohammad et al. [92] reported a decrease in bond strength values when testing both adhesives, Elkassas et al. [132] reported an increase when using the self-etch and Ercan et al. [152] reported a decrease when using the self-etch. Two other authors [154,158] tested two self-etch adhesives and both also reported an overall decrease in bond strength results.

Prati et al. [138] tested four etch-and-rinse adhesives and reported a decrease in bond-strength for two of them when using a 1.5% NaOCl solution. Souza et al. [149] also tested four adhesives, two etch-and-rinse, one self-etch and one universal and reported a decrease in adhesive forces for only one of the etch-and-rinse.

In addition, Wuang et al. [110] reported that when testing a 5% NaOCl solution without phosphoric acid the results were unfavourable but when using both there was a positive outcome regarding bond strength. Also, Cha et al. [66] tested the influence of the use of NaOCl followed or not by a washing step and reported better results when the disinfectant was followed by washing.

EDTA Based Solutions

Of all the ten studies [12,23,39,84,100,110,118,130,132,147] evaluating the effects of the EDTA based solutions on adhesion, only Wang et al. [110] reported a decrease in bond strength associated with the use of such solutions. The authors tested a concentration of 15%.

Green Tea Extract/EGCG/Catechin

The green tea extract was evaluated by two studies [57,67] (at concentrations of 0.05% [57] and 2% [67]), the EGCG was evaluated by four studies [56,67,136,137] (at concentrations ranging from 0.02% [56,136] to 5% [137]) and the catechin by only one study [137]. Of all of the studies evaluating these solutions, only Santiago et al. [56] and Sun et al. [136] reported a decrease in bond strength associated with the use of EGCG but only in a 0.5% concentration.

Ethanol

Of all the four studies [26,59,61,75] evaluating the effects of ethanol as a disinfectant agent, only OZSOY et al. [26] described a decrease in bond strength but only when testing in caries-affected dentin.

Ozone

The use of ozone as a cavity disinfectant was reported in five of the included studies [26,48,93,133,143]. Garcia et al. [93] and Ercan et al. [133] did not report a decrease in bond strength regarding restorations performed in ozone disinfected cavities. However, Rodrigues et al. [48] reported a decrease in bond strength when the ozone was either used after or before the phosphoric acid application, describing a higher decrease in the latter. Dalkilic et al. [143] reported similar results as well as Ozsoy et al. [26] either in healthy or infected dentin.

Boric Acid

Boric acid was reported in two of the studies [133,157]. Akturk et al. [157], which evaluated its effects on adhesion when using it before a self-etch or a universal adhesive system, reported that there was a decrease in bond strength only when the boric acid was used before the universal system. However, Ercan et al. [133] reported a decrease in bond strength when testing the boric acid with a self-etch adhesive.

Iodine Based Solutions

Silva et al. [127], Sharma et al. [147] and Suma et al. [155] evaluated the effect of iodine based solutions on adhesion and all authors reported a decrease in bond strength in the majority of test conditions.

Hydrogen Peroxide

Two studies [152,154] evaluated the effects of 3% hydrogen peroxyde as a cavity disinfectant. Ercan et al. [152] evaluated the use of hydrogen peroxide when using an etch-and-rinse and a self-etch adhesive system. The authors described a decrease in bond strength associated with the use of the self-etch system. Reddy et al. [154] tested two self-etch systems and reported a decrease in bond strength for both adhesives.

Benzalkonic Chloride

Tekçe et al. [100] evaluated the effect of 1% benzalkonic chloride on adhesion 24 h and 12 months after the adhesive procedure. The authors reported a decrease in bond strength only at 12 months.

Other Disinfectant Methods

Urushiol [66,70], doxycycline [103,153], minocycline [104,153], glutaraldehyde based solutions (Gluma®) [13,146], hypochlorous acid [25,47], ferrous sulfate [57], proanthocyanidin [53], ozonated water [93], grape seed extract [109], silver/zinc/titanium nanoparticles [73,108] were all evaluated in only one or two studies and none of the products was associated with a decrease in bond strength in none of the test conditions. The effects of BioPure (Dentsply Sirona, York, PA, USA), a tetracycline based solution, were only reported by Elkassas et al. [132], who tested two distinct adhesive systems. The authors described an increase in bond strength after cavity disinfection with BioPure when used with an etch-and-rinse adhesive system. However, when using a self-etch system the bond strength decreased.

2.3. Clinical Studies

2.3.1. Study Characteristics

Seven clinical studies were included in this review and relevant information regarding each one can be found in Supplementary Materials S2. All of the studies are classified as RCT being that four have a split-mouth design [159–162] and one is a pilot study [163].

The earliest study was published in 2006 [163] and the most recent one was published in 2020 [164].

The number of included participants ranged from 11 [163] to 42 [159,160] patients over 20 years old. The sample size ranged from 41 [163] to 169 [160] non-carious cervical lesions. All authors used relative field isolation except for Torres et al. [162] who used absolute field isolation. The follow-up period ranged from 6 months [160] to 5 years [162].

2.3.2. Interventions—Cavity Disinfection Methods

All of the clinical studies selected for inclusion in this review were analyzed and four of them reported the use of 2% chlorhexidine [159–161,165], two reported the use of 10% NaOCl [162,163] and one reported the use of the diode laser [164].

2.3.3. Clinical Outcomes

Several clinical aspects were assessed and used to withdraw conclusions regarding the effects of each disinfectant. All the clinical studies evaluated postoperative sensitivity, retention of the restoration and marginal discoloration. Presence or absence of secondary caries was evaluated by all authors except for Montagner et al. [160] and Favetti et al. [159]. Marginal integrity/adaptation was also evaluated by all except for Saboia et al. [163]. Pulp vitality was only evaluated in four of the included studies [159–162]. Other clinical aspects such as preoperative sensitivity [165], wear [165], clinical success [160,161], dental integrity [159,160], periodontal health [160] and survival rate [159] were only evaluated in one or two studies.

2.3.4. Effects of Interventions—Outcomes

Dutra-Correa et al. [165], Montagner et al. [160] and Favetti et al. [159] tested a 2% chlorhexidine solution and didn't find statistical differences at baseline and after follow-up regarding the clinical aspects assessed in each study. Saboia et al. [163] tested NaOCl at a 10% concentration and also reported no statistical differences regarding clinical outcomes either at 12 or 24 months. Torres et al. [162] also tested a 10% NaOCl solution and Akarsu et al. [164] tested a diode laser and both reported worse clinical results when comparing to baseline after 5 years and 18 months, respectively, however without statistically significant differences.

The only study presenting with statistically significant differences was the one by Sartori et al. [161] in which 2% chlorhexidine was evaluated. There was a statistically significant difference between the 12 and the 36 months recall for the experimental group regarding marginal discoloration. This difference was also verified between the baseline and the 36 months recall for both experimental and control groups. There was also a statistical difference between the baseline and the 36-month follow-up for the experimental group regarding retention and clinical success.

Of the four studies [159–162] evaluating pulp vitality, all reported a 100% success rate. The same percentage applied for the results regarding periodontal health, evaluated in only one study [160].

2.4. In Situ Studies

Only two in situ studies [61,122] were included in this review.

Simões et al. [61] evaluated the influence of chlorhexidine and ethanol on bond degradation. The authors compared the bond strength of an etch-and-rinse adhesive system at 24 h and 6 months (in vitro study) and in an in situ cariogenic challenge with nine patients aged 20–50 years. The bond strength values of the sticks submitted to the in situ cariogenic challenge were similar to those reported for the in vitro experience and the authors concluded that chlorhexidine and ethanol did not affect the bond strength.

Gunaydin et al. [122] evaluated the effect of 2% chlorhexidine on the immediate and aged dentin bond strength of one etch-and-rinse adhesive system and three self-etch adhesive systems in an in vitro and in an in situ experiment. For the in situ study, 40 patients aged 23–28 years were enrolled. Regardless of the adhesive system used, chlorhexidine treated groups exhibited lower immediate bond strength values and higher aged dentin bond strength values.

2.5. Assessment of Methodological Quality

The results of the studies' quality assessment are presented in Figure 2.

Figure 2. Methodological quality assessment of the included clinical (RCT) using the Revised Cochrane risk-of-bias tool for randomized trials (RoB 2).

Three studies presented an overall classification of "low" risk of bias while four studies presented with "some concerns". These four studies all presented with insufficient information regarding the first domain as regards to the randomization process in which all of them failed to mention whether or not the allocation sequence was concealed until participants were enrolled and assigned to interventions. All of the other domains in all the included studies were well described.

3. Discussion

A cavity disinfectant must be bactericidal and/or bacteriostatic, biocompatible and easy to acquire and handle. It needs to be capable of correctly disinfecting the cavity but without compromising dentin bond strength. Its effect depends on each disinfectants characteristics but also on the type of substrate, adhesive system and restorative material used [90,108,126].

Dental substrates play an important role in the performance of adhesive systems, since the morphological and chemical-mechanical characteristics of healthy dentin are different from those of caries affected dentin [138]. The intertubular dentin of a caries affected substrate is partially demineralized, resulting in a softer and more porous structure, which compromises the adhesive strength [25,27,28]. Moreover, differences between superficial and deep dentin are also identified. Superficial dentin, composed mainly of intertubular dentin, has a higher percentage of collagen and a smaller number of dentinal tubules. The deep dentin, close to the pulp region, is formed mainly by dentinal tubules and presents a reduced percentage of intertubular dentin, mainly after acid etching [32,36,78,136]. As so, deep dentin is more hydrophilic, making disinfectants much more efficient in superficial dentin. In fact, several authors confirmed that adhesion to superficial dentin was significantly higher than that in deep dentin [30–33,37,38].

In most of the selected in vitro studies, the samples were placed in a storage medium before being submitted to adhesive resistance tests, in order to simulate the clinical aging of a material overtime. The ISO/TS 11405:2015 (Dental materials – testing of adhesion to tooth structure) [166] gives guidance on substrate selection, storage and handling of samples for quality testing of the adhesive bond between restorative materials and tooth structure. This ISO suggests distilled water or a 0.5% chloramine solution as good storage media for a maximum of one week after which the samples should be kept in distilled water at the temperature of 4 °C or under −5 °C. No other chemical agents should be used since it might affect absorption, adsorption, diffusion, and dissolution, and consequently alter the physical properties of dentin [167]. Furthermore, the longer the storage time, the worse the mechanical properties of the teeth (such as decreased microhardness and negative influence on bond strength) [168–170].

Although most studies reported the use of distilled water or chloramine as a storage medium, following the recommendations of the ISO, there were several authors using other solutions, such as thymol, which was used in 48 studies [15,17,18,25,28,35,36,39–79].

Given the degenerative changes that take place in dentin proteins, after teeth extraction the ISO/TS 11405:2015 [166] states that when it is not possible to perform experimental procedures immediately after teeth extraction, these should be performed in a time period not superior to 6 months. After the conclusion of all restorative procedures, the samples should be kept in water (ISO 3696:1987, grade 3 [171]) at a temperature of 23 °C.

The ISO/TS 11405:2015 [166] also states that ideally premolars and permanent molars should be used being also preferable to use third molars from individuals with ages ranging from 16–40 years. Almost all authors stated the use of premolars and/or permanent molars except for Yazici et al. [19] that only indicated the use of human teeth without specification on tooth type. However, it was not possible to obtain information regarding the age of the patients.

Regarding disinfectants, there are many available products in the market. In this review only studies that tested, at most, one cavity disinfection method per experimental group were included as a way to try to perceive each disinfectant's true effect regarding bond strength alterations. In addition, experimental disinfectants were not considered.

3.1. In Vitro Studies

Chlorhexidine has a broad spectrum of antibacterial action, specially against gram-positive bacteria, and is used in different medical fields. Of all the species implicated in dental caries, epidemiologic evidence associates *Streptococcus mutans* as the main initiator of dental caries and it is known that chlorhexidine has a strong antibacterial capacity against it. Chlorhexidine also has the ability to inhibit the formation of the acquired pellicle, acting as an antiplaque agent [172–178].

As for the studies included in this review, most of the studies reported a maintenance or even an increase in bond strength values when chlorhexidine was used before the adhesive system, regardless of the concentration tested. Although some authors have studied the use of different adhesive systems (etch-and-rinse, self-etch and universal) after disinfection of the cavity with chlorhexidine, there seems to be no differences between them.

The enzymatic degradation of the collagen matrix by host-derived enzymes plays a significant role in destroying adhesive interfaces. Chlorhexidine is able to inhibit those collagen-degrading enzymes (namely matrix metalloproteinases (MMP) and cysteine cathepsins) which may justify the positive results [179–181]. On the other hand, chlorhexidine seems to have the ability to remove the loose smear debris and to increase the surface energy of dentin, which increases the wetting ability of primers [132,182].

The results are mainly positive and the pretreatment with chlorhexidine demonstrated an adequate preservation of adhesion to dentin, which makes it a safe option for cavity disinfection.

Another possible cavity disinfection method is laser irradiation. According to some authors, irradiation of the dentin surface results in improved microhardness, increased resistance to demineralization and decreased dentin permeability, minimizing bacterial access to the pulp [20,40,88].

However, the results regarding the use of lasers as cavity disinfectants are not consistent, with almost half of the studies reporting that lasers exert a negative effect on the mineral and organic components of dentin, impairing adhesion force [78,183,184].

The Erbium-doped yttrium aluminum garnet (Er:YAG) laser is widely used in different medical fields and has the highest water absorbency which makes it well absorbed in biological tissues containing water. This laser was the most studied one but the results are contradictory, regardless of the laser parameters and the adhesive system used. Dunn et al. [83] reported large peritubular areas with scaling and flaking aspects as well as a defective hybridization between dentin and composite resin when the surface was irradiated before acid-etching, which may justify the negative results. The residual thermal energy can also influence bonding by modifying the hydroxyapatite crystals, denaturing the collagen fibers and excessively dehydrating dentin [31,33,89,145,185,186].

Since laser ablation promotes opening of the dentinal tubules, some authors tested it as an alternative to acid-etching. However, bond strength values decreased in all studies [19,29,37,38,43,44,54,58,78,80,83,94,95,116,117,119,125,140,141,144]. One possible explanation for the negative results is that the Er:YAG laser doesn't demineralize the peritubular dentin, which may hamper the formation of a hybrid layer and resin tags [31,187].

Regarding the other lasers included in this analysis (Er,Cr:YSGG, CO_2, Nd:YAP, diode, Excimer: KrF and ArF, femtosecond, Ti:sapphire) the number of articles that reported results on them is very small. Since different lasers, with different parameters and under different conditions are used, it is not possible to draw conclusions.

Given the significant lack of scientific evidence, the use of lasers as cavity disinfectants should be avoided.

Some materials, such as fluoridated agents, are indicated for dentin pretreatment for different purposes [55]. These products help to diminish the development of secondary caries by decreasing dentin solubility and enhancing dentin remineralization [55,74]. Fluoridated agents such as silver diamine fluoride, ammonium hexafluorsilicate, sodium fluoride, titanium tetrafluoride (TiF_4) and Riva Star were included in this review and its effects as cavity disinfectants prior to adhesive procedures and the influence in bond strength values were analysed. The results were mostly positive with reports of maintenance or even increase the bond strength in four out of six studies—sodium fluoride [69] and titanium tetrafluoride [55,74,106].

Metal fluorides, especially titanium tetrafluoride, have become popular due to their unique interaction with dental hard tissues [106]. In fact, TiF_4 was the most studied fluoridated disinfectant agent, with a total of three studies which reported mostly positive results regarding bond strength. Its beneficial effects can be attributed to the increase of fluoride uptake, which can chemically reduce demineralization of dental hard tissues, but also to the formation of an acid-stable surface layer, referred to as glaze-like layer [188], which provides mechanical protection of the surface, forming a fine layer of titanium containing material, covering the surface and occluding the dentinal tubules [189,190]. However, one of the erosions' inhibiting characteristics of TiF_4 depends on its method of application on dentin. It seems that application with a microbrush leads to surface wear rather than allowing the formation of the glaze-like surface layer [191]. In all of the three included studies which tested TiF_4, the product was actively applied with a microbrush for

60 s, which may possibly justify why pretreatment with TiF_4 had no negative effects on bond strength.

Nevertheless, there was only a total of six studies regarding the use of fluoridated agents as cavity disinfectants and only three evaluating TiF_4, which makes it unadvisable to recommend its use until further studies are conducted on the matter to prove its feasibility.

Sodium hypochlorite is one of the most commonly used cavity disinfectants in clinical practice due to its antibacterial action and wettability property [132]. It also has the highest antimicrobial activity against anaerobic bacteria, as well as against *Streptococcus mutans* [66,192,193].

As regards to its use in restorative protocols, sodium hypochlorite has been considered for use prior to adhesive procedures as a cavity disinfectant. However, as for the studies' results, there is a lack of consensus since some authors [64,74,129,130] reported positive results with maintenance or increase of bond strength values and others [18,45,70,106,150] reported a decrease. These differences might be attributed to differences in experimental protocols such as different NaOCl concentrations, adhesive systems and dentin substrates. Also, the decreased bond strength values may also be due to the oxygen released by NaOCl molecules as it may inhibit adhesive polymerization and compromise the mechanical performance of the bonding interfaces [32].

The substrate plays an important role on the adhesion of current adhesive systems and, in this case, four studies [30,32,34,35] evaluated the effects of sodium hypochlorite demineralization on deep and superficial dentin. Although one study [34] did not find differences when testing deep nor superficial dentin when using a 5% NaOCl solution, two other studies [32,35] revealed better results when testing a 10% NaOCl solution in superficial dentin. Since there are morphological differences between these two substrates, this might occur due to a decrease in the area of intertubular dentin available for bonding in deep dentin [30]. Furthermore, failure in removing all residual water confined deep into demineralized surfaces induces the formation of poorly polymerized polymer chains [34]. In addition, lower concentrations and times of application promote a continuous and slow denaturation of dentin collagen, leading to the formation of a gel layer on the demineralized dentin, preventing the diffusion of adhesive monomers. As the concentration and the time of application increases, denaturation progresses to degradation and complete dissolution of collagen, increasing the porosity of the dentin surface and diffusion of adhesive monomers through demineralized dentin [194]. This may also justify the results since a higher concentration of NaOCl—10%—was used in those two studies. Only one study [30] reported an increase in bond strength when testing in deep dentin with a 5% NaOCl solution. According to the authors, the total or partial removal of the collagen layer in the deep dentin by sodium hypochlorite may facilitate the penetration of the adhesive and avoid adhesive failures on deep cavities.

Some studies also evaluated the use of sodium hypochlorite in combination with different adhesive systems (etch-and-rinse, self-etch and/or universal) but there was no obvious consensus amongst authors since results were very incoherent making it impossible to draw conclusions regarding this matter.

There is no clear evidence on the effects on bond strength when using sodium hypochlorite as a cavity disinfectant prior to adhesive procedures and so caution is required until further studies assure its effects.

Ethylenediamine tetraacetic acid (EDTA) is an organic compound, capable of chelating calcium ions and selectively removing hydroxyapatite without entering deeply into the dentinal tubules [110]. It is an MMP inhibitor solution capable of increasing the longevity of the adhesive interface, dissolving the mineral components of dentin without altering the stability of the organic matrix and without causing collagen denaturation [23,84].

Nine [12,23,39,84,100,118,130,132,147] out of the ten studies evaluating the effects of EDTA reported maintenance or even an increase on mean bond strength despite the adhesive system used (etch-and-rinse, self-etch or universal). Although there is a need for further studies, the use of EDTA based solutions as cavity disinfectants prior to adhesive procedures seems to be a favorable alternative.

Ethanol is another possible alternative as a cavity disinfection method since when applied to dentin cavities it has the ability to expel water from dentin, keeping the collagen network distended. The interfibrillar spaces in the collagen matrix are filled with ethanol and the exposed collagen fibers are involved, preventing its degradation by MMP, thus providing a better substrate for resin placement and monomer infiltration [26,75]. It is also able to create a more hydrophobic environment, reducing water absorption over time, which is a key factor in the degradation of the adhesive bond [75]. This is in line with the results from the included studies since out of all the four studies [26,59,61,75] evaluating the effects of ethanol, only Ozsoy et al. [26] described a decrease in bond strength but only when ethanol was tested in caries-affected dentin. This substrate has a high degree of porosity due to mineral loss and partial or total tubular obstruction in the intertubular area which may justify these results. In addition, adhesive systems cannot infiltrate as deeply into demineralized intertubular structures as acids can, resulting in lower bond strength values than when testing in healthy dentin [26].

Since a decrease in bond strength was only observed when testing in caries-affected dentin [26] and since this is a more complex substrate for adhesion but also the main purpose of using cavity disinfectants, more studies should be conducted testing ethanol in caries-affected dentin to assess its feasibility when attempting to disinfect this substrate. Even though there is limited information about the use of ethanol as a cavity disinfectant, from the available results it looks like a promising alternative.

Ozone is a naturally occurring compound of three oxygen atoms. It is a strong oxidizing agent, with antibacterial activity capable of disrupting the cell wall and cytoplasmic membrane of microorganisms [157], which allows it to efficiently eliminate bacteria, fungi, protozoa, and viruses [48,143].

Ozone has also been proposed as a cavity disinfection method prior to restorative procedures since due to its oxidative and antimicrobial activity, it has the capacity to oxidize proteins in carious lesions and consequently diffusing and depositing calcium and phosphate ions into the demineralized dental tissues, hence leading to remineralization [48, 143]. However, the results of the five included studies regarding the use of ozone are inconsistent since two authors [93,133] did not report a decrease in bond strength and the other three authors [26,48,143] reported opposite results even when testing either in healthy or infected dentin [26,143]. This might be due to the fact that ozone is an unstable molecule and rapidly dissociates into oxygen molecules which will react with free radicals and may cause the inhibition of polymerization of adhesive systems and thus reduce bond strength [26,48]. Therefore, the decrease in bond strength values may have been caused by the presence of residual oxygen molecules after the use of ozone. However, this is not in accordance with the results by Rodrigues et al. [48] who reported a higher decrease when ozone was used before rather than after the use of phosphoric acid which was immediately followed by the adhesive system. The authors stated that the results could only be justified by some chemical interaction that might have occurred between ozone and phosphoric acid, when the acid was used after the ozone, which interfered directly with the formation of the hybrid layer and might therefore affect the bond strength [48].

The use of ozone as a cavity disinfection method is still a matter of discussion since there is yet little available information on its effects on bond strength.

Some other disinfectant agents such as urushiol [66,70], tetracyclines [103,104,132,153], glutaraldehyde based solutions [13,146], hypochlorous acid [25,47], ferrous sulfate [57], proanthocyanidin [53], ozonated water [93], grape seed extract [109], silver/zinc/titanium nanoparticles [73,108], aloe vera [65,134] and green tea extract/EGCG/catechin [56,57,67, 136,137] were included in this review, although to a lesser extent. Even though none of these products was associated with a decrease on bond strength in the majority of the test conditions, only a very limited number of studies regarding each disinfectant were included. On the other hand, other disinfectants such as boric acid [133,157], benzalkonic chloride [100], iodine based solutions [127,147,155] and hydrogen peroxide [152,154] were not only evaluated in very few studies but also all of them reported a decrease in bond strength in most of the experimental scenarios. Being that, the use of these products should be handled with caution since there is insufficient scientific evidence to support its use as cavity disinfectants.

In addition, there are several studies experimenting with different adhesive systems, given the different methodological approaches and reported results it is not possible to draw conclusions.

The development of new quality in vitro studies is essential. Although there is a lack of a standardized protocol, the experiments should account in detail for storage media and conditions, studied dental substrates, control and test groups, outcome measurement, and limited variables (e.g., the use of more than a cavity disinfectant per experimental group doesn't allow the evaluation of its true effect on bond strength).

3.2. Clinical Studies

Concerning the clinical studies, the results were mainly positive for all tested disinfectants. Only Sartori et al. [161] reported worse results with statistically significant differences when testing a 2% chlorhexidine solution. The authors reported a statistically significant difference between the 12 and the 36 months recall for the experimental group regarding marginal discoloration and also between the baseline and the 36 months recall for both experimental and control groups. One possible explanation is the hypothesis that the repetitive cyclic of parafunctional loadings might have induced a failure in the cervical region of the restoration which might have generated micro-cracks at the restoration margin [161]. There was also a statistical difference between the baseline and the 36-month follow-up for the experimental group regarding retention and clinical success. These results might be due to the fact that despite chlorhexidine could decelerate the rate of resin-dentin bonds degradation by inhibition of MMP, it could not prevent the hydrolytic breakdown of polymers that constitute those bonds. Also, another aspect that could not be clarified in the study by Sartori et al. [161] is the origin of the adhesive failures, whether the negative results were due to a defect within the resin or the dentin matrix part of the hybrid layers.

Even though results concerning these disinfectants were mostly positive there is a clear need for further clinical studies regarding these and other disinfectant alternatives and with longer follow-up periods. According to the American Dental Association [195], it is required a retention rate of at least 90% of the restorations placed after 18 months to obtain full acceptance. Given this guideline, the study by Dutra-Correa et al. [165] achieved full acceptance since after 18 months the retention rate was over 90% when testing a 2% chlorhexidine solution. Torres et al. [162] reported a 97% retention rate after 18 months but not after 3 nor 5 years. Saboia et al. [163] also reported a 90% retention rate after 24 months but only for one of the tested adhesive systems.

According to the American Dental Association guidelines [195], non-carious cervical lesions are ideal models for testing the bonding of restorative materials to dental tissue. This is due to the non-carious loss of dental hard tissue having the most of the bonding area in dentinal tissue with only a small incisal/occlusal margin in enamel, they do not have a retentive shape and do not require preparation before restoration. However, the causes of the diminished longevity of non-carious class V restorations are still poorly understood, in contrast with other restorations [159–161] since they are known to have a multifactorial etiology and the prognosis may be significantly affected by several factors related to the material, patient and the environment [160]. The oral environment presents a challenge to the longevity of adhesive resistance since there are several factors that can influence the long term success of restorations such as temperature changes, masticatory load cycles, water absorption and pH fluctuations [196]. It is also important to consider that these kind of lesions might be due to characteristics of an individual patient (for example, parafunctional habits) [159–161] and so it is important to consider patient's individual characteristics since these might affect the survival of restorations.

In order to reduce the impact of these variations on the studies, a good alternative might be to use a split-mouth design which was used in four of the included studies [159–162]. Also, different teeth cavity configurations [161] and each lesions' different clinical characteristics [159,160] should also be taken into account since these might turn substrates more complex to adhesive procedures [159,160]—enamel, dentin or sclerotic dentin [159,160,165]. In the study by Montagner et al. [160] some cavity variables affected the retention of the cervical restorations regardless of the treatment since deeper and wider lesions presented statistically more retention failure than the others. On the other hand, the effect of cavity configuration and location of the restoration margin in relation to the gingival margin were not significant factors for restoration retention in the study by Favetti et al. [159].

A summary of the available evidence on each cavity disinfectant is presented in Table 2.

Table 2. Summary table of the available evidence on the use of cavity disinfectants.

Disinfectant	Does the Available Evidence Support Its Use as a Cavity Disinfectant?
Aloe vera	No (limited number of studies)
Benzalkonic chloride	No
Boric acid	No
Chlorhexidine	Yes
EDTA	Yes (positive results but limited number of studies)
Ethanol	No (limited number of studies)
Ferrous sulfate	No (limited number of studies)
Fluoride	Yes (positive results but limited number of studies)
Glutaraldehyde	No (limited number of studies)
Grape seed extract	No (limited number of studies)
Green tea extract/EGCG/catechin	No (limited number of studies)
Hydrogen peroxide	No
Hypochlorous acid	No (limited number of studies)
Iodine	No
Laser	No
Ozonated water	No (limited number of studies)
Ozone	No
Proanthocyanidin	No (limited number of studies)
Silver/zinc/titanium nanoparticles	No (limited number of studies)
Sodium hypochlorite	No
Tetracyclines	No (limited number of studies)
Urushiol	No (limited number of studies)

The main limitations of the included studies concerns the randomization process in which four of the studies [161,163–165] failed to mention whether or not the allocation sequence was concealed until participants were enrolled and assigned to interventions. Also, the field isolation is an important aspect to consider when performing adhesive procedures and since all authors used relative field isolation except for Torres et al. [162], who used absolute field isolation, this is also an important limitation. Restoration losses of non-carious cervical lesions are more prevalent in the posterior region due to stress intensity and moisture contamination in the cervical region which justifies the importance of absolute field isolation [164].

As so, there is a clear need for further clinical studies regarding this topic that allow the adoption of strict inclusion criteria, the generation of random sequences and the use of blind participants, examiners and evaluators. There is also a need for longer follow-up periods, in order to assess whether these products can be incorporated into restorative protocols or not.

4. Materials and Methods

This systematic review was registered on the International Prospective Register of Systematic Reviews (PROSPERO) platform (temporary ID: 199614) and designed according to PRISMA methodology (Preferred *Reporting Items for Systematic Reviews and Meta-Analysis*).

4.1. Inclusion and Exclusion Criteria

In vitro, in situ and clinical studies evaluating the effect of applying, at most, one cavity disinfection method per experimental group and evaluating the effect of the disinfection method on dentin adhesion of commercially available conventional adhesives and composites were included. Only in vitro and in situ studies reporting results for bond strength tests, presented in the form of mean and standard deviation, were considered.

Clinical studies on postoperative sensitivity, marginal pigmentation and adaptation, restoration loss, secondary caries, microinfiltration and pulp vitality were included.

Studies with no control group, in deciduous teeth, in bleached teeth or in teeth with endodontic treatment were excluded. Studies evaluating the adhesion of posts, cements, sealants, brackets or glass ionomer, studies that used manipulated adhesives (experimental or mixed with other solutions) and studies evaluating root or pulp chamber walls were also excluded. Regarding the laser, the studies that used it to prepare the cavity and not only as a disinfection method were also excluded. Previous contamination of teeth with hemostatic agents, saliva or blood was also considered an exclusion criterion.

All review articles, cell or animal studies, letters, case reports, abstracts and comments were excluded.

4.2. Search Strategy

An electronic search was carried out in the Cochrane Library (www.cochranelibrary.com), PubMed (www.ncbi.nlm.nih.gov/pubmed) and Web of Science (www.webofscience.com) databases, by articles published until August the 8th of 2020, without restrictions on the type, region or year of publication. Only articles in English, Portuguese and Spanish were included. Articles that weren't available online were excluded.

For the research, the terms MeSH "Dentin", "Disinfection", "Anti-Bacterial Agents", "Chlorhexidine", "Sodium Hypochlorite", "Lasers", "Ozone", "Ethanol", "Aloe" and "Adhesives" were used. The search keys used in the different databases are found in Table 3.

4.3. Study Selection

The eligibility of the initially selected articles was evaluated by reading titles, abstracts and full text by two reviewers, independently. Any disagreement was discussed and the opinion of a third reviewer was obtained when necessary.

4.4. Data Extraction

Selected articles were read independently by two reviewers. During data extraction, two Microsoft® Excel (Microsoft, Washington, WA, USA) tables were elaborated. The first, for the in vitro studies, included the parameters: name of the authors, year of publication, groups (n), storage solutions and materials used, and results (bond strength). The second table, for the in vivo studies, included the parameters: name of the authors, year of publication, type of study, groups (n), follow-up period, materials used, and results. Any disagreement was discussed and the opinion of a third reviewer was obtained when necessary.

Table 3. Search keys used in the different databases. The asterisk represents any group of characters.

Database	Research
Cochrane Library	#1 MeSH descriptor: [Dentin] explodes all trees #2 dentin #3 cavity #4 MeSH descriptor: [Disinfection] in all MeSH products #5 disinfect* #6 antibacteria* #7 MeSH descriptor: [Anti-Bacterial Agents] explodes all trees #8 chlorhexidine #9 MeSH descriptor: [Chlorhexidine] explodes all trees #10 "sodium hypochlorite" #11 MeSH descriptor: [Sodium Hypochlorite] explodes all trees #12 laser #13 MeSH descriptor: [Lasers] explodes all trees #14 ozone #15 MeSH descriptor: [Ozone] explodes all trees #16 ethanol #17 MeSH descriptor: [Ethanol] explodes all trees #18 "aloe vera" #19 MeSH descriptor: [Aloe] explodes all trees #20 #1 OR #2 OR #3 #21 #4 OR #5 OR #6 OR #7 OR #8 OR #9 OR #10 #11 OR #13 OR #12 #14 OR #15 OR #16 OR #17 OR #18 OR #19 #22 adhesion #23 adhesive #24 MeSH descriptor: [Adhesives] in all MeSH products #25 bond strength #26 #22 OR #23 OR #24 OR #25 #27 #20 AND #21 and #26 #28 #20 AND #21
Web of Science	TS = ((dentin[MeSH Terms] OR dentin OR cavity) AND (disinfect* OR antibacteria* OR chlorhexidine OR "sodium hypochlorite" OR laser OR ozone OR ethanol OR "aloe vera") AND ("bond strength" OR adhesion or adhesive))
Pubmed	(dentin[MeSH Terms] OR dentin OR cavity) AND (disinfection[MeSH Terms] OR disinfect* OR antibacteria* OR agents, antibacterial[MeSH Terms] OR chlorhexidine[MeSH Terms] OR chlorhexidine OR "sodium hypochlorite" OR sodium hypochlorite[MeSH Terms] OR laser OR lasers[MeSH Terms] OR ozone OR ozone[MeSH Terms] OR ethanol OR ethanol[MeSH Terms] OR "aloe vera" OR aloe[MeSH Terms]) AND ("bond strength" OR adhesion OR adhesive OR adhesives[MeSH Terms])

4.5. Quality Assessment of the Included Clinical Studies

The methodological quality assessment of included clinical studies (randomized controlled trials, RCT) was assessed using the Revised Cochrane risk-of-bias tool for randomized trials (RoB 2) [197] by two independent reviewers. Briefly, five domains were evaluated: (D1) risk of bias arising from the randomization process; (D2) risk of bias due to deviations from the intended interventions; (D3) Risk of bias due to missing outcome data; (D4) Risk of bias in measurement of the outcome; (D5) Risk of bias in selection of the reported results. From this evaluation, each study may vary its overall classification regarding bias risk as "low", "high" or "some concerns".

5. Conclusions

A variety of different products is available for cavity disinfection prior to adhesive procedures. However, there are only a few that have been tested to a proper extent and with proved in vitro and clinical viability.

Chlorhexidine is a popular disinfectant and it was possible to conclude that it is a safe option for cavity disinfection since results are mainly positive with an adequate preservation of adhesion to dentin. Other disinfectants such as EDTA and ethanol may be promising alternatives but there is a clear need for further studies to safely suggest their use as cavity disinfectants. Also, further research is needed to clarify not only the effect of cavity disinfectants in bond strength but also their efficacy against cariogenic bacteria, their application times, products' concentration, their use before or after acid-etching and their combination with different adhesive systems and dental substrates.

Supplementary Materials: Can be found at https://www.mdpi.com/1422-0067/22/1/353/s1.

Author Contributions: Conceptualization, A.C., I.A. and E.C.; methodology, A.C., I.A., A.P., C.M.M., G.S.; software, A.C., I.A.; validation, A.C., I.A., A.P., C.M.M., G.S., M.M.F., E.C.; data extraction and analysis, A.C., I.A., B.R., I.M., J.S., E.C.; writing-original draft preparation, A.C., I.A., B.R., J.S.; writing-review and editing, A.C., I.A., A.P., C.M.M., G.S., M.M.F., E.C.; supervision, A.C., I.A., E.C. All authors have read and agreed to the published version of the manuscript.

Funding: This research received no external funding.

Institutional Review Board Statement: Not applicable.

Informed Consent Statement: Not applicable.

Data Availability Statement: The data presented in this study are available in Supplementary Materials S1 and S2.

Conflicts of Interest: The authors declare no conflict of interest

Abbreviations

CHX	Chlorhexidine
DT	Diamond tip
DTUS	Ultrasonic diamond tip
EDTA	Ethylenediamine tetraacetic acid
EGCG	Epillocatechin gallate
MSP	Medium short pulse
PBS	Phosphate buffer saline
QSP	Quantum square pulse
TC	Thermocyling

References

1. Okada, E.M.P.; Ribeiro, L.N.S.; Stuani, M.B.S.; Borsatto, M.C.; Fidalgo, T.K.D.S.; De Paula-Silva, F.W.G.; Küchler, E.C. Effects of chlorhexidine varnish on caries during orthodontic treatment: A systematic review and meta-analysis. *Braz. Oral Res.* **2016**, *30*, e115. [CrossRef]
2. Cao, C.Y.; Mei, M.L.; Li, Q.L.; Lo, E.C.M.; Chu, C.H. Methods for biomimetic remineralization of human dentine: A systematic review. *Int. J. Mol. Sci.* **2015**, *16*, 4615–4627. [CrossRef]
3. Pinna, R.; Maioli, M.; Eramo, S.; Mura, I.; Milia, E. Carious affected dentine: Its behaviour in adhesive bonding. *Aust. Dent. J.* **2015**, *60*, 276–293. [CrossRef] [PubMed]
4. Kidd, E.A.M. How "clean" must a cavity be before restoration? *Caries Res.* **2004**, *38*, 305–313. [CrossRef] [PubMed]
5. Demarco, F.F.; Corrêa, M.B.; Cenci, M.S.; Moraes, R.R.; Opdam, N.J.M. Longevity of posterior composite restorations: Not only a matter of materials. *Dent. Mater.* **2012**, *28*, 87–101. [CrossRef] [PubMed]
6. Mjör, I.A.; Gordan, V.V. Failure, Repair, Refurbishing and Longevity of Restorations. *Oper. Dent.* **2002**, *27*, 528–534. [PubMed]
7. Mjör, I.A.; Moorhead, J.E.; Dahl, J.E.; Haslum, N. Reasons for replacement of restorations in permanent teeth in general dental practice. *Int. Dent. J.* **2000**, *50*, 361–366. [CrossRef] [PubMed]
8. Chisini, L.A.; Collares, K.; Cademartori, M.G.; de Oliveira, L.J.C.; Conde, M.C.M.; Demarco, F.F.; Corrêa, M.B. Restorations in primary teeth: A systematic review on survival and reasons for failures. *Int. J. Paediatr. Dent.* **2018**, *28*, 123–139. [CrossRef] [PubMed]

9.   De Menezes, L.R.; da Silva, E.O.; Maurat da Rocha, L.V.; Ferreira Barbosa, I.; Rodrigues Tavares, M. The use of clays for chlorhexidine controlled release as a new perspective for longer durability of dentin adhesion. *J. Mater. Sci. Mater. Med.* **2019**, *30*, 132. [CrossRef]

10.   Bin-Shuwaish, M.S. Effects and effectiveness of cavity disinfectants in operative dentistry: A literature review. *J. Contemp. Dent. Pract.* **2016**, *17*, 867–879. [CrossRef]

11.   Miranda, C.; Silva, G.V.; Vieira, M.D.; Costa, S.X.S. Influence of the chlorhexidine application on adhesive interface stability: literature review. *Rev. Sul-Bras. Odontol.* **2014**, *11*, 276–285.

12.   Gwinnett, A.J. Effect of Cavity Disinfection on Bond Strength to Dentin. *J. Esthet. Restor. Dent.* **1992**, *4*, 11–13. [CrossRef] [PubMed]

13.   Davalloo, R.; Tavangar, M.; Ebrahimi, H.; Darabi, F.; Mahmoudi, S. In Vitro Comparative Evaluation of Newly Produced Desensitizer and Chlorhexidine and Gluma on Bond Strength and Bond Longevity of Composite to Dentin. *J. Dent.* **2019**, *21*, 111–118. [CrossRef]

14.   Al Deeb, L.; Bin-Shuwaish, M.S.; Abrar, E.; Naseem, M.; Al-Hamdan, R.S.; Maawadh, A.M.; Al Deeb, M.; Almohareb, T.; Al Ahdal, K.; Vohra, F.; et al. Efficacy of chlorhexidine, Er Cr YSGG laser and photodynamic therapy on the adhesive bond integrity of caries affected dentin. An in-vitro study. *Photodiagnosis Photodyn. Ther.* **2020**, *31*, 101875. [CrossRef]

15.   Mapar, M.; Moalemnia, M.; Kalantari, S. Comparison of the effect of Chlorhexidine and collagen cross-linking agent (Quercus Extract) on the tensile bond strength of composite to dentin. *Med. Sci.* **2020**, *24*, 1168–1175.

16.   Fernandes, G.L.; Strazzi-Sahyon, H.B.; Suzuki, T.Y.U.; Briso, A.L.F.; dos Santos, P.H. Influence of chlorhexidine gluconate on the immediate bond strength of a universal adhesive system on dentine subjected to different bonding protocols: An in vitro pilot study. *Oral Heal. Prev. Dent.* **2020**, *18*, 71–76. [CrossRef]

17.   Geng Vivanco, R.; Cardoso, R.S.; Sousa, A.B.S.; Chinelatti, M.A.; de Vincenti, S.A.F.; Tonani-Torrieri, R.; Pires-de-Souza, F. de C.P. Effect of thermo-mechanical cycling and chlorhexidine on the bond strength of universal adhesive system to dentin. *Heliyon* **2020**, *6*, e03871. [CrossRef]

18.   Nima, G.; Cavalli, V.; Bacelar-Sá, R.; Ambrosano, G.M.B.; Giannini, M. Effects of sodium hypochlorite as dentin deproteinizing agent and aging media on bond strength of two conventional adhesives. *Microsc. Res. Tech.* **2020**, *83*, 186–195. [CrossRef]

19.   Yazici, E.; Gurgan, S.; Gutknecht, N.; Imazato, S. Effects of erbium: Yttrium-aluminum-garnet and neodymium: Yttrium-aluminum-garnet laser hypersensitivity treatment parameters on the bond strength of self-etch adhesives. *Lasers Med. Sci.* **2010**, *25*, 511–516. [CrossRef]

20.   Koshiro, K.; Inoue, S.; Niimi, K.; Koase, K.; Sano, H. Bond strength and SEM observation of CO2 laser irradiated dentin, bonded with simplified-step adhesives. *Oper. Dent.* **2005**, *30*, 170–179.

21.   Rolla, J.N.; Mota, E.G.; Oshima, H.M.S.; Burnett, L.H.; Spohr, A.M. Nd:YAG laser influence on microtensile bond strength of different adhesive systems for human dentin. *Photomed. Laser Surg.* **2006**, *24*, 730–734. [CrossRef] [PubMed]

22.   Gan, J.; Liu, S.; Zhou, L.; Wang, Y.; Guo, J.; Huang, C. Effect of Nd:YAG laser irradiation pretreatment on the long-term bond strength of etch-and-rinse adhesive to dentin. *Oper. Dent.* **2016**, *42*, 62–72. [CrossRef] [PubMed]

23.   Erhardt, M.C.G.; Osorio, R.; Toledano, M. Dentin treatment with MMPs inhibitors does not alter bond strengths to caries-affected dentin. *J. Dent.* **2008**, *36*, 1068–1073. [CrossRef] [PubMed]

24.   Mobarak, E.H. Effect of chlorhexidine pretreatment on bond strength durability of caries-affected dentin over 2-year aging in artificial saliva and under simulated intrapulpal pressure. *Oper. Dent.* **2011**, *36*, 649–660. [CrossRef] [PubMed]

25.   Kunawarote, S.; Nakajima, M.; Foxton, R.M.M.; Tagami, J. Effect of pretreatment with mildly acidic hypochlorous acid on adhesion to caries-affected dentin using a self-etch adhesive. *Eur. J. Oral Sci.* **2011**, *119*, 86–92. [CrossRef]

26.   Ozsoy, A.; Erdemir, U.; Yucel, T.; Yildiz, E. Effects of cavity disinfectants on bond strength of an etch-and-rinse adhesive to water- or ethanol-saturated sound and caries-affected dentin. *J. Adhes. Sci. Technol.* **2015**, *29*, 2551–2564. [CrossRef]

27.   Lenzi, T.L.; Tedesco, T.K.; Soares, F.Z.M.; Loguercio, A.D.; de O Rocha, R. Chlorhexidine does not increase immediate bond strength of etch-and-rinse adhesive to caries-affected dentin of primary and permanent teeth. *Braz. Dent. J.* **2012**, *23*, 438–442. [CrossRef]

28.   Kucukyilmaz, E.; Savas, S.; Akcay, M.; Bolukbasi, B. Effect of silver diamine fluoride and ammonium hexafluorosilicate applications with and without Er:YAG laser irradiation on the microtensile bond strength in sound and caries-affected dentin. *Lasers Surg. Med.* **2016**, *48*, 62–69. [CrossRef]

29.   Cersosimo, M.C.P.; Matos, A.B.; Couto, R.S.D.; Marques, M.M.; de Freitas, P.M. Short-pulse Er:YAG laser increases bond strength of composite resin to sound and eroded dentin. *J. Biomed. Opt.* **2016**, *21*, 48001. [CrossRef]

30.   Toledano, M.; Perdigão, J.; Osorio, E.; Osorio, R. Influence of NaOCl deproteinization on shear bond strength in function of dentin depth. *Am. J. Dent.* **2002**, *15*, 252–255.

31.   Ceballos, L.; Toledano, M.; Osorio, R. Bonding to Er-YAG-laser- treated Dentin. *J. Dent. Res.* **2002**, *81*, 119–122. [CrossRef]

32.   Uceda-Gómez, N.; Reis, A.; Carrilho, M.R.D.O.; Loguercio, A.D.; Rodriguez Filho, L.E. Effect of sodium hypochlorite on the bond strength of an adhesive system to superficial and deep dentin. *J. Appl. Oral Sci.* **2003**, *11*, 223–228. [CrossRef] [PubMed]

33.   Gonçalves, M.; Corona, S.A.M.; Palma-Dibb, R.G.; Pécora, J.D. Influence of pulse repetition rate of Er:YAG laser and dentin depth on tensile bond strength of dentin-resin interface. *J. Biomed. Mater. Res. Part A* **2007**, *86*, 477–482. [CrossRef] [PubMed]

34.   Aguilera, F.S.; Osorio, R.; Osorio, E.; Moura, P.; Toledano, M. Bonding efficacy of an acetone/based etch-and-rinse adhesive after dentin deproteinization. *Med. Oral Patol. Oral Cir. Bucal* **2012**, *17*, 649–654. [CrossRef]

35. Montagner, A.F.; Skupien, J.A.; Borges, M.F.; Krejci, I.; Bortolotto, T.; Susin, A.H. Effect of sodium hypochlorite as dentinal pretreatment on bonding strength of adhesive systems. *Indian J. Dent. Res.* **2015**, *26*, 416–420. [CrossRef]

36. Bueno, M.D.F.T.; Basting, R.T.; Turssi, C.P.; França, F.M.G.; Amaral, F.L.B. Effect of 2% chlorhexidine digluconate application and water storage on the bond strength to superficial and deep dentin. *J. Adhes. Sci. Technol.* **2015**, *29*, 1258–1267. [CrossRef]

37. Alaghehmand, H.; Nasrollah, F.N.; Nokhbatolfoghahaei, H.; Fekrazad, R. An in vitro comparison of the bond strength of composite to superficial and deep dentin, treated with Er:YAG laser irradiation or acid-etching. *J. Lasers Med. Sci.* **2016**, *7*, 167–171. [CrossRef]

38. Karadas, M.; Çağlar, İ. The effect of Er:YAG laser irradiation on the bond stability of self-etch adhesives at different dentin depths. *Lasers Med. Sci.* **2017**, *32*, 967–974. [CrossRef]

39. Say, E.C.; Koray, F.; Tarim, B.; Soyman, M.; Gülmez, T. In vitro effect of cavity disinfectants on the bond strength of dentin bonding systems. *Quintessence Int.* **2004**, *35*, 56–60.

40. Franke, M.; Taylor, A.W.; Lago, A.; Fredel, M.C. Influence of Nd:YAG laser irradiation on an adhesive restorative procedure. *Oper. Dent.* **2006**, *31*, 604–609. [CrossRef]

41. Malta, D.A.M.P.; Kreidler, M.A.M.; Villa, G.E.; De Andrade, M.F.; Fontana, C.R.; Lizarelli, R.F.Z. Bond strength of adhesive restorations to Er: YAG laser-treated dentin. *Laser Phys. Lett.* **2007**, *4*, 153–156. [CrossRef]

42. Campos, E.A.; Correr, G.M.; Leonardi, D.P.; Barato-Filho, F.; Gonzaga, C.C.; Zielak, J.C. Chlorhexidine diminishes the loss of bond strength over time under simulated pulpal pressure and thermo-mechanical stressing. *J. Dent.* **2009**, *37*, 108–114. [CrossRef] [PubMed]

43. Sierpinsky, L.M.G.; Lima, D.M.; Candido, M.S.M.; Bagnato, V.S.; Porto-Neto, S.T. Microtensile bond strength of different adhesive systems in dentin irradiated with Er:YAG laser. *Laser Phys. Lett.* **2008**, *5*, 547–551. [CrossRef]

44. Yazici, A.R.; Karaman, E.; Ertan, A.; Özgünaltay, G.; Dayangaç, B. Effect of different pretreatment methods on dentin bond strength of a one-step self-etch adhesive. *J. Contemp. Dent. Pract.* **2009**, *10*, 41–48. [CrossRef] [PubMed]

45. Baseggio, W.; Consolmagno, E.C.; De Carvalho, F.L.N.; Ueda, J.K.; Schmitt, V.L.; Formighieri, L.A.; Naufel, F.S. Effect of deproteinization and tubular occlusion on microtensile bond strength and marginal microleakage of resin composite restorations. *J. Appl. Oral Sci.* **2009**, *17*, 462–466. [CrossRef] [PubMed]

46. Eugénio, S.; Osorio, R.; Sivakumar, M.; Vilar, R.; Monticelli, F.; Toledano, M. Bond Strength of an Etch-and-Rinse Adhesive to KrF Excimer Laser-Treated Dentin. *Photomed. Laser Surg.* **2010**, *28*, 97–102. [CrossRef]

47. Kunawarote, S.; Nakajima, M.; Shida, K.; Kitasako, Y.; Foxton, R.M.M.; Tagami, J. Effect of dentin pretreatment with mild acidic HOCl solution on microtensile bond strength and surface pH. *J. Dent.* **2010**, *38*, 261–268. [CrossRef]

48. Rodrigues, P.C.F.C.F.; Souza, J.B.B.; Soares, C.J.J.; Lopes, L.G.G.; Estrela, C. Effect of ozone application on the resindentin microtensile bond strength. *Oper. Dent.* **2011**, *36*, 537–544. [CrossRef]

49. Nguyen, D.; Chang, K.; Hedayatollahnajafi, S.; Staninec, M.; Chan, K.; Lee, R.; Fried, D. High-speed scanning ablation of dental hard tissues with a $\lambda = 9.3$ μm CO_2 laser: Adhesion, mechanical strength, heat accumulation, and peripheral thermal damage. *J. Biomed. Opt.* **2011**, *16*, 71410. [CrossRef]

50. Carvalho, A.O.; Reis, A.F.; De Oliveira, M.T.; De Freitas, P.M.; Aranha, A.C.C.; Eduardo, C.D.P.; Giannini, M. Bond strength of adhesive systems to Er,Cr: YSGG laser-irradiated dentin. *Photomed. Laser Surg.* **2011**, *29*, 747–752. [CrossRef]

51. De Castro, F.L.A.; Andrade, M.F.; Hebling, J.; Lizarelli, R.F.Z. Nd: YAG laser irradiation of etched/unetched dentin through an uncured two-step etch-and-rinse adhesive and its effect on microtensile bond strength. *J. Adhes. Dent.* **2012**, *14*, 137–145. [CrossRef]

52. Gerhardt-Szep, S.; Werelius, K.; De Weerth, F.; Heidemann, D.; Weigl, P. Influence of femtosecond laser treatment on shear bond strength of composite resin bonding to human dentin under simulated pulpal pressure. *J. Biomed. Mater. Res. Part B Appl. Biomater.* **2012**, *100*, 177–184. [CrossRef]

53. Verma, R.; Singh, U.P.; Tyagi, S.P.; Nagpal, R.; Manuja, N. Long-term bonding effectiveness of simplified etch-and-rinse adhesives to dentin after different surface pre-treatments. *J. Conserv. Dent.* **2013**, *16*, 367–370. [PubMed]

54. De Oliveira, M.T.; Reis, A.F.; Arrais, C.A.G.; Cavalcanti, A.N.; Aranha, A.C.C.; De Paula Eduardo, C.; Giannini, M. Analysis of the interfacial micromorphology and bond strength of adhesive systems to Er:YAG laser-irradiated dentin. *Lasers Med. Sci.* **2013**, *28*, 1069–1076. [CrossRef] [PubMed]

55. Bridi, E.C.; Amaral, F.L.B.; Turssi, C.P.; Basting, R.T. Influence of dentin pretreatment with titanium tetrafluoride and self-etching adhesive systems on microtensile bond strength. *Am. J. Dent.* **2013**, *26*, 121–126. [PubMed]

56. Santiago, S.L.; Osorio, R.; Neri, J.R.; De Carvalho, R.M.; Toledano, M. Effect of the flavonoid epigallocatechin-3-gallate on resin-dentin bond strength. *J. Adhes. Dent.* **2013**, *15*, 535–540. [CrossRef]

57. Zheng, P.; Zaruba, M.; Attin, T.; Wiegand, A. Effect of different matrix metalloproteinase inhibitors on microtensile bond strength of an etch-and-rinse and a self-etching adhesive to dentin. *Oper. Dent.* **2014**, *40*, 80–86. [CrossRef]

58. Shirani, F.; Birang, R.; Malekipour, M.R.; Hourmehr, Z.; Kazemi, S. Shear bond strength of resin composite bonded with two adhesives: Influence of Er: YAG laser irradiation distance. *Dent. Res. J. (Isfahan)* **2014**, *11*, 689–694.

59. Manso, A.P.; Grande, R.H.M.; Bedran-Russo, A.K.; Reis, A.; Loguercio, A.D.; Pashley, D.H.; Carvalho, R.M. Can 1% chlorhexidine diacetate and ethanol stabilize resin-dentin bonds? *Dent. Mater.* **2014**, *30*, 735–741. [CrossRef]

60. Castro, F.L.A.; Carvalho, J.G.; Andrade, M.F.; Saad, J.R.C.; Hebling, J.; Lizarelli, R.F.Z. Bond strength of composite to dentin: Effect of acid etching and laser irradiation through an uncured self-etch adhesive system. *Laser Phys.* **2014**, *24*, 1. [CrossRef]

61. Simões, D.M.S.; Basting, R.T.; Amaral, F.L.B.; Turssi, C.P.; França, F.M.G. Influence of chlorhexidine and/or ethanol treatment on bond strength of an etch-and-rinse adhesive to dentin: An in vitro and in situ study. *Oper. Dent.* **2014**, *39*, 64–71. [CrossRef] [PubMed]

62. Galafassi, D.; Scatena, C.; Colucci, V.; Rodrigues, A.L.; Campos Serra, M.; Milori Corona, S.A. Long-term chlorhexidine effect on bond strength to Er: YAG laser irradiated-dentin. *Microsc. Res. Tech.* **2014**, *77*, 37–43. [CrossRef] [PubMed]

63. Maenosono, R.M.A.; Bim Júnior, O.; Duarte, M.A.N.H.; Palma-Dibb, R.G.U.; Wang, L.; Ishikiriama, S.K. Diode laser irradiation increases microtensile bond strength of dentin. *Braz. Oral Res.* **2015**, *29*, 1–5. [CrossRef] [PubMed]

64. Chauhan, K.; Basavanna, R.S.; Shivanna, V. Effect of bromelain enzyme for dentin deproteinization on bond strength of adhesive system. *J. Conserv. Dent.* **2015**, *18*, 360–363. [CrossRef] [PubMed]

65. Sinha, D.J.; Jaiswal, N.; Vasudeva, A.; Garg, P.; Tyagi, S.P.; Chandra, P. Comparative Evaluation of the Effect of Chlorhexidine and Aloe Barbadensis Miller (Aloe Vera) on Dentin Stabilization Using Shear Bond Testing. *J Conserv Dent* **2016**, *19*, 406–409. [CrossRef]

66. Cha, H.S.; Shin, D.H. Antibacterial capacity of cavity disinfectants against Streptococcus mutans and their effects on shear bond strength of a self-etch adhesive. *Dent. Mater. J.* **2016**, *35*, 147–152. [CrossRef]

67. Gerhardt, K.M.F.; Oliveira, C.A.R.; França, F.M.G.; Basting, R.T.; Turssi, C.P.; Amaral, F.L.B. Effect of epigallocatechin gallate, green tea extract and chlorhexidine application on long-term bond strength of self-etch adhesive to dentin. *Int. J. Adhes. Adhes.* **2016**, *71*, 23–27. [CrossRef]

68. Balloni, E.C.P.; Do Amaral, F.L.B.; França, F.M.G.; Turssi, C.P.; Basting, R.T. Influence of chlorhexidine in cavities prepared with ultrasonic or diamond tips on microtensile bond strength. *J. Adhes. Sci. Technol.* **2016**, *31*, 1133–1141. [CrossRef]

69. Neri, J.R.; de Nojosa, J.S.; Yamauti, M.; Mendonça, J.S.; Santiago, S.L. Pretreatment with sodium fluoride maintains dentin bond strength of a two-step self-etch adhesive after thermal stressing. *J. Adhes. Dent.* **2017**, *19*, 517–523. [CrossRef]

70. Kim, B.R.; Oh, M.H.; Shin, D.H. Effect of cavity disinfectants on antibacterial activity and microtensile bond strength in class I cavity. *Dent. Mater. J.* **2017**, *36*, 368–373. [CrossRef]

71. Rechmann, P.; Bartolome, N.; Kinsel, R.; Vaderhobli, R.; Rechmann, B.M.T. Bond strength of etch-and-rinse and self-etch adhesive systems to enamel and dentin irradiated with a novel CO2 9.3 µm short-pulsed laser for dental restorative procedures. *Lasers Med. Sci.* **2017**, *32*, 1981–1993. [CrossRef] [PubMed]

72. Alaghehmad, H.; Mansouri, E.; Esmaili, B.; Bijani, A.; Nejadkarimi, S.; Rahchamani, M. Effect of 0.12% chlorhexidine and zinc nanoparticles on the microshear bond strength of dentin with a fifth-generation adhesive. *Eur. J. Dent.* **2018**, *12*, 105–110. [CrossRef] [PubMed]

73. Jowkar, Z.; Shafiei, F.; Asadmanesh, E.; Koohpeima, F. Influence of silver nanoparticles on resin-dentin bond strength durability in a self-etch and an etch-and-rinse adhesive system. *Restor. Dent. Endod.* **2019**, *44*, 1–9. [CrossRef] [PubMed]

74. Sharafeddin, F.; Haghbin, N. Comparison of Bromelain Enzyme, Sodium Hypochlorite, and Titanium Tetrafluoride on Shear Bond Strength of Restorative Composite to Dentin: An in vitro Study. *J. Dent. (Shiraz Iran)* **2019**, *20*, 264–270. [CrossRef]

75. Hussein, A.A.A.; Al-Shamma, A.M.W. Effect of Chlorhexidine and/or Ethanol Pre-bonding Treatment on the Shear Bond Strength of Resin Composite to Dentin. *Int. J. Med. Res. Heal. Sci.* **2019**, *8*, 150–159.

76. Kasraei, S.; Yarmohamadi, E.; Ranjbaran Jahromi, P.; Akbarzadeh, M. Effect of 940nm Diode Laser Irradiation on Microtensile Bond Strength of an Etch and Rinse Adhesive (Single Bond 2) to Dentin. *J. Dent. (Shiraz Iran)* **2019**, *20*, 30–36. [CrossRef]

77. Perdigao, J.; Denehy, G.E.; Swift, E.J. Effects of chlorhexidine on dentin surfaces and shear bond strengths. *Am. J. Dent.* **1994**, *7*, 81–84.

78. Ramos, R.P.; Chimello, D.T.; Chinelatti, M.A.; Nonaka, T.; Pécora, J.D.; Palma Dibb, R.G. Effect of Er:YAG laser on bond strength to dentin of a self-etching primer and two single-bottle adhesive systems. *Lasers Surg. Med.* **2002**, *31*, 164–170. [CrossRef]

79. De Castro, F.L.A.; De Andrade, M.F.; Duarte Júnior, S.L.L.; Vaz, L.G.; Ahid, F.J.M. Effect of 2% chlorhexidine on microtensile bond strength of composite to dentin. *J. Adhes. Dent.* **2003**, *5*, 129–138. [CrossRef]

80. Visuri, S.R.; Gilbert, J.L.; Wright, D.D.; Wigdor, H.A.; Walsh, J.T.J. Shear strength of composite bonded to Er: YAG laser-prepared dentin. *J. Dent. Res.* **1996**, *75*, 599–605. [CrossRef]

81. Gürgan, S.; Bolay, Ş.; Kiremitçi, A. Effect of disinfectant application methods on the bond strength of composite to dentin. *J. Oral Rehabil.* **1999**, *26*, 836–840. [CrossRef] [PubMed]

82. Gonçalves, M.; Milori Corona, S.A.; Borsatto, M.C.; Gomes Silva, P.C.; Djalma Pécora, J. Tensile bond strength of dentin-resinous system interfaces conditioned with Er:YAG laser irradiation. *J. Clin. Laser Med. Surg.* **2002**, *20*, 89–93. [CrossRef]

83. Dunn, W.J.; Davis, J.T.; Bush, A.C. Shear bond strength and SEM evaluation of composite bonded to Er:YAG laser-prepared dentin and enamel. *Dent. Mater.* **2005**, *21*, 616–624. [CrossRef]

84. Osorio, R.; Erhardt, M.C.G.; Pimenta, L.A.F.; Osorio, E.; Toledano, M. EDTA treatment improves resin-dentin bonds' resistance to degradation. *J. Dent. Res.* **2005**, *84*, 736–740. [CrossRef] [PubMed]

85. Abo, T.; Asmussen, E.; Uno, S.; Tagami, J. Short- and long-term in vitro study of the bonding of eight commercial adhesives to normal and deproteinized dentin. *Acta Odontol. Scand.* **2006**, *64*, 237–243. [CrossRef] [PubMed]

86. Loguercio, A.D.; Stanislawczuk, R.; Polli, L.G.; Costa, J.A.; Michel, M.D.; Reis, A. Influence of chlorhexidine digluconate concentration and application time on resin-dentin bond strength durability. *Eur. J. Oral Sci.* **2009**, *117*, 587–596. [CrossRef]

87. Stanislawczuk, R.; Amaral, R.C.; Zander-Grande, C.; Gagler, D.; Reis, A.; Loguercio, A.D. Chlorhexidine-containing acid conditioner preserves the longevity of resin-dentin bonds. *Oper. Dent.* **2009**, *34*, 481–490. [CrossRef]

88. Çelik, Ç.; Özel, Y.; Bağiş, B.; Erkut, S. Effect of laser irradiation and cavity disinfectant application on the microtensile bond strength of different adhesive systems. *Photomed. Laser Surg.* **2010**, *28*, 267–272. [CrossRef]
89. Ferreira, L.S.; Apel, C.; Francci, C.; Simoes, A.; Eduardo, C.P.; Gutknecht, N. Influence of etching time on bond strength in dentin irradiated with erbium lasers. *Lasers Med. Sci.* **2010**, *25*, 849–854. [CrossRef]
90. Mohammad, E.E.C.; Amir, A.; Soodabeh, K.; Ahmad, A. Effect of chlorhexidine on the shear bond strength of self-etch adhesives to dentin. *African J. Biotechnol.* **2011**, *10*, 10054–10057. [CrossRef]
91. Stanislawczuk, R.; Reis, A.; Loguercio, A.D. A 2-year in vitro evaluation of a chlorhexidine-containing acid on the durability of resin-dentin interfaces. *J. Dent.* **2011**, *39*, 40–47. [CrossRef] [PubMed]
92. Mohammad, E.E.C.; Mehdi, A.K.; Soodabeh, K.; Mohammadreza, H.M. Effect of sodium hypochlorite on the shear bond strength of fifth- and seventh-generation adhesives to coronal dentin. *African J. Biotechnol.* **2011**, *10*, 12697–12701. [CrossRef]
93. Garcia, E.J.; Serrano, A.P.M.; Urruchi, W.I.; Deboni, M.C.; Reis, A.; Grande, R.H.M.; Loguercio, A.D. Influence of ozone gas and ozonated water application to dentin and bonded interfaces on resin-dentin bond strength. *J. Adhes. Dent.* **2012**, *14*, 363–370. [CrossRef]
94. Firat, E.; Gurgan, S.; Gutknecht, N. Microtensile bond strength of an etch-and-rinse adhesive to enamel and dentin after Er:YAG laser pretreatment with different pulse durations. *Lasers Med. Sci.* **2012**, *27*, 15–21. [CrossRef] [PubMed]
95. Baraba, A.; Dukić, W.; Chieffi, N.; Ferrari, M.; Anić, I.; Miletić, I. Influence of different pulse durations of Er:YAG Laser based on variable square pulse technology on microtensile bond strength of a self-etch adhesive to dentin. *Photomed. Laser Surg.* **2013**, *31*, 116–124. [CrossRef]
96. Ustunkol, I.; Yazici, A.R.; Gorucu, J.; Dayangac, B. Influence of laser etching on enamel and dentin bond strength of Silorane System Adhesive. *Lasers Med. Sci.* **2015**, *30*, 695–700. [CrossRef]
97. Zhou, L.; Wang, Y.; Yang, H.; Guo, J.; Tay, F.R.; Huang, C. Effect of chemical interaction on the bonding strengths of self-etching adhesives to deproteinised dentine. *J. Dent.* **2015**, *43*, 973–980. [CrossRef]
98. Da Silva, E.M.; Glir, D.H.; Gill, A.W.M.C.; Giovanini, A.F.; Furuse, A.Y.; Gonzaga, C.C. Effect of chlorhexidine on dentin bond strength of two adhesive systems after storage in different media. *Braz. Dent. J.* **2015**, *26*, 642–647. [CrossRef]
99. Koizumi, H.; Hamama, H.H.; Burrow, M.F. Effect of a silver diamine fluoride and potassium iodide-based desensitizing and cavity cleaning agent on bond strength to dentine. *Int. J. Adhes. Adhes.* **2016**, *68*, 54–61. [CrossRef]
100. Tekçe, N.; Tuncer, S.; Demirci, M.; Balci, S. Do matrix metalloproteinase inhibitors improve the bond durability of universal dental adhesives? *Scanning* **2016**, *38*, 535–544. [CrossRef]
101. Ruschel, V.C.; Malta, D.A.M.P.; Monteiro, S. Dentin bond strength of an adhesive system irradiated with an Nd:YAG laser. *Laser Phys.* **2016**, *26*, 116101. [CrossRef]
102. Kusdemir, M.; Çetin, A.R.; Özsoy, A.; Toz, T.; Öztürk Bozkurt, F.; Özcan, M. Does 2% chlorhexidine digluconate cavity disinfectant or sodium fluoride/hydroxyethyl methacrylate affect adhesion of universal adhesive to dentin? *J. Adhes. Sci. Technol.* **2016**, *30*, 13–23. [CrossRef]
103. Oliveira, H.; Tedesco, T.K.; Rodrigues-Filho, L.E.; Soares, F.Z.M.; Rocha, O.R. Doxycycline as a matrix metalloproteinase inhibitor to prevent bond degradation: The effect of acid and neutral solutions on dentin bond strength. *Gen. Dent.* **2016**, *64*, 14–17.
104. Loguercio, A.D.; Stanislawczuk, R.; Malaquias, P.; Gutierrez, M.F.; Bauer, J.; Reis, A. Effect of minocycline on the durability of dentin bonding produced with etch-and- rinse adhesives. *Oper. Dent.* **2016**, *41*, 511–519. [CrossRef]
105. Loguercio, A.D.; Hass, V.; Gutierrez, M.F.; Luque-Martinez, I.V.; Szesz, A.; Stanislawczuk, R.; Bandeca, M.C.; Reis, A. Five-year effects of chlorhexidine on the in vitro durability of resin/dentin interfaces. *J. Adhes. Dent.* **2016**, *18*, 35–42. [CrossRef]
106. Sharafeddin, F.; Koohpeima, F.; Razazan, N. The Effect of Titanium Tetrafluoride and Sodium Hypochlorite on the Shear Bond Strength of Methacrylate and Silorane Based Composite Resins: An In-Vitro Study. *J. Dent. (Shiraz Iran)* **2017**, *18*, 82–87.
107. Kucukyilmaz, E.; Botsali, M.S.; Korkut, E.; Sener, Y.; Sari, T. Effect of Different Modes of Erbium:yttrium Aluminum Garnet Laser on Shear Bond Strength To Dentin. *Niger. J. Clin. Pract.* **2017**, *20*, 1277–1282.
108. Jowkar, Z.; Farpour, N.; Koohpeima, F.; Mokhtari, M.J.; Shafiei, F. Effect of silver nanoparticles, zinc oxide nanoparticles and titanium dioxide nanoparticles on microshear bond strength to enamel and dentin. *J. Contemp. Dent. Pract.* **2018**, *19*, 1405–1412. [CrossRef]
109. Silva, A.C.; Melo, P.; Ferreira, J.; Oliveira, S.; Gutknecht, N. Influence of grape seed extract in adhesion on dentin surfaces conditioned with Er,Cr:YSGG laser. *Lasers Med. Sci.* **2019**, *34*, 1493–1501. [CrossRef]
110. Wang, J.; Song, W.; Zhu, L.; Wei, X. A comparative study of the microtensile bond strength and microstructural differences between sclerotic and Normal dentine after surface pretreatment. *BMC Oral Health* **2019**, *19*, 216. [CrossRef]
111. Breschi, L.; Cammelli, F.; Visintini, E.; Mazzoni, A.; Carrilho, M.; Cadenaro, M.; Foulger, S.; Tay, F.R. Influence of chlorhexidine concentration on the durability of etch-and-rinse dentin bonds: A 12-month in vitro study. *J. Adhes. Dent.* **2009**, *11*, 191–198. [PubMed]
112. Ramos, A.C.B.; Esteves-Oliveira, M.; Arana-Chavez, V.E.; De Paula Eduardo, C. Adhesives bonded to erbium: Yttrium-aluminum-garnet laser-irradiated dentin: Transmission electron microscopy, scanning electron microscopy and tensile bond strength analyses. *Lasers Med. Sci.* **2010**, *25*, 181–189. [CrossRef] [PubMed]
113. Dalli, M.; Ercan, E.; Zorba, Y.O.; Ince, B.; Şahbaz, C.; Bahşi, E.; Çolak, H. Effect of 1% chlorhexidine gel on the bonding strength to dentin. *J. Dent. Sci.* **2010**, *5*, 8–13. [CrossRef]

114. Sano, K.; Tonami, K.I.; Ichinose, S.; Araki, K. Effects of ArF excimer laser irradiation of dentin on the tensile bonding strength to composite resin. *Photomed. Laser Surg.* **2012**, *30*, 71–76. [CrossRef] [PubMed]
115. Saraceni, C.H.; Liberti, E.; Navarro, R.S.; Cassoni, A.; Kodama, R.; Oda, M. Er:YAG-laser and sodium hypochlorite influence on bond to dentin. *Microsc. Res. Tech.* **2013**, *76*, 72–78. [CrossRef]
116. Jiang, Q.; Chen, M.; Ding, J. Comparison of tensile bond strengths of four one-bottle self-etching adhesive systems with Er:YAG laser-irradiated dentin. *Mol. Biol. Rep.* **2013**, *40*, 7053–7059. [CrossRef]
117. Ribeiro, C.F.; de Paiva Gonçalves, S.E.; Yui, K.C.K.; Borges, A.B.; Barcellos, D.C.; Brayner, R. Dentin Bond Strength: Influence of Er:YAG and Nd:YAG Lasers. *Int. J. Periodontics Restor. Dent.* **2013**, *33*, 373–377. [CrossRef]
118. Kasraei, S.; Azarsina, M.; Khamverdi, Z. Effect of Ethylene diamine tetra acetic acid and sodium hypochlorite solution conditioning on microtensile bond strength of one-step self-etch adhesives. *J. Conserv. Dent.* **2013**, *16*, 243–246. [CrossRef]
119. Portillo, M.; Lorenzo, M.C.; Moreno, P.; García, A.; Montero, J.; Ceballos, L.; Fuentes, M.V.; Albaladejo, A. Influence of Er:YAG and Ti:sapphire laser irradiation on the microtensile bond strength of several adhesives to dentin. *Lasers Med. Sci.* **2015**, *30*, 483–492. [CrossRef]
120. Heredia, A.; Ferreira da Silva, D.F.; Gluer Carracho, H.; Borges, G.A.; Spohr, A.M. Influence of Nd:YAG laser on the durability of resin-dentin bonds. *J. Laser Appl.* **2015**, *27*, 42004. [CrossRef]
121. Pucci, C.R.; Barbosa, N.R.; Bresciani, E.; Yui, K.C.K.; Filomena, M.R.L.H.; Barcellos, D.C.; Torres, C.R.G. Influence of dentin deproteinization on bonding degradation: 1-year results. *J. Contemp. Dent. Pract.* **2016**, *17*, 985–989. [CrossRef] [PubMed]
122. Gunaydin, Z.; Yazici, A.R.; Cehreli, Z.C. In vivo and in vitro effects of chlorhexidine pretreatment on immediate and aged dentin bond strengths. *Oper. Dent.* **2016**, *41*, 258–267. [CrossRef] [PubMed]
123. Bravo, C.; Sampaio, C.S.S.; Hirata, R.; Puppin-Rontani, R.M.M.; Mayoral, J.R.R.; Giner, L. In-vitro Comparative Study of the use of 2 % Chlorhexidine on Microtensile Bond Strength of Different Dentin Adhesives: A 6 Months Evaluation. *Int. J. Morphol.* **2017**, *35*, 893–900. [CrossRef]
124. Bravo, C.; Sampaio, C.S.; Hirata, R.; Puppin-Rontani, R.M.; Mayoral, J.R.; Giner, L. Effect of 2 % Chlorhexidine on Dentin Shear Bond Strength of Different Adhesive Systems: A 6 Months Evaluation. *Int. J. Morphol.* **2017**, *35*, 1140–1146. [CrossRef]
125. Trevelin, L.T.; Da Silva, B.T.F.; De Freitas, P.M.; Matos, A.B. Influence of Er:YAG laser pulse duration on the long-term stability of organic matrix and resin-dentin interface. *Lasers Med. Sci.* **2019**, *34*, 1391–1399. [CrossRef]
126. El-housseiny, A.A.; Jamjoum, H. The effect of caries detector dyes and a cavity cleanising agent on Composite Resin Bonding To Enamel and Dentin. *J. Clin. Pediatric Dent.* **2000**, *25*, 57–64. [CrossRef]
127. Da Silva, N.R.F.A.; Calamia, C.S.; Coelho, P.G.; Carrilho, M.R.D.O.; De Carvalho, R.M.; Caufield, P.; Thompson, P. Van Effect of 2% iodine disinfecting solution on bond strength to dentin. *J. Appl. Oral Sci.* **2006**, *14*, 399–404. [CrossRef]
128. Bengtson, C.R.G.; Bengtson, A.L.; Bengtson, N.G.; Turbino, M.L. Efeito da Clorexidina 2% na Resistência de União de Dois Sistemas Adesivos à Dentina Humana. *Pesq. Bras. Odontoped Clin. Integr.* **2008**, *8*, 51–56. [CrossRef]
129. Fawzy, A.S.; Amer, M.A.; El-Askary, F.S. Sodium hypochlorite as dentin pretreatment for etch-and-rinse single-bottle and two-step self-etching adhesives: Atomic force microscope and tensile bond strength evaluation. *J. Adhes. Dent.* **2008**, *10*, 135–144. [CrossRef]
130. Sauro, S.; Mannocci, F.; Toledano, M.; Osorio, R.; Pashley, D.H.; Watson, T.F. EDTA or H3PO4/NaOCl dentine treatments may increase hybrid layers' resistance to degradation: A microtensile bond strength and confocal-micropermeability study. *J. Dent.* **2009**, *37*, 279–288. [CrossRef]
131. Chou, J.C.; Chen, C.C.; Ding, S.J. Effect of Er,Cr:YSGG laser parameters on shear bond strength and microstructure of dentine. *Photomed. Laser Surg.* **2009**, *27*, 481–486. [CrossRef] [PubMed]
132. Elkassas, D.W.; Fawzi, E.M.; El Zohairy, A. The effect of cavity disinfectants on the micro-shear bond strength of dentin adhesives. *Eur. J. Dent.* **2014**, *8*, 184–190. [CrossRef] [PubMed]
133. Ercan, E.; Çolak, H.; Hamidi, M.M.; İbrahimov, D.; Gulal, E. Can Dentin Surfaces Be Bonded Safely With Ozone and Boric Acid? *Ozone Sci. Eng.* **2015**, *37*, 556–562. [CrossRef]
134. Sinha, D.J.; Jandial, U.A.; Jaiswal, N.; Singh, U.P.; Goel, S.; Singh, O. Comparative evaluation of the effect of different disinfecting agents on bond strength of composite resin to dentin using two-step self-etch and etch and rinse bonding systems: An in-vitro study. *J. Conserv. Dent.* **2018**, *21*, 424–427. [CrossRef] [PubMed]
135. Sacramento, P.A.; De Castilho, A.R.F.; Banzi, É.C.F.; Puppi-Rontani, R.M. Influence of cavity disinfectant and adhesive systems on the bonding procedure in demineralized dentin—a one-year in vitro evaluation. *J. Adhes. Dent.* **2012**, *14*, 575–583. [CrossRef] [PubMed]
136. Sun, Q.; Gu, L.; Quan, J.; Yu, X.; Huang, Z.; Wang, R.; Mai, S. Epigallocatechin-3-gallate enhance dentin biomodification and bond stability of an etch-and-rinse adhesive system. *Int. J. Adhes. Adhes.* **2018**, *80*, 115–121. [CrossRef]
137. Kalaiselvam, R.; Ganesh, A.; Rajan, M.; Kandaswamy, D. Evaluation of bioflavonoids on the immediate and delayed microtensile bond strength of self-etch and total-etch adhesive systems to sound dentin. *Indian J. Dent. Res.* **2018**, *29*, 133–136. [CrossRef]
138. Prati, C.; Chersoni, S.; Pashley, D.H. Effect of removal of surface collagen fibrils on resin-dentin bonding. *Dent. Mater.* **1999**, *15*, 323–331. [CrossRef]
139. Huang, M.S.; Li, M.T.; Huang, F.M.; Ding, S.J. The effect of thermocycling and dentine pre-treatment on the durability of the bond between composite resin and dentine. *J. Oral Rehabil.* **2004**, *31*, 492–499. [CrossRef]

140. Ramos, R.P.; Chinelatti, M.A.; Chimello, D.T.; Borsatto, M.C.; Pécora, J.D.; Palma-Dibb, R.G. Bonding of self-etching and total-etch systems to Er:YAG laser-irradiated dentin. Tensile bond strength and scanning electron microscopy. *Braz. Dent. J.* **2004**, *15*, SI-9–SI-20.

141. De Carvalho, R.C.R.; De Freitas, P.M.; Otsuki, M.; Eduardo, C.D.P.; Tagami, J. Micro-shear bond strength of Er:YAG-laser-treated dentin. *Lasers Med. Sci.* **2008**, *23*, 117–124. [CrossRef] [PubMed]

142. Chang, Y.E.; Shin, D.H. Effect of chlorhexidine application methods on microtensile bond strength to dentin in class i cavities. *Oper. Dent.* **2010**, *35*, 618–623. [CrossRef] [PubMed]

143. Dalkilic, E.E.; Arisu, H.D.; Kivanc, B.H.; Uctasli, M.B.; Omurlu, H. Effect of different disinfectant methods on the initial microtensile bond strength of a self-etch adhesive to dentin. *Lasers Med. Sci.* **2012**, *27*, 819–825. [CrossRef]

144. Davari, A.; Sadeghi, M.; Bakhshi, H. Shear Bond Strength of an Etch-and-rinse Adhesive to Er:YAG Laser- and/or Phosphoric Acid-treated Dentin. *J. Dent. Res. Dent. Clin. Dent. Prospects* **2013**, *7*, 67–73. [CrossRef] [PubMed]

145. Corona, S.A.M.; Atoui, J.A.; Chimello, D.T.; Borsatto, M.C.; Pecora, J.D.; Dibb, R.G.P. Composite resin's adhesive resistance to dentin: Influence of Er:YAG laser focal distance variation. *Photomed. Laser Surg.* **2005**, *23*, 229–232. [CrossRef]

146. Chiang, Y.S.; Chen, Y.L.; Chuang, S.F.; Wu, C.M.; Wei, P.J.; Han, C.F.; Lin, J.C.; Chang, H.T. Riboflavin-ultraviolet-A-induced collagen cross-linking treatments in improving dentin bonding. *Dent. Mater.* **2013**, *29*, 682–692. [CrossRef] [PubMed]

147. Sharma, V.; Kumar, S.; Rampal, P. Shear bond strength of composite resin to dentin after application of cavity disinfectants—SEM study. *Contemp. Clin. Dent.* **2011**, *2*, 155. [CrossRef]

148. Rayar, S.; Sadasiva, K.; Singh, P.; Thomas, P.; Senthilkumar, K.; Jayasimharaj, U. Department Effect of 2% Chlorhexidine on Resin Bond Strength and Mode of Failure Using Two Different Adhesives on Dentin: An In Vitro Study. *J. Pharm. Bioallied Sci.* **2019**, *11*, S325–S330. [CrossRef]

149. Souza, F.B.; Silva, C.H.V.; Dibb, R.G.P.; Delfino, C.S.; Beatrice, L.C.S. Bonding performance of different adhesive systems to deproteinized dentin: Microtensile bond strength and scanning electron microscopy. *J. Biomed. Mater. Res. Part B Appl. Biomater.* **2005**, *75*, 158–167. [CrossRef]

150. Saboia, V.P.A.; Nato, F.; Mazzoni, A.; Orsini, G.; Putignano, A.; Giannini, M.; Breschi, L. Adhesion of a Two-step Etch-and-Rinse Collagen-depleted Dentin. *J. Adhes Dent.* **2008**, *10*, 419–422.

151. Hedayatollahnajafi, S.; Staninec, M.; Watanabe, L.; Lee, C.; Fried, D. Dentin bond strength after ablation using a CO 2 laser operating at high pulse repetition rates. *Lasers Dent. XV* **2009**, *7162*, 71620F. [CrossRef]

152. Ercan, E.; Erdemir, A.; Zorba, Y.O.; Eldeniz, A.U.; Dalli, M.; İnce, B.; Kalaycioglu, B. Effect of Different Cavity Disinfectants on Bond Strength of a Posterior Composite. *J. Adhes Dent.* **2009**, *11*, 343–346. [CrossRef] [PubMed]

153. Stanislawczuk, R.; Da Costa, J.A.; Polli, L.G.; Reis, A.; Loguercio, A.D. Effect of tetracycline on the bond performance of etch-and-rinse adhesives to dentin. *Braz. Oral Res.* **2011**, *25*, 459–465. [CrossRef]

154. Reddy, M.S.C.; Mahesh, M.C.; Bhandary, S.; Pramod, J.; Shetty, A. Evaluation of effect of different cavity disinfectants on shear bond strength of composite resin to dentin using two-step self-etch and one-step self-etch bondingsystems: A comparative in vitro study. *J. Contemp. Dent. Pract.* **2013**, *14*, 275–280. [CrossRef] [PubMed]

155. Suma, N.K.; Shashibhushan, K.K.; Reddy, S. Effect of Dentin Disinfection with 2% Chlorhexidine Gluconate and 0.3% Iodine on Dentin Bond Strength: An in vitro Study. *Int. J. Clin. Pediatr. Dent.* **2017**, *10*, 223–228. [CrossRef] [PubMed]

156. Zabeu, G.S.; Maenossono, R.M.; Scarcella, C.R.; Brianezzi, L.F.F.; Palma-Dibb, R.G.; Ishikiriama, S.K. Effect of diode laser irradiation on the bond strength of polymerized nonsimplified adhesive systems after 12 months of water storage. *J. Appl. Oral Sci.* **2019**, *27*. [CrossRef]

157. Akturk, E.; Bektas, O.O.; Ozkanoglu, S.; Akin, E.G.G. Do Ozonated Water and Boric Acid Affect the Bond Strength to Dentin in Different Adhesive Systems? *Niger. J. Clin. Pract.* **2019**, *22*, 1758–1764. [CrossRef]

158. Ebrahimi-Chaharom, M.E.; Kimyai, S.; Mohammadi, N.; Oskoee, P.A.; Daneshpuy, M.; Bahari, M. Effect of sodium ascorbate on the bond strength of all-in-one adhesive systems to NaOCl-treated dentin. *J. Clin. Exp. Dent.* **2015**, *7*, e595–e599. [CrossRef]

159. Favetti, M.; Schroeder, T.; Montagner, A.F.; Correa, M.B.; Pereira-Cenci, T.; Cenci, M.S. Effectiveness of pre-treatment with chlorhexidine in restoration retention: A 36-month follow-up randomized clinical trial. *J. Dent.* **2017**, *60*, 44–49. [CrossRef]

160. Montagner, A.F.; Perroni, A.P.; Corrêa, M.B.; Masotti, A.S.; Pereira-Cenci, T.; Cenci, M.S. Effect of pre-treatment with chlorhexidine on the retention of restorations: A randomized controlled trial. *Braz. Dent. J.* **2015**, *26*, 234–241. [CrossRef]

161. Sartori, N.; Stolf, S.C.; Silva, S.B.; Lopes, G.C.; Carrilho, M. Influence of chlorhexidine digluconate on the clinical performance of adhesive restorations: A 3-year follow-up. *J. Dent.* **2013**, *41*, 1188–1195. [CrossRef]

162. Torres, C.R.G.; Barcellos, D.C.; Batista, G.R.; Pucci, C.R.; Antunes, M.J.S.; De La Cruz, D.B.; Borges, A.B. Five-year clinical performance of the dentine Deproteinization Technique in non-carious cervical lesions. *J. Dent.* **2014**, *42*, 816–823. [CrossRef] [PubMed]

163. Saboia, V.P.A.; Almeida, P.C.; Ritter, A.V.; Swift, E.J.; Pimenta, L.A.F. 2-Year clinical evaluation of sodium hypochlorite treatment in the restoration of non-carious cervical lesions: A pilot study. *Oper. Dent.* **2006**, *31*, 530–535. [CrossRef] [PubMed]

164. Akarsu, S.; Karademir, S.A.; Ertaş, S.A. The Effect of Diode Laser Application on Restoration of Non Carious Cervical Lesion: Clinical Follow Up. *Niger J Clin Pr* **2020**, *23*, 165–171.

165. Dutra-Correa, M.; Saraceni, C.H.C.; Ciaramicoli, M.T.; Kiyan, V.H.; Queiroz, C.S. Effect of chlorhexidine on the 18-month clinical performance of two adhesives. *J. Adhes. Dent.* **2013**, *15*, 287–292. [CrossRef] [PubMed]

166. *ISO/TS 11405:2015 Dental Materials—Testing of Adhesion to Tooth Structure*; International Organisation for Standardization: Geneva, Switzerland, 2015.
167. Secilmis, A.; Dilber, E.; Gokmen, F.; Ozturk, N.; Telatar, T. Effects of storage solutions on mineral contents of dentin. *J. Dent. Sci.* **2011**, *6*, 189–194. [CrossRef]
168. Santana, F.R.; Pereira, J.C.; Pereira, C.A.; Fernandes Neto, A.J.; Soares, C.J. Influence of method and period of storage on the microtensile bond strength of indirect composite resin restorations to dentine. *Braz. Oral Res.* **2008**, *22*, 352–357. [CrossRef]
169. Aydin, B.; Pamir, T.; Baltaci, A.; Orman, M.N.; Turk, T. Effect of storage solutions on microhardness of crown enamel and dentin. *Eur. J. Dent.* **2015**, *9*, 262–266. [CrossRef]
170. Tosun, G.; Sener, Y.; Sengun, A. Effect of storage duration/solution on microshear bond strength of composite to enamel. *Dent. Mater. J.* **2007**, *26*, 116–121. [CrossRef]
171. *ISO 3696:1987—Water for Analytical Laboratory Use—Specification and Test Methods*; International Organisation for Standardization: Geneva, Switzerland, 1987.
172. Coelho, A.; Paula, A.; Carrilho, T.; Silva, M.J.; Botelho, F.C.E. Chlorhexidine mouthwash as an anticaries agent—a systematic review. *Quintessence Int.* **2017**, *48*, 585–591.
173. Lim, K.S.; Kam, P.C.A. Chlorhexidine.-pharmacology and clinical applications. *Anaesth Intensive Care* **2008**, *36*, 502–512. [CrossRef] [PubMed]
174. Marsh, P.D.; Keevil, C.W.; McDermid, A.S.; Williamson, M.I.; Ellwood, D.C. Inhibition by the antimicrobial agent chlorhexidine of acid production and sugar transport in oral streptococcal bacterial. *Arch. Oral Biol.* **1983**, *28*, 233–240. [CrossRef]
175. Iwami, Y.; Schachtele, C.F.; Yamada, T. Mechanism of inhibition of glycolysis in Streptococcus mutans NCIV 11723 by chlorhexidine. *Oral Microbiol. Immunol.* **1995**, *10*, 360–364. [CrossRef] [PubMed]
176. Ximenes, M.; Cardoso, M.; Astorga, F.; Arnold, R.; Pimenta, L.A.; de S Viera, R. Antimicrobial activity of ozone and NaF-chlorhexidine on early childhood caries. *Braz. Oral Res.* **2017**, *31*, 1–10. [CrossRef]
177. Nomura, R.; Inaba, H.; Matayoshi, S.; Yoshida, S.; Matsumi, Y.; Matsumoto-Nakano, M.; Nakano, K. Inhibitory effect of a mouth rinse formulated with chlorhexidine gluconate, ethanol, and green tea extract against major oral bacterial species. *J. Oral Sci.* **2020**, *62*, 206–211. [CrossRef]
178. Türkün, M.; Türkün, L.S.; Kalender, A. Effect of Cavity Disinfectants on the Sealing Ability of Nonrinsing Dentin-Bonding Resins. *Yearb. Dent.* **2006**, *2006*, 10–11. [CrossRef]
179. Zhou, J.; Tan, J.; Yang, X.; Xu, X.; Li, D.; Chen, L. MMP-Inhibitory effect of chlorhexidine applied in a self-etching adhesive. *J. Adhes. Dent.* **2011**, *13*, 111–115.
180. Osorio, R.; Yamauti, M.; Osorio, E.; Ruiz-Requena, M.E.; Pashley, D.; Tay, F.; Toledano, M. Effect of dentin etching and chlorhexidine application on metalloproteinase-mediated collagen degradation. *Eur. J. Oral Sci.* **2011**, *119*, 79–85. [CrossRef]
181. Montagner, A.F.; Sarkis-Onofre, R.; Pereira-Cenci, T.; Cenci, M.S. MMP Inhibitors on Dentin Stability: A Systematic Review and Meta-analysis. *J. Dent. Res.* **2014**, *93*, 733–743. [CrossRef]
182. Meiers, J.C.; Kresin, J.C. Cavity disinfectants and dentin bonding. *Oper Dent.* **1996**, *21*, 153–159.
183. Goncalves, M.; Corona, S.A.M.; Borsatto, M.C.; Pecora, J.D.; Dibb, R.G.P. Influence of pulse frequency Er:YAG laser on the tensile bond strength of a composite to dentin. *Am. J. Dent.* **2005**, *18*, 165–167. [PubMed]
184. Matsumotoa, K.; Hossain, M.; Tsuzuki, N.; Yamada, Y. Morphological and compositional changes of human dentin aftermEr:YAG laser irradiation. *J. Oral Laser Appl.* **2003**, *3*, 115–130.
185. Martínez-Insua, A.; Da Silva Dominguez, A.; Rivera, F.G.; Santana-Penín, U.A. Differences in bonding to acid-etched or Er:YAG-laser-trated enamel and dentin surfaces. *J. Prosthet Dent.* **2000**, *84*, 280–288. [CrossRef] [PubMed]
186. De Munck, J.; Van Meerbeek, B.; Yudhira, R.; Lambrechts, P.; Vanherle, G. Micro- tensile bond strength of two adhesives to Er:YAG-lased vs bur-cut enamel and dentin. *Eur. J. Oral. Sci.* **2002**, *110*, 322–329. [CrossRef] [PubMed]
187. Schein, M.T.; Bocangel, J.S.; Nogueira, G.E.C.; Schein, P.A.L. SEM evaluation of the interaction pattern between dentin and resin after cavity preparation using ER: YAG laser. *J. Dent.* **2003**, *5712*, 127–135. [CrossRef]
188. Büyükyilmaz, T.; Ogaard, B.; Rølla, G. The resistance of titanium tetrafluoride-treated human enamel to strong hydrochloric acid. *Eur. J. Oral Sci.* **1997**, *105*, 473–477. [CrossRef]
189. Hals, E.; Tveit, A.B.; Tötdal, B.; Isrenn, R. Effect of NaF, TiF4 and APF solutions on root surfaces in vitro, with specialreference to uptake of F. *Caries Res.* **1981**, *15*, 468–476. [CrossRef]
190. Wefel, J.S.; Harless, J.D. The effect of topical fluoride agents on fluoride uptake and surface morphology. *J. Dent Res.* **1981**, *60*, 1842–1848. [CrossRef]
191. Wiegand, A.; Magalhães, A.C.; Sener, B.; Waldheim, E.; Attin, T. TiF(4) and NaF at pH 1.2 but not at pH 3.5 are able to reduce dentin erosion. *Arch Oral Biol.* **2009**, *54*, 790–795. [CrossRef]
192. Arslan, S.; Ozbilge, H.; Kaya, E.G.; Er, O. In vitro antimicrobial activity of propolis, BioPure MTAD, sodium hypochlorite, and chlorhexidine on. *Saudi Med. J.* **2011**, *32*, 479–483.
193. Estrela, C.; Estrela, C.R.A.; Decurcio, D.A.; Hollanda, A.C.B.; Silva, J.A. Antimicrobial efficacy of ozonated water, gaseous ozone, sodium hypochlorite and chlorhexidine in infected human root canals. *Int. Endod. J.* **2007**, *40*, 85–93. [CrossRef] [PubMed]
194. Osorio, R.; Ceballos, L.; Tay, F.; Cabrerizo-Vilchez, M.A.; Toledano, M. Effect of sodium hypochlorite on dentin bonding with a polyalkenoic acid-containing adhesive system. *J. Biomed. Mater. Res.* **2002**, *60*, 316–324. [CrossRef] [PubMed]

195. American Dental Association Council on Scientific Affairs Acceptance program guidelines: Dentin and enamel adhesive materials. *Am. Dent. Assoc. Counc. Sci. Aff. Chicago Am. Dent. Assoc.* **2001**, *12*, 1–12.
196. Gurgan, S.; Kiremitci, A.; Cakir, F.Y.; Gorucu, J.; Alpaslan, T.; Yazici, E.; Gutknecht, N. Shear bond strength of composite bonded to Er,Cr:YSGG laser-prepared dentin. *Photomed. Laser Surg.* **2008**, *26*, 495–500. [CrossRef] [PubMed]
197. Sterne, J.A.C.; Savović, J.; Page, M.J.; Elbers, R.G.; Blencowe, N.S.; Boutron, I.; Cates, C.J.; Cheng, H.-Y.; Corbett, M.S.; Eldridge, S.M.; et al. RoB 2: A revised tool for assessing risk of bias in randomised trials. *BMJ* **2019**, *366*. [CrossRef] [PubMed]

International Journal of
Molecular Sciences

MDPI

Article

Silver Nanoparticles from Oregano Leaves' Extracts as Antimicrobial Components for Non-Infected Hydrogel Contact Lenses

Anastasia Meretoudi [1], Christina N. Banti [1,*], Panagiotis K. Raptis [1], Christina Papachristodoulou [2], Nikolaos Kourkoumelis [3], Aris A. Ikiades [2], Panagiotis Zoumpoulakis [4], Thomas Mavromoustakos [5] and Sotiris K. Hadjikakou [1,6,*]

[1] Inorganic Chemistry Laboratory, Department of Chemistry, University of Ioannina, 45110 Ioannina, Greece; ameretoudi1996@gmail.com (A.M.); panagiwtisraptis93@yahoo.gr (P.K.R.)

[2] Department of Physics, University of Ioannina, 45110 Ioannina, Greece; xpapaxri@uoi.gr (C.P.); ikiadis@uoi.gr (A.A.I.)

[3] Medical Physics Laboratory, Medical School, University of Ioannina, 45110 Ioannina, Greece; nkourkou@uoi.gr

[4] Laboratory of Chemistry, Analysis & Design of Food Processes, Department of Food Science and Technology, University of West Attica, 11635 Attica, Greece; pzoump@eie.gr

[5] Organic Chemistry Laboratory, Department of Chemistry, University of Athens Greece, 15571 Athens, Greece; tmavrom@chem.uoa.gr

[6] Institute of Materials Science and Computing, University Research Center of Ioannina (URCI), 45110 Ioannina, Greece

* Correspondence: cbanti@uoi.gr (C.N.B.); shadjika@uoi.gr (S.K.H.); Tel.: +30-26510-08374 (S.K.H.)

Citation: Meretoudi, A.; Banti, C.N.; Raptis, P.K.; Papachristodoulou, C.; Kourkoumelis, N.; Ikiades, A.A.; Zoumpoulakis, P.; Mavromoustakos, T.; Hadjikakou, S.K. Silver Nanoparticles from Oregano Leaves' Extracts as Antimicrobial Components for Non-Infected Hydrogel Contact Lenses. *Int. J. Mol. Sci.* **2021**, *22*, 3539. https://doi.org/10.3390/ijms22073539

Academic Editor: Iolanda Francolini

Received: 12 March 2021
Accepted: 25 March 2021
Published: 29 March 2021

Publisher's Note: MDPI stays neutral with regard to jurisdictional claims in published maps and institutional affiliations.

Abstract: The oregano leaves' extract (ORLE) was used for the formation of silver nanoparticles (AgNPs(ORLE)). ORLE and AgNPs(ORLE) (2 mg/mL) were dispersed in polymer hydrogels to give the pHEMA@ORLE_2 and pHEMA@AgNPs(ORLE)_2 using hydroxyethyl–methacrylate (HEMA). The materials were characterized by X-ray fluorescence (XRF) spectroscopy, X-ray powder diffraction analysis (XRPD), thermogravimetric differential thermal analysis (TG-DTA), derivative thermogravimetry/differential scanning calorimetry (DTG/DSC), ultraviolet (UV-Vis), and attenuated total reflection mode (ATR-FTIR) spectroscopies in solid state and UV–Vis in solution. The crystallite size value, analyzed with XRPD, was determined at 20 nm. The antimicrobial activity of the materials was investigated against Gram-negative bacterial strains *Pseudomonas aeruginosa* (*P. aeruginosa*) and *Escherichia coli* (*E. coli*). The Gram-positive ones of the genus of *Staphylococcus epidermidis* (*S. epidermidis*) and *Staphylococcus aureus* (*S. aureus*) are known to be involved in microbial keratitis by the means of inhibitory zone (IZ), minimum inhibitory concentration (MIC), and minimum bactericidal concentration (MBC). The IZs, which developed upon incubation of *P. aeruginosa*, *E. coli*, *S. epidermidis*, and *S. aureus* with paper discs soaked in 2 mg/mL of AgNPs(ORLE), were 11.7 ± 0.7, 13.5 ± 1.9, 12.7 ± 1.7, and 14.3 ± 1.7 mm. When the same dose of ORLE was administrated, the IZs were 10.2 ± 0.7, 9.2 ± 0.5, 9.0 ± 0.0, and 9.0 ± 0.0 mm. The percent of bacterial viability when they were incubated over the polymeric hydrogel discs of pHEMA@AgNPs(ORLE)_2 was interestingly low (66.5, 88.3, 77.7, and 59.6%, respectively, against of *P. aeruginosa*, *E. coli*, *S. epidermidis*, and *S. aureus*) and those of pHEMA@ORLE_2 were 89.3, 88.1, 92.8, and 84.6%, respectively. Consequently, pHEMA@AgNPs(ORLE)_2 could be an efficient candidate toward the development of non-infectious contact lenses.

Keywords: silver nanoparticles; oregano leaves' extract; antimicrobial materials; hydrogels; contact lens

1. Introduction

Contact lenses are made of hydrogels such as poly–2–hydroxyethylmethacrylate (pHEMA) [1,2]. The pHEMA is a synthetic biocompatible hydrogel material for contact lens' development, which was approved by the Food and Drug Administration (FDA) in 1971 [3]. The advantages of the use of pHEMA in contact lenses arise from its water and oxygen permeability [1,2]. However, bacteria such as *Pseudomonas aeruginosa, Escherichia coli, Staphylococcus epidermidis,* and *Staphylococcus aureus* are often colonizing contact lens' materials [4,5]. Thus, contact lens may be the etiology of bacterial infections such as Microbial Keratitis (MK). This initiates the research for novel material's' development, which exhibits the benefits of typical contact lenses without the peril for severe bacterial infections. The dispersion of antimicrobial agents leads to loading of antibacterial properties in the polymeric material [6,7]. Silver(I) ions have been pursued as additives to contact lenses for their antimicrobial efficacy [8]. However, the major drawback of the ionic silver usage is the inactivation through complexation with chloride anions of the human fluids and their precipitation. Silver nanoparticles, on the other hand, are a valuable alternative [8]. The silver nanoparticles show efficient antimicrobial property compared to other salts due to their extremely large surface area, which provides better contact between the nanoparticles and the cells of the microorganisms [9].

While essential oils, such as *Origanum vulgare,* exhibit antibacterial, antifungal, antiviral, anti-inflammatory, and antioxidant properties [10,11], plant extracts are used for chemical synthetic procedure of nanomaterials as reducing or stabilizing agents [12]. The type of the plant extracts affects the characteristics of the produced nanoparticles since different plant extracts contain different concentrations and combinations of organic reducing agents [12]. Thus, AgNPs from ethanolic oregano leaves' extracts exhibit stronger antimicrobial activity against Gram-positive (*Staphylococcus aureus, Bacillus subtilis*) strains than against Gram-negative (*Escherichia coli, Salmonella typhimurium*) strains [13]. Moreover, silver nanoparticles from aqueous oregano plant extract have shown antibacterial and antifungal activities against *Shigella sonnei, Micrococcus luteus, Escherichia coli, Aspergillus flavus, Alternaria alternate, Paecilomyces variotii,* and *Phialophora alba* [14]. AgNPs from essential oil of oregano possessed antimicrobial activity against *S. aureus* [10].

Aiming for the formation of new, non-infectious materials for sterile antimicrobial contact lenses' development, we report here the synthesis of AgNPs(ORLE) obtained from oregano leaves extract (ORLE). Consequently, they were incorporated in pHEMA hydrogels. The materials pHEMA@ORLE_2 and pHEMA@AgNPs(ORLE)_2 were characterized by XRF, XRPD, TG-DTA, DTG/DSC, UV-Vis, ATR–FTIR, and refractive index in solid state and UV–Vis spectroscopy in solution. The antimicrobial activity of the materials was evaluated against the Gram-negative bacterial strains *P. aeruginosa* and *E. coli* and the Gram-positive ones, *S. epidermidis* and *S. aureus*, which are involved in microbial contamination of contact lenses, leading to ocular infections.

2. Results

2.1. General Aspects

In order to evaluate the antimicrobial efficiency of contact lens with antimicrobial components, a solution of AgNPs(ORLE) (2 mg/mL) was dispersed in pHEMA (Scheme 1) during its polymerization procedure. Consequently, the pHEMA@AgNPs(ORLE)_2 was derived. Similarly, oregano leaves' extract (ORLE) was isolated and used for the preparation of pHEMA@ORLE_2. The pHEMA, pHEMA@ORLE_2, and pHEMA@AgNPs(ORLE)_2 discs of 10-mm diameter were cut, cleaned from monomers, and stored either in sterilized NaCl 0.9% *w/w* solution or they were dried at 50 °C (Figure S1).

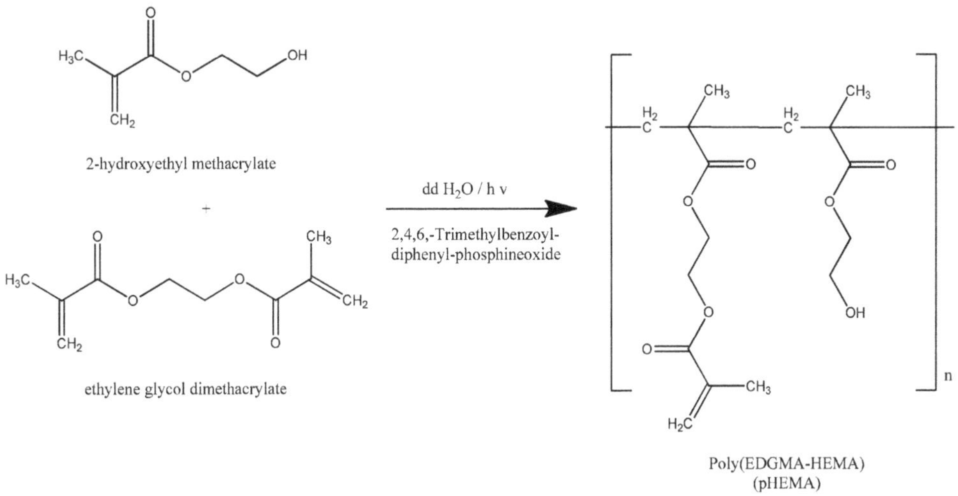

Scheme 1. Preparation reaction of poly-hydroxyethyl–methacrylate. (pHEMA).

2.2. Solid State

2.2.1. Refractive Index

The refractive indexes of stored discs in saline solution exhibit almost the same refractive index (1.434 (pHEMA), 1.434 (pHEMA@ORLE_2), and 1.433 (pHEMA@AgNPs(ORLE)_2)). These values are in accordance with the corresponding one of the pHEMA@platinum–nanoparticles (1.424–1.436) [8] and pHEMA@AGMNA–1 (AGMNA= {[Ag$_6$(μ_3-HMNA)$_4$(μ_3-MNA)$_2$]$^{2-}$·[(Et$_3$NH)$^+$]$_2$·(DMSO)$_2$·(H$_2$O)]}) (1.436) [7], although the ideal hydrogels should have a refractive index value matching the range of 1.372–1.381 [3]. However, the fabricated hydrogel is highly transparent, with refractive indexes ranging from 1.42 to 1.45 in the spectra range from 400 to 800 nm [15,16].

2.2.2. X-ray Fluorescence Spectroscopy

The XRF spectrum of the AgNPs(ORLE) and pHEMA@AgNPs(ORLE)_2 powders confirmed the presence of Ag. Moreover, the Ag Kα X-ray emission was used for quantitative determination of Ag in the AgNPs(ORLE). The content of silver in AgNPs(ORLE) was determined at 37 $\pm$ 5% *w/w*. (Figure 1). However, the silver content in the case of pHEMA@AgNPs(ORLE)_2 was unable to be determined due to its low quantity, which was close to the limits of the method.

2.2.3. X-ray Powder Diffraction Analysis (XRPD)

To perform crystallite size analysis of AgNPs(ORLE) with XRPD (Figure 2), the different contributions to specimen broadening were separated. The observed diffraction profile was the result of the of a specimen profile (f) and the aberrations introduced by the diffractometer. Therefore, the shape of the instrument line profile (g) due to the non-ideal optical contribution of the diffractometer and the wavelength distribution of the radiation should be known. Using a corundum (Al$_2$O$_3$) standard, the g was convoluted with f, in order to obtain a calculated instrument line profile, which was then applied to the measured profile of our sample (h). The pseudo-Voigt shape function was used, and the curve shape parameter, the skewness, and the kα_2 contribution were allowed to be refined during the peak fitting. The peak width was defined as the full width at half maximum (FWHM), knowing that the peak width varied inversely with crystallite size. Microstrain broadening analysis was not performed and, therefore, results might be less accurate at larger 2-theta angles. The profiles of the three Bragg peaks at 38.2, 64.5, and 77.5 were calculated and the following crystallite sizes were obtained: 210, 197, and 194 Å. The similar obtained values showed minimal crystalline strain, yielding an average crystallite size value of approximately 20 nm.

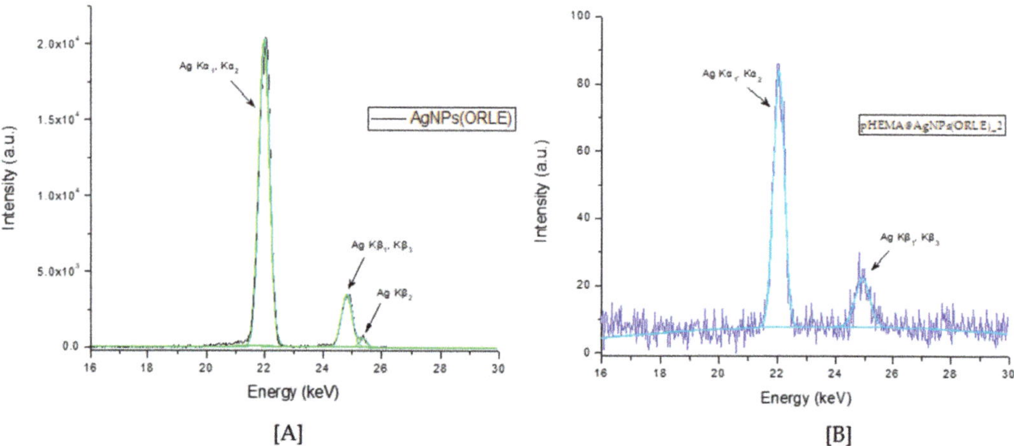

Figure 1. The Ag Kα radiation emitted from AgNPs(ORLE) (**A**) and pHEMA@AgNPs(ORLE)_2 (**B**) Oregano leaves extract (ORLE).

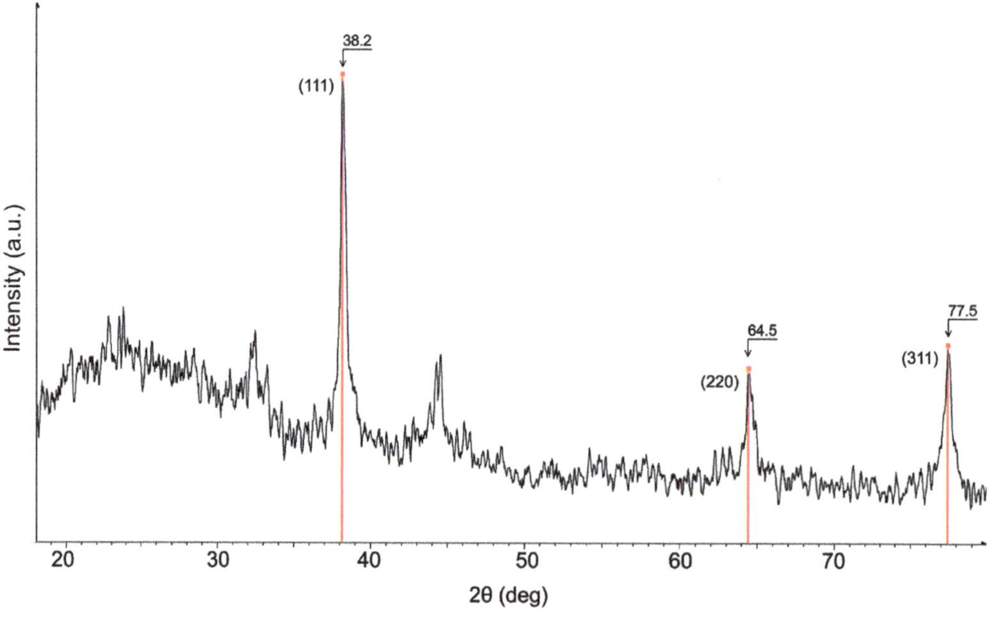

Figure 2. Powder X-ray diagram of sample AgNPs(ORLE) with the characteristic Bragg peaks of Ag nanoparticles.

2.2.4. Thermogravimetric Analysis of AgNPs(ORLE), pHEMA, and pHEMA@AgNPs(ORLE)_2

Differential Scanning Calorimetry (DSC)

In order to clarify whether AgNPs(ORLE) interact with pHEMA in solid state leading to a mixture or a composite material, DSC studies were carried out on the powder of dry pHEMA and pHEMA@AgNPs(ORLE)_2 (Figure 3). An endothermic transition at 455.3 °C in the DSC thermo-diagram of pHEMA@AgNPs(ORLE)_2 was observed at higher temperature at 459.7°C in the free pHEMA, suggesting the formation of a composite material rather than a mixture of precursors. The formation of the composite material was further supported by the sharp endothermic transition at 443.8 °C in the diagram of pHEMA@AgNPs(ORLE)_2, which is absent in the corresponding diagrams of AgNPs(ORLE) and pHEMA, respectively (Figure 3).

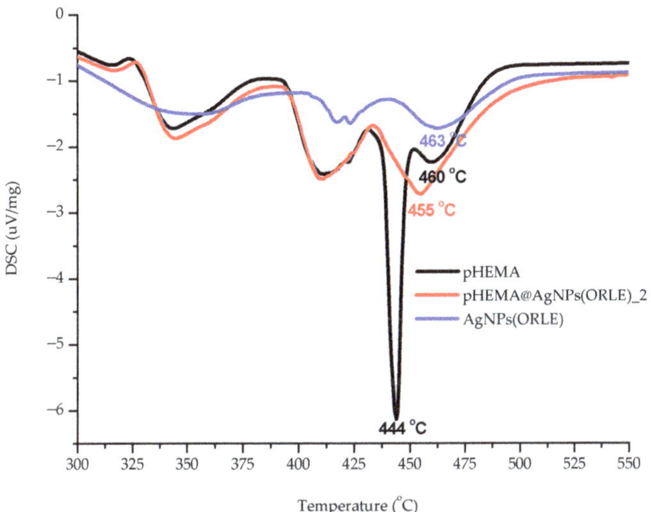

Figure 3. Differential scanning calorimetry (DSC) thermo-diagrams of pHEMA, AgNPs(ORLE), and pHEMA@AgNPs(ORLE)_2.

Thermal Decomposition of AgNPs(ORLE) and The Composite pHEMA@AgNPs(ORLE)_2

TG-DTA analysis was performed under air on the powder of dry AgNPs(ORLE) and pHEMA@AgNPs(ORLE)_2 with increasing the temperature at a rate of 10 °C min^{-1} from ambient up to 500 °C (Figure 4). The composite pHEMA@AgNPs(ORLE)_2 decomposed with five endothermic steps at 50–221, 221–298, 298–376, 376–430 and 430–527 °C with total mass loss 96.0%. The AgNPs(ORLE) decomposed with four endothermic steps at 50–218, 218–393, 393–424, and 424–463 °C with total mass loss of 96.5% (Figure 4).

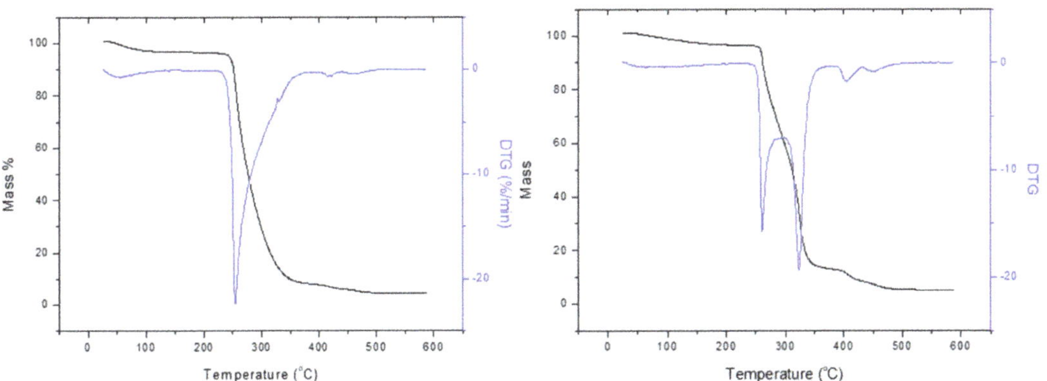

Figure 4. TGA diagrams of AgNPs(ORLE) (**left**) and pHEMA@AgNPs(ORLE)_2 (**right**).

2.2.5. Ultraviolet-Visible Spectroscopy (UV–Vis)

Solid State UV–Vis Spectra

The dispersion of the AgNPs(ORLE) in pHEMA within pHEMA@AgNPs(ORLE)_2 was verified by the solid state UV–Vis spectroscopy (Figure 5). The characteristic broad band at 330 nm in the solid state UV–Vis spectrum of pHEMA@AgNPs(ORLE)_2, which is absent in the corresponding one of free pHEMA, suggested the incorporation of the AgNPs(ORLE) in it.

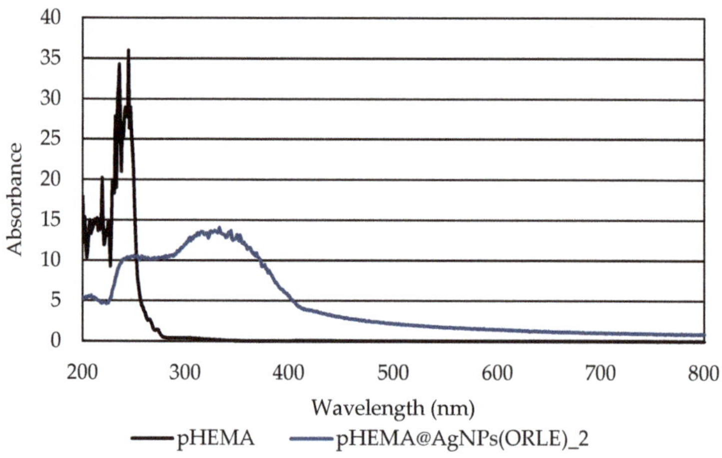

Figure 5. UV solid state of pHEMA and pHEMA@AgNPs(ORLE)_2.

Solution State Spectra

Since absorption band in UV–Vis spectra of AgNPs between 400–500 nm has been assigned to the collective resonance of electrons at the surface of the silver nanoparticles, these bands were used for the determination of silver nanoparticles' size [17,18]. The UV–Vis spectrum of AgNPs(ORLE) in double distilled (dd) water solution (0.4 mg/mL) (Figure 6) exhibited a broad band at 479 nm, which correlated to nanoparticles' size of ~65 nm [19]. It is mentioned here that the size of 20 nm, which was determined by XRPD spectroscopy (see above), refers to crystallite sizes and not to solubilized particles studied here by UV–Vis spectroscopy.

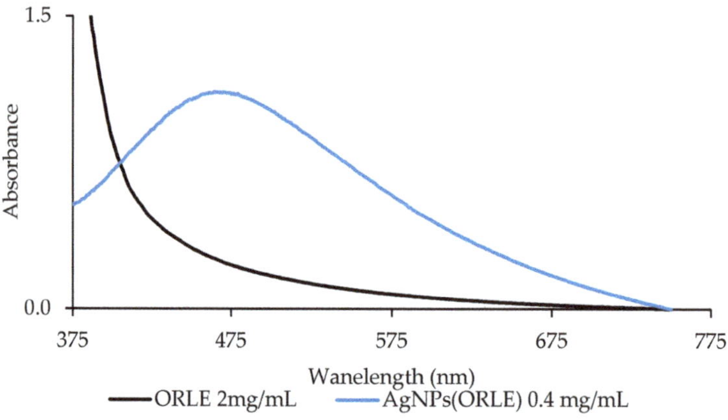

Figure 6. UV–Vis spectrum of AgNPs(ORLE) in double-distilled (dd) water solution.

2.3. Antibacterial Activity of ORLE and AgNPs(ORLE)

2.3.1. Determination of the Inhibition Zone (IZ) through Agar Disc-Diffusion Method

The agar disc-diffusion method was employed in order to survey the sensitivity of the microorganism to the antibacterial agent studied here [20]. The diameters of the bacterial growth inhibition zones when *P. aeruginosa, E. coli, S. epidermidis,* and *S. aureus* were treated by ORLE and AgNPs(ORLE) with a dose of 2 mg/mL upon their incubation for 20 h are summarized in Table 1 (Figure S2). The bacteria showed higher sensitivity to the AgNPs(ORLE) than ORLE, which rose up to 1.6-fold higher in the case of in the case of *S. aureus*. The microbe strains were classified into three categories according to the size of IZ, caused by an antimicrobial agent in their agar dilution culture: (1) strains, where the agent causes IZ $\geq$17 mm, were susceptible; (ii) those where the agent created IZ between 13 to 16 mm (13 $\leq$ IZ $\leq$ 16 mm) were intermediate; and (3) in those where the agent caused IZ $\leq$ 12 mm, the microbes were considered as resistant strains [21]. Therefore, the strains studied here (*P. aeruginosa, E. coli, S. epidermidis,* and *S. aureus*) responded intermediately to AgNPs(ORLE), while they were considered as resistant strains toward ORLE [21].

Table 1. Inhibition Zones (IZs). Minimum Inhibitory Concentrations (MICs), Minimum Bactericidal Concentrations (MBCs), Biofilm Elimination Concentrations (BECs) of ORLE, and AgNPs(ORLE) against *P. aeruginosa, E. coli, S. epidermidis,* and *S. aureus.*

Material Title	*P. aeruginosa*	*E. coli*	*S. epidermidis*	*S. aureus*
IZ (mm)				
AgNPs(ORLE) 2 mg/mL	13.1 ± 1.6	12.3 ± 0.7	12.7 ± 1.7	14.8 ± 1.1
ORLE 2 mg/mL	10.2 ± 0.7	9.3 ± 0.5	ND	ND
pHEMA@AgNPs(ORLE)_2	10.3 ± 0.7	ND	11.0 ± 1.9	10.3 ± 0.7
pHEMA@ORLE_2	ND	ND	ND	ND
Bacteria Viability (%)				
pHEMA@AgNPs(ORLE)_2	66.5	88.3	77.7	59.6
pHEMA@ORLE_2	89.3	88.1	92.8	84.6
MIC (µg/mL)				
AgNPs(ORLE)	139.5 ± 17.4	124.3 ± 12.9	272.2 ± 14.1	>300
ORLE	>300	>300	>300	>300
MBC (µg/mL)				
AgNPs(ORLE)	135.7 ± 35.2	>300	>300	>300
ORLE	>300	>300	>300	>300
BEC (µg/mL)				
AgNPs(ORLE)	945 ± 72	–	–	>1000

ND = Not Developed (IZ) was developed. ORLE = oregano leaves extract. IZ= Inhibition Zone. MIC= Minimum Inhibitory Concentration, MBC= Minimum Bactericidal Concentration, BEC= Biofilm Elimination Concentrations.

The antimicrobial activity of AgNPs(ORLE) against the microbes prompted us to load it into pHEMA for the development of new non-infectious contact lens pHEMA@AgNPs(ORLE)_2. The pHEMA@ORLE_2 was also prepared for comparison. The inhibition zones of pHEMA@AgNPs(ORLE)_2 against *P. aeruginosa, E. coli, S. epidermidis.* and *S. aureus* microbes suggested mild antimicrobial activity (Table 1, Figure 7). Moreover, no inhibition zones were developed when pHEMA@ORLE_2 or pHEMA was used against all tested strains (Figure S3). Since the agar diffusion test showed that pHEMA@AgNPs(ORLE)_2 discs, with diameter of 10 mm, which contained 2 mg/mL AgNPs(ORLE), developed shorter IZs against *P. aeruginosa, E. coli, S. epidermidis,* and *S. aureus* (10.3 ± 0.7, not developed (ND), 11.0 ± 1.9, and 10.3 ± 0.7 mm, respectively) than soaked paper discs, (diameter of 9 mm) with a solution of AgNPs(ORLE) (2 mg/mL) (13.1 ± 1.6, 12.3 ± 0.7, 12.7 ± 1.7, and 14.8 ± 1.1 mm, respectively), a minor releasing of the discs' ingredient could be concluded.

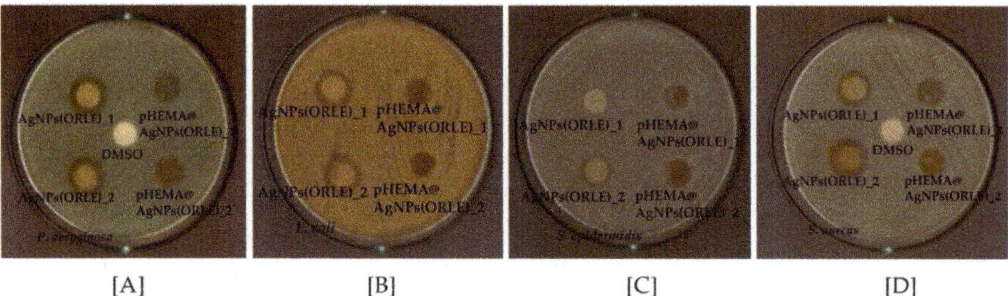

[A]　　　　　　[B]　　　　　　[C]　　　　　　[D]

Figure 7. Inhibition zones of pHEMA and pHEMA@AgNPs(ORLE)_2 against *Pseudomonas aeruginosa* (*P. aeruginosa*) (**A**), *Escherichia coli* (*E. coli*) (**B**), *Staphylococcus epidermidis* (*S. epidermidis*) (**C**), and *Staphylococcus aureus* (*S. aureus*) (**D**).

Thus, pHEMA, pHEMA@ORLE_2, and pHEMA@AgNPs(ORLE)_2 discs were placed in tests tubes that contained 5×10^5 cfu/mL of *P. aeruginosa*, *E. coli*, *S. epidermidis*, and *S. aureus* microbes. The calculated bacterial% viability of *P. aeruginosa*, *E. coli*, *S. epidermidis*, and *S. aureus* upon their incubation with pHEMA@AgNPs(ORLE)_2 discs for 20 h was 66.5, 88.3, 77.7, and 59.6%, respectively (Figure 8). On the contrary, no or meaningless influence in the bacterial viability was observed against these bacterial strains, upon their treatment by discs of pHEMA@ORLE_2 or pure pHEMA (Table 1, Figure 8).

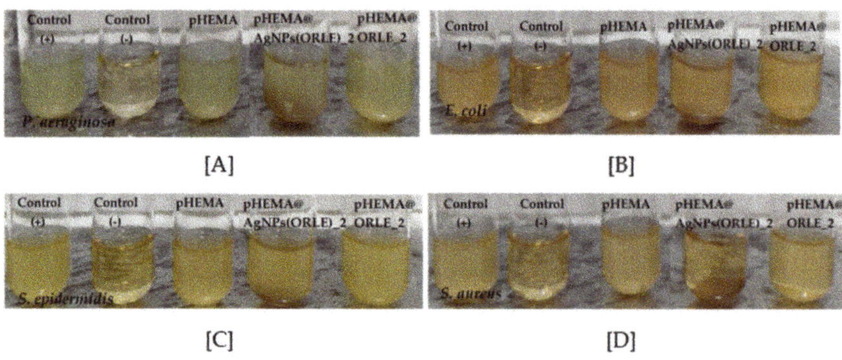

[A]　　　　　　　　　　[B]

[C]　　　　　　　　　　[D]

Figure 8. Bacteria viability of pHEMA, pHEMA@ORLE_2, and pHEMA@AgNPs(ORLE)_2 and against *P. aeruginosa* (**A**), *E. coli* (**B**), *S. epidermidis* (**C**), and *S. aureus* (**D**).

2.3.2. Effects on Biofilm Formation

The adhesion of bacterial cells to surfaces is the initial stage of the formation of biofilm. Thus, bacterial colonies are attached in a surface and they are protected in a polysaccharides' matrix. The necessity of the surgical removal of the infected tissue, which is then followed in the untreated cases, contains high risk for corneal blindness. The biofilm elimination can be achieved by applying metallodrugs [22].

The removal of biofilm was also assessed using crystal violet assay [23]. The discs of pHEMA@AgNPs(ORLE)_2 eliminated the biofilm of *P. aeruginosa* by 22%, while no inhibitory activity was observed against the biofilm of *S. aureus* (Figure S4).

2.3.3. Minimum Inhibitory (MIC) and Minimum Bactericidal (MBC) Concentrations

The antimicrobial potency of ORLE and AgNPs(ORLE) against Gram-negative (*P. aeruginosa* and *E. coli*) and Gram-positive (*S. epidermidis* and *S. aureus*) was evaluated by the mean of MIC, which is the lowest concentration needed for the inhibition of the bacterial growth, and the MBC, which is the lowest concentration of an antibacterial agent that can eliminate 99.9% of the bacterial inoculum [23,24]. The MIC and MBC values of ORLE and AgNPs(ORLE) against microbes studied here are summarized in Table 1 (Figures S5–S10). ORLE was inactive against the strains used.

2.3.4. Effects on Biofilm Formation by AgNPs(ORLE)

The effect of AgNPs(ORLE) against biofilm formation of *P. aeuroginosa* and *S. aureus* was studied by the biofilm elimination concentration (BEC) using crystal violet assay [7,23]. The BEC was defined as the concentration required to achieve at least a 99.9% reduction in the viability of biofilm bacteria. The composite material inhibited the biofilm formation, reducing 100% of its biomass, at 945 ± 72 µg/mL, against *P. aeuroginosa,* while it was not determined for *S aureus* for concentrations up to 1000 µg/mL (Figure S11). The BEC value of ciprofloxacin, a known antibiotic use for the bacterial keratitis, was 221 µg/mL toward *P. aeuroginosa* [6].

2.3.5. In Vitro Toxicity against Normal Human Corneal Epithelial Cells (HCECs)

The in vitro toxicity of ORLE and AgNPs(ORLE) was evaluated against normal human corneal epithelial cells (HCECs) after their incubation for a period of 48 h. The Inhibitory Concentration of the 50% of the cells (IC_{50}) values for ORLE and AgNPs(ORLE) were higher than 200 µg/mL, suggesting low toxicity toward normal cells. The cell viability of HCECs upon their incubation with discs of pHEMA@ORLE_2 and pHEMA@AgNPs(ORLE)_2 was decreased at 90.0 ± 9.5 and $54.9 \pm 7.1\%$, respectively, in respect of the cells incubated with pHEMA discs.

2.3.6. In Vivo Toxicity Evaluation by Brine Shrimp Artemia Salina

Artemia salina is a zooplanktonic crustacean [25,26] and the nauplii of the brine shrimp are considered as a simple and suitable model system for acute toxicity tests [26,27]. The assay can be correlated with the toxicity data of rodents and humans and shows a good correlation with cytotoxicity tests, making these measurements suitable as preliminary results [26,28,29]. The lethality was noted in terms of deaths of larvae.

The (%) mortality of *Artemia salina* larvae upon their incubation with ORLE and AgNPs(ORLE) at the concentrations of 150–300 µg/mL are summarized in Table 2. The mortality rate (%) of brine shrimp larvae incubated with ORLE and AgNPs(ORLE) rose from 4.0 and 7.1% (150 µg/mL) up to 78% at 300 µg/mL for both ORLE and AgNPs(ORLE) (Table 2). No mortality of brine shrimp larvae was found upon incubation of pHEMA, pHEMA@ORLE_2, and pHEMA@AgNPs(ORLE)_2 for 24 h, indicating their non-toxic behavior.

Table 2. Percent mortality of *Artemia salina* larvae in increasing concentrations of solutions with ORLE and AgNPs(ORLE).

Tested Concentration (µg/mL)	% Mortality	
	ORLE	AgNPs(ORLE)
150	4.0 ± 3.3	7.1 ± 3.4
200	33.9 ± 9.9	36.6 ± 11.6
250	34.5 ± 14.1	26.4 ± 11.4
300	78.4 ± 13.4	78.8 ± 23.1

3. Materials and Methods

3.1. Materials and Instruments

All solvents used were of reagent grade. Tryptone and soy peptone were purchased from Biolife, Milano, Italy. Agar was purchased from Sigma-Aldrich St. Louis, MO, USA product of Spain. Sodium cloride, D(+)–Glucose, and di potassium hydrogen phosphate trihydrate were purchased from Merck, Darmstadt, Germany. DMSO was purchased from Riedel–de Haen (Seelze, Germany). Melting points were measured in open tubes with a Stuart Scientific apparatus and are uncorrected. A UV–1600 PC series spectrophotometer of VWR international GmbH, Darmstadt, Germany was used to obtain electronic absorption spectra. ATR-FTIR spectra in the region of 4000–370 cm^{-1} were obtained with a Cary 670 FTIR spectrometer, Agilent Technologies Agilent Technologies. XRF measurement was carried out using an Am–241 radio isotopic source (exciting radiation 59.5 keV).

3.2. Preparation of ORLE and AgNPs(ORLE)

Dry oregano leaves (4 g) were refluxed using a Soxhlet apparatus with 50 mL dd water for 3 h. The ORLE extract was then obtained with filtration of the initial extract using Whatman No. 1 filter paper. In a beaker (250 mL), 10 mL ORLE and 170 mg $AgNO_3$ (1 mmol) were added. The beaker was placed in a microwave oven (700 W) for 2 min. Consequently, the solution was placed in the sonicator (ultrasonic) for 10 min. The solution was centrifugal at 6000 rpm for 20 min and the AgNPs(ORLE) were collected.

ORLE: yield 0.2 g, brown powder, melting point: 180–184 °C (change of color); ATR-IR (cm^{-1}): 3855 w, 3265 w, 1590 m, 1427, 1266 w, 1079, 1026, 989, 730, 685, 605, 564, 526, 505, 476, 455, and 426; UV–Vis (ddH_2O): λmax = 356 nm.

AgNPs(ORLE): yield 0.0232 g; content of Ag in AgNPs(ORLE): 37 ± 5% *w/w*; melting point: >250 °C IR (cm^{-1}): 3135 w, 2641, 2324 w, 1699 w, 1594 m, 1504 w, 1442 w, 1225, 1145, 1011, 807 w, 766 w, 7368 w, 692 w, 664 w, 630 w, 595 w, 570 w, 543 w, 509 w, 487, 473 w, 448, 431 s, and 410 s; UV–Vis (ddH_2O): λmax = 479 nm.

3.3. Synthesis of pHEMA@ORLE_2 and pHEMA@AgNPs(ORLE)_2

Hydrogels of ORLE or AgNPs(ORLE) dispersed in pHEMA were obtained as follows: 2.7 mL of HEMA were mixed with 2 mL of double-distilled water (ddw), which contained ORLE or AgNPs(ORLE) (2 mg/mL) and 10 μL of ethyleneglycol dimethacrylate (EGDMA). The solution was then degassed by bubbling with nitrogen for 15 min. Trimethylbenzoyldiphenyl-phosphineoxide (TPO) initiator (6 mg) was added to the solution and mixed for 5 min at 800 rpm. The solution was poured into the mold and was then placed under a UV mercury lamp (λmax = 280 nm), 15 watts, for photopolymerization for 40 min. Unreacted monomers were removed by immersing the gel in water for 15 min. Discs with 10-mm diameter were cut and were washed by immersion in water, NaCl 0.9%, and HCl 0.1 M, and again in water. The discs were then dried at 40 °C until no weight change would occur. The yield of dry pHEMA, pHEMA@ORLE_2, and pHEMA@AgNPs(ORLE)_2 was 1.669 and 1.683 g, respectively.

The pHEMA@ORLE_2: light-brown color; IR (cm^{-1}): 624 w, 480 w, 428 w, UV–Vis (solid state): λmax = 237 nm

The pHEMA@AgNPs(ORLE)_2: dark-brown color; Ag content in pHEMA@AgNPs (ORLE): 0.21 ± 0.04% *w/w*; IR (cm^{-1}) 624 w, 511 m, 480 w, 424 w; UV–Vis (solid phase): λmax = 332 nm.

3.4. Refractive Indexes

The values of refractive indexes of the lenses were measured with an Abbe refractometer (NAR–1T, Atago Co., Ltd., Tokyo, Japan) at 24 °C.

3.5. X-ray Fluorescence Spectroscopy

XRF measurement was carried out using an Am–241 radioisotopic source (exciting radiation 59.5 keV). For the detection of X-ray fluorescence, a Si (Li) detector was used. The measuring time was chosen so as to collect ~2000 data on the weaker Kα peak.

3.6. X-ray Powder Diffraction (XRPD)

The study of the samples by using the X-ray powder diffraction was accomplished by a diffraction meter D8 AdvanceBruker, department of Physics, University of Ioannina. Radiation CuKa (40 kV, 40 mA, $\lambda K\alpha$) and the monochromator system of diffracted beam were used. The X-ray powder diffraction patterns were measured in the area of 2θ angles between $2°$ and $80°$, using a rotation step $0.02°$ and time of 2 sec per step. All samples measured with the above diffraction meter were in fine-grained powder form.

3.7. Thermogravimetric Differential Thermal Analysis (TG-DTA), Differential Scanning Calorimetry (DTG/DSC)

For the DTA/TG measurements, a DTG/TG NETZSCH STA 449C was used. For the measurements, the samples were placed inside a platinum capsule on one side of the thermal scale while on the other side a-alumina was used as reference sample. The speed of temperature increase was 10 °C/min at a temperature range of 25–500 °C and the measurements took place in the air. All the measured samples were in fine-grained powder form.

3.8. Bacterial Strains

The bacterial strains of *P. aeruginosa* (PAO1), *Escherichia coli* Dh5a (*E. coli*), *S. aureus* (ATCC® 25923™), and *S. epidermidis* (ATCC® 14990™) were adopted in the experiments. The bacterial strain *P. aeruginosa* PAO1 was kindly offered from Prof. A. Koukou (Laboratory of Biochemistry, University of Ioannina-Greece). The biological experiments were performed in triplicates. The values were evaluated by the statistical analysis software package included in the MS Office excel.

3.9. Effects on the Growth of Microbial Strains

The procedure was performed as previously reported [6,7,23]. Briefly, bacterial strains plated onto trypticase soy agar were incubated at 37 °C for 18–24 h. Three to five isolated colonies of the same morphological appearance were selected from a fresh agar plate using a sterile loop and transferred into a tube containing 2 mL of sterile saline solution. The optical density at 620 nm was adjusted to 0.1, which corresponded to 10^8 cfu/mL [6,7,23].

For the evaluation of MIC, the inoculum size for broth dilution was 5×10^5 cfu/mL. The culture solution was treated with ORLE and AgNPs(ORLE) (20–300 µg/mL).

For the evaluation of MBC, the bacteria were initially cultivated in the presence of ORLE and AgNPs(ORLE) in broth culture for 20 h. The MBC values were determined in duplicate, by subculturing 4 µL of the broth an agar plate [6,7,23].

In order to evaluate the viability of microbes on pHEMA, pHEMA@ORLE_2, and pHEMA@AgNPs(ORLE)_2, discs of the materials were placed in the test tubes, which contained 5×10^5 cfu/mL of *P. aeruginosa*, *E. coli*, *S. epidermidis*, and *S. aureus* microbes [6,7,23]. The optical densities of the supernatant solutions were then measured to give the % viability of microbes after incubation for 18–24 h [6,7,23].

3.10. Removal of Biofilm, Using Crystal Violet Assay

Bacteria with a density of 1.3×10^6 cfu/mL were inoculated into Luria–Bertani agar (LB agar) medium for *P. aeruginosa* or tryptic soy broth for *S. aureus* (total volume = 1500 µL) and cultured for 20 h at 37 °C. Afterwards, the content of each tube was carefully removed, the tubes were washed with 1 mL 0.9% saline dilution, and 2 mL of broth were added. The negative control contained only broth. Then, the bacteria were incubated with ORLE, AgNPs(ORLE), pHEMA, pHEMA@ORLE_2, and pHEMA@AgNPs(ORLE)_2 for 20 h at 37 °C. The content of each tube was aspirated and was washed three times with 1 mL methanol and 2 mL 0.9% saline and left to dry. Then, the tubes were stained for 15 min with crystal violet solution (0.1% *w/v*). Excess stain was rinsed off with 1 mL methanol and 2 mL 0.9% saline solution and, afterwards, 3 mL 0.9% saline solution. The tubes were left to dry for 24 h and the bounded crystal violet was released by adding 30% glacial acetic acid. The optical density of the solution yielded was then measured at 550 nm, to give the biofilm biomass [6,7,23].

3.11. Determination of the Inhibition Zone (IZ)

The procedure was performed as previously reported [6,7,23]. Agar plates were inoculated with a standardized inoculum (10^8 cfu/mL) of the microorganisms. Discs of pHEMA or pHEMA@ORLE_2 and pHEMA@AgNPs(ORLE)_2 with 10-mm diameter or ORLE, AgNPs(ORLE) were placed on the agar surface and the Petri plates were incubated for 20 h.

3.12. Sulforhodamine B Assay

Initially, the HCECs were seeded in a 24-well plate in a density of 7.5×10^4 cells. After 24 h of cell incubation, ORLE, AgNPs(ORLE), the discs of pHEMA, pHEMA@ORLE_2, and pHEMA@AgNPs(ORLE)_2 were added in the wells. After 24-h incubation of HCECs with the discs, the discs were removed, the culture medium was aspirated, and the cells were fixed with 300 µL of 10% cold trichloroacetic acid (TCA). The plate was left for 30 min at 4 °C, washed five times with deionized water, and left to dry at room temperature for at least 24 h. Subsequently, 300 µL of 0.4% (*w/v*) sulforhodamine B (SRB) (Sigma) in 1% acetic acid solution was added to each well and left at room temperature for 20 min. SRB was removed, and the plate was washed five times with 1% acetic acid before air drying. Bound SRB was solubilized with 1 mL of 10 mM unbuffered Tris-base solution. Absorbance was read in a 24-well plate reader at 540 nm [6,7,23].

3.13. Evaluation of Toxicity with Brine Shrimp Assay

Brine shrimp assay was performed by a method previously described [3,7,26]. One-g cysts were initially hydrated in freshwater for one hour in a separating funnel or cone-shaped container. Seawater was prepared by dissolving 17 g of sea salt in 500 mL of distilled water [7,26,30]. The container was facilitated with good aeration for 48 h at room temperature and under continuous illumination. After hatching, nauplii released from the eggshells were collected at the bright side of the cone (near the light source) by using a micropipette. The larvae were isolated from the eggs by aliquoting them in a small beaker containing NaCl 0.9% [7,26,30]. An aliquot (0.1 mL) containing about 6 to 10 nauplii was introduced to each well of a 24-well plate and ORLE, AgNPs(ORLE), or one disc of pHEMA or pHEMA@ORLE_2 and pHEMA@AgNPs(ORLE)_2 were added in each well. The final volume of each well was 1 mL with NaCl 0.9%. The brine shrimps were observed at the interval time of 24 h, using a stereoscope. Larvae were considered dead if they did not exhibit any internal or external movement in 10 s of observation. Each experiment was repeated three times.

4. Conclusions

This work aimed for the development of sterilized and non-infectious contact lens. For this purpose, oregano leaves' extract (ORLE) was used for the formation of silver nanoparticles AgNPs(ORLE). AgNPs(ORLE) is an efficient disinfectant. The microbial strains *P. aeruginosa*, *E. coli*, *S. epidermidis*, and *S. aureus* involved in microbial keratitis responded intermediately to AgNPs(ORLE). The dispersion of AgNPs(ORLE) in pHEMA resulted in the composite material pHEMA@AgNPs(ORLE)_2. The percent of bacterial viability of *P. aeruginosa*, *E. coli*, *S. epidermidis*, and *S. aureus* upon their incubation with pHEMA@AgNPs(ORLE)_2 discs decreased by 33.5, 11.7, 22.3, and 40.4%, respectively, while it was also preventing the formation of biofilm on their surface. The cell viability of HCECA upon their incubation with pHEMA@AgNPs(ORLE)_2 discs for 48 h decreased by $54.9 \pm 7.1\%$. No mortality of brine shrimp larvae was found upon their incubation with pHEMA@AgNPs(ORLE)_2 for 24 h, indicating non-toxic in vivo behavior. Thus, pHEMA@AgNPs(ORLE)_2 is an effective candidate for the development of new non-infectious contact lens.

Supplementary Materials: The following are available online at https://www.mdpi.com/article/10 .3390/ijms22073539/s1. Figure S1. Dry discs of hydrogels pHEMA@ORLE_2 (left) and pHEMA@ AgNPs(ORLE)_2 (right). Figure S2. Bacterial growth inhibition zones developed in *P. aeruginosa* (A), *E. coli* (B), *S. epidermidis* (C), and *S. aureus* (D) by ORLE and AgNPs(ORLE) with doses of 1 or 2 mg/mL upon their incubation for 20 h. Figure S3. Bacterial growth inhibition zones developed in *P. aeruginosa* (A), *E. coli* (B), *S. epidermidis* (C), and *S. aureus* (D) by ORLE and pHEMA@ORLE with doses of 1 or 2 mg/mL upon their incubation for 20 h. Figure S4. Removal of preformed biofilm of *P. aeruginosa* (A) and *S. aureus* (B) caused pHEMA, pHEMA@ORLE_2, and pHEMA@AgNPs(ORLE)_2 discs. Figure S5. Minimum inhibitory concentration of ORLE (A) and AgNPs(ORLE) (B) against *P. aeruginosa*. Figure S6. minimum inhibitory concentration of ORLE (A) and AgNPs(ORLE) (B) against *E. coli*. Figure S7. Minimum inhibitory concentration of ORLE (A) and AgNPs(ORLE) (B) against *S. epidermidis*. Figure S8. Minimum inhibitory concentration of ORLE (A) and AgNPs(ORLE) (B) against *S. aureus*. Figure S9. Minimum bactericidal concentration of AgNPs(OLRE) against *P. aeruginosa* (A), *E. coli* (B), *S. epidermidis* (C), and *S. aureus* (D). Figure S10. Minimum bactericidal concentration of ORLE against *P. aeruginosa* (A), *E. coli* (B), *S. epidermidis* (C), and *S. aureus* (D). Figure S11. The biofilm growth of *P. aeruginosa* (A) and *S. aureus* (B), under increasing concentrations of AgNPs(ORLE) stained by crystal violet.

Author Contributions: Conceptualization, S.K.H.; investigation, A.M., P.K.R., C.N.B., N.K., C.P., A.A.I., P.Z., and T.M.; methodology, C.N.B. and S.K.H.; supervision, S.K.H.; validation, S.K.H.; writing—original draft, C.N.B. and S.K.H.; writing—review and editing, C.N.B. and S.K.H. All authors have read and agreed to the published version of the manuscript.

Funding: This research was co-financially supported by the European Union and Greek national funds through the Operational Program Competitiveness, Entrepreneurship and Innovation, under the call RESEARCH—CREATE—INNOVATE (project code: T1EDK–02990).

Institutional Review Board Statement: Not applicable.

Informed Consent Statement: Informed consent was obtained from all subjects involved in the study.

Acknowledgments: (1) The International Graduate Program in "Biological Inorganic Chemistry", which operates at the University of Ioannina within the collaboration of the Departments of Chemistry of the Universities of Ioannina, Athens, Thessaloniki, Patras, Crete, and the Department of Chemistry of the University of Cyprus (http://bic.chem.uoi.gr/BIC-En/index-en.html). (2) This research has been co-financed by the European Union and Greek national funds through the Operational Program Competitiveness, Entrepreneurship and Innovation, under the call RESEARCH—CREATE— INNOVATE (project code: T1EDK–02990).

Conflicts of Interest: The authors declare no conflict of interest.

Abbreviations

ATR–FT–IR	Attenuated Total Reflection mode
BEC	Biofilm Elimination Concentration
DTG/DSC	Differential Scanning Calorimetry
E. coli	*Escherichia coli*
FWHM	Full Width at Half Maximum
HCEC	Normal Human Corneal Epithelial Cells
IZ	Inhibitory Zone
MBC	Minimum Bactericidal Concentration
MIC	Minimum Inhibitory Concentration
MK	Microbial Keratitis
P. aeruginosa	*Pseudomonas aeruginosa*
pHEMA	poly–2–hydroxyethylmethacrylate
S. aureus	*Staphylococcus aureus*
S. epidermidis	*Staphylococcus epidermidis*
TD–DTA	Thermogravimetric Differential Thermal Analysis
UV–Vis	Ultra violet
XRF	X-ray fluorescence spectroscopy
XRPD	X-ray powder diffraction analysis

References

1. Musgrave, C.S.A.; Fang, F. Contact Lens Materials: A Materials Science Perspective. *Materials* **2019**, *12*, 261. [CrossRef]
2. Wichterle, O.; Lim, D. Hydrophilic Gels for Biological Use. *Nature* **1960**, *185*, 117–118. [CrossRef]
3. Caló, E.; Khutoryanskiy, V.V. Biomedical applications of hydrogels: A review of patents and commercial products. *Eur. Polym. J.* **2015**, *65*, 252–267. [CrossRef]
4. Schaefer, F.; Bruttin, O.; Zografos, L.; Guex-Crosier, Y. Bacterial keratitis: A prospective clinical and microbiological study. *Br. J. Ophthalmol.* **2001**, *85*, 842–847. [CrossRef] [PubMed]
5. Stapleton, F.; Carnt, N. Contact lens–related microbial keratitis: How have epidemiology and genetics helped us with pathogenesis and prophylaxis. *Eye* **2012**, *26*, 185–193. [CrossRef]
6. Milionis, I.; Banti, C.B.; Sainis, I.; Raptopoulou, C.P.; Psycharis, V.; Kourkoumelis, N.; Hadjikakou, S.K. Silver ciprofloxacin (CIPAG): A successful combination of chemically modified antibiotic in inorganic–organic hybrid. *J. Biol. Inorg. Chem.* **2018**, *23*, 705–723. [CrossRef] [PubMed]
7. Rossos, A.K.; Banti, C.N.; Kalampounias, A.; Papachristodoulou, C.; Kordatos, K.; Zoumpoulakis, P.; Mavromoustakos, T.; Kourkoumelis, N.; Hadjikakou, S.K. pHEMA@AGMNA-1: A novel material for the development of antibacterial contact lens. *Mater. Sci. Eng. C* **2020**, *111*, 110770. [CrossRef]
8. Xiao, A.; Dhand, C.; Leung, C.M.; Beuerman, R.W.; Ramakrishna, S.; Lakshminarayanan, R. Strategies to Design Antimicrobial Contact Lenses and Contact Lens Cases. *J. Mater. Chem. B* **2018**, *6*, 2171–2186. [CrossRef]
9. Rai, M.; Yadav, A.; Gade, A. Silver nanoparticles as a new generation of antimicrobials. *Biotechnol. Adv.* **2009**, *27*, 76–83. [CrossRef]
10. Maciel, M.V.D.O.B.; da Rosa Almeida, A.; Machado, M.H.; Elias, W.C.; da Rosa, C.G.; Teixeira, G.L.; Noronha, C.M.; Bertoldi, F.C.; Nunes, M.R.; de Armas, R.D.; et al. Green synthesis, characteristics and antimicrobial activity of silver nanoparticles mediated by essential oils as reducing agents, Biocatal. *Agric. Biotechnol.* **2020**, *28*, 101746.
11. Rodriguez-Garcia, I.; Silva-Espinoza, B.A.; Ortega-Ramirez, L.A.; Leyva, J.M.; Siddiqui, M.W.; Cruz-Valenzuela, M.R.; Gonzalez-Aguilar, G.A.; Ayala-Zavala, J.F. Oregano Essential Oil as an Antimicrobial and Antioxidant Additive in Food Products. *Crit. Rev. Food Sci. Nutr.* **2016**, *56*, 1717–1727. [CrossRef] [PubMed]
12. Rahimi-Nasrabadi, M.; Mahdi Pourmortazavi, S.; Ataollah Sadat Shandiz, S.; Ahmadi, F.; Batooli, H. Green synthesis of silver nanoparticles using Eucalyptus leucoxylon leaves extract and evaluating the antioxidant activities of extract. *Nat. Prod. Res.* **2014**, *28*, 1964–1969. [CrossRef] [PubMed]
13. Hambardzumyan, S.; Sahakyan, N.; Petrosyan, M.; Jawad Nasim, M.; Jacob, C.; Trchounian, A. Origanum vulgare L. extract–mediated synthesis of silver nanoparticles, their characterization and antibacterial activities. *AMB Express* **2020**, *10*, 162. [CrossRef]
14. Shaik, M.R.; Khan, M.; Kuniyil, M.; Al-Warthan, A.; Alkhathlan, H.Z.; Siddiqui, M.R.H.; Shaik, J.P.; Ahamed, A.; Mahmood, A.; Khan, M.; et al. Plant-Extract-Assisted Green Synthesis of Silver Nanoparticles Using Origanum vulgare L. *Extract and Their Microbicidal Activities. Sustainability* **2018**, *10*, 913.
15. Varikooty, J.; Keir, N.; Woods, C.A.; Fonn, D. Measurement of the refractive index of soft contact lenses during wear. *Eye Contact Lens* **2010**, *36*, 2–5. [CrossRef]
16. Childs, A.; Li, H.; Lewittes, D.; Dong, B.; Liu, W.; Shu, X.; Sun, C.; Zhang, H.F. Fabricating customized hydrogel contact lens. *Sci. Rep.* **2016**, *6*, 34905. [CrossRef]
17. Abdel-Halim, E.S.; Alanazi, H.H.; Al-Deyab, S.S. Utilization of hydroxypropyl carboxymethyl cellulose in synthesis of silver nanoparticles. *Int. J. Biol. Macromol.* **2015**, *75*, 467–473. [CrossRef] [PubMed]
18. Garza-Navarro, M.A.; Aguirre-Rosales, J.A.; Llanas-Vázquez, E.E.; Moreno-Cortez, I.E.; Torres-Castro, A.; González-González, V. Totally ecofriendly synthesis of silver nanoparticles from aqueous dissolutions of polysaccharides. *Int. J. Polym. Sci.* **2013**, 436021. [CrossRef]
19. Oldenburg, S.J. Silver Nanoparticles: Properties and Applications, Sigma–Aldrich. Available online: www.sigmaaldrich.com/technical--documents/articles/materials--science/nanomaterials/silver--nanoparticles.html.
20. Matuschek, E.; Brown, D.F.J.; Kahlmeter, G. Development of the EUCAST disk difusion antimicrobial susceptibility testing method and its implementation in routine microbiology laboratories. *Clin. Microbiol. Infect.* **2014**, *20*, o255–o266. [CrossRef] [PubMed]
21. Clinical and Laboratory Standards Institute (CLSI). *Performance Standards for Antimicrobial Susceptibility Testing;* Approved Standard, 25th Informational Supplement, CLSI document M100-S25; Clinical and Laboratory Standards Institute: Wayne, PA, USA, 2015.
22. Doroshenko, N.; Rimmer, S.; Hoskins, R.; Garg, P.; Swift, T.; Spencer, H.L.M.; Lord, R.M.; Katsikogianni, M.; Pownall, D.; MacNeil, S.; et al. Antibiotic functionalised polymers reduce bacterial biofilm and bioburden in a simulated infection of the cornea. *Biomater. Sci.* **2018**, *6*, 2101–2109. [CrossRef] [PubMed]
23. Karetsi, V.A.; Banti, C.N.; Kourkoumelis, N.; Papachristodoulou, C.; Stalikas, C.D.; Raptopoulou, C.P.; Psycharis, V.; Zoumpoulakis, P.; Mavromoustakos, T.; Sainis, I.; et al. An Efficient Disinfectant, Composite Material {SLS@[Zn3(CitH)2]} as Ingredient for Development of Sterilized and Non Infectious Contact Lens. *Antibiotics* **2019**, *8*, 213. [CrossRef] [PubMed]
24. Wiegand, I.; Hilpert, K.; Hancock, R.E.W. Agar and broth dilution methods to determine the minimal inhibitory concentration (MIC) of antimicrobial substances. *Nat. Protoc.* **2008**, *3*, 163–175. [CrossRef] [PubMed]
25. Zhu, B.; Zhu, S.; Li, J.; Hui, X.; Wang, G.-X. The developmental toxicity, bioaccumulation and distribution of oxidized single walled carbon nanotubes in Artemia salina. *Toxicol. Res.* **2018**, *7*, 897–906. [CrossRef]

26. Banti, C.N.; Hadjikakou, S.K. Evaluation of Toxicity with Brine Shrimp Assay. *Bio Protoc.* **2020**, *11*, e3895. [CrossRef]
27. Trompeta, A.F.; Preiss, I.; Ben-Ami, F.; Benayahu, Y.; Charitidis, C.A. Toxicity testing of MWCNTs to aquatic organisms. *RSC Adv.* **2019**, *9*, 36707–36716. [CrossRef]
28. da Silveira Carvalho, J.M.; de Morais Batista, A.H.; Nogueira, N.A.P.; Holanda, A.K.M.; de Sousa, J.R.; Zampieri, D.; Bezerra, M.J.B.; Barreto, F.S.; de Moraes, M.O.; Batista, A.A.; et al. A biphosphinic ruthenium complex with potent anti–bacterial and anti–cancer activity. *New J. Chem.* **2017**, *41*, 13085–13095. [CrossRef]
29. Živković, M.B.; Matić, I.Z.; Rodić, M.V.; Novaković, I.T.; Sladić, D.M.; Krstić, N.M. Synthesis, characterization and in vitro cytotoxic activities of new steroidal thiosemicarbazones and thiadiazolines. *RSC Adv.* **2016**, *6*, 34312–34333. [CrossRef]
30. Ketikidis, I.; Banti, C.N.; Kourkoumelis, N.; Tsiafoulis, C.G.; Papachristodoulou, C.; Kalampounias, A.G.; Hadjikakou, S.K. Conjugation of Penicillin–G with Silver(I) Ions Expands Its Antimicrobial Activity against Gram Negative Bacteria. *Antibiotics* **2020**, *9*, 25. [CrossRef]

International Journal of
Molecular Sciences

Article

Efficient Photodynamic Killing of Gram-Positive Bacteria by Synthetic Curcuminoids

Sung-Jen Hung [1,2]**, Yi-An Hong** [1,3]**, Kai-Yu Lin** [4]**, Yi-Wen Hua** [4]**, Chia-Jou Kuo** [4]**, Anren Hu** [1,3,*]**, Tzenge-Lien Shih** [4,*] **and Hao-Ping Chen** [1,5,6,*]

[1] Institute of Medical Sciences, Tzu Chi University, Hualien 97004, Taiwan;
    md.hong@msa.hinet.net (S.-J.H.); amyhung840809@gmail.com (Y.-A.H.)
[2] Department of Dermatology, Hualien Tzu Chi Hospital, Hualien 97002, Taiwan
[3] Department of Laboratory Medicine and Biotechnology, Tzu Chi University, Hualien 97004, Taiwan
[4] Department of Chemistry, Tamkang University, New Taipei City 25137, Taiwan;
    s930608eagle@gmail.com (K.-Y.L.); a0987719278@gmail.com (Y.-W.H.); b24631839@gmail.com (C.-J.K.)
[5] Department of Biochemistry, Tzu Chi University, Hualien 97004, Taiwan
[6] Integration Center of Traditional Chinese and Modern Medicine, Hualien Tzu Chi Hospital,
    Hualien 97002, Taiwan
*   Correspondence: anren@gms.tcu.edu.tw (A.H.); tlshih@mail.tku.edu.tw (T.-L.S.);
    hpchen@mail.tcu.edu.tw (H.-P.C.); Tel.: +886-3-856-5301 (ext. 2335) (A.H.); +886-2-8631-5024 (T.-L.S.);
    +886-3-856-5301 (ext. 2433) (H.-P.C.)

Received: 27 October 2020; Accepted: 22 November 2020; Published: 27 November 2020

Abstract: In our previous study, we have demonstrated that curcumin can efficiently kill the anaerobic bacterium *Propionibacterium acnes* by irradiation with low-dose blue light. The curcuminoids present in natural plant turmeric mainly include curcumin, demethoxycurcumin, and bisdemethoxycurcumin. However, only curcumin is commercially available. Eighteen different curcumin analogs, including demethoxycurcumin and bisdemethoxycurcumin, were synthesized in this study. Their antibacterial activity against Gram-positive aerobic bacteria *Staphylococcus aureus* and *Staphylococcus epidermidis* was investigated using the photodynamic inactivation method. Among the three compounds in turmeric, curcumin activity is the weakest, and bisdemethoxycurcumin possesses the strongest activity. However, two synthetic compounds, (1*E*,6*E*)-1,7-bis(5-methylthiophen-2-yl)hepta-1,6-diene-3,5-dione and (1*E*,6*E*)-1,7-di(thiophen-2-yl)hepta-1,6-diene-3,5-dione, possess the best antibacterial activity among all compounds examined in this study. Their chemical stability is also better than that of bisdemethoxycurcumin, and thus has potential for future clinical applications.

Keywords: bisdemethoxycurcumin; curcumin; curcuminoid; demethoxycurcumin; photodynamic inactivation; *Staphylococcus aureus*; *Staphylococcus epidermis*

---

1. Introduction

The emergence of drug-resistant bacteria has brought challenges to global public health and clinical treatments [1]. For all antibiotics currently used, a corresponding drug-resistant bacteria can be found [2]. The development of a new generation of antibiotics has become an increasingly important issue. However, progress in developing new antibiotics is dramatically slow [3]. More recently, antimicrobial photodynamic therapy (aPDT) appears to be a promising alternative approach and may become a new antimicrobial method [4]. Unlike traditional antibiotics, aPDT uses a photosensitizer or a nontoxic photoactivatable dye, visible light, and reactive oxygen to generate reactive oxygen species, like singlet oxygen or superoxide, to kill bacteria.

The antimicrobial activity of methylene blue, toluidine blue, rose bengal [5,6], indocyanine green [7], curcumin [8,9], and chlorin [10] induced by PDT has been reported previously. More recently, a synthetic compound, TTPy, has been proven to completely kill Gram-positive bacteria, namely *Staphylococcus aureus* and *Staphylococcus epidermidis,* under white light (60 mW/cm^2) for 15 min [11]. However, as reported by our group previously, curcumin, a natural cooking spice isolated from *Curcuma longa* L rhizome, could kill the anaerobic Gram-positive bacteria *Propionibacterium acnes,* entirely under the irradiation of blue light (3 mW/cm^2) for only 1 min [9]. Curcumin, therefore, appears to be an attractive aPDT agent.

Curcuminoids in natural plant turmeric include curcumin (compound **3**), demethoxycurcumin (compound **4**), and bisdemethoxycurcumin (compound **5**) [12,13]. Curcumin is the primary form among them. At present, neither demethoxycurcumin nor bisdemethoxycurcumin is commercially available. Therefore, in contrast to curcumin, the biological activities of demethoxycurcumin and bisdemethoxycurcumin are relatively unknown. To further explore and improve the aPDT properties of curcumin, demethoxycurcumin, bisdemethoxycurcumin, and fifteen curcumin analogs (compounds **6–19** in Figure 1) were synthesized. The aPDT activities of the aforementioned compounds against Gram-positive bacteria *S. aureus* and *S. epidermidis* were investigated in this study.

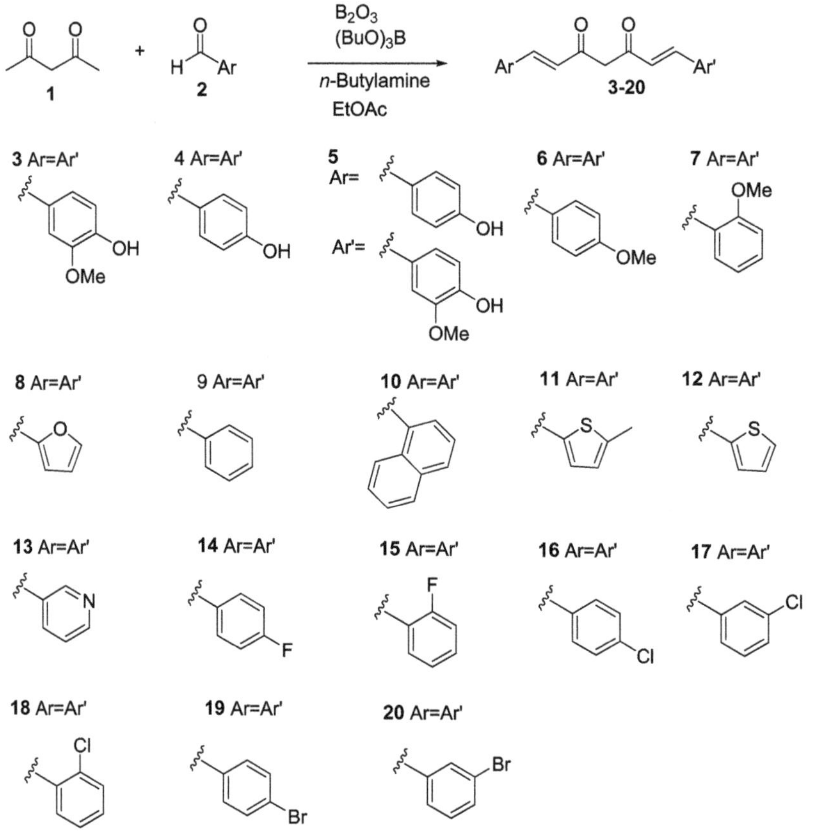

Figure 1. Chemical synthesis of curcuminoids **3–20**.

2. Results and Discussion

2.1. Chemical Synthesis of Compounds 3–20

Synthesis of symmetric curcuminoids **3**, **4**, and **6–20** followed Pabon's method [14] (Figure 1). All of the starting materials were commercially available and inexpensive. One equivalent of

2,4-pentanedione was treated with two equivalents of corresponding aldehydes using B_2O_3 and $(BuO)_3B$ as complexing agents (see experimental). In contrast, the asymmetric curcuminoid **5** was applied to the strategy mentioned above, except one equivalent of aldehyde (Ar or Ar′) was added first. Notably, the subsequent aldehyde was added slowly via a syringe pump to afford a better yield of **5**. The NMR spectra of synthetic compounds are included in the Supplementary Figures S1–S36.

2.2. Antimicrobial Activity of Compound **3**–**20** with Blue Light Irradiation

As shown in Figure 2, the antibacterial activities of compounds **3**–**20** against *S. epidermid* is were investigated. Compounds **4**, **5**, **8**, **11**, and **12** were the most effective among the eighteen compounds. The antibacterial activity of curcumin (compound **3**) was relatively weak, with a killing rate: 14.1%. In contrast to the previous report, the killing rate of curcumin against *P. acnes* is nearly 100% under similar experimental conditions. The possible reason for this difference is that *P. acnes* is an anaerobic bacterium, whereas *S. epidermidis* is aerobic. This result indicated that the antibacterial activity of demethoxycurcumin (compound **4**) and bisdemethoxycurcumin (compound **5**) was higher than that of curcumin (compound **3**), the primary isomer form in plant turmeric, under aerobic conditions.

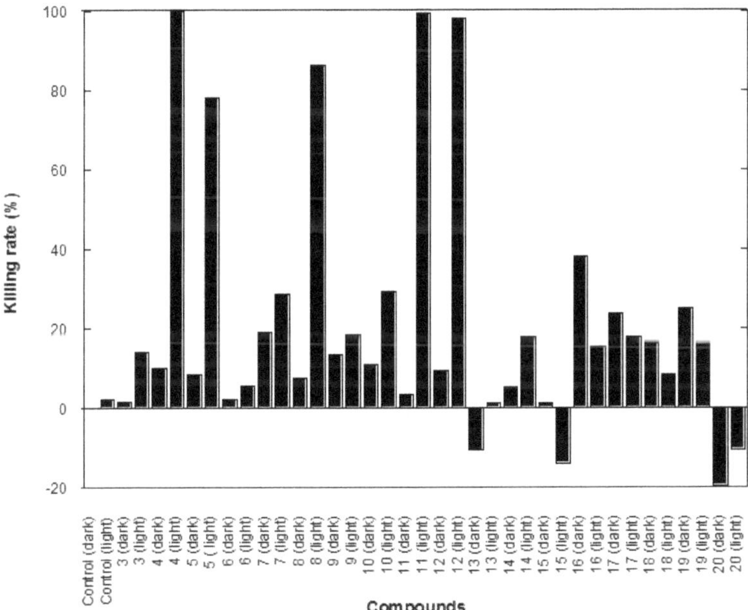

Figure 2. Bacterial killing activities of curcumin analogs on aerobic bacterium *Staphylococcus epidermidis*.

Our previous results showed that curcumin's photolytic products include vanillin, camphor, and acenaphthylene [9]. This result suggests that the formation of radicals is involved in this photolytic process. Generally, the antibacterial activity of compounds with halogen atom attached to the arene (compounds **14**–**20**) was low. Because the halogen atom is an electron-withdrawing group, this result implies that halogen's attachment on those curcumin analogs is not conducive to these compounds' photolysis. Previous studies have shown that curcumin binds effectively to the liposomal bilayer and locates preferentially in the hydrophobic acyl chain region [15]. Compounds **14**–**20** with halogen-substituted molecules should be much more hydrophobic than curcumin, altering the interactions with the bacterial lipid bilayer.

Different working concentrations of the compounds and bacterial strains were then used to compare the antibacterial activity of compounds **3**, **4**, **5**, **8**, **11**, and **12** (Table 1). Compounds **3**, **4**, and **5** are present in natural plant turmeric. It is interesting to note that synthetic compounds **8**, **11**, and **12** contain a hetero five-membered ring group. When the bacterial strain was switched to the other Gram-positive bacterium, *S. aureus*, the antibacterial activity of compounds **5** and **8** was significantly reduced. Furthermore, the concentration of the compounds **4**, **11**, and **12** was lowered to 0.5 ppm (Table 1). Thus, compounds **11** and **12** were the most effective among all the compounds tested in this study. The antibacterial activity of compound **11** on the Gram-negative bacterium *Escherichia coli* was also examined. The killing rate in the experimental and control groups was 18.1% and 17.1%, respectively, even when the working concentration of compound **11** was enhanced to 2 ppm. This result is in accordance with the previous report [11]. The synthetic compound TTPy can photodynamically kill Gram-positive bacteria *S. aureus* and *S. epidermidis*, but not the Gram-negative bacterium *E. coli*. All these results might come from the differences in cell envelop structures between Gram-positive and Gram-negative bacteria.

Table 1. The killing efficiency of compounds **3**, **4**, **5**, **8**, **11**, and **12** against *Staphylococcus aureus* and *S. epidermidis* with 1 min blue light irradiation.

Bacterial Strain	*S. aureus*	*S. epidermidis*	*S. epidermidis*
Working Concentration	1 ppm	1 ppm	0.5 ppm
Control (in dark)	N/A	N/A	N/A
Control (with BL irradiation)	2.9 ± 2.2	7.7 ± 10.0	7.6 ± 4.8
3 (in dark)	12.0 ± 12.0	2.6 ± 12.4	
3 (with BL irradiation)	18.6 ± 6.3	14.9 ± 1.3	
4 (in dark)	19.3 ± 13.6	−5.0 ± 19.6	5.9 ± 7.0
4 (with BL irradiation)	100 ± 0	98.5 ± 1.5	22.3 ± 3.2
5 (in dark)	17.1 ± 2.0	12.5 ± 4.4	
5 (with BL irradiation)	31.0 ± 2.5	71.1+ 9.8	
8 (in dark)	0.7 ± 6.2	4.3 ± 8.8	
8 (with BL irradiation)	27.1 ± 18.0	91.8 ± 7.3	
11 (in dark)	−2.0 ± 5.2	6.7 ± 4.1	26.0 ± 11.3
11 (with BL irradiation)	99.7 ± 0.3	99.8 ± 0.2	97.3 ± 0.7
12 (in dark)	13.5 ± 3.9	4.2 ± 4.7	7.9 ± 9.0
12 (with BL irradiation)	100 ± 0	99.7 ± 0.3	87.8 ± 12.2

All experiments were performed in triplicate. All data are expressed as the mean ± standard deviation. BL: blue light.

*2.3. SEM Observation of Microbial Membrane Disruption after the Treatment of Compound **11** and Irradiation with Blue Light*

Our previous results showed that curcumin could disrupt *P. acnes* cell membranes after irradiation with blue light under anaerobic conditions by SEM [9]. Neither *S. aureus* nor *S. epidermidis* could be efficiently killed by curcumin under aerobic conditions in this study (Table 1), even though a previous report indicated that curcumin inhibited the growth of multi-resistant *S. aureus* by irradiation with LED for as long as 20 min [16]. SEM also examined the compound **11**-treated and blue light-irradiated *S. epidermidis* under aerobic conditions in this study. As shown in Figure 3, the bacterial cell membrane integrity was disrupted, and cellular morphology was altered. While the blue light irradiation time increases from 1 min to 5 min, the cell membrane damage also significantly increases.

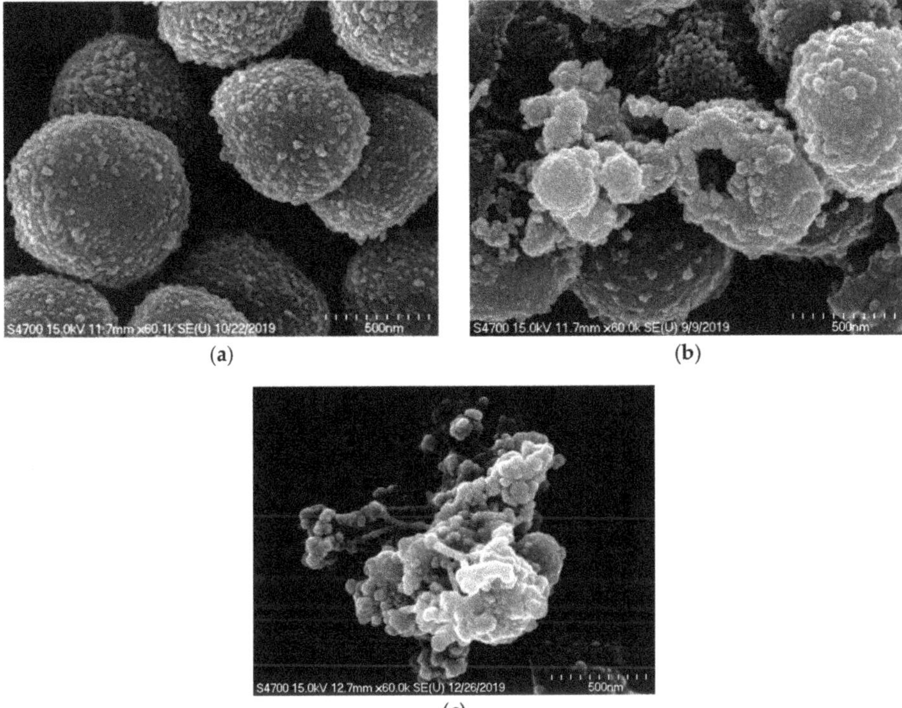

Figure 3. Scanning electron microscopy analysis of *Staphylococcus epidermidis* irradiated with blue light in the presence of 1 ppm compound **11**. (×60,000) The cell wall surface of *S. epidermidis* was severely damaged after the treatment. (**a**) Before irradiation with blue light, (**b**) Irradiation with blue light for 1 min, and (**c**) Irradiation with blue light for 5 min.

2.4. Chemical Stability of Compounds 4, 11, and 12

Curcumin easily undergoes autoxidation reactions in liquid at neutral-basic and alkaline pH [17]. The absorption spectra of compounds **4**, **11**, and **12** in the DMSO stock solution were recorded after storage at room temperature in the dark for 48 h. Their absorption spectra were recorded and shown in Figure 4. The maximum absorbance of compound **4** (λmax = 426 nm), **11** (λmax = 440 nm), and **12** (λmax = 426 nm) decreased 6.4%, 0.8%, and 1.3%, respectively. The color change of compounds **11** and **12** was not obvious. These results suggest that the chemical stability of compounds **11** and **12** is better than that of compound **4**. The NMR spectra of the degraded compound **4** were included in the Supplementary Figure S37.

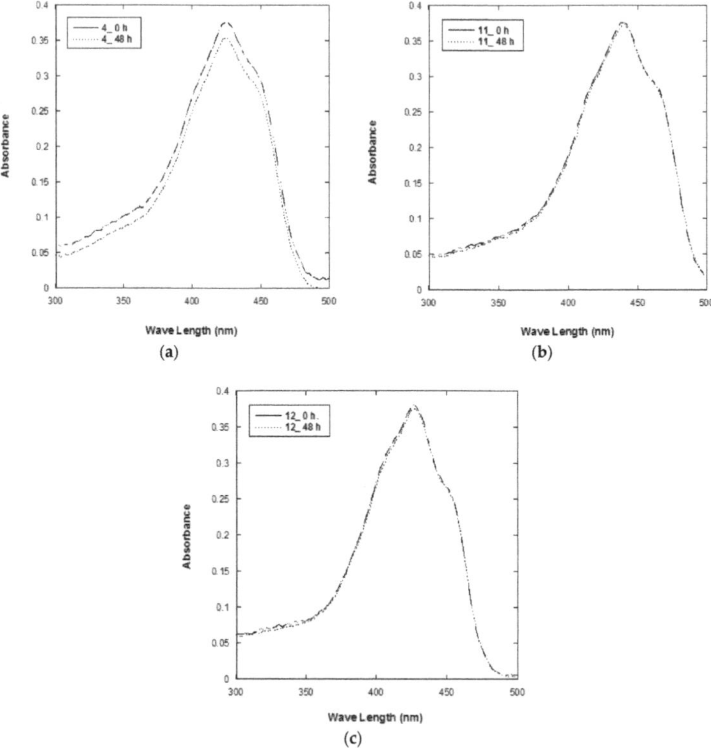

Figure 4. Absorption spectra of (**a**) compound **4** (20 ppm), (**b**) compound **11** (20 ppm), and (**c**) compound **12** (20 ppm), before and after storage in the dark for 48 h.

3. Materials and Methods

3.1. Synthesis of the Curcumin Analogs 3–20

All chemicals were purchased from Sigma-Aldrich (Shanghai, China) or Alfa-Aesar (Heysham, Lancashire, UK) companies and used without further purification. ^{1}H and ^{13}C NMR data were recorded on a Bruker 600 Ultrashield NMR spectrophotometer (Bruker, New Taipei City, Taiwan). The chemical shifts were reported in part per million (ppm) with the designated deuterium solvent relative to the residual solvent as internal standard (CDCl$_3$, ^{1}H: 7.26 ppm; ^{13}C: 77.0 ppm.; CD$_3$OD, ^{1}H: 4.78 ppm; ^{13}C: 49.15 ppm). Purification by flash column chromatography (SiliaFlash® P60, 40–63 μm 60 Å, SiliCycle® Inc., Quebec City, QC, Canada) was performed on 230–400 mesh SiO$_2$. The melting points were measured by a MP-2D apparatus (Fargo, New Taipei City, Taiwan) and not corrected. The mass data were obtained from JEOL JNS-700 (Akishima, Tokyo, Japan) by either EI or FAB and Bruker UltraFlex II for ESI (Bruker, New Taipei City, Taiwan).

3.1.1. General Procedure in Preparation of Compounds **3**, **4**, and **6–20**

A mixture of acetylacetone (1.00 equiv.) and B$_2$O$_3$ (0.50 equiv.) in EtOAc (0.250 M) was heated at 50 °C for 30 min, followed by the addition of aldehyde (2.00 equiv.) and (BuO)$_3$B (4.00 equiv.) in EtOAc (1.0 M), which was stirred at at 25 °C for 30 min before being added into the aforementioned solution. The resulting mixture was heated at 50 °C for 30 min, followed by the slow addition of *n*-butylamine (0.50 equiv.) in EtOAc (0.80 M), and then heated at 80 °C until reaction completion as indicated by TLC indication. Once the reaction was completed, HCl (1.0 N) was added and stirred for 30 min and then diluted with EtOAc and H_2O. The organic layer was separated, dried by $MgSO_4$, filtrated, and concentrated.

3.1.2. (1*E*,6*E*)-1,7-Bis(4-hydroxy-3-methoxyphenyl)hepta-1,6-diene-3,5-dione (**3**)

Vanillin (0.500 g, 3.29 mmol). Purification by flash column chromatography (EtOAc:Hexane = 1:3–1:1; EtOAc:Hexane = 1:2, R_f = 0.4) afforded **3** (0.2179 g, 0.266 mmol) as a yellow solid. Yield: 36%. Mp 182–186 °C ^{1}H NMR (600 MHz, CD_3OD): δ7.56(d, *J* = 15.7 Hz, 2H), 7.20 (s, 2H), 7.09 (d, *J* = 7.9 Hz, 2H), 6.80 (d, *J* = 8.1 Hz, 2H), 6.61 (d, *J* = 15.7 Hz, 2H), 3.90 (s, 6H). ^{13}C NMR (150 MHz, CD_3OD): δ185.0, 184.8, 161.2, 150.5, 149.5, 142.2, 142.0, 131.3, 128.7, 128.1, 124.2, 122.3, 122.1, 117.0, 116.7, 111.9, 56.6. HRMS (FAB) calculated for $C_{21}H_{20}O_6$ ([M]$^+$): 368.1260. Found: 368.1261.

3.1.3. (1*E*,6*E*)-1,7-Bis(4-hydroxyphenyl)hepta-1,6-diene-3,5-dione (**4**)

4-Hydroxybenzaldehyde (0.500 g, 4.10 mmol). Purification by flash column chromatography (EtOAc:Hexane = 1:3–1:1; EtOAc:Hexane = 1:3, R_f = 0.3) afforded **4** (0.218 g, 0.592 mmol) as a red solid. Yield: 42%. Mp 232–236 °C. ^{1}H NMR (600 MHz, CD_3OD): δ7.55 (d, *J* = 15.8 Hz, 2H), 7.50 (dd, *J* =7.8, 4H), 6.81 (d, *J* = 7.8 Hz, 4H), 6.57 (d, *J* = 15.8 Hz, 4H). ^{13}C NMR (150 MHz, CD_3OD): δ184.8, 161.1, 141.9, 131.2, 128.0, 122.0, 117.0. HRMS (FAB) calculated for $C_{19}H_{16}O_4$ ([M]$^+$): 308.1049. Found: 308.1049.

3.1.4. (1*E*,6*E*)-1-(4-hydroxy-3-methoxyphenyl)-7-(4-hydroxyphenyl)Hepta-1,6-diene-3,5-dione (**5**)

Followed the general procedure except the vanillin was used half equivalent relative to acetyl acetone. The resulting mixture was purified by flash column chromatography (EtOAc:Hexane = 1:2–1:1) to afford an intermediate as a yellow solid in 28% yield. This yellow solid (0.200 g, 0.850 mmol) was applied the general procedure and used the equivalent amount of 4-hydroxybenzaldehyde (0.207 g, 1.7 mmol). At the end of reaction time, purification by flash column chromatography (EtOAc:Hexane = 1:3–1:1; EtOAc:Hexane = 1:2, R_f = 0.3) afforded **5** (0.090 g, 0.266 mmol) as a red solid. Yield: 31%. Mp 172–174 °C. ^{1}H NMR (600 MHz, CD_3OD): δ7.53 (d, *J* = 15.8 Hz, 2H), 7.52 (d, *J* = 15.8 Hz, 2H), 7.44 (d, *J* = 8.4 Hz, 1H), 7.15 (s, 1H), 7.06 (d, *J* = 8.2 Hz, 1H), 6.80–6.78 (m, 3H), 6.57 (d, *J* = 15.8 Hz, 1H),6.54 (d, J = 15.8 Hz, 1H), 4.86 (s, 2H), 3.87 (s, 3H). ^{13}C NMR (150 MHz, CD_3OD): δ185.0, 184.8, 161.2, 150.5, 149.5, 142.2, 142.0, 131.3, 128.7, 124.2, 122.4, 122.1, 117.0, 116.7, 111.9, 56.6. HRMS (ESI) calculated for $C_{20}H_{19}O_5$ ([M+H]$^+$): 339.1232. Found: 339.1227.

3.1.5. (1*E*,6*E*)-1,7-Bis(4-methoxyphenyl)hepta-1,6-diene-3,5-dione (**6**)

p-Anisaldehyde (0.447 g, 3.282 mmol). Purification by flash column chromatography (EtOAc:Hexane = 1:5–1:1; EtOAc:Hexane = 1:3, R_f = 0.3) afforded **6** (0.379 g, 1.128 mmol) as a red solid. Yield: 69%. Mp 163–165 °C. ^{1}H NMR (600 MHz, $CDCl_3$): δ7.62 (d, *J* = 15.8 Hz, 2H), 7.51 (d, *J* = 8.8 Hz, 2H), 6.91 (d, *J* = 8.8 Hz, 2H), 6.50 (d, *J* = 15.8 Hz, 2H), 5.79 (s, 1H), (s, 1H), 3.84 (s, 6H). ^{13}C NMR (150 MHz, $CDCl_3$): δ183.3, 161.3, 140.1, 129.7, 127.8, 121.8, 114.4, 101.3, 55.4. HRMS (ESI) calculated for $C_{21}H_{21}O_4$ ([M+H]$^+$): 337.1440. Found: 337.1434.

3.1.6. (1*E*,6*E*)-1,7-Bis(2-methoxyphenyl)hepta-1,6-diene-3,5-dione (**7**)

2-Methoxybenzaldehyde (2.723 g, 20.000 mmol). Purification by flash column chromatography (EtOAc:Hexane = 1:3–1:1; EtOAc:Hexane = 1:2, R_f = 0.3) afforded **7** (1.033 g, 3.071 mmol) as a yellow solid. Yield: 31%. Mp 114–116 °C. ^{1}H NMR (600 MHz, $CDCl_3$): δ7.99 (d, *J* = 16.0 Hz, 2H), 7.55 (d, *J* = 7.6 Hz, 2H), 7.34 (t, *J* = 7.8 Hz, 2H), 6.97 (t, *J* = 7.5 Hz, 2H), 6.92 (d, *J* = 8.2 Hz, 2H), 6.72 (d, *J* = 16.0 Hz, 2H), 5.88 (s, 1H), 3.90 (s, 6H). ^{13}C NMR (150 MHz, $CDCl_3$): δ183.8, 158.4, 135.7, 131.2, 128.6, 124.8, 124.1, 120.7, 111.2, 101.5, 55.5. HRMS (FAB) calculated for $C_{21}H_{20}O_4$ ([M]$^+$): 336.1362. Found: 336.1359.

3.1.7. (1*E*,6*E*)-1,7-Di(furan-2-yl)hepta-1,6-diene-3,5-dione (**8**)

2-Furaldehyde (0.510 g, 5.308 mmol). Purification by flash column chromatography (EtOAc:Hexane = 1:10–1:5; EtOAc:Hexane = 1:10, R_f = 0.4) afforded **8** (0.0691 g, 0.270 mmol) as an orange-yellow solid. Yield: 10%. Mp 128–129 °C. ^{1}H NMR (600 MHz, CDCl$_3$): δ7.48 (s, 2H), 7.40 (d, *J* = 15.5 Hz, 2H), 6.60 (d, *J* = 3.4 Hz, 2H), 6.51 (d, *J* = 15.5 Hz, 2H), 6.40 (dd, *J* = 2.9, 1.4 Hz, 2H), 5.74 (s, 1H). ^{13}C NMR (150 MHz, CDCl$_3$): δ182.7, 151.7, 144.7, 126.8, 121.8, 114.8, 112.5, 102.3. HRMS (FAB) calculated for C$_{15}$H$_{12}$O$_4$ ([M]$^+$): 256.0736. Found: 256.0736.

3.1.8. (1*E*,6*E*)-1,7-Diphenylhepta-1,6-diene-3,5-dione (**9**)

Benzaldehyde (2.122 g, 20.000 mmol). Purification by flash column chromatography (EtOAc:Hexane = 1:3–1:1; EtOAc:Hexane = 1:2, R_f = 0.8) afforded **9** (2.016 g, 7.300 mmol) as a yellow solid. Yield: 73%. Mp 154–155 °C. ^{1}H NMR (600 MHz, CDCl$_3$): δ7.67 (d, *J* = 15.9 Hz, 2H), 7.56 (d, *J* = 6.6 Hz, 4H), 7.42–7.36 (m, 6H). 6.54 (d, *J* = 15.8 Hz, 2H), 5.86 (s, 1H). ^{13}C NMR (150 MHz, CDCl$_3$): δ183.3, 140.6, 135.0, 130.1, 128.9, 128.1, 124.1, 101.8. HRMS (FAB) calculated for C$_{19}$H$_{16}$O$_2$ ([M]$^+$): 276.1150. Found: 276.1150.

3.1.9. (1*E*,6*E*)-1,7-Di(naphthalen-1-yl)hepta-1,6-diene-3,5-dione (**10**)

1-Naphthaldehyde (0.500 g, 3.201 mmol). Purification by flash column chromatography (EtOAc:Hexane = 1:10–1:5; EtOAc:Hexane = 1:10, R_f = 0.5) afforded **10** (0.156 g, 0.4115 mmol) as a yellow solid. Yield: 26%. Mp 177–180 °C. ^{1}H NMR (600 MHz, CDCl$_3$): δ8.55 (d, *J* = 15.6 Hz, 2H), 8.27 (d, *J* = 8.4 Hz, 2H), 7.90 (t, *J* = 8.2 Hz, 4H), 7.82 (d, *J* = 7.2 Hz, 2H), 7.60 (td, *J* = 8.2, 1.1 Hz, 2H), 7.55 (td, *J* = 8.0, 1.0 Hz, 2H), 7.51 (t, *J* = 7.6 Hz, 2H), 6.76 (d, *J* = 15.6 Hz, 2H), 5.95 (s, 1H). ^{13}C NMR (150 MHz, CDCl$_3$):δ183.3, 137.5, 133.8, 132.4, 131.6, 130.5, 128.7, 126.9, 126.6, 126.3, 125.5, 124.9, 123.5, 102.2. HRMS (FAB) calculated for C$_{27}$H$_{20}$O$_2$ ([M]$^+$): 376.1463. Found: 376.1466.

3.1.10. (1*E*,6*E*)-1,7-Bis(5-methylthiophen-2-yl)hepta-1,6-diene-3,5-dione (**11**)

5-Methylthiophene-2-carboxaldehyde (2.524 g, 20.003 mmol). Purification by flash column chromatography (EtOAc:Hexane = 1:4–1:1; EtOAc:Hexane = 1:3, R_f = 0.6) afforded **11** (1.138 g, 3.601 mmol) as a brown solid. Yield: 36%. Mp 140–141 °C. ^{1}H NMR (600 MHz, CDCl$_3$): δ7.67 (d, *J* = 15.3 Hz, 2H), 7.05 (d, *J* = 3.3 Hz, 2H), 6.71 (d, *J* = 3.2 Hz, 2H), 6.26 (d, *J* = 15.4 Hz, 2H), 5.68 (s, 1H), 2.50 (s, 6H). ^{13}C NMR (150 MHz, CDCl$_3$): δ182.7, 144.0, 138.6, 133.3, 131.5, 126.7, 121.7, 101.4, 15.8. HRMS (FAB) calculated for C$_{17}$H$_{16}$O$_2$S$_2$ ([M]$^+$): 316.0592. Found: 316.0593.

3.1.11. (1*E*,6*E*)-1,7-Di(thiophen-2-yl)hepta-1,6-diene-3,5-dione (**12**)

2-Thiophenecarboxaldehyde (1.116 g, 9.951 mmol). Purification by flash column chromatography (CH$_2$Cl$_2$:Hexane = 3:1–20:1; CH$_2$Cl$_2$:Hexane = 1:1, R_f = 0.5) afforded **12** (0.317 g, 1.101 mmol) as a brown solid. Yield: 22%. Mp 195–197 °C. ^{1}H NMR (600 MHz, CDCl$_3$): δ7.75 (d, *J* = 15.4 Hz, 2H), 7.38 (d, *J* = 5.0 Hz, 2H), 7.26 (d, *J* = 4.2 Hz, 2H), 7.06 (dd, *J* = 5.0, 3.6 Hz, 2H), 6.41 (d, *J* = 15.4 Hz, 2H), 5.74 (s, 1H). ^{13}C NMR (150 MHz, CDCl$_3$): δ182.7, 140.5, 133.1, 130.9, 128.4, 123.0, 101.8. HRMS (FAB) calculated for C$_{15}$H$_{12}$O$_2$S$_2$ ([M]$^+$): 288.0279. Found: 288.0279.

3.1.12. (1*E*,6*E*)-1,7-Di(pyridin-3-yl)hepta-1,6-diene-3,5-dione (**13**)

3-Pyridinecarboxaldehyde (1.000 g, 9.340 mmol). Purification by flash column chromatography (EtOAc:Hexane = 1:5–1:1; EtOAc:Hexane = 1:5, R_f = 0.5) afforded **13** (0.602 g, 2.35 mmol) as a brown solid. Yield: 47%. Mp 180–181 °C. ^{1}H NMR (600 MHz, CDCl$_3$): δ8.79 (s, 2H), 8.60 (d, *J* = 4.1 Hz, 2H), 7.86 (d, *J* = 7.8 Hz, 2H), 7.66 (d, *J* = 15.9 Hz, 2H), 7.34 (dd, *J* = 7.7, 4.9 Hz, 2H), 6.70 (d, *J* = 15.9 Hz, 2H), 5.89 (s, 1H). ^{13}C NMR (150 MHz, CDCl$_3$): δ182.8, 150.8, 149.7, 137.2, 134.3, 130.7, 125.8, 123.8, 102.2. HERMS (FAB) calculated for C$_{15}$H$_{12}$O$_4$ ([M]$^+$): 256.0736. Found: 256.0736.

3.1.13. (1*E*,6*E*)-1,7-Bis(4-fluorophenyl)hepta-1,6-diene-3,5-dione (**14**)

4-Fluorobenzaldehyde (1.240 g, 9.991 mmol). Purification by flash column chromatography (EtOAc:Hexane = 1:3–1:1; EtOAc:Hexane = 1:3, R_f = 0.7) afforded **14** (0.312 g, 0.998 mmol) as a pale-yellow solid. Yield: 20%. Mp 172–173 °C. ^{1}H NMR (600 MHz, CDCl$_3$): δ7.63 (d, J = 15.8 Hz, 2H), 7.55 (d, J = 5.5 Hz, 2H), 7.54 (d, J = 5.5 Hz, 2H), 7.09 (d, J = 8.5 Hz, 2H), 7.06 (d, J = 11.5 Hz, 2H), 6.54 (d, J = 15.7 Hz, 2H), 5.81 (s,1H). ^{13}C NMR (150 MHz, CDCl$_3$): δ183.1, 163.8 ($^1J_{\text{C-F}}$ = 249.7 Hz), 139.4, 131.2. 129.9 ($^3J_{\text{C-F}}$ = 8.3 Hz), 123.7, 116.1 ($^2J_{\text{C-F}}$ = 21.8 Hz), 101.8. HERMS (FAB) calculated for C$_{19}$H$_{14}$F$_2$O$_2$ ([M]$^+$): 312.0962. Found: 312.0963.

3.1.14. (1*E*,6*E*)-1,7-Bis(2-fluorophenyl)hepta-1,6-diene-3,5-dione (**15**)

2-Fluorobenzaldehyde (0.408 g, 3.290 mmol). Purification by flash column chromatography (EtOAc:Hexane = 1:10–1:1; EtOAc:Hexane = 1:2, R_f = 0.6) afforded **15** (0.323 g, 1.04 mmol) as a yellow solid. Yield: 63%. Mp 100–102 °C. ^{1}H NMR (600 MHz, CDCl$_3$): δ7.78 (d, J = 16.1 Hz, 2H), 7.57 (td, J = 7.6, 1.6 Hz, 2H), 7.36–7.33 (m, 2H), 7.18 (t, J = 7.7 Hz, 2H), 7.11 (dd, J = 9.0, 8.3 Hz, 2H), 6.76 (d, J = 16.1 Hz, 2H), 5.90 (s, 1H). ^{13}C NMR (150 MHz, CDCl$_3$): δ183.3, 161.5 ($^1J_{\text{C-F}}$ = 252.0 Hz), 133.4, 131.4 ($^3J_{\text{C-F}}$ = 7.5 Hz), 129.2, 126.6 ($^3J_{\text{C-F}}$ = 6.0 Hz), 124.4 ($^4J_{\text{C-F}}$ = 3.0 Hz), 116.2 ($^2J_{\text{C-F}}$ = 11.0 Hz), 116.2 ($^2J_{\text{C-F}}$ = 21.0 Hz). HRMS (ESI) calculated for C$_{19}$H$_{15}$F$_2$O$_2$ ([M+H]$^+$): 313.1040. Found: 313.1038.

3.1.15. (1*E*,6*E*)-1,7-Bis(4-chlorophenyl)hepta-1,6-diene-3,5-dione (**16**)

4-Chlorobenzaldehyde (2.811 g, 20.000 mmol). Purification by flash column chromatography (EtOAc:Hexane = 1:4–1:1; EtOAc:Hexane = 1:4, R_f = 0.6) afforded **16** (1.800 g, 5.21 mmol) as a yellow solid. Yield: 52%. Mp 165–166 °C. ^{1}H NMR (600 MHz, CDCl$_3$): δ7.61 (d, J = 15.8 Hz, 2H), 7.49 (d, J = 8.4 Hz, 4H), 7.37 (d, J = 8.4 Hz, 4H), 6.59 (d, J = 15.8 Hz, 2H), 5.83 (s, 1H). ^{13}C NMR (150 MHz, CDCl$_3$): δ183.0, 129.3, 136.0, 133.4, 129.3, 129.2, 124.5, 102.2. HRMS (FAB) calculated for C$_{19}$H$_{14}$Cl$_2$O$_2$ 344.0317. Found: 344.0317.

3.1.16. (1*E*,6*E*)-1,7-Bis(3-chlorophenyl)hepta-1,6-diene-3,5-dione (**17**)

3-Chlorobenzaldehyde (2.811 g, 20.000 mmol). Purification by flash column chromatography (EtOAc:Hexane = 1:5–1:1; EtOAc:Hexane = 1:1, R_f = 0.6) afforded **17** (1.417 g, 4.107 mmol) as an amorphous yellow solid. Yield: 41%. Mp 153–154 °C. ^{1}H NMR (600 MHz, CDCl$_3$): δ7.57 (d, J = 15.8 Hz, 2H), 7.52 (s, 2H), 7.39 (d, J = 7.0 Hz, 2H), 7.35–7.25 (m, 4H), 6.60 (d, J = 15.8 Hz, 2H), 5.83 (s, 1H). ^{13}C NMR (150 MHz, CDCl$_3$): δ182.9, 139.1, 136.7, 134.9, 130.1, 129.9, 127.6, 126.4, 125.2, 102.3. HRMS (ESI) calculated for C$_{19}$H$_{15}$Cl$_2$O$_2$ ([M+H]$^+$): 345.0449. Found: 345.0448.

3.1.17. (1*E*,6*E*)-1,7-Bis(2-chlorophenyl)hepta-1,6-diene-3,5-dione (**18**)

2-Chlorobenzaldehyde (3.390 g, 24.116 mmol). Purification by flash column chromatography (EtOAc:Hexane = 1:4–1:1; EtOAc:Hexane = 1:4, R_f = 0.6) afforded **18** (0.757 g, 2.194 mmol) as a yellow solid. Yield: 18%. Mp 147–148 °C. ^{1}H NMR (600 MHz, CDCl$_3$): δ8.06 (d, J = 15.8 Hz), 7.67–7.64 (m, 2H), 7.43–7.41 (m, 2H), 7.32–7.27 (m, 4H), 6.62 (d, J = 15.8 Hz, 2H), 5.91 (s, 1H). ^{13}C NMR (150 MHz, CDCl$_3$): δ183.1, 136.5, 135.1, 133.1, 130.8, 130.3, 127.5, 127.0, 126.5, 101.7. HRMS (ESI) calculated for C$_{19}$H$_{15}$Cl$_2$O$_2$ ([M+H]$^+$): 345.0449. Found: 345.0444.

3.1.18. (1*E*,6*E*)-1,7-Bis(4-bromophenyl)hepta-1,6-diene-3,5-dione (**19**)

4-Bromobenzaldehyde (0.609 g, 3.292 mmol). Purification by flash column chromatography (EtOAc:Hexane = 1:10–1:1; EtOAc:Hexane = 1:2, R_f = 0.6) afforded **19** (0.429 g, 0.993 mmol) as a yellow solid. Yield: 60%. Mp 233–235 °C. ^{1}H NMR (600 MHz, CDCl$_3$): δ7.60 (d, J = 15.8 Hz, 2H), 7.53 (d, J = 8.4 Hz, 4H), 7.42 (d, J = 8.4 Hz, 4H), 6.61 (d, J = 15.8 Hz, 2H), 5.83 (s, 1H). ^{13}C NMR (150 MHz, CDCl$_3$): δ183.0, 139.4, 133.9, 132.2, 129.5, 124.6, 124.4, 102.1. HRMS (ESI) calculated for C$_{19}$H$_{15}$Br$_2$O$_2$ ([M+H]$^+$): 432.9439. Found: 432.9434.

3.1.19. (1*E*,6*E*)-1,7-Bis(3-bromophenyl)hepta-1,6-diene-3,5-dione (**20**)

3-Bromobenzaldehyde (0.500 g, 2.702 mmol). Purification by flash column chromatography (EtOAc:Hexane = 1:10–1:3; EtOAc:Hexane = 1:10, R_f = 0.4) afforded **20** (0.429 g, 0.993 mmol) as a yellow solid. Yield: 74%. Mp 152–154 °C. ^{1}H NMR (600 MHz, CDCl$_3$): δ7.71 (t, *J* = 1.7 Hz, 2H), 7.58 (d, *J* = 15.8 Hz, 2H), 7.51–7.49 (m, 2H), 7.47 (d, *J* = 7.9 Hz, 2H), 7.28 (d, *J* = 7.9 Hz, 2H), 6.21 (d, *J* = 16.2 Hz, 2H), 5.84 (s, 1H). ^{13}C NMR (150 MHz, CDCl$_3$): δ182.9, 139.1, 137.1, 132.9, 130.6, 130.4, 126.9, 125.3, 123.1, 103.3. HRMS (ESI) calculated for C$_{19}$H$_{15}$Br$_2$O$_2$ ([M+H]$^+$): 432.9434. Found: 432.9437.

3.2. Photodynamic Antibacterial Studies

The photo-irradiation system for the microbial viability experiments was reported previously [9]. The blue light intensity was 3.0 mW/cm^2 using a DC 5V power supply. The LED (Vetalux Company, Tainan, Taiwan) emission spectra were from 410 to 510 nm with λmax = 462 nm. *S. epidermidis* TCU-1 BCRC 81267 and *S. aureus* subsp. aureus TCU-2 BCRC 81268 were obtained from the Bioresource Collection and Research Center, Hsinchu, Taiwan. *Escherichia coli* was provided by Professor Kai-Chih Chang (Department of Laboratory Medicine and Biotechnology, Tzu Chi University, Taiwan). All bacterial strains were cultured in LB medium (BD Biosciences, San Jose, CA, USA) at 37 °C until OD$_{600}$ reached 1.0. The number of bacteria was about 10^9 CFU/mL.

Curcumin and its analogs were dissolved in 100% DMSO (Sigma-Aldrich, Shanghai, China), and the concentration of this stock was 2000 ppm. These DMSO stocks were diluted with LB medium. A total of 2 mL bacterial cultures was treated with 0.5 or 1 ppm of curcumin and its synthetic derivatives, and irradiated with 3.0 mW/cm^2 of blue light for 1 min (equivalent to radiant exposure of 0.18 J/cm^2). The cultures were then serially diluted before streaking and spreading on LB agar plates. After incubation at 37 °C overnight, the microbial colonies were counted, and the killing ratio was calculated as follows:

$$\text{Killing ratio } (\%) = \left\{ 1 - \frac{T_{(\text{CFU/mL})}}{C_{(\text{CFU/mL})}} \right\} \times 100\% \tag{1}$$

where T is the colony number of the curcumin and its synthetic derivatives-treated group, and C is the colony number of the control group (DMSO only) without light irradiation.

3.3. Scanning Electron Microscope (SEM) Observation of Microbial Membrane Disruption

After the treatment of compound **11** and blue light irradiation, the surface morphological changes in *S. epidermidis* cells were examined using Hitachi S-4700 SEM (Hitachi, Tokyo, Japan). The preparation for SEM samples was described previously [9]. A total of 2 mL 10^9 CFU/mL bacterial culture was treated with 1 ppm compound **11** and irradiated with blue light for 1 or 5 min.

3.4. Chemical Stability of Compounds *4*, **11**, and **12**

The 20 ppm DMSO solutions of compounds **4**, **11**, and **12** were prepared and stored in the dark at room temperature for 48 h. Before and after storage, the UV-visible spectra of compounds **1**, **8**, and **9** were recorded in the wavelength range of 220–750 nm.

3.5. Statistical Analysis

The experiments were performed in triplicate, and the data are expressed as mean ± standard deviation of three individual experiments. The data were assessed by analysis of variance (ANOVA) using SPSS Statistics (IBM, Armonk, NY, USA). $p < 0.05$ was considered significant.

4. Conclusions

The antibacterial activity of eighteen curcumin analogs against Gram-positive aerobic bacteria *S. aureus* and *S. epidermidis* was investigated by the photodynamic inactivation method. The antibacterial activity of all analogs containing halogen atom (compounds **14** to **20**) was low. The reason for this is still not clear. Two compounds, (1*E*,6*E*)-1,7-bis(5-methylthiophen-2-yl)hepta-1,6-diene-3,5-dione (compound **11**) and (1*E*,6*E*)-1,7-di(thiophen-2-yl)hepta-1,6-diene-3,5-dione (compound **12**), had the strongest antibacterial activity. Their chemical stability was also better than that of natural curcuminoids. Because natural curcuminoids are easily oxidized in solution, this feature makes these two compounds potentially useful for future clinical applications.

Supplementary Materials: The following are available online at http://www.mdpi.com/1422-0067/21/23/9024/s1.

Author Contributions: Synthesis of curcumin and its analogs, K.-Y.L., Y.-W.H. and C.-J.K.; antimicrobial activity assay, SEM, and stability test for curcumin analogs, S.-J.H. and Y.-A.H.; data curation, A.H., T.-L.S. and H.-P.C.; conceptualization and design of the study, H.-P.C.; writing—original draft, H.-P.C. and T.-L.S.; writing—review and editing, H.-P.C.; supervision, A.H., T.-L.S. and H.-P.C. All authors have read and agreed to the published version of the manuscript.

Funding: This research was funded by the Ministry of Science and Technology, Taiwan, R.O.C., grant number MOST 108-2113-M320-001, and Hualien Tzu Chi Hospital, grant number TCRD 109-77, to A.H.

Acknowledgments: We thank the Electron Microscopy Laboratory at Tzu Chi University for technical assistance with the SEM.

Conflicts of Interest: The authors declare no conflict of interest.

Abbreviations

aPDT	antimicrobial photodynamic therapy
BL	Blue light
SEM	Scanning Electron Microscope

References

1. Vivas, R.; Barbosa, A.A.T.; Dolabela, S.S.; Jain, S. Multidrug-Resistant Bacteria and Alternative Methods to Control Them: An Overview. *Microb. Drug Resist.* **2019**, *25*, 890–908. [CrossRef] [PubMed]

2. Medina, E.; Pieper, D.H. Tackling Threats and Future Problems of Multidrug-Resistant Bacteria. *Fungal Physiol. Immunopathog.* **2016**, *398*, 3–33. [CrossRef]

3. Hesterkamp, T. Antibiotics Clinical Development and Pipeline. *Curr. Top. Microbiol.* **2016**, *398*, 447–474. [CrossRef]

4. Nakonechny, F.; Firer, M.A.; Nitzan, Y.; Nisnevitch, M. Intracellular Antimicrobial Photodynamic Therapy: A Novel Technique for Efficient Eradication of Pathogenic Bacteria. *Photochem. Photobiol.* **2010**, *86*, 1350–1355. [CrossRef] [PubMed]

5. Tegos, G.P.; Hamblin, M.R. Phenothiazinium Antimicrobial Photosensitizers Are Substrates of Bacterial Multidrug Resistance Pumps. *Antimicrob. Agents Chemother.* **2006**, *50*, 196–203. [CrossRef] [PubMed]

6. Cieplik, F.; Etabenski, L.; Ebuchalla, W.; Emaisch, T. Antimicrobial photodynamic therapy for inactivation of biofilms formed by oral key pathogens. *Front. Microbiol.* **2014**, *5*, 405. [CrossRef] [PubMed]

7. Wang, X.-H.; Peng, H.-S.; Yang, W.; Ren, Z.-D.; Liu, X.-M.; Liu, Y.-A. Indocyanine green-platinum porphyrins integrated conjugated polymer hybrid nanoparticles for near-infrared-triggered photothermal and two-photon photodynamic therapy. *J. Mater. Chem. B* **2017**, *5*, 1856–1862. [CrossRef] [PubMed]

8. Araújo, N.C.; Fontana, C.R.; Gerbi, M.E.M.; Bagnato, V.S. Overall-Mouth Disinfection by Photodynamic Therapy Using Curcumin. *Photomed. Laser Surg.* **2012**, *30*, 96–101. [CrossRef] [PubMed]

9. Yang, M.-Y.; Chang, K.-C.; Chen, L.-Y.; Hu, A. Low-dose blue light irradiation enhances the antimicrobial activities of curcumin against Propionibacterium acnes. *J. Photochem. Photobiol. B Biol.* **2018**, *189*, 21–28. [CrossRef] [PubMed]

10. Embleton, M.L.; Nair, S.P.; Cookson, B.D.; Wilson, M. Selective lethal photosensitization of methicillin-resistant Staphylococcus aureus using an IgG-tin (IV) chlorin e6 conjugate. *J. Antimicrob. Chemother.* **2002**, *50*, 857–864. [CrossRef] [PubMed]

11. Kang, M.; Zhou, C.; Wu, S.; Yu, B.; Zhang, Z.; Song, N.; Lee, M.M.S.; Xu, W.; Xu, F.; Wangb, D.; et al. Evaluation of Structure–Function Relationships of Aggregation-Induced Emission Luminogens for Simultaneous Dual Applications of Specific Discrimination and Efficient Photodynamic Killing of Gram-Positive Bacteria. *J. Am. Chem. Soc.* **2019**, *141*, 16781–16789. [CrossRef] [PubMed]

12. Zhang, J.; Jinnai, S.; Ikeda, R.; Wada, M.; Hayashida, S.; Nakashima, K. A Simple HPLC-fluorescence Method for Quantitation of Curcuminoids and Its Application to Turmeric Products. *Anal. Sci.* **2009**, *25*, 385–388. [CrossRef] [PubMed]

13. Esatbeyoglu, T.; Huebbe, P.; Ernst, I.M.A.; Chin, D.; Wagner, A.E.; Rimbach, G. Curcumin-From Molecule to Biological Function. *Angew. Chem. Int. Ed.* **2012**, *51*, 5308–5332. [CrossRef] [PubMed]

14. Pabon, H.J.J. A synthesis of curcumin and related compounds. *Recl. Trav. Chim. Pays-bas* **1964**, *83*, 379–386. [CrossRef]

15. Karewicz, A.; Bielska, D.; Gzyl-Malcher, B.; Kepczynski, M.; Lach, R.; Nowakowska, M. Interaction of curcumin with lipid monolayers and liposomal bilayers. *Colloids Surf. B Biointerfaces* **2011**, *88*, 231–239. [CrossRef] [PubMed]

16. Freitas, M.A.; Pereira, A.H.; Pinto, J.G.; Casas, A.; Ferreira-Strixino, J. Bacterial viability after antimicrobial photodynamic therapy with curcumin on multiresistant Staphylococcus aureus. *Futur. Microbiol.* **2019**, *14*, 739–748. [CrossRef] [PubMed]

17. Wang, Y.-J.; Pan, M.-H.; Cheng, A.-L.; Lin, L.-I.; Ho, Y.-S.; Hsieh, C.-Y.; Lin, J.-K. Stability of curcumin in buffer solutions and characterization of its degradation products. *J. Pharm. Biomed. Anal.* **1997**, *15*, 1867–1876. [CrossRef]

Publisher's Note: MDPI stays neutral with regard to jurisdictional claims in published maps and institutional affiliations.

International Journal of
Molecular Sciences

Article

In Vitro Evaluation of Gentamicin or Vancomycin Containing Bone Graft Substitute in the Prevention of Orthopedic Implant-Related Infections

Alessandro Bidossi [1,†], Marta Bottagisio [1,*,†], Nicola Logoluso [2] and Elena De Vecchi [1]

[1]   IRCCS Istituto Ortopedico Galeazzi, Laboratory of Clinical Chemistry and Microbiology, 20161 Milan, Italy;
     alessandro.bidossi@grupposandonato.it (A.B.); elena.devecchi@grupposandonato.it (E.D.V.)

[2]   IRCCS Istituto Ortopedico Galeazzi, Department of Reconstructive Surgery of Osteo-Articular Infections
     C.R.I.O. Unit, 20161 Milan, Italy; nicola.logoluso@grupposandonato.it

*   Correspondence: marta.bottagisio@grupposandonato.it; Tel.: +39-02-6621-4886

†   These authors equally contributed.

Received: 17 November 2020; Accepted: 2 December 2020; Published: 4 December 2020

Abstract: Antibiotic-loaded bone graft substitutes are attractive clinical options and have been used for years either for prophylaxis or therapy for periprosthetic and fracture-related infections. Calcium sulfate and hydroxyapatite can be combined in an injectable and moldable bone graft substitute that provides dead space management with local release of high concentrations of antibiotics in a one-stage approach. With the aim to test preventive strategies against bone infections, a commercial hydroxyapatite/calcium sulfate bone graft substitute containing either gentamicin or vancomycin was tested against *Staphylococcus aureus*, *Staphylococcus epidermidis* and *Pseudomonas aeruginosa*, harboring different resistance determinants. The prevention of bacterial colonization and biofilm development by selected microorganisms was investigated along with the capability of the eluted antibiotics to select for antibiotic resistance. The addition of antibiotics drastically affected the ability of the selected strains to adhere to the tested compound. Furthermore, both the antibiotics eluted by the bone graft substitutes were able to negatively impair the biofilm maturation of all the staphylococcal strains. As expected, *P. aeruginosa* was significantly affected only by the gentamicin containing bone graft substitutes. Finally, the prolonged exposure to antibiotic-containing sulfate/hydroxyapatite discs did not lead to any stable or transient adaptations in either of the tested bacterial strains. No signs of the development of antibiotic resistance were found, which confirms the safety of this strategy for the prevention of infection in orthopedic surgery.

Keywords: prosthetic joint infection; fracture-related infection; bone graft substitute; revision arthroplasty; local antibiotic

---

1. Introduction

In the 1960s, the first total knee and hip replacements were implanted in patients, heralding the starting point of a revolution in orthopedic surgery [1]. Today, total joint replacement and internal fixation of fractures represent two of the most successful interventions in modern medicine, improving the quality of life of and fostering longevity in an elderly population. The rising number of orthopedic implants is unfortunately associated with an increase in implant-related infections (IRI), which represents a significant clinical, public and economic burden. IRI may be divided into periprosthetic joint infections (PJI) [2–4] and fracture-related infections (FRI) [5,6], both presenting two different clinical and surgical challenges.

The most commonly isolated bacteria causing IRI are *Staphylococcus aureus* and coagulase-negative staphylococci, which account for more than 60% of PJI, followed by other clinically relevant microbes such as streptococci, enterococci, Gram-negative rods and anaerobes [7]. Most of them can form a biofilm, which is defined as an aggregate of microbial cells embedded within a slimy self-produced matrix [8]. Single free-floating bacterial cells start aggregating and attaching to implant surfaces. An immature biofilm continues growing in cellular density and, later on, extracellular components promote microbial aggregation and stimulate the development of the slimy extracellular matrix, conferring extreme resistance to the action of immune cells and antimicrobials [9]. In addition, bacterial resistance and tolerance to antimicrobial agents [10] play a key role in infection persistence and therapeutic failure [11]. The cornerstones of effective eradication of the bacterial IRI are the resection of necrotic tissue by surgical debridement and antibiotic therapy [12,13]. Systemic antibiotic therapy, however, often offers only low local concentrations, depending on bioavailability and tissue penetration, and can cause severe adverse events. The additional use of local antibiotics can achieve high doses of active drugs exceeding the minimum inhibitory concentrations (MICs) [11] exclusively at the site of infection, thereby optimizing therapeutic efficacy and minimizing the risk of systemic toxicity [14].

Different degradable biomaterials have been suggested as antibiotic carriers [15]. Specifically, antibiotic-containing bone graft substitutes (BGS) were designed to promote and protect the bone healing process. CERAMENT™ is a resorbable biphasic biomaterial with a powder composition of 40% hydroxyapatite and 60% calcium sulfate, which offers a specific balance between material dissolution and bone regeneration. Premanufactured with either gentamicin (CERAMENT™|G) or vancomycin (CERAMENT™|V), it is aimed at ensuring antibacterial protection during the bone regeneration process.

Pre-clinical and clinical investigations of the prefabricated gentamicin-eluting BGS reported a prolonged release of high drug concentrations [16–18], which demonstrated that it was effective in vitro against sessile bacteria [19]. The clinical impact of the BGS has been reported in the literature [20–22], proving its efficacy and safety in the management of orthopedic infections. However, in vitro data concerning the activity of the vancomycin-loaded BGS are needed to support the clinical use of this formulation. Hence, the aim of this study was to further investigate in vitro the ability of the antibiotic-eluting hydroxyapatite/calcium sulfate BGS to prevent bacterial adhesion and biofilm formation by clinically relevant microorganisms, which were isolated from PJI and FRI, together with the evaluation of a possible selection for bacterial resistance to the eluted antibiotics.

2. Results

2.1. Bacterial Adhesion on Material Surface

2.1.1. Staphylococcus aureus

Results observed when testing *S. aureus* adhesion on the different materials are summarized in Figure 1a. The clinical isolate susceptible to all antibiotics (SauS) showed greater affinity to the surface of unloaded ceramic bone void filler (CBVF) than to sandblasted titanium (ST), with a significant increase in the number of attached cells concordant with the incubation time.

After incubation for 120 min, 24 ± 25.3 CFU/mL were found on ST vs. 73.3 ± 42.5 CFU/mL on CBVF. Ceramic filler containing gentamicin (CG) and vancomycin (CV) completely inhibited the adhesion of SauS at all time points. Different from SauS, the glycopeptide-intermediate *S. aureus* (GISA) strain showed no significant difference in the adhesion to ST or CBVF. At 120 min, 48 ± 45.8 CFU/mL were found on ST vs. 38.6 ± 27.4 CFU/mL on CBVF. CG and CV were still able to prevent the adhesion of GISA at all time points completely (0 CFU/mL).

At 120 min, ST exposed a significantly higher attachment of gentamicin-resistant *S. aureus* (SauG) than CBVF or CG (857.3 ± 211 vs. 242.6 ± 92.5 vs. 320 ±178.4 CFU/mL, respectively). Indeed, CG could not inhibit the adhesion of the gentamicin-resistant SauG (300 CFU/mL); only CV was able to completely impede SauG adhesion (0 CFU/mL).

In contrast to all other strains, methicillin-resistant *S. aureus* (MRSA) showed a higher adhesion to CG with respect to ST and CBVF (96 ± 37.1 vs. 29.3 ± 28.1 vs. 48 ± 37.6 CFU/mL, respectively) at

120 min. As observed with the other staphylococcal strains, CV completely inhibited MRSA adhesion (0 CFU/mL).

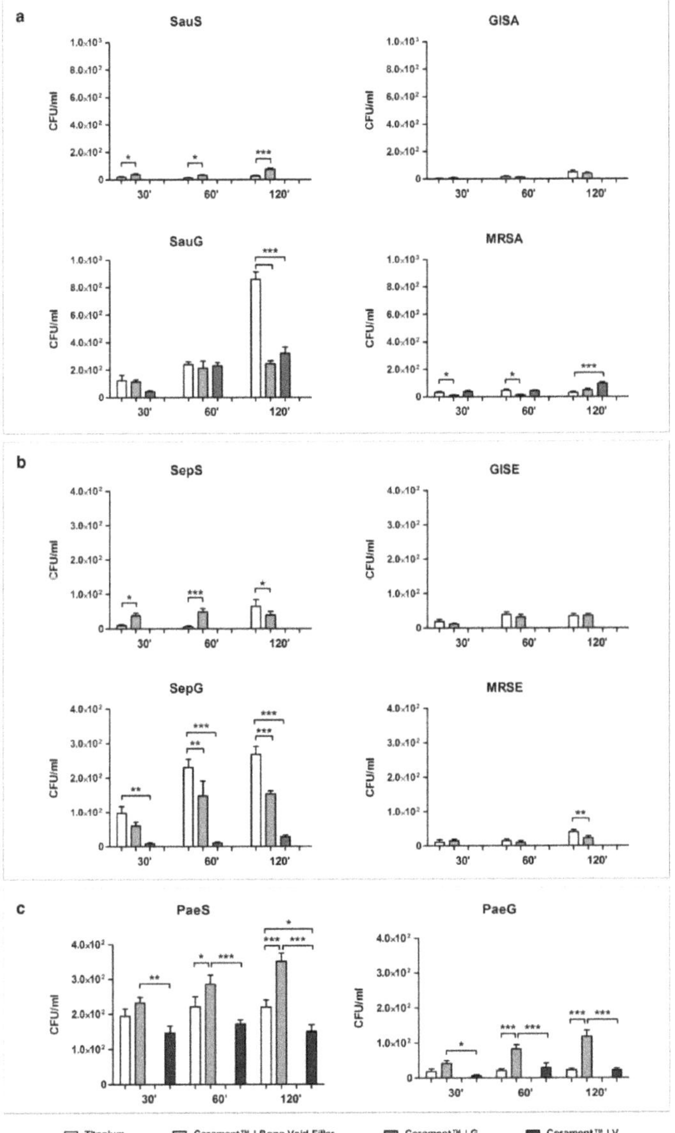

Figure 1. Bacterial adhesion on material surface as determined by plating and colony counting. (**a**) *Staphylococcus aureus*, (**b**) *Staphylococcus epidermidis* and (**c**) *Pseudomonas aeruginosa* adhesion on sandblasted titanium (ST), ceramic bone void filler (CBVF), ceramic filler containing gentamicin (CG) and ceramic filler containing vancomycin (CV) after 30, 60 and 120 min of incubation. Statistical significance: $p < 0.05$ (*), $p < 0.01$ (**) and $p < 0.001$ (***).

2.1.2. *Staphylococcus epidermidis*

The *S. epidermidis*-susceptible strain (SepS), the gentamicin-resistant strain (SepG) and methicillin-resistant *S. epidermidis* (MRSE) showed a significantly higher adhesion to ST than to CBVF after 120 min (Figure 1b). In glycopeptide-intermediate clinical isolate (GISE) no significant differences in the adhesion to ST or CBVF were found.

Similar to the observation on *S. aureus*, CG and CV were able to inhibit the adhesion of SepS and GISA at all time points (0 CFU/mL). Interestingly, CG significantly impaired the adhesion of the resistant SepG if compared with CBVF (28 ± 21.1 vs. 153.3 ± 36 CFU/mL) and even completely inhibited the adhesion of MRSE at all time points. CV was able to prevent the colonization of the disc's surface of all the tested *S. epidermidis* strains, regardless of the antibiotic susceptibility profile.

2.1.3. *Pseudomonas aeruginosa*

Both the gentamicin-susceptible (PaeS) and the gentamicin-resistant (PaeG) *P. aeruginosa* strains demonstrated a higher surface affinity to CBVF than to ST, regardless of their antibiotic susceptibility profile (Figure 1c). In comparison to CBVF, CV was able to significantly decrease the adhesion of PaeS (150.6 ± 74 vs. 350.6 ± 91.9 CFU/mL at 120 min) and PaeG (22.6 ± 19.8 vs. 118.6 ± 66.5 CFU/mL at 120 min) at all time points. CG was able to completely prevent the adhesion not only of PaeS, but also of the gentamicin-resistant PaeG at all time points.

2.2. *Biofilm Formation*

In the analysis of biofilm formation by confocal laser scan microscopy (CLSM), all tested bacterial isolates demonstrated a significantly higher biofilm biomass production on ST than on CBVF (Figure 2). Specifically, the biofilm biomass was $9.0 \times 10^7 \pm 3.9 \times 10^6$ μm^3 on ST vs. $9.2 \times 10^7 \pm 2.0 \times 10^6$ μm^3 on CBVF for MRSA, $6.8 \times 10^7 \pm 2.3 \times 10^7$ μm^3 on ST vs. $1.6 \times 10^7 \pm 5.2 \times 10^6$ μm^3 on CBVF for MRSE and $3.5 \times 10^7 \pm 1.2 \times 10^6$ μm^3 on ST vs. $5.2 \times 10^6 \pm 4.5 \times 10^5$ μm^3 on CBVF for PaeG, respectively. Representative 3D reconstructions of mature biofilm and surface structures of the analyzed substrates are depicted in Figures 3 and 4.

2.2.1. *Staphylococcus aureus*

CSLM analysis allowed quantification of the biofilm biomass at $\sim 6.2 \times 10^6 \pm 1.2 \times 10^6$ μm^3 on CBVF for SauS (Figure 5a). The biofilm biomass was significantly reduced to $\sim 1.2 \times 10^6 \pm 7.4 \times 10^4$ μm^3 on CG and to $\sim 1.4 \times 10^6 \pm 3.9 \times 10^5$ μm^3 on CV. In MRSA and GISA, similar patterns were found with a significant reduction of the biofilm biomass on CG or CV. For SauG, only CV was able to significantly decrease the biofilm production, whereas CG did not show a significant reduction. However, as reported in Table 1 and depicted in Figure 5a, CG displayed a markedly increased mortality rate of MRSA and GISA, as compared to CV (55.3% and 48.9% vs. 38.6% and 20.6%, respectively).

2.2.2. *Staphylococcus epidermidis*

The CLSM results for the *S. epidermidis* strains are comparable to the results observed for *S. aureus*. SepS, GISE, SepG and MRSE demonstrated a significantly lower biofilm-forming capacity on CG or CV, as compared to CBVF.

For example, MRSE biofilm biomass was $\sim 1.6 \times 10^7 \pm 5.2 \times 10^6$ μm^3 on CBVF, $\sim 1.8 \times 10^6 \pm 2.7 \times 10^5$ μm^3 on CG and $\sim 2.2 \times 10^6 \pm 3.7 \times 10^5$ μm^3 on CV, respectively. Notably, CG was able to significantly hinder biofilm formation even in the gentamicin-resistant SepG ($1.1 \times 10^7 \pm 1.2 \times 10^6$ μm^3 on CBVF vs. $3.6 \times 10^6 \pm 1.4 \times 10^6$ μm^3 on CG). When considering the bactericidal activity against sessile cells, CG displayed an increased mortality rate of the MRSE and GISE strains, similar to that observed for *S. aureus*, whereas CV induced a notable reduction of vital cells towards SepG (Table 1, Figure 5b).

2.2.3. *Pseudomonas aeruginosa*

CV was not able to suppress biofilm development of PaeS and PaeG (Figure 5c). Biofilm biomass of PaeS was $5.1 \times 10^6 \pm 1.6 \times 10^6$ μm^3 on CBVF vs. $5.3 \times 10^6 \pm 4.2 \times 10^5$ μm^3 on CV. However, CV seemingly induced a deeper stress when compared to CBVF, since the sessile biomass of both *P. aeruginosa* strains displayed an increased percentage of dead cells (Table 1, Figure 5c). As expected, CG displayed greater activity in preventing *P. aeruginosa* biofilm formation by significantly reducing the biomass of both isolates, being more effective against PaeS ($5.1 \times 10^6 \pm 1.6 \times 10^6$ μm^3 on CBVF vs. $2.1 \times 10^6 \pm 2.9 \times 10^5$ μm^3 on CG) than against PaeG ($5.2 \times 10^6 \pm 4.5 \times 10^5$ μm^3 on CBVF vs. $2.7 \times 10^6 \pm 1.7 \times 10^5$ μm^3 on CG).

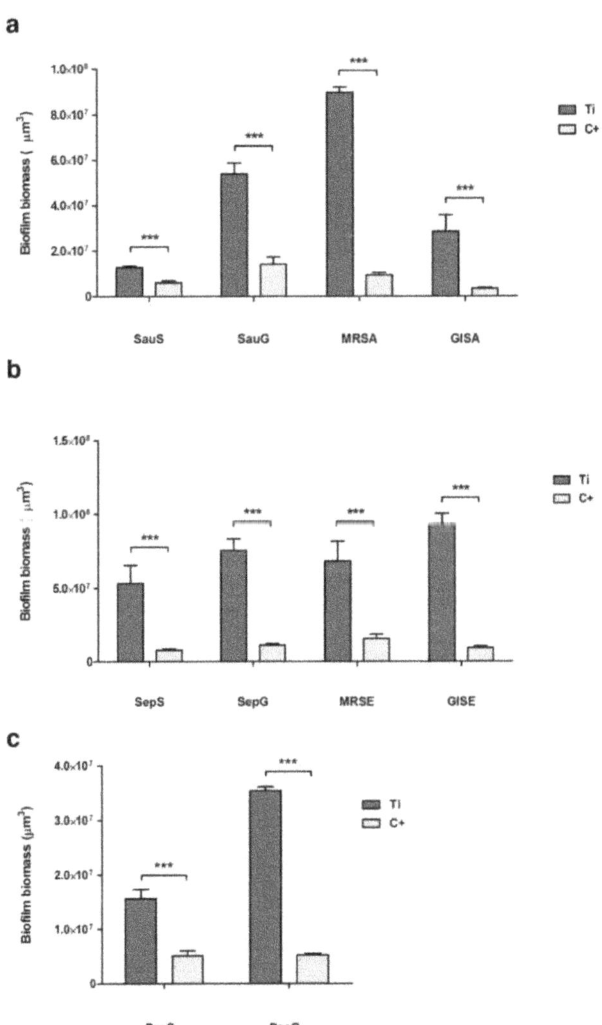

Figure 2. Biofilm biomass secreted by (**a**) *Staphylococcus aureus*, (**b**) *Staphylococcus epidermidis* and (**c**) *Pseudomonas aeruginosa* isolates on ST and CBVF. Statistical significance: $p < 0.001$ (***).

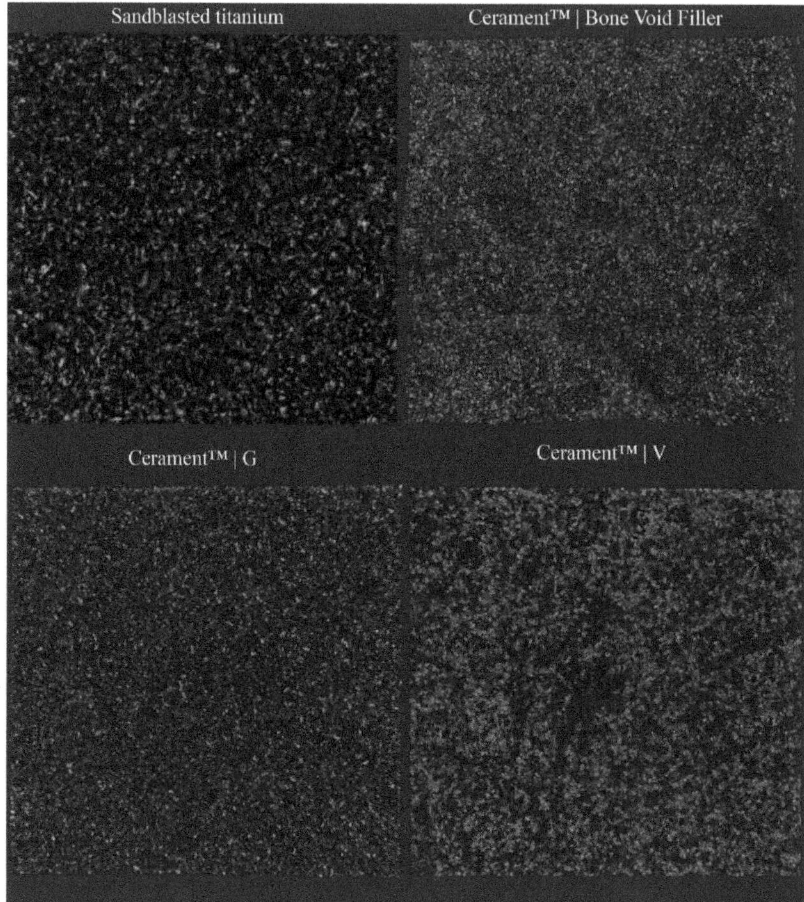

Figure 3. Representative confocal laser scan microscopy (CLSM) 3D images of sessile methicillin-resistant *Staphylococcus aureus* on ST, CBVF, CG and CV stained with Filmtracer™ LIVE/DEAD™ Biofilm Viability Kit. Magnification: 20×.

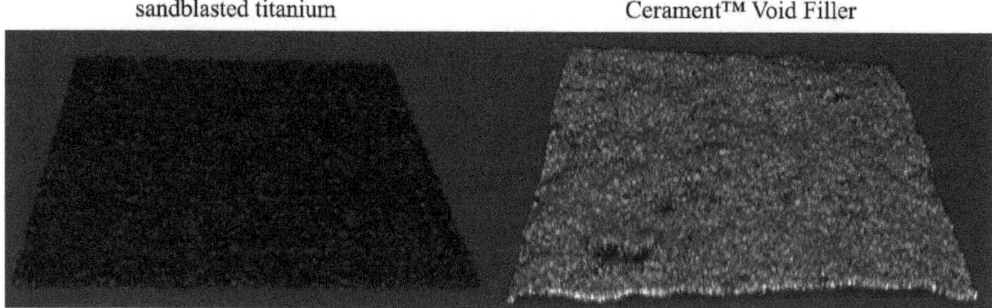

Figure 4. Representative CLSM 3D images of the surface characteristics of ST and CBVF. Magnification: 20×.

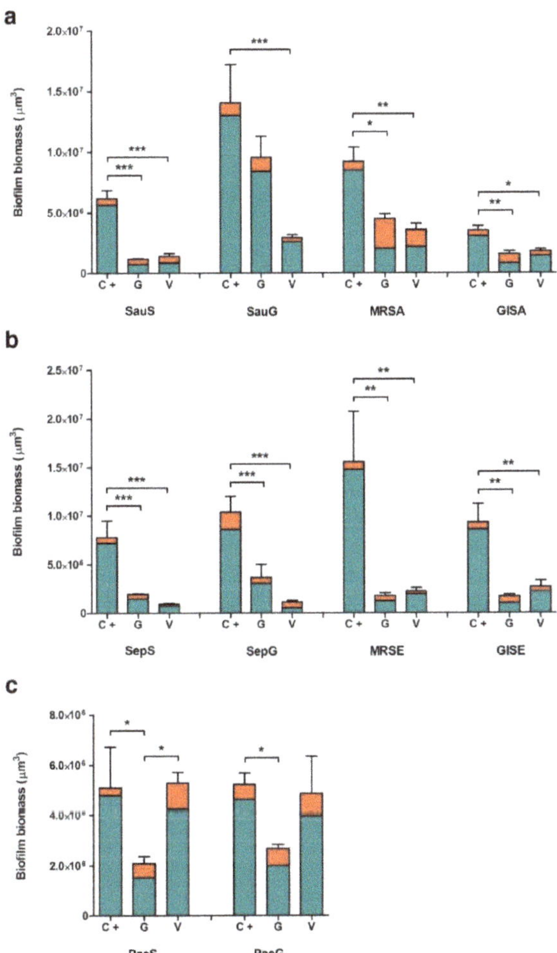

Figure 5. Ratio of live:dead cells (green:red) of (**a**) *Staphylococcus aureus*, (**b**) *Staphylococcus epidermidis* and (**c**) *Pseudomonas aeruginosa* isolates in biofilm formed on CBVF, CG and CV. Statistical significance: $p < 0.05$ (*), $p < 0.01$ (**) and $p < 0.001$ (***).

Table 1. Mortality rate by CLSM analysis. Data are expressed as percentage (%) of dead cells against the whole biofilm biomass. CBVF: Cerament^{TM} Bone Void Filler; CG: Cerament^{TM} G; CV: Cerament^{TM} V.

		CBVF	**CG**	**CV**
	SauS	8.6	36.7	39.5
S. aureus	SauG	7.7	12.1	11.6
	MRSA	7.9	55.3	38.6
	GISA	14.2	48.9	20.6
	SepS	7.7	26.2	20.6
S. epidermidis	SepG	15.8	16.8	52.3
	MRSE	5.0	31.2	12.4
	GISE	7.8	40.3	18.2
P. aeruginosa	PaeS	6.0	26.8	19.6
	PaeG	11.4	25.2	18.6

2.3. Resistance Selection to Eluted Vancomycin and Gentamicin

In the tested experimental conditions, no adaptation in the presence of CG or CV was observed by means of the zone of inhibition (ZOI) measurements on a solid agar medium. When repeatedly exposed to a freshly prepared antibiotic-loaded disc, all the tested strains displayed uniform ZOIs with no colonies of viable bacteria inside the inhibition halo.

When in the presence of a new disc, all the strains of *S. aureus* and *P. aeruginosa* maintained a comparable susceptibility profile (ZOI diameter) at all the time points, as depicted in Figure 6a,c, respectively (Δ ZOI = ± 2 mm between T1 and T7), with the exception of the MRSA strain, which showed a slight decrease in susceptibility to gentamicin after seven consecutive cycles (ZOI 10.3 ± 0.6 mm and 4.3 ± 0.6 mm at T1 and T7, respectively). In contrast, *S. epidermidis* strains, except for SepG, displayed an increased ZOI diameter after the second CG cycle (14.7 ± 0.6 mm vs. 19.3 ± 0.6 mm for SepS, 14.3 ± 0.6 mm vs. 17.3 ± 1.2 mm for MRSE and 14.3 ± 0.6 mm vs. 17 ± 1 mm for GISE at T1 and T2, respectively) (Figure 6b). Nevertheless, the increased susceptibility profile proved to be transitory as the growth–inhibition diameters reduced during the following cycles, reaching values like those observed at T1 (Δ ZOI = ± 2 mm between T1 and T7). All other differences observed were not significant.

When the same disc was used from T1 to T7, all the tested staphylococcal strains showed an absence of inhibition halo at the later time points in the presence of CV, comprising both the glycopeptide-intermediate strains (GISA and GISE). Otherwise, in the presence of CG, the gentamicin-resistant strains (SauG and SepG) were not inhibited from T4 and T3, respectively. The absence of any stable or transient adaptation was then confirmed by measuring MIC values for all the tested clinical isolates exposed to both gentamicin and vancomycin released from antibiotic-loaded discs. No significant changes in the susceptibility profile were observed. Indeed, MIC fluctuations recorded never exceeded 1-fold dilution (Tables 2 and 3).

Table 2. Resistance profile of the tested strains to gentamicin before (T0) and after (T7) exposure to CG and after cycle on antibiotic-free medium (T14). Numbers indicate gentamicin concentrations (μg/mL). (=), exposure to the same CG disc; (≠), exposure to a fresh CG disc.

		T0	T7 (=)	T14 (=)	T7 (≠)	T14 (≠)
S. aureus	SauS	0.5	0.125–1	0.5	0.25–0.5	0.25
	MRSA	0.5	0.125–1	0.5	1	1
	SauG	256	256–512	256–512	512	512
	GISA	0.25	0.25–0.5	0.25	0.5–1	0.5
S. epidermidis	SepS	0.125	0.125–0.25	0.125	0.125–0.25	0.125–0.25
	MRSE	0.25	0.125–0.25	0.125–0.25	0.25	0.125–0.25
	SepG	128	256	128	128–256	128–256
	GISE	0.125	0.125	0.125	0.125	0.125
P. aeruginosa	PaeS	0.125	0.125	0.125	0.25	0.125–0.125
	PaeG	16	8	8–16	8	8–16

Table 3. Resistance profile of the tested strains to vancomycin before (T0) and after (T7) exposure to CV and after cycle on antibiotic-free medium (T14). Numbers indicate vancomycin concentrations (μg/mL). (=), exposure to the same CG disc; (≠), exposure to a fresh CV disc.

		T0	T7 (=)	T14 (=)	T7 (≠)	T14 (≠)
S. aureus	SauS	0.25	0.125–0.25	0.25	0.5	0.25
	MRSA	1	0.5–1	0.5	2	1–2
	SauG	0.5	0.5–1	0.5	1	0.5
	GISA	4	4	2–4	4	2–4
S. epidermidis	SepS	1	2	1	1–2	1
	MRSE	0.125	0.125	0.125	0.125	0.125
	SepG	1	1–2	1	2	2
	GISE	4	4–8	4	4–8	4
P. aeruginosa	PaeS	>1024	>1024	>1024	>1024	>1024
	PaeG	>1024	>1024	>1024	>1024	>1024

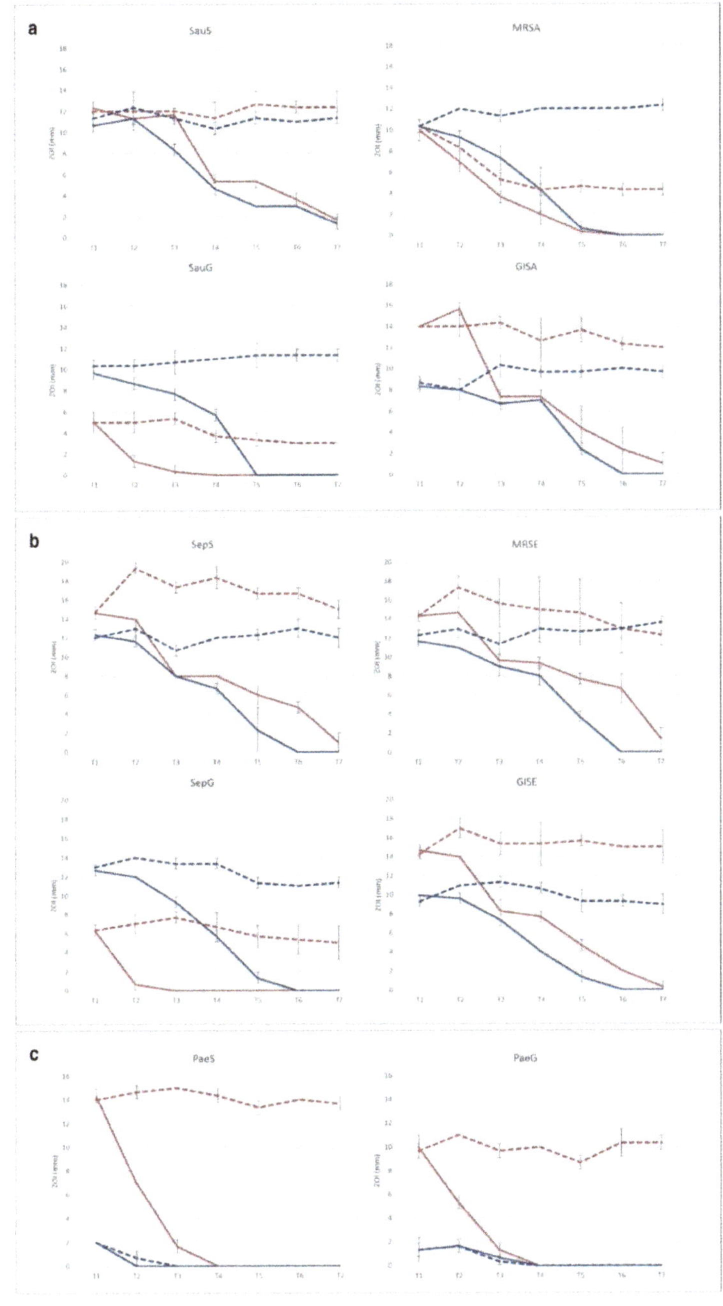

Figure 6. Zone of inhibition (ZOI) of (**a**) *Staphylococcus aureus*, (**b**) *Staphylococcus epidermidis* and (**c**) *Pseudomonas aeruginosa* growth on agar plates before (T0) and after (T7) exposure to CG/CV. ZOI (mm ± SD) are plotted against time (days). (=), exposure to the same CG/CV disc; (≠), exposure to a fresh CG/CV disc.

3. Discussion

One of the most feared complications in orthopedic surgery is the bacterial colonization of abiotic surfaces causing PJI and FRI, occurring in 1% to 2% of the patients receiving orthopedic implants [23]. Antibiotic-eluting BGS represent an attractive clinical option in the management of PJI and FRI. Gentamicin-loaded ceramic fillers have been extensively employed in clinics and their successful use reported in the literature [20–22,24–28]. The increasing emergence of aminoglycoside resistance has led to the development of a new BGS formulation enriched with vancomycin, an additional option alongside gentamicin in the orthopedic field [29]. Although these are promising strategies, there is a lack of in vitro data supporting the safe use of both CG and CV in the treatment of biofilm-related infections. Indeed, their impact on bacterial adhesion, biofilm formation and selection of resistance has not been systematically evaluated in vitro.

3.1. Bacterial Adhesion

Clinically relevant strains of *S. aureus*, *S. epidermidis* and *P. aeruginosa* were tested for their ability to adhere in the first two hours of contact to solidified calcium sulfate/hydroxyapatite discs, to evaluate a possible contamination upon surgical implantation. This scenario can be expected in a contaminated open fracture (e.g., Gustilo–Anderson IIIb), where, even after the initial extensive debridement and irrigation, bacteria may persist and attach to bone fragments or internal fixation devices. When adhesion to unloaded CBVF was compared to ST, staphylococcal isolates displayed a heterogeneous trend, which was mainly similar or lower to that of ST, with the exception of the two susceptible strains. On the contrary, all the *P. aeruginosa* isolates showed a greater affinity for CBVF discs than for ST under the tested experimental conditions. It is well known that this reversible event not only varies according to the physical characteristics of the implant, but also relies on differences between bacterial species and even strains. Indeed, the surface hydrophilicity of the material, the presence of functional groups and electric charge, as well as the physicochemical characteristics of the microbial membrane modulate initial physicochemical interactions influencing early bacterial adhesion [30]. In this context, the use of antibiotic-loaded CG or CV disks drastically affected the ability of the selected strains to adhere to the tested compounds. CG inhibited the adhesion of the gentamicin-susceptible bacterial strains (SauS, GISA, SepS and GISE) and reduced the bacterial load of the gentamicin-resistant staphylococci (SauG, SepG, MRSA and MRSE), but could not completely impede their adhesion. Since gentamicin is active against Gram-negative bacteria, CG inhibited the attachment of PaeS and PaeG, whereas CV completely avoided the adhesion of all staphylococci strains. As expected, CV was not able to impair the adhesion of the Gram-negative *P. aeruginosa* strains.

In the clinical setting, this finding could be used in the prophylaxis of infection in contaminated open fractures. If a bone void or fracture gap is present during the initial debridement, it could be filled with CG. Since open fractures might be contaminated by Gram-positive and Gram-negative bacteria [31], the shown effect of CG on *P. aeruginosa* might be beneficial in this setting. Aljawadi and co-workers recently presented their clinical results on the use of CG in addition to systemic antibiotics in Gustilo–Anderson IIIb open fractures in a single-stage "Fix and Flap" procedure in 80 patients [24]. The limb salvage rate in this challenging patient group was 96.25% and the infection rate was 1.25%. A randomized controlled trial for CG as part of the surgical repair for open tibial diaphyseal fractures is on its way and recruitment of 200 patients has been finalized (www.clincaltrials.gov, Identifier: NCT02820363).

3.2. Evaluation of Biofilm Formation

The ability to form biofilm on the tested substrates was evaluated by CLSM. Despite the porous structure of CBVF, all the tested strains displayed a significantly greater biomass volume on sandblasted titanium than on CBVF.

Interestingly, biofilm formation by all the tested staphylococcal strains was negatively affected by the presence of antibiotic-eluting BGS, highlighting the efficacy of the initial burst release of an antimicrobial even in such an in vitro setting conducted under ideal conditions, which does not accurately recapitulate in vivo conditions. This experimental design does not consider important aspects such as the lack of surrounding tissues, the physiological supply of nutrients, cellular resorption and immune system action, which might concur in impairing bacterial contamination and help to reduce the number of bacteria on the site of contamination. The high bacterial inoculum (much greater than that of a possible contamination event, which is likely to happen from few cells) and a nutrient-rich growth medium represent a "worst case scenario", which understandably supports our results.

CG was able to significantly suppress biofilm formation of all staphylococci strains, except for SauG and both *P. aeruginosa* strains. CV reduced biofilm formation of all staphylococci, including MRSA and MRSE, but had no impact on Gram-negative *P. aeruginosa* biofilm biomass. The observed increase of mortality rate of cells encased within the biofilm matrix could be ascribed to the effects of the high vancomycin concentration, which might cause the perturbation of cell shape, as observed on Gram-negative strains by others [32].

To date, clinical outcomes of CG and CV have been published in PJI [22], FRI [25,27], chronic Osteomyelitis (cOM) [20,21], and Diabetic Foot Osteomyelitis (DFO) [26,28]. McNally and co-workers found a reduction of the recurrence rate of cOM of 4% in a one-stage procedure compared to common reports in the literature of about 10% [21]. Briefly, they evaluated the use of CG in the management of deep bone infection in a one-stage procedure. Considering the unknown bacterial resistance pattern, the broader Gram-negative and Gram-positive spectrum and the unavailability of CV at the time of the study, CG proved effective in the treatment of chronic Osteomyelitis, even in the presence of gentamicin-resistant clinical isolates.

The results of our in vitro investigation showed that CV might be beneficial in the management of staphylococcal infections. Selecting between CG and CV is only possible if the causative bacteria have been identified in the first stage of a two-stage procedure. Logoluso et al. used a two-stage approach in the management of PJI, where CG was used in 13 patients and CV in 7 (only staphylococcal and enterococcal infections), finding only one recurrence of PJI in their series of 20 consecutive cases [22]. In FRI, the recurrence rate was reported in 7.7% [25] and 8.5% of cases [27], and in DFO in 10% [26].

3.3. Resistance Selection by Eluted Vancomycin and Gentamicin

There is an ongoing discussion if local antibiotics can induce resistance or lead to a selection of resistant bacteria. The slow penetration rate of antibiotics through the biofilm matrix may give time to the cells to adapt phenotypically and genotypically. Sessile bacteria encased within the biomass are able to incorporate eDNA from the biofilm matrix, and in vitro tests suggested that horizontal gene transfer may even be 10,000 times more likely in biofilms compared to their planktonic counterparts [33–36]. Furthermore, cells in a biofilm are inherently prone to spontaneous mutations, possibly due to an increased response to stress and consequent DNA damage [37,38]. In the orthopedic field, this topic is gaining particular interest with regard to reduced staphylococcal susceptibility to glycopeptide antibiotics, which are commonly employed in the treatment of orthopedic infections because of their noteworthy bone penetration and the diffusion of methicillin-resistant strains [39]. While all ZOIs of gentamicin-resistant strains (SauG, SepG and PaeG) were notably inferior to those of non-resistant strains, such marked difference was not observed in the presence of vancomycin for the glycopeptide-intermediate strains. This result confirms that aminoglycoside resistance in staphylococci is derived from the acquisition of a gene (i.e., aph(3′)-IIIa), whereas resistance to glycopeptide mainly consists of a reduced susceptibility by accumulation of point mutations [40]. However, it is worth noting that glycopeptide-intermediate staphylococci behavior in the presence of CV was like that of the other strains.

For all the aforementioned reasons, it is extremely important to test whether the exposure to antibiotic-loaded biomaterials might lead to the acquisition of a resistance profile. The results obtained from the present study suggest that prolonged exposure either to the same or new antibiotic-eluting hydroxyapatite/calcium sulfate discs did not lead to stable or transient adaptations in either of the tested bacterial strains.

Despite the aforementioned limitations of the study, the emergence of antibiotic-resistant strains in the presence of an antibiotic-enriched compound in a clinical scenario seems extremely unlikely considering the initial burst of high concentrations of active drugs and the subsequent retention of a sustained amount of antibiotics, as also reported from previous studies [19]. Furthermore, the combined use of both gentamicin and vancomycin-eluting compounds might be considered for clinical cases presenting high-risk of infection recurrence due to polymicrobial contamination. This low-risk profile, together with the clinically confirmed regenerative properties of the tested resorbable materials [20,41], confirms further the pertinence of antibiotic-containing BGS as prophylactic and treatment options and proves them potentially safe for clinical application in orthopedic infection surgery [42].

4. Materials and Methods

4.1. Tested Bacterial Strains and Biomaterials

Clinically relevant bacterial strains from patients with IRI were isolated at the Laboratory of Clinical Chemistry and Microbiology of the IRCCS Galeazzi Orthopedic Institute and were used in this study. Biofilm-producing strains of *S. aureus* (one susceptible strain, SauS; one gentamicin-resistant strain, SauG; one methicillin-resistant strain, MRSA; one glycopeptide-intermediate strain, GISA), *S. epidermidis* (one susceptible strain, SepS; one gentamicin-resistant strain, SepG; one methicillin-resistant strain, MRSE; one glycopeptide-intermediate strain, GISE) and *P. aeruginosa* (one gentamicin-susceptible strain, PaeS and one gentamicin-resistant strain, PaeG) were selected. The bacterial ability to adhere and to produce biofilm, as well as the emergence of antibiotic resistance in vitro were tested on the following substrates: sandblasted titanium (ST; used as growth control) (Geass, Pozzuolo del Friuli, Italy), CERAMENT^{TM}|BONE VOID FILLER, a biphasic degradable bone graft substitute composed of 40% hydroxyapatite and 60% calcium sulfate (without antibiotic, used as growth control), CERAMENT^{TM}|G, containing 17.5 mg gentamicin/mL paste and CERAMENT^{TM}|V, containing 66 mg vancomycin/mL paste. Sterile kits of the products were supplied by BONESUPPORT AB (Lund, Sweden) and mixed according to the instructions for use provided by the manufacturer.

4.2. Evaluation of Bacterial Adhesion on Material Surface

The effects of different formulations of the calcium sulfate/hydroxyapatite bone graft substitute on bacterial adhesion were tested. To standardize the experimental conditions, 6 × 2.5 mm discs of unloaded CBVF and CG and CV were produced by pouring freshly prepared injectable pastes into a custom-made mold of polydimethylsiloxane (PDMS, Farnell, Lainate, Italy). The theoretical concentration of gentamicin in each disc was approximately 1.23 mg, whereas that of vancomycin was approximately 4.66 mg per disc. Discs were placed in a 96-well microplate containing 180 µL of Brain Heart Infusion (BHI) broth (BioMérieux, Craponne, France). Each well was inoculated with 20 µL of a bacterial overnight culture to obtain a final density of 1.0×10^6 CFU/mL. To evaluate the first phases of bacterial attachment upon tested substrates, after 30, 60 and 120 min of incubation at 37 °C, the discs were rinsed three times with sterile saline to remove non-adherent bacteria [43,44]. Then, discs were immersed in 200 µL of a 0.1% *w/v* dithiothreitol (DTT; Sigma-Aldrich, Milan, Italy) solution and mechanically stirred for 15 min at room temperature to detach bacteria adhering to the discs [45]. Finally, 10 µL from each well was serially diluted and drop-plated on Tryptic Soy Agar (TSA; Sigma-Aldrich, Milan, Italy) plates and incubated at 37 °C for 24 h for colony forming units (CFU) count [46]. The adhesion tests were performed in triplicates for each strain.

4.3. Analysis of Biofilm Formation by CLSM

The ability of the tested clinical strains to form biofilm on the antibiotic-containing and unloaded biphasic bone graft substitutes was evaluated using CLSM [47]. Discs were placed in a 24-well microplate with 950 µL of BHI broth (Biomérieux, Craponne, France) and 50 µL of 1.0×10^7 CFU/mL bacterial suspension. After 48 h of incubation at 37 °C, discs were rinsed three times with sterile saline to remove non-adherent bacteria and stained with Filmtracer™ LIVE/DEAD™ Biofilm Viability Kit (Thermo Fisher Diagnostics, Monza, Italy), according to the manufacturer's instructions. Briefly, the staining solution was prepared by adding 3 µL of SYTO9 and 3 µL of propidium iodide to 1 mL of filter-sterilized water. Samples were incubated with 20 µL of staining solution at room temperature in the dark for 15 min. Afterwards, samples were washed with sterile saline and examined with an upright TCS SP8 (Leica Microsystems CMS GmbH, Mannheim, Germany) using a 20× dry objective (HC PL FLUOTAR 20×/0.50 DRY). Images from at least three randomly selected areas were acquired for each sample. The obtained images were processed with Las X (Leica Microsystems CMS GmbH, Mannheim, Germany) and analyzed with Fiji software (Fiji, ImageJ, Wayne Rasband National Institutes of Health). The test was performed in triplicates for each strain.

4.4. Determination of Minimum Inhibitory Concentration

The MIC for each strain was determined by the broth microdilution method, following the guidelines of the European Committee on Antimicrobial Susceptibility Testing [48]. A microbial suspension was prepared for each bacterial strain in Mueller–Hinton (MHB, Millipore, Milan, Italy) broth to an optical density equal to a standard turbidity of 0.5 McFarland ($\sim 1.5 \times 10^8$ CFU/mL). After obtaining a microbial load of 5×10^6 CFU/mL, 10 µL of each suspension was inoculated in a 96-well microplate containing 90 µL of a serial 2-fold dilution of either gentamicin (Sigma-Aldrich, Milan, Italy) or vancomycin (VWR, Milan, Italy). The MIC was defined as the lowest antibiotic concentration that inhibited the bacterial growth after 24 h of incubation at 37 °C, as assessed by the unaided eye. The experiment was performed in triplicates.

4.5. Evaluation of Selection for Bacterial Resistance after Exposure to Eluted Antibiotics

The emergence of bacterial resistance to the antibiotics released from the drug-loaded formulations was investigated by the modified Kirby–Bauer method as reported by others [49]. Briefly, either gentamicin or vancomycin-eluting discs were placed on TSA plates previously seeded with a bacterial suspension corresponding to a standard turbidity of 0.5 McFarland ($\sim 5 \times 10^8$ CFU/mL) (T0). After incubation at 37 °C for 24 h, the zone of inhibition (ZOI) was measured and bacterial cells growing on the edge thereof were subcultured on a fresh TSA plate for 7 days (from T1 to T7). Specifically, two different procedures were applied: (i) the same exhausted antibiotic-containing disc was used throughout the whole experiment (up to T7); (ii) the antibiotic-containing substrate was replaced with a fresh disc every 24 h for 7 days. The ZOI was measured and recorded daily. After overnight incubation in antibiotic-free BHI broth, bacteria were further subcultured on antibiotic-free TSA plates for an additional 7 days (T14) to assess the stability of the potentially acquired resistance. MIC values were determined before (T0) and after exposure to the eluted antibiotics (T7) and after the last cycle on antibiotic-free TSA plates (T14). The acquisition of bacterial resistance to an antimicrobial compound was defined as ≥4-fold increase in MIC values [50].

5. Conclusions

From the results obtained in this in vitro study, it can be concluded that CV efficacy is comparable to CG in the treatment of Gram-positive biofilm and has been proven to be effective against gentamicin-resistant strains. The use of antibiotic-containing hydroxyapatite/calcium sulfate bone graft substitutes could be a feasible strategy in the prevention of bacterial colonization and biofilm development by microorganisms and may prove to be a promising approach for dead-space management in the treatment of cOM, FRI, DFO and PJI, always combined with systemic antibiotic therapy.

Author Contributions: Conceptualization, N.L. and E.D.V.; methodology, A.B. and M.B.; formal analysis, A.B. and M.B.; data curation, A.B. and M.B.; writing—original draft preparation, A.B. and M.B.; writing—review and editing, N.L. and E.D.V.; visualization, M.B.; supervision, E.D.V.; funding acquisition, E.D.V. All authors have read and agreed to the published version of the manuscript.

Funding: This study was partly sponsored by BONESUPPORT AB and partially funded by the Italian Ministry of Health (RC Project).

Acknowledgments: The authors would like to thank Michael Diefenbeck and Maria Eugenia Butini for technical and scientific support.

Conflicts of Interest: The authors declare no conflict of interest. The funders had no role in the design of the study; in the collection, analyses, or interpretation of data; in the writing of the manuscript; or in the decision to publish the results.

Abbreviations

BGS	Bone Graft Substitute
IRI	Implant-related Infection
PJI	Prosthetic Joint Infection
FRI	Fracture-related Infection
cOM	Chronic Osteomyelitis
DFO	Diabetic Foot Osteomyelitis
MIC	Minimum Inhibitory Concentrations
CBVF	Ceramic Bone Void Filler
CG	Ceramic Filler Containing Gentamicin
CV	Ceramic Filler Containing Vancomycin
CLSM	Confocal Laser Scan Microscopy
ZOI	Zone of Inhibition

References

1. Kapadia, B.H.; Berg, R.A.; Daley, J.A.; Fritz, J.; Bhave, A.; Mont, M.A. Periprosthetic joint infection. *Lancet* **2016**, *387*, 386–394. [CrossRef]
2. Izakovicova, P.; Borens, O.; Trampuz, A. Periprosthetic joint infection: Current concepts and outlook. *EFORT Open Rev.* **2019**, *4*, 482–494. [CrossRef] [PubMed]
3. Osmon, D.R.; Berbari, E.F.; Berendt, A.R.; Lew, D.; Zimmerli, W.; Steckelberg, J.M.; Rao, N.; Hanssen, A.; Wilson, W.R.; Infectious Diseases Society of America. Diagnosis and management of prosthetic joint infection: Clinical practice guidelines by the Infectious Diseases Society of America. *Clin. Infect. Dis.* **2013**, *56*, e1–e25. [CrossRef] [PubMed]
4. Zimmerli, W.; Trampuz, A.; Ochsner, P.E. Prosthetic-joint infections. *N. Engl. J. Med.* **2004**, *351*, 1645–1654. [CrossRef]
5. Govaert, G.A.M.; Kuehl, R.; Atkins, B.L.; Trampuz, A.; Morgenstern, M.; Obremskey, W.T.; Verhofstad, M.H.J.; McNally, M.A.; Metsemakers, W.J.; Fracture-Related Infection Consensus Group. Diagnosing Fracture-Related Infection: Current Concepts and Recommendations. *J. Orthop. Trauma* **2020**, *34*, 8–17. [CrossRef]
6. Metsemakers, W.J.; Morgenstern, M.; McNally, M.A.; Moriarty, T.F.; McFadyen, I.; Scarborough, M.; Athanasou, N.A.; Ochsner, P.E.; Kuehl, R.; Raschke, M.; et al. Fracture-related infection: A consensus on definition from an international expert group. *Injury* **2018**, *49*, 505–510. [CrossRef]
7. Drago, L.; De Vecchi, E.; Bortolin, M.; Zagra, L.; Romano, C.L.; Cappelletti, L. Epidemiology and Antibiotic Resistance of Late Prosthetic Knee and Hip Infections. *J. Arthroplast.* **2017**, *32*, 2496–2500. [CrossRef]
8. Bjarnsholt, T.; Ciofu, O.; Molin, S.; Givskov, M.; Hoiby, N. Applying insights from biofilm biology to drug development—Can a new approach be developed? *Nat. Rev. Drug Discov.* **2013**, *12*, 791–808. [CrossRef]
9. Hall-Stoodley, L.; Stoodley, P. Evolving concepts in biofilm infections. *Cell. Microbiol.* **2009**, *11*, 1034–1043. [CrossRef]

10. Brauner, A.; Fridman, O.; Gefen, O.; Balaban, N.Q. Distinguishing between resistance, tolerance and persistence to antibiotic treatment. *Nat. Rev. Microbiol.* **2016**, *14*, 320–330. [CrossRef]

11. Lalidou, F.; Kolios, G.; Drosos, G.I. Bone infections and bone graft substitutes for local antibiotic therapy. *Surg. Technol. Int.* **2014**, *24*, 353–362. [PubMed]

12. Metsemakers, W.J.; Fragomen, A.T.; Moriarty, T.F.; Morgenstern, M.; Egol, K.A.; Zalavras, C.; Obremskey, W.T.; Raschke, M.; McNally, M.A.; Fracture-Related Infection (FRI) Consensus Group. Evidence-Based Recommendations for Local Antimicrobial Strategies and Dead Space Management in Fracture-Related Infection. *J. Orthop. Trauma* **2020**, *34*, 18–29. [CrossRef] [PubMed]

13. Trampuz, A.; Zimmerli, W. Diagnosis and treatment of implant-associated septic arthritis and osteomyelitis. *Curr. Infect. Dis. Rep.* **2008**, *10*, 394–403. [CrossRef] [PubMed]

14. Winkler, H. Treatment of chronic orthopaedic infection. *EFORT Open Rev.* **2017**, *2*, 110–116. [CrossRef]

15. Inzana, J.A.; Schwarz, E.M.; Kates, S.L.; Awad, H.A. Biomaterials approaches to treating implant-associated osteomyelitis. *Biomaterials* **2016**, *81*, 58–71. [CrossRef]

16. Stravinskas, M.; Horstmann, P.; Ferguson, J.; Hettwer, W.; Nilsson, M.; Tarasevicius, S.; Petersen, M.M.; McNally, M.A.; Lidgren, L. Pharmacokinetics of gentamicin eluted from a regenerating bone graft substitute: In vitro and clinical release studies. *Bone Jt. Res.* **2016**, *5*, 427–435. [CrossRef]

17. Stravinskas, M.; Nilsson, M.; Horstmann, P.; Petersen, M.M.; Tarasevicius, S.; Lidgren, L. Antibiotic Containing Bone Substitute in Major Hip Surgery: A Long Term Gentamicin Elution Study. *J. Bone Jt. Infect.* **2018**, *3*, 68–72. [CrossRef]

18. Stravinskas, M.; Nilsson, M.; Vitkauskiene, A.; Tarasevicius, S.; Lidgren, L. Vancomycin elution from a biphasic ceramic bone substitute. *Bone Jt. Res.* **2019**, *8*, 49–54. [CrossRef]

19. Butini, M.E.; Cabric, S.; Trampuz, A.; Di Luca, M. In vitro anti-biofilm activity of a biphasic gentamicin-loaded calcium sulfate/hydroxyapatite bone graft substitute. *Colloids Surf. B Biointerfaces* **2018**, *161*, 252–260. [CrossRef]

20. Ferguson, J.; Athanasou, N.; Diefenbeck, M.; McNally, M. Radiographic and Histological Analysis of a Synthetic Bone Graft Substitute Eluting Gentamicin in the Treatment of Chronic Osteomyelitis. *J. Bone Jt. Infect.* **2019**, *4*, 76–84. [CrossRef]

21. McNally, M.A.; Ferguson, J.Y.; Lau, A.C.; Diefenbeck, M.; Scarborough, M.; Ramsden, A.J.; Atkins, B.L. Single-stage treatment of chronic osteomyelitis with a new absorbable, gentamicin-loaded, calcium sulphate/hydroxyapatite biocomposite: A prospective series of 100 cases. *Bone Jt. J.* **2016**, *98*, 1289–1296. [CrossRef]

22. Logoluso, N.; Drago, L.; Gallazzi, E.; George, D.A.; Morelli, I.; Romano, C.L. Calcium-Based, Antibiotic-Loaded Bone Substitute as an Implant Coating: A Pilot Clinical Study. *J. Bone Jt. Infect.* **2016**, *1*, 59–64. [CrossRef] [PubMed]

23. Trampuz, A.; Widmer, A.F. Infections associated with orthopedic implants. *Curr. Opin. Infect. Dis.* **2006**, *19*, 349–356. [CrossRef] [PubMed]

24. Aljawadi, A.; Islam, A.; Jahangir, N.; Niazi, N.; Ferguson, Z.; Sephton, B.; Elmajee, M.; Reid, A.; Wong, J.; Pillai, A. Adjuvant Local Antibiotic Hydroxyapatite Bio-Composite in the management of open Gustilo Anderson IIIB fractures. Prospective Review of 80 Patients from the Manchester Ortho-Plastic Unit. *J. Orthop.* **2020**, *18*, 261–266. [CrossRef]

25. Drampalos, E.; Mohammad, H.R.; Pillai, A. Augmented debridement for implant related chronic osteomyelitis with an absorbable, gentamycin loaded calcium sulfate/hydroxyapatite biocomposite. *J. Orthop.* **2020**, *17*, 173–179. [CrossRef]

26. Niazi, N.S.; Drampalos, E.; Morrissey, N.; Jahangir, N.; Wee, A.; Pillai, A. Adjuvant antibiotic loaded bio composite in the management of diabetic foot osteomyelitis—A multicentre study. *Foot* **2019**, *39*, 22–27. [CrossRef]

27. Pesch, S.; Hanschen, M.; Greve, F.; Zyskowski, M.; Seidl, F.; Kirchhoff, C.; Biberthaler, P.; Huber-Wagner, S. Treatment of fracture-related infection of the lower extremity with antibiotic-eluting ceramic bone substitutes: Case series of 35 patients and literature review. *Infection* **2020**, *48*, 333–344. [CrossRef]

28. Hutting, K.; van Netten, J.; Dening, J.; Cate, W.T.; van Baal, J. Surgical debridement and gentamicin-loaded calcium sulphate/hydroxyapatite bone void filling to treat diabetic foot osteomyelitis. *Diabet. Foot J.* **2019**, *22*, 22–27.

29. Labby, K.J.; Garneau-Tsodikova, S. Strategies to overcome the action of aminoglycoside-modifying enzymes for treating resistant bacterial infections. *Future Med. Chem.* **2013**, *5*, 1285–1309. [CrossRef]

30. Carniello, V.; Peterson, B.W.; van der Mei, H.C.; Busscher, H.J. Physico-chemistry from initial bacterial adhesion to surface-programmed biofilm growth. *Adv. Colloid Interface Sci.* **2018**, *261*, 1–14. [CrossRef]

31. Torbert, J.T.; Joshi, M.; Moraff, A.; Matuszewski, P.E.; Holmes, A.; Pollak, A.N.; O'Toole, R.V. Current bacterial speciation and antibiotic resistance in deep infections after operative fixation of fractures. *J. Orthop. Trauma* **2015**, *29*, 7–17. [CrossRef] [PubMed]

32. Stokes, J.M.; French, S.; Ovchinnikova, O.G.; Bouwman, C.; Whitfield, C.; Brown, E.D. Cold Stress Makes Escherichia coli Susceptible to Glycopeptide Antibiotics by Altering Outer Membrane Integrity. *Cell Chem. Biol.* **2016**, *23*, 267–277. [CrossRef] [PubMed]

33. Hausner, M.; Wuertz, S. High rates of conjugation in bacterial biofilms as determined by quantitative in situ analysis. *Appl. Environ. Microbiol.* **1999**, *65*, 3710–3713. [CrossRef] [PubMed]

34. Madsen, J.S.; Burmolle, M.; Hansen, L.H.; Sorensen, S.J. The interconnection between biofilm formation and horizontal gene transfer. *FEMS Immunol. Med Microbiol.* **2012**, *65*, 183–195. [CrossRef] [PubMed]

35. Molin, S.; Tolker-Nielsen, T. Gene transfer occurs with enhanced efficiency in biofilms and induces enhanced stabilisation of the biofilm structure. *Curr. Opin. Biotechnol.* **2003**, *14*, 255–261. [CrossRef]

36. Savage, V.J.; Chopra, I.; O'Neill, A.J. Staphylococcus aureus biofilms promote horizontal transfer of antibiotic resistance. *Antimicrob. Agents Chemother.* **2013**, *57*, 1968–1970. [CrossRef]

37. Boles, B.R.; Singh, P.K. Endogenous oxidative stress produces diversity and adaptability in biofilm communities. *Proc. Natl. Acad. Sci. USA* **2008**, *105*, 12503–12508. [CrossRef]

38. Driffield, K.; Miller, K.; Bostock, J.M.; O'Neill, A.J.; Chopra, I. Increased mutability of Pseudomonas aeruginosa in biofilms. *J. Antimicrob. Chemother.* **2008**, *61*, 1053–1056. [CrossRef]

39. Vaudaux, P.; Ferry, T.; Uckay, I.; Francois, P.; Schrenzel, J.; Harbarth, S.; Renzoni, A. Prevalence of isolates with reduced glycopeptide susceptibility in orthopedic device-related infections due to methicillin-resistant Staphylococcus aureus. *Eur. J. Clin. Microbiol. Infect. Dis.* **2012**, *31*, 3367–3374. [CrossRef]

40. Wang, Y.; Li, X.; Jiang, L.; Han, W.; Xie, X.; Jin, Y.; He, X.; Wu, R. Novel Mutation Sites in the Development of Vancomycin- Intermediate Resistance in Staphylococcus aureus. *Front. Microbiol.* **2016**, *7*, 2163. [CrossRef]

41. Hettwer, W.; Horstmann, P.F.; Bischoff, S.; Gullmar, D.; Reichenbach, J.R.; Poh, P.S.P.; van Griensven, M.; Gras, F.; Diefenbeck, M. Establishment and effects of allograft and synthetic bone graft substitute treatment of a critical size metaphyseal bone defect model in the sheep femur. *APMIS Acta Pathol. Microbiol. Immunol. Scand.* **2019**, *127*, 53–63. [CrossRef] [PubMed]

42. Kyriacou, H.; Kamaraj, A.; Khan, W.S. Developments in Antibiotic-Eluting Scaffolds for the Treatment of Osteomyelitis. *Appl. Sci.* **2020**, *10*, 2244. [CrossRef]

43. Azeredo, J.; Azevedo, N.F.; Briandet, R.; Cerca, N.; Coenye, T.; Costa, A.R.; Desvaux, M.; Di Bonaventura, G.; Hebraud, M.; Jaglic, Z.; et al. Critical review on biofilm methods. *Crit. Rev. Microbiol.* **2017**, *43*, 313–351. [CrossRef] [PubMed]

44. Bidossi, A.; Bottagisio, M.; De Grandi, R.; De Vecchi, E. Ability of adhesion and biofilm formation of pathogens of periprosthetic joint infections on titanium-niobium nitride (TiNbN) ceramic coatings. *J. Orthop. Surg. Res.* **2020**, *15*, 90. [CrossRef]

45. De Vecchi, E.; Bottagisio, M.; Bortolin, M.; Toscano, M.; Lovati, A.B.; Drago, L. Improving the Bacterial Recovery by Using Dithiothreitol with Aerobic and Anaerobic Broth in Biofilm-Related Prosthetic and Joint Infections. *Adv. Exp. Med. Biol.* **2017**, *973*, 31–39. [CrossRef]

46. Herigstad, B.; Hamilton, M.; Heersink, J. How to optimize the drop plate method for enumerating bacteria. *J. Microbiol. Methods* **2001**, *44*, 121–129. [CrossRef]

47. Bidossi, A.; De Grandi, R.; Toscano, M.; Bottagisio, M.; De Vecchi, E.; Gelardi, M.; Drago, L. Probiotics Streptococcus salivarius 24SMB and Streptococcus oralis 89a interfere with biofilm formation of pathogens of the upper respiratory tract. *BMC Infect. Dis.* **2018**, *18*, 653. [CrossRef]

48. EUCAST Guidelines. Available online: http://www.eucast.org/fileadmin/src/media/PDFs/EUCAST_files/MIC_testing/Edis5.1_broth_dilution.pdf (accessed on 3 December 2020).

49.   Karr, J.C.; Lauretta, J.; Keriazes, G. In vitro antimicrobial activity of calcium sulfate and hydroxyapatite (Cerament Bone Void Filler) discs using heat-sensitive and non-heat-sensitive antibiotics against methicillin-resistant Staphylococcus aureus and Pseudomonas aeruginosa. *J. Am. Podiatr. Med. Assoc.* **2011**, *101*, 146–152. [CrossRef]

50.   Drago, L.; De Vecchi, E.; Nicola, L.; Tocalli, L.; Gismondo, M.R. In vitro selection of resistance in Pseudomonas aeruginosa and Acinetobacter spp. by levofloxacin and ciprofloxacin alone and in combination with beta-lactams and amikacin. *J. Antimicrob. Chemother.* **2005**, *56*, 353–359. [CrossRef]

Publisher's Note: MDPI stays neutral with regard to jurisdictional claims in published maps and institutional affiliations.

International Journal of
Molecular Sciences

Article

Oxo-Titanium(IV) Complex/Polymer Composites—Synthesis, Spectroscopic Characterization and Antimicrobial Activity Test

Piotr Piszczek [1],* , Barbara Kubiak [1], Patrycja Golińska [2] and Aleksandra Radtke [1],*

[1] Faculty of Chemistry, Nicolaus Copernicus University in Toruń, Gagarina 7, 87-100 Toruń, Poland; basiak0809@gmail.com

[2] Faculty of Biological and Veterinary Sciences, Nicolaus Copernicus University in Toruń, Lwowska 1, 87-100 Toruń, Poland; golinska@umk.pl

* Correspondence: piszczek@umk.pl (P.P.); aradtke@umk.pl (A.R.); Tel.: +48-607-883-357 (P.P.); +48-600-321-294 (A.R.)

Received: 2 November 2020; Accepted: 15 December 2020; Published: 18 December 2020

Abstract: The emergence of a large number of bacterial strains resistant to many drugs or disinfectants currently used contributed to the search of new, more effective antimicrobial agents. In the presented paper, we assessed the microbiocidal activity of tri- and tetranuclear oxo-titanium(IV) complexes (TOCs), which were dispersed in the poly(methyl methacrylate) (PMMA) matrix. The TOCs were synthesized in reaction to $Ti(OR)_4$ (R = ^iPr, ^iBu) and HO_2CR' (R' = 4-$PhNH_2$ and 4-PhOH) in a 4:1 molar ratio at room temperature and in Ar atmosphere. The structure of isolated oxo-complexes was confirmed by IR and Raman spectroscopy and mass spectrometry. The antimicrobial activity of the produced composites (PMMA + TOCs) was estimated against Gram-positive (*Staphylococcus aureus* ATCC 6538 and *S. aureus* ATCC 25923) and Gram-negative (*Escherichia coli* ATCC 8739 and *E. coli* ATCC 25922) bacteria and yeasts of *Candida albicans* ATCC 10231. All produced composites showed biocidal activity against the bacteria. Composites containing $\{Ti_4O_2\}$ cores and the $\{Ti_3O\}$ core stabilized by the 4-hydroxybenzoic ligand showed also high activity against yeasts. The results of investigations carried out suggest that produced (PMMA + TOCs) composites, due to their microbiocidal activity, could find an application in the elimination of microbial contaminations in various fields of our lives.

Keywords: antimicrobial activity; oxo-titanium(IV) complexes; polymer-inorganic composites; physicochemical properties; thermal properties

1. Introduction

The unique physical and chemical properties of materials based on titanium dioxide, i.e., photocatalytic activity, hydrophilicity, or strong absorption of UV radiation, contribute to their wide-application inter alia in various fields of chemical, cosmetics, biomedical, pharmaceutical, and disinfecting applications [1–5]. The photocatalytic activity of TiO_2 nanoparticles or nanocoatings, associated with the electron transfer between the valence band and the conduction one (the O2p-Ti3d charge transfer transition), which can take place in the presence of photons of wavelengths smaller than 350 nm (this corresponds to ca. 5% of the daylight), is of particular importance [3,4,6]. The above-mentioned phenomenon initiates the oxidation and reduction reactions, which proceed on the surface of TiO_2 nanoparticles/nanocoatings and lead to the formation of the reactive oxygen species (ROS) responsible for the antimicrobial activity of these materials, among others [3]. Aiming to extend the absorption range of TiO_2 in daylight, attention was paid to the oxo-titanium(IV) complexes (TOCs), which contain $\{Ti_aO_b\}$ cores of different architectures in their structures and their potential use as antimicrobial and photocatalytic active substrates [7–11]. These compounds are studied for the possibility of their use for the photoinduced degradation of organic dyes or water splitting [8,12]. In the case of TOCs, the electron transfer takes place also between the $2p$ electron shell of the oxygen atom and the $3d$ shell of the titanium atom; however, the presence of ligands, which stabilize the oxo-titanium(IV) complex cluster, act as photosensitizers, and in this way, the radiation can be absorbed in a wider range [13,14]. By placing the ligands in the coordination sphere (e.g., carboxylate ligands (-O$_2$CR′)), the orbitals HOMO-LUMO of ligands and the $\{Ti_aO_b\}$ core can be mixed. As a result, the distance for electrons to move is smaller. The charge transfer from the ligands to the cluster core is possible due to its low-energy absorption band [15,16]. The photocatalytic activity of the oxo-titanium(IV) complexes, of the different $\{Ti_aO_b\}$ core architectures and possessing various carboxylic ligands, was most often assessed based on the photodegradation processes of organic dye aqueous solutions, such as methylene orange (MO), methylene blue (MB), and rhodamine B (RB) [16–19]. The samples, during the photocatalytic experiments were mostly irradiated by UV light in the range 315–400 nm; however, also, visible light was used. Kim et al. studied the oxo-Ti(IV) complexes with the $\{Ti_6O_6\}$ core, and stabilized by such carboxylate ligands as 4-aminobenzoic, 4-amino-2-fluoro-benzoic, 4-amino-2-chloro-benzoic, 4-amino-3-chloro-benzoic, and 5-dichlorobenzoic acids, the anions were especially interesting. They showed a direct dependence of oxo-complex photocatalytic activity versus the way of -O$_2$CR′ group functionalization. This effect is closely related to the change in the distance between HOMO and LUMO orbitals in the tested compounds [19]. The results of our earlier studies on the photocatalytic activity of oxo-clusters with $\{Ti_3O\}$ and $\{Ti_4O_2\}$ cores also exhibited a significant influence in the way carboxylate ligands functionalize [20–22]. According to these works, the oxo-Ti(IV) complexes stabilized by 9-fluorenecarboxylate and 4-aminobenzonate ligands revealed the best photocatalytic properties in the photodegradation of the methylene blue (MB) solution [20,21].

An interesting issue that may be related to the photocatalytic activity of oxo-Ti(IV) complexes are their antimicrobial properties. Nowadays, searching for new antimicrobial materials results from the necessity of their wide use in everyday life and from their importance in the public health system. Analyzing the literature data, we turned our attention on the antibacterial activity of compounds that contain metal-oxo cores, e.g., oxo-iron(III) complexes [23] or polyoxometalates [24], in their structures. The antibacterial assay of the (Fe$_3$O(PhCO$_2$)$_6$(MeOH)$_3$)(NO$_3$)(MeOH)$_2$ cluster showed a significant growth inhibition of *Bacillus cereus* MTCC 1272, *Staphylococcus epidermidis* MTCC 3086, and *Salmonella typhimurium* MTCC 98 but not of *Escherichia coli* MTCC 723 [23]. The results of studies on the antibacterial properties of polyoxometalates against *Moraxella catarrhalis* seem to be particularly interesting. It has been revealed that microbiocidal activity of these types of compounds mainly depends on their composition; metal-oxide anion core shape and size; and the type of the central metal in MO$_6$ unit (M = Mo, V, and W) [24]. Antimicrobial properties have been found in the mixed ligand titanium dioxide complex, which was produced from TiO$_2$ nanoparticles and 8-hydroxyquinoline and glycine as the ligands [25]. This complex showed good antifungal activity against *Alternaria alternata*, *Rhizoctania solani*, *Helminthosporium oryzae*, *Fusarium oxysporum*, *Curvularia lunata*, and *Aspergillus niger*, as well as antibacterial activity against *Salmonella*, *E. coli*, *S. epidermis*, and *Enterococcus faecalis*. However, this compound did not show biocidal activity against *Aspergillus fumigatus*, *Aspergillus terreus*, *Trichoderma viride*, and *Cladosporium herbarium* [25]. The investigations of Kaushal et al. revealed the promising biological activities of titanium complexes synthesized by reacting to TiCl$_4$ with Schiff bases (SBs) [26,27]. The synthesized complexes were tested for their antimicrobial activity against pathogenic bacterial strains i.e., *B. cereus* MTCC 6728, *Micrococcus luteus* MTCC 1809, *S. aureus* MTCC 3160, *S. epidermidis* MTCC 3086, *Aeromonas hydrophila* MTCC 1739, *Aclaligenes faecalis* MTCC 126, *Shigella sonnei* MTCC 2957, *Klebsiella pneumoniae* MTCC 3384, *Pseudomonas aeruginosa* MTCC 1035, and *S. typhimurium* MTCC 1253. The (TiCl$_2$(SB)$_2$) complexes showed higher antimicrobial activity than their parent Schiff bases [27]. The analysis of the literature reports have shown that the antimicrobial activity of multinuclear oxo-titanium(IV) complexes have not been fully explored so far. Attention should be drawn to the results of the study by Svensson et al., which revealed the antibacterial activity of the oxo-complexes that consisted of {Ti$_4$O$_2$} cores stabilized by two triclosan ligands against *S. aureus* [28]. However, in this case, the antibacterial activity of the hydrolyzed form of synthesized complex was studied by the authors.

It should be noted that all the above-mentioned antimicrobial agents can be used in liquid form, which distinguishes them from TOCs systems, which are discussed in the presented paper. The studied compounds (TOCs) containing {Ti$_3$O} and {Ti$_4$O$_2$} cores due to their hydrophobic natures and their resistance to hydrolysis processes [20–22] were dispersed in a poly(methyl methacrylate) (PMMA) matrix (i.e., films (PMMA + TOCs) were produced). The aim of our investigations was the synthesis of a (PMMA + TOCs) composite characterized by very high surface antimicrobial activity. PMMA is an important material, which, due to sufficient mechanical properties and minimal inflammatory response, is used in different fields of our lives—inter alia, in dentistry [29]. An important direction of the research related to the use of PMMA-based materials is their modification aimed to improving their antimicrobial activity. This effect can be achieved by the addition of inorganic agents such as silver nanoparticles, titanium dioxide, a mixture of titanium dioxide and silicon dioxide nanoparticles, and anion powder (Na$_2$SiO$_3$) [29–34]. In this paper, the results of the modification of the PMMA films by the introduction of TOC micro-grains are discussed. The influence of TOC structures on the physicochemical properties and photocatalytic and microbicidal activity of the PMMA + TOC systems was assessed.

2. Results

Studied oxo-complexes (TOCs) were isolated from a the reaction mixture of appropriate titanium alkoxides and organic acids (4:1 alkoxide: acid molar ratio) at room temperature (RT) and at an inert gas atmosphere (Ar, Schlenk line), according to earlier described procedures (Table 1) [20–22]. 4-aminobenzoic acid (HOOC-4-PhNH$_2$) and 4-hydroxobenzoic acid (HOOC-4-PhOH) were used in these reactions. The type of the titanium alkoxide (OR; R = ^{i}Pr and ^{i}Bu) is the factor that directly influences the {Ti$_a$O$_b$} core architecture.

Table 1. The compositions of the reaction mixtures. THF: tetrahydrofuran.

4:1 Molar Ratio		Isolated	
Ti(OR)$_4$	HOOCR′	Solid Product	Solvent
Ti(O^{i}Pr)$_4$	HOOC-4-PhNH$_2$	(1)	THF/^{i}PrOH
Ti(O^{i}Pr)$_4$	HOOC-4-PhOH	(2)	THF/^{i}PrOH
Ti(O^{i}Bu)$_4$	HOOC-4-PhNH$_2$	(3)	Toluene
Ti(O^{i}Bu)$_4$	HOOC-4-PhOH	(4)	THF/^{i}BuOH

Unfortunately, the weak quality of the formed crystals that caused the structure of isolated solid reaction products were determined based on the analysis of IR, Raman spectra (Figures 1 and 2 and Table 2), and mass spectrometry ones (Table 3).

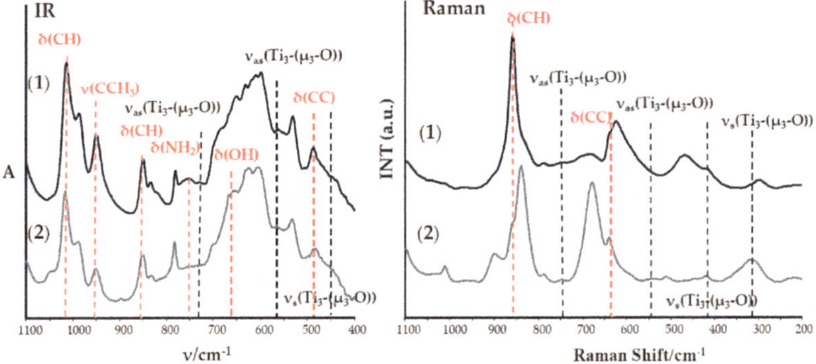

Figure 1. IR and Raman spectra of the (1) and (2) complexes.

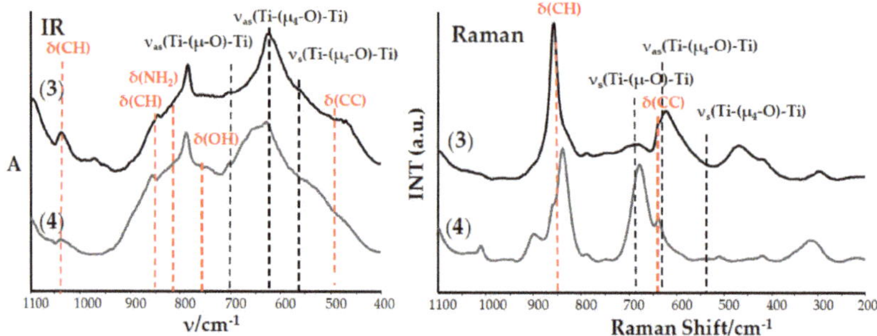

Figure 2. IR and Raman spectra of the (3) and (4) complexes.

The vibrational spectra analysis of the (1)–(4) compounds (Table 2) confirmed the presence of the coordinated carboxylate ligands 1514–1536 cm^{-1} ν_{as}(COO)) containing 1,4-substituted Ph groups 780–784 cm^{-1} (ν(CH), IR spectra) functionalized by NH_2 or OH groups (3100–3500 cm^{-1}) and, also, alkoxide groups 1014–1036 cm^{-1} and 948–988 cm^{-1} (ν(Ti-OR)) in their structures. The identification of normal vibrations bands, which descended from titanium-oxide bridge modes, was the spectral proof of the formed {Ti_aO_b} core type. For this purpose, the results of our previous DFT calculations were used [20–22]. The medium bands at 483 and 483 cm^{-1} in the IR spectra of (1) and (2) and weak and very weak bands, which were found at 719–728 cm^{-1}, 539–564 cm^{-1}, and 342–419 cm^{-1} in the IR and Raman spectra of these compounds, were attributed to vibrations of {Ti_3-(μ_3-O)} cores (Figure 1 and Table 2).

Table 2. Results of the vibrational spectra studies of the (1)–(3) complexes (the band intensity: strong (s), middle (m), weak (w), very weak (vw).

Modes	(1) IR	(1) R	(2) IR	(2) R	(3) IR	(3) R	(4) IR	(4) R
ν(OH) ν(NH$_2$)	3200–3450	3162 (w)	3300–3500		3200–3450	3175 (w)	3300–3500	
ν(CC) (Ph)	1622 (m) 1601 (m)	1603 (s)	1622 (m) 1603 (m)	1595 (s)	1621 (m) 1603 (m)	1602 (s)	1622 (m) 1603 (m)	1594 (s)
δ(NH$_2$)	1587 (m)				1589 (m)			
ν_{as}(COO)	1533 (m) 1516 (m)	1524 (m)	1536 (m) 1516 (m)	1519 (w)	1523 (m)	1524 (m)	1536 (m) 1510 (w)	1518 (w)
ν_s(COO)	1462 (w)	1436 (m)	1464 (w)	1450 (w)	1497 (m)	1459 (w)	1498 (m)	1445 (w)
δ(NH$_2$)	1362 (m)	1306 (m)			1299 (m)	1300 (m)		
ν(C-O) + δ(CC)	1014 (s)		1016 (m)		1036 (m)		1016 (m)	
ν(Ti-OR) + ν(CC) + δ(CH)	986 (m) 949 (m)	856 (m)	988 (m) 951 (m)	838 (m)	971 (m) 948 (w)	856 (m)	988 (m) 951 (w)	838 (m)
ν(CH) 1,4-Ph	780 (m)		782 (m)		784 (s)		782 (m)	
ν(Ti$_3$-(μ_3-O))	750 (w) 728 (w)	747 (vw) 719 (w)	755 (vw) 727 (vw)	748 (vw)				
ν(Ti$_2$-(μ-O))					703 (m)	690 (vw)	702 (m)	687 (vw)
ν(Ti$_4$-(μ_4-O))					639 (m)	623 (w)	634 (m)	610 (vw)
δ(CCC)	683 (w) 666 (m) 622 (m)	622 (m) 605 (w)	701 (w) 638 (m) 629 (w)	678 (m) 638 (w)				
ν(Ti$_3$-(μ_3-O))	561 (w)		564 (w)	539 (vw)				
ν(Ti$_4$-(μ_4-O))			532 (w)	510 (w)	532 (m)	548 (vw)	556 (m)	536 (vw)
ν(Ti$_3$-(μ_3-O))	486 (m)	417 (w)	483 (m)	419 (w)				
ν(Ti$_3$-(μ_3-O))		342 (vw)		350 (vw)				

Analysis of the electrospray ionization-mass spectrometry (ESI-MS) spectra of (**1**) and (**2**) confirm the presence of peaks, which can be assigned to the fragmentation ions containing $\{Ti_3\text{-}(\mu_3\text{-}O)\}$ cores (Table 3). Considering these data, we can state that the trinuclear oxo-Ti(IV) complexes, i.e., $[Ti_3O(O^iPr)_8(O_2C\text{-}4\text{-}PhNH_2)_2]$ (**1**) and $[Ti_3O(O^iPr)_8(O_2C\text{-}4\text{-}PhOH)_2]$ (**2**) were isolated from the 4:1 reaction mixture of $Ti(O^iPr)_4$ and HO_2CR'; $R' = 4\text{-}PhNH_2$ and 4-PhOH using the tetrahydrofuran (THF)/Pr^iOH (1:1 mixture) as a solvent. The bands, which were detected at 702 and 703 cm^{-1} and 687 and 690 cm^{-1} in the IR and Raman spectra of (**3**) and (**4**), were attributed to stretching modes of $Ti_4(\mu_4\text{-}O)$ bridges, whereas bands at 634–639 cm^{-1} and 532–556 cm^{-1} (IR) and 687–690 cm^{-1} and 536–548 cm^{-1} (Raman) were assigned to vibrations of $Ti_4\text{-}(\mu\text{-}O)$ bridges of $\{Ti_4O_2\}$ cores (Figure 2 and Table 2). Additionally, in these cases, the presence of peaks attributed to fragmentation ions containing $\{Ti_4O_2\}$ cores in MS spectra of (**3**) and (**4**) confirms the tetranuclear structures of these complexes (Table 2). The results obtained indicate the isolation of the tetranuclear Ti(IV) oxo-complexes $[Ti_4O_2(O^iBu)_{10}(O_2C\text{-}4\text{-}PhNH_2)_2]$ (**3**) and $[Ti_4O_2(O^iBu)_{10}(O_2C\text{-}4\text{-}PhOH)_2]$ (**4**) from the mother liquor composed of a 4:1 mixture of $Ti(O^iBu)_4$ and the above-mentioned organic acids and a 1:1 mixture of THF and Bu^iOH as a solvent.

Table 3. The results of electrospray ionization-mass spectrometry (ESI-MS) spectra studies of the (**1**)–(**4**) complexes. TOCs: tetranuclear oxo-titanium(IV) complexes.

TOCs	m/z	Fragmentation Ion	Intensity (%)
$[Ti_3O(O^iPr)_8(O_2C\text{-}4\text{-}PhNH_2)_2]$ (**1**)	768	$(Ti_3O(O^iPr)_8(O_2C\text{-}4\text{-}PhNH_2))^+$	8
	550	$(Ti_3O(O^iPr)_2(O_2C\text{-}4\text{-}PhNH_2))^+$	36
$[Ti_3O(O^iPr)_8(O_2C\text{-}4\text{-}PhOH)_2]$ (**2**)	847	$(Ti_3O(O^iPr)_7(O_2C\text{-}4\text{-}PhOH)_2)^+$	26
	769	$(Ti_3O(O^iPr)_8(O_2C\text{-}4\text{-}PhOH))^+$	90
$[Ti_4O_2(O^iBu)_{10}(O_2C\text{-}4\text{-}PhNH_2)_2]$ (**3**)	1080	$(Ti_4O_2(O^iBu)_8(O_2C\text{-}4\text{-}PhNH_2)_2)^+$	5
	881	$(Ti_4O_2(O^iBu)_9)^+$	5
$[Ti_4O_2(O^iBu)_{10}(O_2C\text{-}4\text{-}PhOH)_2]$ (**4**)	1229	$(Ti_4O_2(O^iBu)_{10}(O_2CPhOH)_2) + H^+$	8
	945	$(Ti_4O_2(O^iBu)_8(O_2CPhOH))^+$	10

2.1. UV–Vis Diffuse Reflectance Spectra (UV-Vis-DRS) of the (1)–(4) Oxo-Complexes and HOMO-LUMO Gap Determination

UV-Vis-DRS spectra of the (**1**)–(**4**) oxo-complexes were registered at room temperature using magnesium oxide as a standard reference (Figure 3). The HOMO-LUMO gap values were determined basing on the Kubelka-Munk (K-M) function versus light energy, i.e., $K = f(h\nu)$, where $K = (1-R)^2/2R$ and R is the reflectance, which was used for the optical band gap determination (Figure 3). According to these data, trinuclear oxo-complexes (**1**) and (**2**) exhibit absorption on the border of the UV and Vis range, $\lambda_{max} = 400$ nm for (**1**) and $\lambda_{max} = 360$ nm for (**2**), and with a sharp absorption edge at ca. 520 nm and 480 nm for complexes (**1**) and (**2**), respectively (Figure 3a). The wide absorption bands with maxima at 405 nm and 390 nm and with the same sharp absorption edge at 500 nm were found in the spectra of (**3**) and (**4**), respectively (Figure 3b). The determined HOMO-LUMO energy gap values (ΔE) revealed significant differences between the oxo-complexes with a different type of stabilized carboxylate groups. The lowest ΔE values were found for (**1**), $\Delta E = 2.0$ eV, and (**3**), $\Delta E = 2.15$ eV, i.e., oxo-complexes containing $-O_2C\text{-}4PhNH_2$ ligands. The HOMO-LUMO energy gap values of complexes that contain the $-O_2C\text{-}4\text{-}PhOH$ groups in their structures were the highest and were equal to $\Delta E = 2.38$ eV for (**2**) and $\Delta E = 2.25$ eV for (**4**) (Figure 3).

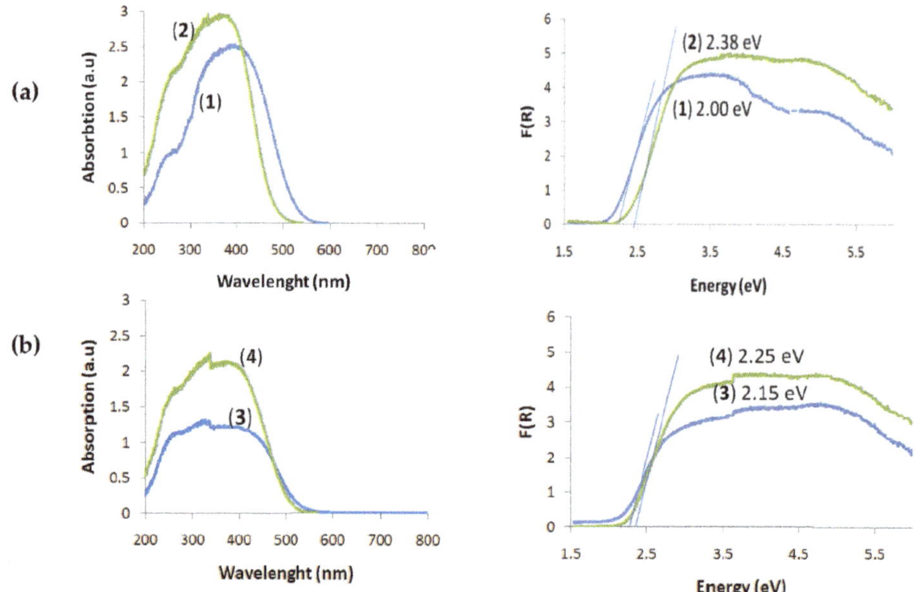

Figure 3. Solid-state UV-Vis-diffuse reflectance spectra (DRS) of the **(1)**–**(4)** micro-grains (the left side) and Kubelka-Munk function versus light energy plot for the HOMO-LUMO gap determination (the right side): (**a**) [Ti$_3$O(O^{i}Pr)$_8$(O$_2$CC$_6$H$_4$NH$_2$)$_2$] (**1**) and [Ti$_3$O(O^{i}Pr)$_8$(O$_2$CC$_6$H$_4$OH)$_2$] (**2**) and (**b**) [Ti$_4$O$_2$(O^{i}Bu)$_{10}$(O$_2$CC$_6$H$_4$NH$_2$)$_2$] (**3**) and [Ti$_4$O$_2$(O^{i}Bu)$_{10}$(O$_2$CC$_6$H$_4$OH)$_2$] (**4**).

2.2. (PMMA + TOC) Composites

The photocatalytic properties of the **(1)**–**(4)** titanium(IV) oxo-complexes (TOCs) and their antimicrobial activity were estimated using (PMMA + TOC) composite foils produced by the dispersion of 20 wt.% TOC micro-grains in the PMMA matrix [21,22]. SEM images, presented in Figure 4, show that composite foils (PMMA + **(1)**, **(2)**, and **(4)**) contain the dispersed TOC grains of diameters 100–300 µm. In the case of (PMMA + **(3)**), fine grains (of diameters 3–5 µm) of the oxo-complex that may form larger aggregates are dispersed in the polymer matrix. The registration of Raman microscopy maps allowed to investigate the TOC grain presence and distribution in the PMMA matrix (Figure 5).

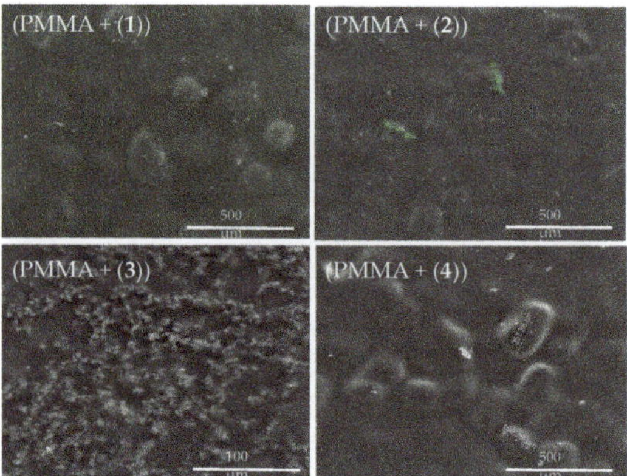

Figure 4. SEM images of poly(methyl methacrylate) (PMMA) + TOCs and TOCs= **(1)**–**(4)** composites dispersed in the PMMA matrix.

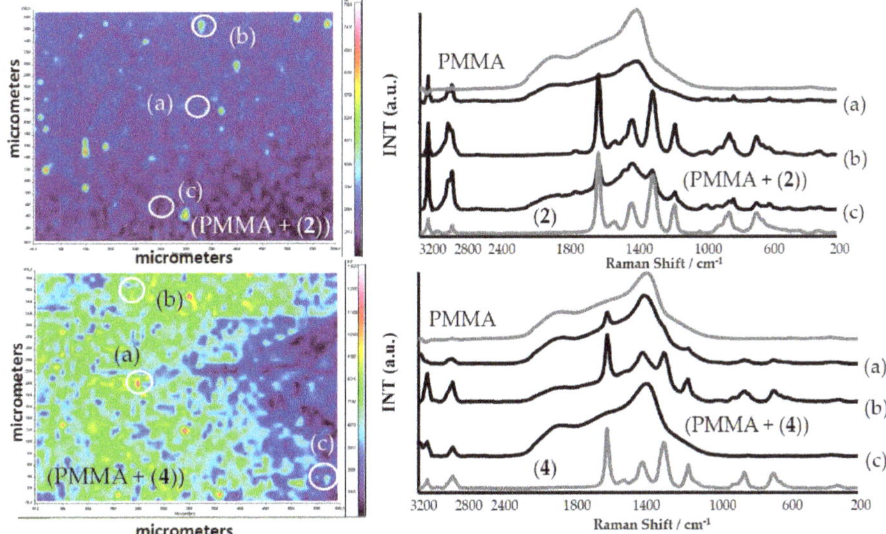

Figure 5. Raman microscopy maps, registered for PMMA + TOC composites (TOCs (**2**) and (**4**)). The Raman spectra, in selected points (a)–(c) on the surface of PMMA + TOCs, are presented alongside.

An analysis of the Raman spectra of composite foils proved the presence of tri- and tetranuclear titanium(IV) oxo-complexes of the earlier determined structure. The use of an ultrasonic bath for the dispersion of TOC grains in the polymer solution allowed the appropriate uniform distribution of the fine grains of TOCs throughout the entire polymer volume (Figure 5). However, probably, the solvent evaporation led to the formation of larger TOC grains and places dominated by the polymer.

2.3. Thermal Analysis of PMMA/TOC Composites

The use of thermogravimetric analysis (TGA) and differential scanning calorimetry (DSC) allowed the thermal characterization of (PMMA + TOCs) composites and the PMMA matrix in the temperature range 30–500 °C at inert atmosphere. The results of these investigations are presented in Table 4 and Figure 6. The analysis of the TGA data of all studied samples revealed a slight weight loss (10–15%), which occurred below 280 °C (I Stage), and the major one (70–85%) between 280 and 450 °C (II Stage). Considering the previous investigations, Stage I can be attributed to the loss of the residual solvent and/or monomer, while Stage II was due the samples' thermal decomposition [35,36].

Table 4. Thermal parameters received from thermogravimetric analysis (TGA) and differential scanning calorimetry (DSC) of the poly(methyl methacrylate) (PMMA) and (PMMA + TOCs) composites (T_g = glass transition temperature, T = evaporation temperature of an unreacted monomer, T_m = melting temperature. T_{max} = temperature in the thermal transition maximum, Δm = thermal transition weight loss).

Composite	DSC			TGA		
	$T_g/°C$	$T/°C$	$T_m/°C$	Stage I $T_{max}/°C/\Delta m/\%$	Stage II $T_{max}/°C/\Delta m/\%$	Solid Residue at 450 °C (%)
PMMA	99.6	150.8	368.6	194.9/12	365.1/85	3
(PMMA+(1))	100.6	-	370.4	188.2/12	367.8/70	18
(PMMA+(2))	101.2	-	375.7	183.9/11	365.4/74	15
(PMMA+(3))	99.0	-	377.8	159.9/15	372.7/76	9
(PMMA+(4))	115.2	-	364.5	171.0/10	374.0/79	11

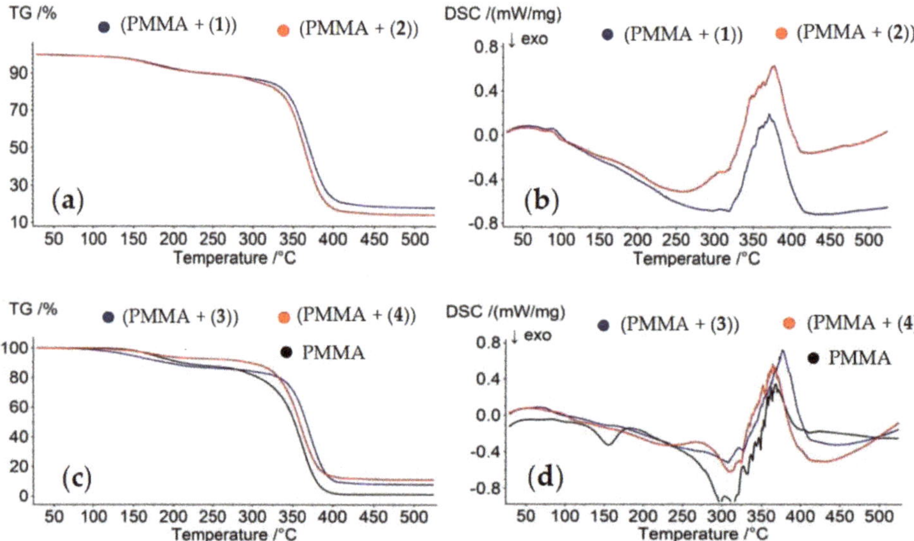

Figure 6. Thermogravimetric curves (TGA) (**a**,**c**) and the differential scanning calorimetry curves of (DSC) (**b**,**d**) of the produced composite materials (PMMA + (**1**)–(**4**)) and PMMA.

An analysis of the DSC thermograms exhibits that the PMMA matrix has a glass transition temperature (T_g) at 99.6 °C, while the exothermic region at 150.8 °C (T), according to previous reports, can be assigned to the evaporation of an unreacted monomer [37] (Figure 6d). The latter effect in the DSC curves of (PMMA + (**1**)–(**4**)) samples was not observed. The endothermic peak at 368.6 °C, which was found on the PMMA DSC thermogram, can be assigned to the PMMA melting temperature (T_m). In the case of composite films enriched with (**1**), (**2**), and (**3**) oxo-complexes, the strong endothermic peaks appear at 370.4 °C, 375.7 °C, and 377.8 °C; this is above the T_m of the pure PMMA matrix. The exception is (PMMA + (**4**)), for which the T_m was registered at about 364.5 °C.

2.4. Estimation of Photocatalytic Activity of the (1)–(4) Oxo-Complexes

The photocatalytic activity of the (PA + TOCs) systems, with TOCs (**1**)–(**4**), was estimated on the basis of the photodecolorization process of the methylene blue (MB) solution irradiated in the UVA range, according to the ISO 10678:2010 procedure [38,39]. The changes of MB concentrations versus irradiation times for the studied (PMMA + TOCs) composites and the reaction observed rate constants are presented in Figure 7 and Table 5. According to these data, the (PMMA + (**3**)) and (PMMA + (**4**)) composites exhibit the best photocatalytic activity. Simultaneously, in the case of the (PMMA + (**1**)) and (PMMA (**2**)) samples, the very weak photocatalytic activity that was noticed ((PMMA + (**1**)) has activity similar to the pure PMMA (difference in absorbance during 180-min measurements, $\Delta A_{180} = 0.011$). The calculated values of ΔA_{180} are presented in Table 5. Moreover, the photodegradation of Rhodamine B of the above-mentioned composite films was carried out under VIS light, and the obtained results are presented in Figure 7 and in Table 5 [40].

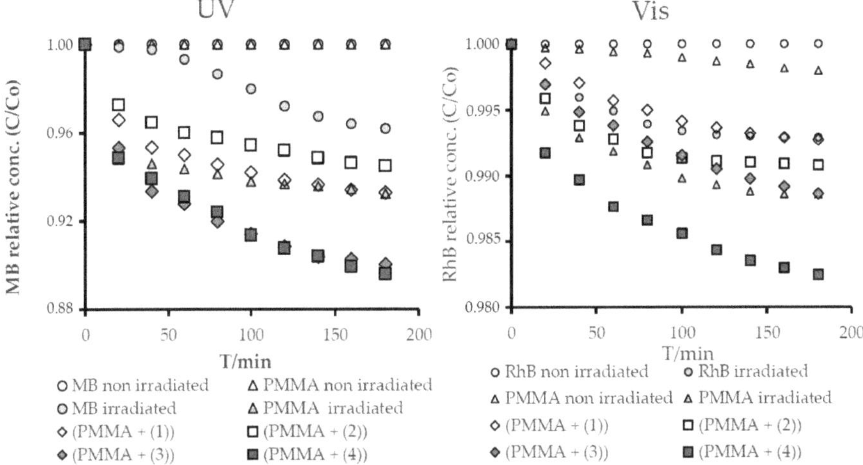

Figure 7. Changes in the concentrations of the methylene blue (MB) and Rhodamine B (RhB) solutions as a function of time for the respective composite materials irradiated with UV and Vis light.

Table 5. Organic dye decolorization percentages and ΔA_{180} parameters for the studied reactions in relation to the composites.

Composite	MB Decolorization [a] (%)	ΔA_{180}	ΔA_{180} in Reference to PMMA
PMMA+(1)	6.70	0.067	−0.001
PMMA+(2)	5.50	0.055	−0.013
PMMA+(3)	9.97	0.100	0.032
PMMA+(4)	10.42	0.104	0.036
PMMA(irradiated)	6.76	0.068	-

Composite	RhB Decolorization [b] (%)	ΔA_{180}	ΔA_{180} in Reference to PMMA
PMMA+(1)	0.73	0.007	−0.004
PMMA+(2)	0.92	0.009	−0.002
PMMA+(3)	1.14	0.011	0
PMMA+(4)	1.75	0.018	0.007
PMMA(irradiated)	1.14	0.011	-

[a] Methylene blue (MB) decolorization at the end of the measurements (t = 180 min). [b] Rhodamine B (RhB) decolorization at the end of the measurements (t = 180 min).

2.5. Antimicrobial Activity of (PMMA + TOCs) Composites

The results of the (PMMA + (**1**)–(**4**)) antimicrobial activity tests against Gram-positive (*S. aureus* ATCC 6538 and *S. aureus* ATCC 25923) and Gram-negative (*E. coli* ATCC 8739 and *E. coli* ATCC 25922) bacteria and yeasts of *Candida albicans* ATCC 10231 of (PMMA + TOCs) composites, where TOCs (**1**)–(**4**) are presented in Table 6. According to these data, all of the (PMMA + TOCs) composites showed high antimicrobial activity against all tested bacteria (*R* between 2.5–6.0). In case of tests performed on the yeasts of *C. albicans* biocidal activity, $R \geq 2$ or 99% was found for PMMA (**2**)–(**4**)—namely, the complex containing the {Ti$_3$O} core and stabilized by the 4-hydroxybenzoic ligand, as well as for complexes with {Ti$_4$O$_2$} cores but not for PMMA (**1**), which is a complex with a {Ti$_3$O} core and stabilized by the -O$_2$C-4-PhNH$_2$ ligand.

Table 6. Antimicrobial activity of the (PMMA +TOCs) systems (R = biocidal rate (%), $R\%$ = reduction percentage).

Microorganisms	PMMA+ (1)		PMMA+ (2)		PMMA+ (3)		PMMA+(4)	
	R	$R\%$	R	$R\%$	R	$R\%$	R	$R\%$
E. coli ATCC 8739	2.5	99.6	6.0	>99.99	6.0	>99.99	6.0	>99.99
E. coli ATCC 25922	3.3	99.95	6.0	>99.99	6.0	>99.99	6.0	>99.99
S. aureus ATCC 6538	6.0	>99.99	6.0	>99.99	6.0	>99.99	6.0	>99.99
S. aureus ATCC 25923	6.0	>99.99	6.0	>99.99	6.0	>99.99	6.0	>99.99
C. albicans ATCC 10231	1.7	85.0	6.0	>99.99	6.0	>99.99	6.0	>99.99

$R \geq 2$ is a biocidal effect when the microbial growth is reduced at least 100 times (99.0%).

3. Discussion

The multinuclear oxo-titanium(IV) complex applicability as a potentially new group of agents with antimicrobial properties was assessed based on carried out investigations. Considering the results of previous works, the oxo-clusters forming the small $\{Ti_aO_b\}$ core structures (i.e., $\{Ti_4O_2\}$ and $\{Ti_3O\}$), and stabilized by -O_2C-4-PhR′ (R′ = -NH_2, -OH) ligands have been chosen to further microbiological experiments [20–23]. The complexes were isolated as micro-grain powders from the reaction mixtures of appropriate titanium alkoxides and organic acids (4:1 molar ratio) at room temperature and in inert atmosphere. The structure of the isolated compounds such as ($Ti_3O(O^iPr)_8(O_2C$-4-PhR′$)_2$) (R′ = -NH_2 (**1**), -OH (**2**)), ($Ti_4O_2(O^iBu)_{10}(O_2C$-4-PhR′$)_2$) (R′ = -NH_2 (**3**), and -OH (**4**)) were determined as a result of IR, Raman, and ESI-MS spectra analyses. Studies of UV-Vis-DRS spectra revealed that the absorption maximum of the (**1**)–(**4**) compounds is shifted towards the visible range, i.e., 360–405 nm, with a sharp absorption edge between 480–520 nm (Figure 3). The results of the previous investigations proved that the $\{Ti_aO_b\}$ cores were absorbed in the UV range, which is due to the O2p-Ti3d charge transfer transition [41]. The shifting of the absorption to the visible range, which was registered in the spectra of the (**1**)–(**4**) compounds, can be explained by the ligand-to-core charge transfer (LCCT) from the -O_2CR′ (R′ = 4-PhNH_2 or 4-PhOH) ligands to the tri- or tetranuclear titanium-oxo cores. A similar effect was discussed by Cui et al. for a group of multinuclear Ti(IV)-oxo complexes stabilized by 4-chlorosalicylate ligands [42]. The designated values of the HOMO-LUMO gap energy for the studied oxo-complexes changes in the row (**1**) < (**3**) < (**4**) < (**2**) (Figure 3). The analysis of these data confirms that the type of carboxylate ligand, which stabilizes the $\{Ti_aO_b\}$ skeleton, is the main factor allowed to effectively control the HOMO-LUMO energy gap of multinuclear oxo-complexes. This effect was highlighted during earlier studies of Ti(IV)-oxo clusters of similar core structures and significant different carboxylate ligands [13,14,21,22,41,42]. The influence of the $\{Ti_aO_b\}$ cluster core size on the HOMO-LUMO energy gap (E_g) is definitely more negligible, e.g., the results of Cui et al.'s investigations revealed that, for the $\{Ti_4O\}$, $\{Ti_{11}O_9\}$, and $\{Ti_{14}O_{12}\}$ clusters stabilized by O^iPr and 4-chlorosilicylate ligands, the E_g values were 3.0, 2.8, and 2.9 eV, i.e., they differed from each other by 0.1 and 0.2 eV [42]. In the case of the studied $\{Ti_3O\}$ ((**1**) and (**2**)) and $\{Ti_4O_2\}$ ((**3**) and (**4**)) clusters, these differences amounted to 0.13 and 0.15 eV, respectively (Figure 3).

The research carried out led to the isolation of micro-grained powders of TOCs containing $\{Ti_3O\}$ ((**1**) and (**2**)) and $\{Ti_4O_2\}$ ((**3**) and (**4**)} cores, in which the absorption maximum was shifted towards the UVA-Vis range in comparison to TiO_2 (UV, λ_{max} = 300–350 nm [43]). The type of the carboxylate ligand, which stabilizes the $\{Ti_aO_b\}$ core, influences the HOMO-LUMO gap values (ΔE), i.e., ΔE (**1**) and (**2**) < ΔE (**3**) and (**4**).

Due to the hydrophobic characters of studied powders, their photocatalytic and microbiocidal activity were estimated using (PMMA + TOCs) composite films of 0.25–0.50-μm thickness. The analysis of the SEM images proved that the uniform dispersion of TOC fine grains (3–5 μm) in the PMMA matrix was found only in the (PMMA + (**3**)) composite, while, in the cases of (PMMA + TOCs and TOC = (**1**), (**2**), and (**4**)) composites, the studied materials contained the dispersed, larger (100–300 μm) TOC grains (Figure 4). The nonuniform distribution of TOCs in composite films revealed the analysis of Raman microscopy maps of the (PMMA + (**2**) and (**4**)) films (Figure 5). The large grains of TOCs are separated from each other by the pure polymer sites and by the areas that are formed by fine grains dispersed in PMMA matrices. The thermal characterization of (PMMA + (**1**)–(**4**)) showed a lack of significant differences between their thermal properties and the PMMA matrix. However, higher thermal decomposition temperatures of composite films versus the pure polymer indicate that the Ti(IV)-oxo complex addition to the PMMA matrix improves its thermal stability. A similar effect was observed for the PMMA matrix containing 20% $BaTiO_3$ [37]. While analyzing the data presented in Table 3, attention should be paid to the differences in the thermal stability of (PMMA + TOCs) composites resulting from the type of added oxo-cluster. The dispersion of trinuclear oxo-complexes in the PMMA matrix slightly improves the thermal stability of the composite in comparison to the pure polymer, i.e., 0.3–2.7 °C, while this difference is clearer and was found to be 7.6–8.9 °C for the tetranuclear ones (Table 4).

For the reason that studied oxo-complexes can be a potential source of the reactive oxygen species (ROS), before the antimicrobial tests, their photocatalytic activity was determined. Luo et al. [18] and Lv et al. [44] demonstrated the promising photocatalytic activity of the oxo-Ti(IV) clusters dispersed in solutions of MO (methyl orange) and MB. Considering the potential application of TOCs (i.e., (**1**)–(**4**)) clusters, as an antimicrobial agent introduced to the polymer matrix (e.g., PMMA), the photocatalytic activity assessment was made for their (PMMA + (**1**)–(**4**)) composites. Our earlier studies drew our attention to the good photocatalytic activity of $(Ti_4O_2(O^iPr)_{10}(O_2C\text{-}4\text{-}PhNH_2)_2)$, which the 20 wt.% addition to the polystyrene (PS) matrix or PMMA was one, clearly improved the photodecolorization effect of methylene blue (MB) solution [21,22]. Additionally, trinuclear systems exhibited good photocatalytic activity, which, in the case of $(Ti_3O(O^iBu)_8(O_2CR')_2)$, (R' = -$C_{13}H_9$, -3-$PhNO_2$) was proven to be better than for the appropriate tetranuclear oxo-Ti(IV) systems [22]. In the case of the (PMMA + (**1**)–(**4**)) films, the best photocatalytic activity revealed the (PMMA + (**4**)) system irradiated with UVA and Visible light for the MB solution and RhB one, respectively. The PMMA film containing uniformly dispersed fine grains of (**3**) showed slightly weaker activity also in the UVA-Vis range. Cui et al. drew attention to the importance of the size of the oxo-cluster and the related specific surface area [42]. In our case, the cluster size was similar, while the sizes of the grains dispersed in the PMMA matrix were different (Figure 4), which can be associated with the influence of the grain surfaces on their photocatalytic activity. It is interesting that a decrease the of Ti_3O core size of the oxo-clusters influences the significant decrease of the (PMMA + TOCs) composites (where the TOC (**1**) and (**2**)) have photocatalytic activity, especially in Vis range, and practically no significant changes in UV activity.

The (PMMA + TOCs) systems were sterilized with the use of UVC light, which, according to our UV–Vis-DRS measurements, did not activate the studied samples. In the next stage, the samples were irradiated with natural indoor light directly before inoculation with microorganisms. Considering the results of our earlier works regarding the photocatalytic activity of the (polymer + TOCs) systems [21,22], we assumed that the antimicrobial mechanism of the action of the TOCs could be similar to that of the TiO_2-assisted ultraviolet treatment (TUV). TUV is a well-known technique used in water disinfection [45,46] or the food industry [47] for the inactivation of microorganisms. The degree of photoinduced deactivation of the microorganisms depends on various treatment parameters, the nature of the microorganisms, and environmental conditions [45,48]. The mechanisms for the bactericidal action of TiO_2 photocatalysis are associated with the generation of reactive oxygen species (ROSs), such as superoxide radical anion ($O_2^-\cdot$), hydrogen peroxide (H_2O_2), and hydroxyl radical (OH·), which cause oxidative damage to living organisms [47,49]. Many authors suggested that the cell membrane is the primary site of ROS attack [50–52]. This attack of the cell membrane by ROS induce lipid peroxidation. The cell membrane damage directly leads to the leakage of minerals, proteins, and genetic materials, which is the root cause of cell death [51]. Apart from the cytoplasmic membrane, the supercoiled plasmid DNA, genomic DNA, and internal organelles are destructed when the bacteria are exposed to TUV [47]. Similar mechanisms of PMMA coated with TiO_2 action against *E. coli* and *S. aureus* were suggested by Su et al. [49]. Moreover, they claimed that a further oxidative attack of internal cellular components accelerates the cell death and, ultimately, results in the decrease of the survival ratio of both tested bacterial strains, namely *S. aureus* and *E. coli* BL21. Furthermore, Pan et al. [53] evaluated the antibacterial properties of nano-trititanate ($H_2Ti_3O_7$ nanomaterial) against *E. coli* MC1061 and observed that the bactericidal capabilities of these nano-trititanates were more significant compared to nano-TiO_2, both with and without exposure to UV light. Various nano-titanates activated with UV light inhibited the bacterial growth in the range of 11.9–57.1%, while these nonirradiated in the range of 25.7–94.7%. The dependence between the photocatalytic properties and antimicrobial activity was also noticed for studied foils of (PMMA + TOCs) composites, containing the 20 wt.% of (**1**)–(**4**) TOCs. The weakest biocidal activity was observed for the (PMMA + (**1**)), which simultaneously showed the worst photocatalytic properties, even irradiated by the light in the UVA-Vis range. The photocatalytic activity of the (PMMA + (**2**)–(**4**)) films was higher, which can explain the strong inhibition of microorganism growth (>99.99%) on their surfaces. It should be noted that the biocidal activity of the (PMMA + TOC) composite films was assessed for 20 wt.% of TOC micro-grain contents due to our earlier studies of (polymer + (**3**)) composite photocatalytic activity (polymer PS and PMMA) [22]. Although ROS generation is the main proposed mechanism of antimicrobial activity of titanium nanomaterials or titanium nanomaterial-containing complexes, the other mechanism of action should be also considered. It is well-known that some nanomaterials, especially metal-based ones, exhibit strong inhibitory effects towards a broad spectrum of bacterial strains. It is claimed that the biocidal effect of these materials, including titanium oxide nanoparticles, results from the "electromagnetic" attraction between the microbe and metal oxide nanoparticles, as microorganisms carry a negative charge while metal oxides carry a positive charge [54]. Overall, metal-based nanoparticles may interact with the sulfur-containing proteins present in the cell envelopes causing irreversible changes in the cell wall structure, resulting in its disruption and affecting the permeability of the cell membrane by altering transport activity through the plasma [55] or disrupting the ATP production [56]. Nanoparticles can further penetrate inside the microbial cells and interact with ribosomes and biomolecules such as proteins, lipids, and DNA, which may alter translation, replication, and other processes in microbial cells. Moreover, the ion release, as a secondary oxidation process, contributes to the biocidal properties of nanomaterials. Metal ions also affect the losses in the ability to replicate DNA, translation process in ribosomes, and protein activities [57–59]. They can affect membrane transport and the release of potassium (K^+) ions from the microbial cells. The increase of membrane permeability caused by both nanomaterials and released ions may lead to the leakage of cellular contents, including ions, proteins, reducing sugars, and, sometimes, cellular energy reservoirs (ATP) [58,60–62]. It should, however, be noted that the above-mentioned mechanism can be the result of the direct contact between the nanoparticles and bacterial cell. In the discussed case, the TOC micro-grains were surrounded by the polymer matrix, thus limiting their mutual contact. Therefore, the received results suggested that a mechanism based on the ROS formation seems to be the most probable.

In further investigations, the lowest but still highly active content of TOCs as an antimicrobial agent in the polymer matrix should be determined. Moreover, the possible cytotoxicity of studied (PMMA + TOCs) composites should be assessed, which is extremely important in the case of the potential biomedical applications of these systems. The antimicrobial activity of nanomaterials prevents bacterial adhesion and biofilm formation. This activity is especially important for medical applications of nanomaterials or nanomaterial-containing composites. Biofilms protect the underlying microbes from antibiotics and host defense mechanisms and, thus, may lead to serious infections. Moreover, the antimicrobial properties of nanomaterials may be used to prevent microbial contaminations that cause food spoilage, the spread of foodborne diseases, and bio-fouling of materials [29,63].

4. Materials and Methods

4.1. Materials

Titanium(IV) isopropoxide (Aldrich, St. Louis, MO, USA), titanium(IV) isobutoxide (Aldrich, St. Louis, MO, USA), 4-aminobenzoic acid (Aldrich, St. Louis, MO, USA), and 4-hydroxybenzoic acid (Aldrich, St. Louis, MO, USA) were purchased commercially and were used without further purification. All solvents used in the synthesis, i.e., tetrahydrofuran (THF), isobutanol (HO^iBu), and isopropanol (HO^iPr) were distilled before their use and stored in argon atmosphere. The processes of Ti(IV) oxo-complexes synthesis were carried out using the standard Schlenk technique in the inert gas atmosphere (Ar) and at room temperature (RT).

4.2. Synthesis of Ti(IV) Oxo-Complexes (TOCs) and (PMMA +TOCs) Composites

The synthesis of $[Ti_3O(O^iPr)_8(O_2CC_6H_4NH_2)_2]$ (**1**): 0.12 g of 4-aminobenzoic acid (0.875 mmol) was added to the solution of 1-mL titanium(IV) isopropoxide (3.5 mmol) in 2 mL of THF/Pr^iOH (1:1), leading to a clear yellow solution. The solution was left for crystallization. The yield based on acid: 68% (0.31 g). Anal. Calc. for $C_{38}H_{68}O_{13}Ti_3N_2$: C, 50.44; H, 7.52; N, 3.10; Ti, 15.93. Found: C, 50.38; H, 7.48; N, 2.99; Ti, 15.96.

The synthesis of $[Ti_3O(O^iPr)_8(O_2CC_6H_4OH)_2]$ (**2**): 0.12 g of 4-hydroxybenzoic acid (0.875 mmol) was added to the solution of 1-mL titanium(IV) isopropoxide (3.5 mmol) in 2 mL of THF/Pr^iOH (1:1), leading to a clear yellow solution. The solution was left for crystallization. The yield based on acid: 74% (0.33 g). Anal. Calc. for $C_{38}H_{66}O_{15}Ti_3$: C, 50.33; H, 7.28; Ti, 15.89. Found: C, 50.81; H, 7.22; Ti, 15.80.

The synthesis of $[Ti_4O_2(O^iBu)_{10}(O_2CC_6H_4NH_2)_2]$ (**3**): complex was synthesized, as reported [21]. 0.12 g of 4-aminobenzoic acid (0.875 mmol) was added to the solution of 1.19-mL titanium(IV) isobutoxide (3.5 mmol) in 2 mL of toluene, leading to a clear yellow solution. The solution was left for crystallization. The yield based on acid: 41% (0.22 g). Anal. Calc. for $C_{54}H_{102}O_{14}Ti_4N_2$: C, 52.86; H, 8.38; N, 2.28; Ti, 15.61. Found: C, 53.14; H, 7.83; N, 2.05; Ti, 15.56.

The synthesis of $[Ti_4O_2(O^iBu)_{10}(O_2CC_6H_4OH)_2]$ (**4**): 0.12 g of 4-hydroxybenzoic acid (0.875 mmol) was added to the solution of 1.19-mL titanium(IV) isobutoxide (3.5 mmol) in 2 mL of THF/Bu^iOH (1:1), leading to a clear yellow solution. The solution was left for crystallization. The yield based on acid: 63% (0.30 g). Anal. Calc. for $C_{54}H_{100}O_{18}Ti_4$: C, 52.77; H, 8.14; Ti, 15.64. Found: C, 51.36; H, 7.92; Ti, 16.01.

The polymer foils containing 20 wt.% of synthesized Ti(IV)oxo-clusters (TOCs) were prepared by an addition of the solution of the TOC solution (ca. 0.25 g of (**1**)–(**4**) TOCs were dispersed in 1 cm^3 of THF) to the poly(methyl methacrylate) (PMMA) solution (1.0 g of PMMA dissolved in 5 cm^3 of THF). The resulting mixtures were stirred in an ultrasonic bath for 120 min; in the next step, they were poured into a glass Petri dish and left for the evaporation of the solvent at RT. The composite foil thickness ca. 50 μm were characterized by Raman and IR spectroscopy and scanning electron microscopy.

4.3. Analytical Procedures

The structures of the isolated solid reaction products (crystals and powders) were confirmed using vibrational spectroscopy methods, i.e., IR spectrophotometry (Perkin Elmer Spectrum 2000 FTIR spectrophotometer ($400–4000\ cm^{-1}$ range, KBr pellets)) and Raman spectroscopy (RamanMicro 200 spectrometer (PerkinElmer, Waltham, MA, USA)). Raman spectra were registered using a laser with the wavelength 785 nm, with a maximum power of 350 mW, in the range $200–3200\ cm^{-1}$, using a 20 × 0.40/FN22 objective lens and an exposure time of 15 s each time. Elemental analyses were performed on an Elemental Analyser vario Macro CHN Elemental Analyser vario Macro CHN Elementar Analysen Systeme GmbH (Elementar, Hanau, Germany). The mass spectra were recorded using the ESI-MS method using a QToFSynapt G2 Si (Waters Corporation, Mundelein, IL, USA) spectrometer. The main measurement parameters: capillary voltage: 2.5–3 kV, source temperature: 110 °C, sampling cone voltage 20–50 V, source offset 40–60 V, desolvation temperature 250–350 °C, cone gas flow: 50.0 dm^3/h, and desolation gas flow: 798.0–900 dm^3/h. The produced PMMA + TOCs foil surfaces were studied using a scanning electron microscope with field emission (SEM, Quanta 3D FEG, Houston, TX, USA). Composite materials underwent thermal treatment (Bruker Optik, Ettingen, Germany) in the range 20–500 °C, with the heating speed of 5 °C/min in the atmosphere of nitrogen.

4.4. HOMO-LUMO Gap Determination

The HOMO-LUMO gap energy values of isolated (**1**)–(**4**) complexes were determined by using diffuse reflectance UV-Vis spectra (UV-VIS-DRS), which were registered between 200 and 800 nm. Jasco V-750 spectrophotometer was used (JASCO Deutschland GmbH, Pfungstadt, Germany). The recorded spectra were evaluated in terms of energy band gap values via Spectra Manager TM CFR software.

4.5. The Photocatalytic Activity Evaluation of (PMMA + TOCs) Composites

The photocatalytic activity of PMMA + TOCs foils (TOCs = (**1**)–(**4**)) was studied by monitoring the degradation processes of MB aqueous solution according to ISO 10678:2010 procedure and, also, Rhodamine B (RhB) solution [38–40]. Foil samples of sizes 10×10 mm were preconditioned by exposure to UVA or Vis light for 30 h. In the next step, foils were placed in quartz cuvettes with both dye solutions (V = 3.5 cm^3 and C = 2.0×10^{-5} M). After 12 h in the dark, the solutions were replaced by the appropriate test of MB and RhB solutions (for both: V = 3.5 cm^3 and C = 1.0×10^{-5} M). The prepared samples were exposed to UVA irradiation (18-W lamp, 340–410-nm range, with a maximum at 365 nm) and Vis light (77-W tungsten halogen lamp, range of 350–2200 nm). All cuvettes were covered with quartz glass panes during irradiation. MB absorbance at 660 nm and RhB absorbance at 554 nm were registered (Metertech SP-830 PLUS, Metertech, Inc., Taipei, Taiwan) after 0, 20, 40, 60, 80, 100, 120, 140, 160, and 180 min of irradiation. Percentage of MB decolorization was calculated using the equation

$$\% \text{ dye decolorization} = ((C_0 - C_t)/C_0) \times 100 = ((A_0 - A_t)/A_0) \times 100 \tag{1}$$

where C_0 is an initial concentration of dye, C_t is a dye concentration at a given time t, and A_0 and A_t are absorbances at 0 and t times [38,64].

The rate of both photodegradation processes was estimated by a simple calculation of a difference in absorbance during 180-min measurements (ΔA_{180}). Such kind of calculation is a result of a low degradation degree of dyes.

4.6. The Evaluation of Antimicrobial Properties of (PMMA + TOCs) Systems

Antimicrobial activity of PMMA + TOCs (TOCs = (**1**)–(**4**)) composite foils ($30 \times 30 \times 0.05$ mm) was studied against Gram-positive (*S. aureus* ATCC 6538 and *S. aureus* ATCC 25923) and Gram-negative (*E. coli* ATCC 8739 and *E. coli* ATCC 25922) bacteria and yeasts of *Candida albicans* ATCC 10231 using method according to the ISO 22196:2011 standard [65]. All strains were purchased from the American Type Culture Collection (Manassas, VA, USA). PMMA +TOC specimens were sterilized using a UVC lamp for 30 min in the laminar hood (Bioquell, Hampshire, UK) and placed in sterile Petri plates. The earlier investigations revealed that the absorption maximum of the (PMMA + TOCs) samples was found at the UVA-Vis border, which enabled the use of UVC radiation in order for their sterilization. Microbial suspension ($1.0–1.8 \times 10^6$ colony-forming units (cfu) $cm^{3/-1}$) prepared in sterile deionized water was placed on the surfaces of the appropriate PMMA + TOCs samples and covered with sterile foil films (polypropylene; PP) and incubated for 24 h at 37 °C in a humid atmosphere. The PP film was then removed and microbial suspension collected into a 2-cm^3 centrifuge tube. Subsequently, serial ten-fold dilutions of each sample were prepared. Aliquots (100 µL) of each dilution were aseptically spread over the surface of Tryptic Soy Agar (TSA, Becton Dickinson, Sparks Glencoe, MD, USA) or Sabouraud Dextrose Agar (SDA, Becton Dickinson) in Petri plates for the bacteria and fungi, respectively. Inoculated samples were incubated at 37 °C for 24 h. Assays were performed in triplicate. After incubation, colony-forming units (cfu) were counted on the agar plates. Control was PMMA inoculated with the test microorganism.

The antimicrobial activity was calculated using Formula (2), according to the ISO 22196:2011 standard [65]:

$$R = U_t - A_t \tag{2}$$

where U_t is the average of the common logarithm of the number of viable bacteria in cells/cm^2 recovered from the untreated test specimens (PMMA) after 24 h, and A_t is the average of the common logarithm of the number of viable bacteria in cells/cm^2 recovered from the treated test specimens (PMMA + TOCs) after 24 h. $R \geq 2$ determines the biocidal activity

The percentage reduction ($R\%$) of the bacterial or fungal growth was calculated using Formula (3):

$$R\% = ((B - A)/B) \times 100 \tag{3}$$

where R is the biocidal rate (%), B—the average number of microorganisms on unmodified PMMA in T_0, and A—the average number of microorganisms on the surfaces of the studied (PMMA + TOCs) composites after 24 h. The 100-times reduction of the number of colonies determines at least 99% of the reduction of microbial growth (biocidal activity).

5. Conclusions

The work carried out led to the production of oxo-titanium(IV) complexes (TOCs), i.e., $(Ti_3O(O^iPr)_8(O_2C\text{-}4\text{-}PhR')_2)$ **(1)** and **(2)** and $(Ti_4O_2(O^iBu)_{10}(O_2C\text{-}4\text{-}PhR')$ **(3)** and **(4)** (R' = -NH$_2$ and -OH), which structures were confirmed by IR, Raman spectroscopy, and ESI-MS spectrometry. The dispersion of 20 wt.% of TOCs **(1)**–**(4)** in the PMMA matrix allowed the formation of (PMMA + TOCs) composite films for which the thermal and photocatalytic properties and biocidal activity were estimated. The obtained results suggest that $(Ti_4O_2(O^iBu)_{10}(O_2C\text{-}4\text{-}PhOH)_2)$ **(4)** can be used for the production of composites (polymer + TOC), which, in the form of films, coatings, or resin additives, can be used as a bactericidal agent in various areas of our lives. It should be noted that the potential use of materials based on PMMA with the addition of TOCs in dentistry would require medical experiments carried out in accordance with the procedures specified by Isola et al. [66,67].

Author Contributions: Conceptualization, P.P.; methodology, P.P., B.K., P.G., and A.R.; formal analysis, P.P., B.K., P.G., and A.R.; investigation, B.K. and P.G.; data curation, P.P. and B.K.; writing—original draft preparation, P.P.; writing—review and editing, P.P. and A.R.; and supervision, P.P. and A.R. All authors have read and agreed to the published version of the manuscript.

Funding: This research received no external funding.

Conflicts of Interest: The authors declare no conflict of interest.

References

1. Hashimoto, K.; Irie, H.; Fujishima, A. TiO$_2$ Photocatalysis: A Historical Overview and Future Prospects. *Jpn. J. Appl. Phys.* **2005**, *44*, 8269–8285. [CrossRef]
2. Chen, X.; Mao, S.S. Titanium Dioxide Nanomaterials: Synthesis, Properties, Modifications, and Applications. *Chem. Rev.* **2007**, *107*, 2891–2959. [CrossRef] [PubMed]
3. Carp, O.; Huisman, C.L.; Reller, A. Photoinduced Reactivity of Titanium Dioxide. *Prog. Solid State Chem.* **2004**, *32*, 33–177. [CrossRef]
4. Fujishima, A.; Zhang, X.X.; Tryk, D.A. TiO$_2$ photocatalysis and related surface phenomena. *Surf. Sci. Rep.* **2008**, *63*, 515–558. [CrossRef]
5. Benedict, J.B.; Coppens, P. The Crystalline Nanocluster Phase as a Medium for Structural and Spectroscopic Studies of Light Absorption of Photosensitizer Dyes on Semiconductor Surfaces. *J. Am. Chem. Soc.* **2010**, *132*, 2938–2944. [CrossRef] [PubMed]
6. Li, N.; Matthews, P.D.; Luo, H.-K.; Wright, D.S. Novel properties and potential applications of functional ligand-modified polyoxotitanate cages. *Chem. Commun.* **2016**, *52*, 11180–11190. [CrossRef]
7. Assi, H.; Mouchaham, G.; Steunou, N.; Devic, T.; Serre, C. Titanium coordination compounds: From discrete metal complexes to metal–organic frameworks. *Chem. Soc. Rev.* **2017**, *46*, 3431–3452. [CrossRef]
8. Benedict, J.B.; Freindorf, R.; Trzop, E.; Cogswell, J. Large Polyoxotitanate Clusters: Well-Defined Models for Pure-Phase TiO$_2$ Structures and Surfaces. *J. Am. Chem. Soc.* **2010**, *132*, 13669–13671. [CrossRef]
9. Coppens, P.; Chen, Y.; Trzop, E. Crystallography and Properties of Polyoxotitanate Nanoclusters. *Chem. Rev.* **2014**, *114*, 9645–9661. [CrossRef]

10. Rozes, L.; Sanchez, C. Titanium oxo-clusters: Precursors for a Lego-like construction of nanostructured hybrid materials. *Chem. Soc. Rev.* **2011**, *40*, 1006–1030. [CrossRef]

11. Su, H.-C.; Wu, Y.-Y.; Hou, J.-L.; Zhang, G.-L.; Zhu, Q.-Y.; Dai, J. Dye molecule bonded titanium alkoxide: A possible new type of dye for sensitized solar cells. *Chem. Commun.* **2016**, *52*, 4072–4075. [CrossRef] [PubMed]

12. Zou, D.-H.; Cui, L.-N.; Liu, P.-Y.; Yang, S.; Zhu, Q.-Y.; Dai, J. Triphenylamine derived titanium oxo clusters: An approach to effective organic–inorganic hybrid dyes for photoactive electrodes. *Chem. Commun.* **2018**, *54*, 9933–9936. [CrossRef] [PubMed]

13. Fan, Y.; Zhang, Y.; Zou, H.-M.; Li, H.-M.; Cui, Y.; Jing, Q.-S.; Lu, H.-T. Modulating the band gap and photoelectrochemical activity of dicarboxylate-stabilized titanium-oxo clusters. *Inorg. C. Acta* **2018**, *482*, 16–22. [CrossRef]

14. Macyk, W.; Szaciłowski, K.; Stochel, G.; Buchalska, M.; Kuncewicz, J.; Łabuz, P. Titanium(IV) complexes as direct TiO$_2$ photosensitizers. *Coord. Chem. Rev.* **2010**, *254*, 2687–2701. [CrossRef]

15. Hong, Z.-F.; Xu, S.-H.; Yan, Z.-H.; Lu, D.-F.; Kong, X.-J.; Long, L.-S.; Zheng, L.-S. A Large Titanium Oxo Cluster Featuring a Well-Defined Structural Unit of Rutile. *Cryst. Growth Des.* **2018**, *18*, 4864–4868. [CrossRef]

16. Su, M.K.; Wu, Y.; Tan, W.; Wang, D.; Yuan, M.H. A monomeric bowl-like pyrogallol[4]arene Ti12 coordination complex. *Chem. Commun.* **2017**, *53*, 9598–9601. [CrossRef]

17. Wu, Y.-Y.; Luo, W.; Wang, Y.-H.; Pu, Y.-Y.; Zhang, X.; You, L.-S.; Dai, J. Titanium–oxo–Clusters with Dicarboxylates: Single-Crystal Structure and Photochromic Effect. *Inorg. Chem.* **2012**, *51*, 8982–8988. [CrossRef]

18. Luo, W.; Ge, G. Two Titanium-oxo-Clusters with Malonateand Succinate Ligands: Single-Crystal Structuresand Catalytic Property. *J. Clust. Sci.* **2016**, *27*, 635–643. [CrossRef]

19. Kim, S.; Sarkar, D.; Kim, Y.; Park, M.H.; Yoon, M.; Kim, Y.; Kim, M. Synthesis of functionalized titanium-carboxylate molecular clusters and their catalytic activity. *J. Ind. Eng. Chem.* **2017**, *53*, 171–176. [CrossRef]

20. Janek, M.; Muzioł, T.M.; Piszczek, P. Trinuclear Oxo-Titanium Clusters: Synthesis, Structure, and Photocatalytic Activity. *Materials* **2019**, *12*, 3195. [CrossRef]

21. Janek, M.; Radtke, A.; Muzioł, T.M.; Jerzykiewicz, M.; Piszczek, P. Tetranuclear Oxo-Titanium Clusters with Different Carboxylate Aromatic Ligands: Optical Properties, DFT Calculations, and Photoactivity. *Materials* **2018**, *11*, 1661. [CrossRef] [PubMed]

22. Janek, M.; Muzioł, T.; Piszczek, P. The structure and photocatalytic activity of the tetranuclear titanium(IV) oxo-complex with 4-aminobenzoate ligands. *Polyhedron* **2018**, *141*, 110–117. [CrossRef]

23. Pathak, S.; Jana, B.; Mandal, M.; Mandal, V.; Ghorai, T.K. Antimicrobial activity study of a μ_3-oxo bridged [Fe$_3$O(PhCO$_2$)$_6$(MeOH)$_3$](NO$_3$)(MeOH)$_2$] cluster. *J. Mol. Struct.* **2017**, *1147*, 480–486. [CrossRef]

24. Gumeroval, N.I.; Al-Sayed, E.; Lukáš Krivosudský, L.; Čipčić-Paljetak, H.; Verbanac, D.; Rompel, A. Antibacterial Activity of Polyoxometalates Against Moraxella catarrhalis. *Front. Chem.* **2018**, *6*, 336. [CrossRef]

25. Kumar, S.; Bhanjana, G.; Kumar, R.; Dilbaghi, N. Synthesis, Characterization and Antimicrobial Activity of Nano Titanium(IV) Mixed Ligand Complex. *Mater. Focus* **2013**, *2*, 475–481. [CrossRef]

26. Kaushal, R.; Kumar, N.; Thakur, A.; Nehra, K.; Awasthi, P.; Kaushal, R.; Arora, S. Synthesis, Spectral Characterization, Antibacterial and Anticancer Activity of some Titanium Complexes. *Anticancer Agents Med. Chem* **2018**, *18*, 739–746. [CrossRef] [PubMed]

27. Kaushal, R.; Thakur, S. Syntheses and Biological screening of Schiff base complexes of Titanium(IV). *Chem. Eng. Trans.* **2013**, *32*, 1801–1806. [CrossRef]

28. Svensson, F.G.; Seisenbaeva, G.A.; Kessler, V.G. Mixed-Ligand Titanium "Oxo Clusters": Structural Insights into the Formation and Binding of Organic Molecules and Transformation into Oxide Nanostructures on Hydrolysis and Thermolysis. *Eur. J. Inorg. Chem.* **2017**, *35*, 4117–4122. [CrossRef]

29. Zafar, M.S. Prosthodontic Applications of PolymethylMethacrylate (PMMA): An Update. *Polymers* **2020**, *12*, 2299. [CrossRef]

30. Liu, M.; Zhang, X.; Zhang, J.; Zheng, Q.; Liu, B. Study on Antibacterial Property of PMMA Denture Base Materials with Negative Ion Powder. *Mater. Sci. Eng.* **2018**, *301*, 012034. [CrossRef]

31. Chen, R.; Han, Z.; Huang, Z.; Karki, J.; Wang, C.; Zhu, B.; Zhang, X. Antibacterial activity, cytotoxicity and mechanical behaviour of nano-enhanced denture base resin with different kinds of inorganic antibacterial agents. *Dent. Mol. J.* **2017**. [CrossRef] [PubMed]

32. Sadagar, A.; Khalil, S.; Kassaee, M.Z.; Shahroudi, A.S.; Pourakbari, B.; Bahador, A. Antimicrobial properties of poly(methyl methacrylate) acrylic resins incorporated with silicon dioxide and titanium dioxide nanoparticles on cariogenic bacteria. *J. Orthod. Sci.* **2016**, *5*, 7–13. [CrossRef]

33. De Mori, A.; Di Grgorio, E.; Kao, A.P.; Tozzi, G.; Barbu, E.; Sanghani-Kerai, A.; Draheim, R.R.; Rolado, M. Antibacterial PMMA Composite Cments with Tunable Thermal and Mechanical Properties. *ACS Omega* **2019**, *4*, 19664–19675. [CrossRef] [PubMed]

34. Kumar, M.; Arun, S.; Upadhyaya, P.; Pugazhenthi, G. Properties of PMMA/clay nanocomposites prepared using various compatibilizer. *Int. J. Mech. Mat. Eng.* **2015**, *10*, 7. [CrossRef]

35. Gałka, P.; Kowalonek, J.; Kaczmarek, H. Thermogravimetric analysis of thermal stability of poly(methylmethacrylate) films modified with photoinitiators. *J. Therm. Anal. Calorim.* **2014**, *115*, 1387–1394. [CrossRef]

36. Lu, P.; Huang, Q.; Mukherjee, A.; Hsieh, Y.-L. Effects of polymer matrices to the formation of silicon carbide (SiC) nanoporous fibers and nanowires under carbothermal reduction. *J. Mater. Chem.* **2011**, *21*, 1005–1012. [CrossRef]

37. Elshereksi, N.W.; Mohamd, S.H.; Arifin, A.; Ishak, Z.A.M. Thermal Characterization of Poly(Methyl Methacrylate) Filled with Barium Titanate as Denture Base Material. *J. Phys. Sci.* **2014**, *25*, 15–27.

38. ISO 10678:2010(E) Standards. Fine Ceramics (Advanced Ceramics, Advanced Technical Ceramics)—Determination of Photocatalytic Activity of Surfaces in an Aqueous Medium by Degradation of Methylene Blue. Available online: https://www.iso.org/standard/46019.html (accessed on 10 June 2020).

39. Ehlert, M.; Radtke, A.; Topolski, A.; Śmigiel, J.; Piszczek, P. The Photocatalytic Activity of Titania Coatings Produced by Electrochemical and Chemical Oxidation of Ti6Al4V Substrate, Estimated According to ISO 10678:2010. *Materials* **2020**, *13*, 2649. [CrossRef]

40. Patrick, W.; Dietmar, S. Photodegradation of rhodamine B In aqueous solution via $SiO_2@TiO_2$ nanospheres. *J. Photochem. Photobiol. A Chem.* **2007**, *185*, 19–25. [CrossRef]

41. Liu, J.-X.; Gao, M.-Y.; Fang, W.-H.; Zhang, L.; Zhang, J. Bandgap Engineering of Titanium-Oxo Clusters: Labile Surface Sites Used for Ligand Substitution and Metal Incorporation. *AngewandeChemie* **2016**, *128*, 5246–5251. [CrossRef]

42. Cui, Y.; Zou, G.-D.; Li, H.-M.; Huang, Y.; Fan, Y. 4-Chlorocalicylate-stabilized titanium-oxo clusters with structures mediated by tetrazole and their photophysical properties. *Polyhedron* **2019**, *157*, 177–182. [CrossRef]

43. Habibi, S.; Jamshidi, M. Sol–gel synthesis of carbon-doped TiO_2 nanoparticles based on microcrystalline cellulose for efficient photocatalytic degradation of methylene blue under visible light. *J. Environ. Technol.* **2020**, *41*, 3233–3247. [CrossRef] [PubMed]

44. Lv, H.-T.; Li, H.-M.; Zou, G.-D.; Cui, Y.; Huang, Y.; Fan, Y. Tioanium-oxo clusters functionalized with cateholate-tye ligands: Modulating the optical properties through charge-transfer transition. *Dalton Trans.* **2018**, *47*, 8158–8163. [CrossRef] [PubMed]

45. Kim, S.; Ghafoor, K.; Lee, J.; Feng, M.; Hong, J.; Lee, D.; Park, J. Bacterial inactivation in water, DNA strand breaking, and membrane damage induced by ultraviolet assisted titanium dioxide photocatalysis. *Water Res.* **2013**, *47*, 4403–4411. [CrossRef]

46. Wu, M.J.; Bak, T.; O'Doherty, P.J.; Moffitt, M.C.; Nowotny, J.; Bailey, T.D.; Kersaitis, C. Photocatalysis of titanium dioxide for water disinfection: Challenges and future perspectives. *Int. J. Photogramm.* **2014**, 973484. [CrossRef]

47. Ramesh, T.; Nayak, B.; Amirbahman, A.; Tripp, C.P.; Mukhopadhyay, S. Application of ultraviolet light assisted titanium dioxide photocatalysis for food safety: A review. *Innov. Food Sci. Emerg. Technol.* **2016**, *38*, 105–115. [CrossRef]

48. Maness, P.; Smolinski, S.; Blake, D.M.; Huang, Z.; Wolfrum, E.J.; Jacoby, W.A. Bactericidal activity of photocatalytic TiO_2 reaction: Toward an understanding of its killing mechanism. *Appl. Environ. Microbiol.* **1999**, *65*, 4094–4098. [CrossRef]

49. Su, W.; Wang, S.; Wang, X.; Fu, X.; Weng, J. Plasma pre-treatment and TiO_2 coating of PMMA for the improvement of antibacterial properties. *Surf. Coat. Technol.* **2010**, *205*, 465–469. [CrossRef]

50. Lin, H.; Xu, Z.; Wang, X.; Long, J.; Su, W.; Fu, X.; Lin, Q. Photocatalytic and antibacterial properties of medical-grade PVC material coated with TiO_2 film. *J. BiomedMater. Res. B Appl. Biomater.* **2008**, *87*, 425–431. [CrossRef]

51. Yu, J.C.; Ho, W.; Lin, J.; Yip, H.; Wong, P.K. Photocatalytic activity, antibacterial effect, and photoinduced hydrophilicity of TiO$_2$ films coated on a stainless steel substrate. *Environ. Sci. Technol.* **2003**, *37*, 2296–2301. [CrossRef]

52. Huang, Z.; Maness, P.C.; Blake, D.M.; Wolfrum, E.J.; Smolinski, S.L.; Jacoby, W.A. Bactericidalmode of titaniumdioxidephotocatalysis. *J. Photochem. Photobiol. A Chem.* **2000**, *130*, 163–170. [CrossRef]

53. Pan, R.; Liu, Y.; Chen, W.; Dawson, G.; Wang, X.; Li, W.; Dong, B.; Zhu, Y. The toxicity evaluation of nano-trititanate with bactericidal properties in vitro. *Nanotoxicology* **2012**, *6*, 327–337. [CrossRef]

54. Jesline, A.; John, N.P.; Narayanan, P.M.; Vani, C.; Murugan, S. Antimicrobial activity of zinc and titanium dioxide nanoparticles against biofilm-producing methicillin-resistant. *Staphylococcus aureus. Appl. Nanosci.* **2015**, *5*, 157–162. [CrossRef]

55. Ghosh, S.; Patil, S.; Ahire, M.; Kitture, R.; Kale, S.; Pardesi, K. Synthesis of silver nanoparticles using *Dioscoreabulbifera* tuberextract and evaluation of its synergistic potential in combination with antimicrobial agents. *Int. J. Nanomed.* **2012**, *7*, 483–496. [CrossRef]

56. Inbaneson, S.J.; Ravikumar, S.; Manikandan, N. Antibacterial potential of silver nanoparticles against isolated urinary tract infectious bacterial pathogens. *Appl. Nanosci.* **2011**, *1*, 231–236. [CrossRef]

57. Morones, J.R.; Elechiguerra, J.L.; Camacho, A.; Ramirez, J.T. The bactericidal effect of silver nanoparticles. *Nanotechnology* **2005**, *16*, 2346–2353. [CrossRef]

58. Chauhan, R.; Kumar, A.; Abraham, J. A biological approach to the synthesis of silver nanoparticles with *Streptomyces* sp. JAR1 and its antimicrobial activity. *Sci. Pharm.* **2013**, *81*, 607–621. [CrossRef]

59. Jung, W.K.; Koo, H.C.; Kim, K.W.; Shin, S.; Kim, S.H.; Park, Y.H. Antibacterial activity and mechanism of action of the silver ion in *Staphylococcus aureus* and *Escherichia coli*. *Appl. Environ. Microbiol.* **2008**, *74*, 2171–2178. [CrossRef]

60. Lok, C.N.; Ho, C.M.; Chen, R.; He, Q.Y.; Yu, W.Y.; Sun, H. Proteomic analysis of the mode of antibacterial action of silver nanoparticles. *J. Proteome Res.* **2006**, *5*, 916–924. [CrossRef]

61. Kim, S.H.; Lee, H.S.; Ryu, D.S.; Choi, S.J.; Lee, D.S. Antibacterial activity of silver-nanoparticles against *Staphylococcus aureus* and *Escherichia coli*. *Korean J. Microbiol. Biotechnol.* **2011**, *39*, 77–85.

62. Li, J.; Rong, K.; Zhao, H.; Li, F.; Lu, Z.; Chen, R. Highly selective antibacterial activities of silver nanoparticles against *Bacillus subtilis*. *J. Nanosci. Nanotechnol.* **2013**, *13*, 6806–6813. [CrossRef] [PubMed]

63. Bajpaia, V.K.; Kamleb, M.; Shuklaa, S.; Mahatoc, D.K.; Chandrad, P.; Hwange, S.K.; Kumarb, P.; Huhe, Y.S.; Han, Y.-K. Prospects of using nanotechnology for food preservation, safety, and security. *J. Food Drug Anal.* **2018**, *26*, 1201–1214. [CrossRef] [PubMed]

64. Xu, H.; Ouyang, S.; Liu, L.; Reunchan, P.; Umezawa, N.; Ye, J. Recent advances in TiO$_2$-based photocatalysis. *J. Mat. Chem. A* **2014**, *2*, 12642–12661. [CrossRef]

65. ISO 22196:2011 Standards. Measurement of Antibacterial Activity on Plastics and Other Non-Porous Surfaces. Available online: https://www.kieferusa.com/wp-content/uploads/2019/07/MRSA-ISO-22196.pdf (accessed on 3 June 2019).

66. Isola, G.; Alibrandi, A.; Currò, M.; Matarese, M.; Ricc, A.S.; Matarese, G.; Ientile, R.; Kocher, T. Evaluation of salivary and serum ADMA levels in patients with periodontal and cardiovascular disease as subclinical marker of cardiovascular risk. *J. Periodontol.* **2020**, *91*, 1076–1084. [CrossRef]

67. Isola, G.; Polizzi, A.; Patini, R.; Ferlito, S.; Alibrandi, A.; Palazzo, G. Association among serum and salivary A. actinomycetemcomitans specific immunoglobulin antibodies and periodontitis. *BMC Oral Health* **2020**, *20*, 283. [CrossRef]

Publisher's Note: MDPI stays neutral with regard to jurisdictional claims in published maps and institutional affiliations.

International Journal of
Molecular Sciences

Article

Valorization of *Gleditsia triacanthos* Invasive Plant Cellulose Microfibers and Phenolic Compounds for Obtaining Multi-Functional Wound Dressings with Antimicrobial and Antioxidant Properties

Ioana Cristina Marinas [1,2], Eliza Oprea [3,*], Elisabeta-Irina Geana [4], Oana Tutunaru [5], Gratiela Gradisteanu Pircalabioru [1], Irina Zgura [6] and Mariana Carmen Chifiriuc [1]

1   Research Institute of the University of Bucharest-ICUB, Microbiology Department, Faculty of Biology, University of Bucharest, 91-95 Spl. Independentei, 050095 Bucharest, Romania; ioana.cristina.marinas@gmail.com (I.C.M.); gratiela87@gmail.com (G.G.P.); carmen.chifiriuc@gmail.com (M.C.C.)
2   National Institute of Research & Development for Food Bioresources—IBA Bucharest, 6 Dinu Vintila Street, 021102 Bucharest, Romania
3   Department of Organic Chemistry, Biochemistry and Catalysis, Faculty of Chemistry, University of Bucharest, 4-12 Regina Elisabeta, 030018 Bucharest, Romania
4   National R&D Institute for Cryogenics and Isotopic Technologies—ICIT Rm. Valcea, 4 Uzinei Street, PO Raureni, 240050 Ramnicu Valcea, Romania; irina.geana@icsi.ro
5   National Institute for Research and Development in Microtechnologies IMT-Bucharest, Erou Iancu Nicolae Street, 126A, 077190 Bucharest, Romania; oana.tutunaru@imt.ro
6   Department of Optical Processes in Nanostructured Materials, National Institute of Materials Physics Atomistilor Street, 405A, 077125 Magurele, Romania; irina.zgura@infim.ro
*   Correspondence: eliza.oprea@g.unibuc.ro; Tel.: +40-723-250-470

Citation: Marinas, I.C.; Oprea, E.; Geana, E.; Tutunaru, O.; Pircalabioru, G.G.; Zgura, I.; Chifiriuc, M.C. Valorization of *Gleditsia triacanthos* Invasive Plant Cellulose Microfibers and Phenolic Compounds for Obtaining Multi-Functional Wound Dressings with Antimicrobial and Antioxidant Properties. *Int. J. Mol. Sci.* **2021**, *22*, 33. https://dx.doi.org/10.3390/ijms22010033

Received: 16 November 2020
Accepted: 17 December 2020
Published: 22 December 2020

Publisher's Note: MDPI stays neutral with regard to jurisdictional claims in published maps and institutional affiliations.

Abstract: *Gleditsia triacanthos* is an aggressive invasive species in Eastern Europe, producing a significant number of pods that could represent an inexhaustible resource of raw material for various applications. The aim of this study was to extract cellulose from the *Gleditsia triacanthos* pods, characterize it by spectrophotometric and UHPLC–DAD-ESI/MS analysis, and use it to fabricate a wound dressing that is multi-functionalized with phenolic compounds extracted from the leaves of the same species. The obtained cellulose microfibers (CM) were functionalized, lyophilized, and characterized by ATR-FTIR and SEM. The water absorption and retention capacity as well as the controlled release of phenolic compounds with antioxidant properties evaluated in temporal dynamics were also determined. The antimicrobial activity against reference and clinical multi-drug-resistant *Escherichia coli*, *Staphylococcus aureus*, *Pseudomonas aeruginosa*, *Acinetobacter baumannii*, *Enterobacter cloacae*, *Candida albicans*, and *Candida parapsilosis* strains occurred immediately after the contact with the tested materials and was maintained for 24 h for all tested microbial strains. In conclusion, the multi-functionalized cellulose microfibers (MFCM) obtained from the reproductive organs of an invasive species can represent a promising alternative for the development of functional wound dressings with antioxidant and antimicrobial activity, as well as being a scalable example for designing cost-effective, circular bio-economy approaches to combat the accelerated spread of invasive species.

Keywords: multi-functionalized cellulose microfibers; controlled release; wound dressing; antimicrobial activity; antioxidant activity

1. Introduction

Micro- and nanomaterials derived from biological macromolecules are used in a wide range of industrial, technological, and biomedical applications, e.g., for adsorption, ultrafiltration, packaging, preservation of historical artefacts, thermal insulators, and fire retardants, energy extraction and storage, acoustics, sensory, controlled delivery of drugs, and especially tissue engineering. Due to their mechanical robustness, hydrophilicity, biocompatibility, and good biodegradability, micro- and nanocellulose have specific functions, such as tissue repair, regeneration, and healing, and they are also antimicrobial nanomaterials with shape memory and intelligent membranes [1,2].

Different types of cellulose with a well-defined and functionalized morphology with different active principles can be used to obtain advanced wound dressings with high performance for preventing wound infections and accelerated wound healing [1]. Previous studies of wood-based nanofibrillated cellulose (NFC) have shown a high capacity to absorb liquids and form translucent films, suggesting the benefits of using cellulose fibers as a dressing. The property of maintaining a moist environment is necessary for the proper healing of chronic wounds, while translucency makes it possible to evaluate the wound progress without requiring the dressing removal [3]. In addition, these nanomaterials can provide various drug delivery properties and other effects, depending on the raw materials they come from [4] and can be used for the design of functional wound dressing or for topical drug delivery, offering several advantages over other administration pathways, including eliminating internal metabolism, minimizing pain, prolonging the drug release, and controlling drug absorption by removing skin bandage [5]. On the other side, Nie (2019) showed that the nanocellulose fibers (NF) have lower thermal stability compared with raw microcellulose fibers (MF) due to the reduced particle size, increased specific surface area, and increased exposed carbon and reactive group activity. The thermal degradation of NF generates low molecular weight segments and breakpoint defects. The same effect was produced after introducing the carboxylate group on the cellulose surface of the nano-sized fiber (prepared by oxidation of TEMPO ((2,2,6,6-tetramethyl-1-piperidinyl)oxidanyl) [2]. Microcellulose fibers can be obtained from different plant species (including invasive ones), such as bamboo [6], *Artemisia vulgaris* [2], and banana plant wastes [7].

Gleditsia triacanthos L. (*Fabaceae*) is a native tree to North America, with an extraordinary tolerance to high temperatures and drought, conditions that paradoxically favor its spread. This species is also very tolerant of industrial and transport emissions with relatively high resistance to soils with extreme pH. The combined clonal and sexual reproduction, the short juvenile period, the high seed production and the high germination rate of the seeds are considered defining traits for the degree of invasiveness of these species [8]. Initially introduced as an ornamental and aromatic species, it spread rapidly, invading different ecoregions, being presently considered one of the most aggressive legume invaders in the world [9]. This species invasiveness poses serious risks for ecosystems biodiversity, structure, and functions, as well as conservation of the protected areas, and requires high costs for its eradication [10]. The selective harvesting of reproductive organs of such invasive species for practical use could stop their uncontrolled spread. Taking into account that plant extracts offer unlimited therapeutic options, one of the sectors that could benefit from the potential use of invasive plants is the biomedical field.

Thus, the therapeutic valorization of these species could be both an ecological solution to stop their uncontrolled spreading and an economic solution, providing low costs raw materials. The purpose of the study was to obtain cellulose microfibers (CM) from the reproductive organs of *Gleditsia triachantos* and functionalize them with active principles (phenolic compounds from *G. triachantos* leaves) with antioxidant and antimicrobial activity in order to obtain functional wound dressings with improved performance in preventing wound infections, therefore, accelerating the wound healing process.

2. Results

2.1. Chemical Characterization of Phenolic Compounds of G. triachantos Leaves

2.1.1. Total Phenolic Compounds (TPC)

The quantification of total phenolic compounds (TPC) was performed using the calibration curve, resulting in a content of 113.62 ± 1.09 mg TPC (GAE, gallic acid equivalents)/g dry plant, and the flavonoid content was 35.24 ± 0.55 mg (QE, quercetin equivalents)/g dry plant for the *G. triachantos* leaves extract.

2.1.2. Ultra-High-Performance Liquid Chromatography Diode Array Detector Electrospray Ionization Tandem Mass Spectrometry (UHPLC–DAD-ESI/MS)

Phenolic compound profile by UHPLC–DAD-ESI/MS (ultra-high-performance liquid chromatography diode array detector electrospray ionization tandem mass spectrometry). The concentrations of phenolic acids (PAs), flavonoids, and flavonoid heterosides are shown in Table S1 of Supplementary Materials.

The main phenolic compounds identified in the alcoholic extract of G. triachantos were vanillic acid (773.88 µg/L), 4-hydroxy benzoic acid (570.37 µg/L), ferulic acid (319.81 µg/L), catechin (34,271.15 µg/L), quercetin (3,007.89 µg/L), and epicatechin (1,431.47 µg/L). The extract was rich in chlorogenic acid (11,984.43 µg/L), cinnamic acid (2,736.42 µg/L, an acid that lacks phenolic groups but is still usually included in the phenolic compounds class), and rutin (1,302.31 µg/L). The following compounds were also evidenced by UHPLC–DAD-ESI/MS: abscisic acid (16.10 µg/L), a sesquiterpenoid plant hormone, and ellagic acid (165.13 µg/L), a dimer of gallic acid.

2.2. Multi-Functionalized Cellulose Microfiber Fabrication

G. triachantos pods were ground into powder (to pass through a 100 mesh sieve) with a structure characterized by ATR-FTIR (Attenuated Total Reflectance—Fourier Transform Infrared) to highlight the specific bands of the three major components (see Section 2.3.1), which are hemi-cellulose (39.07% ± 1.49%), cellulose (29.78% ± 1.28%), and lignin (27.71% ± 0.75%) [11]. Then, the wood powder was subjected to purification by successive extractions to remove the extractable substances, hemicellulose (toluene and ethanol), and lignin. The lignocellulosic material required six cycles of delignification. The sample was taken from each cycle to quantify the total lignin content by a similar protocol to that used to determine the hydroxycinnamic acid content. The delignification process was continued for two more cycles after the lignin content was below the detection limit. After the last delignification cycle, the aqueous suspension of the obtained cellulose fibers was washed until no chloride ions were found, quantified by the Mohr titrimetric method. After extraction, the freeze-dried cellulosic material was characterized to highlight the lack of lignin content and compared to CM functionalized with phenolic compounds extracted from the leaves of the same species.

2.3. Characterization of Cellulose Microfibers (CM) and Multi-Functionalized Cellulose Microfibers (MFCM)

2.3.1. ATR-FTIR Analysis

The ATR-FTIR spectra of the studied materials, CM and multi-functionalized cellulose microfibers (MFCM), are presented in Figure 1. According to Danial (2015) [12], the lack of the C=O band in both spectra that should have been observed at the wave number ≈1733 cm^{-1} highlights the successful removal of hemicelluloses. The complete removal of lignin was given by the lack of bands at ≈1509 cm^{-1} and ≈1592 cm^{-1}, bands attributed to the aromatic C=C vibrations in the plane as well as the lack of the band located at 1264 cm^{-1}, which would have been assigned C–O–C specific to ether bonds from lignin. The C–C band breathing rings (≈1158 cm^{-1}) and the C–O–C glycosidic ether band (≈1100 cm^{-1}) were attributed to the presence of cellulose. The band at ≈2900 cm^{-1}, observed in all samples, corresponded to the C–H stretching vibration that reflects the organic content in general [13].

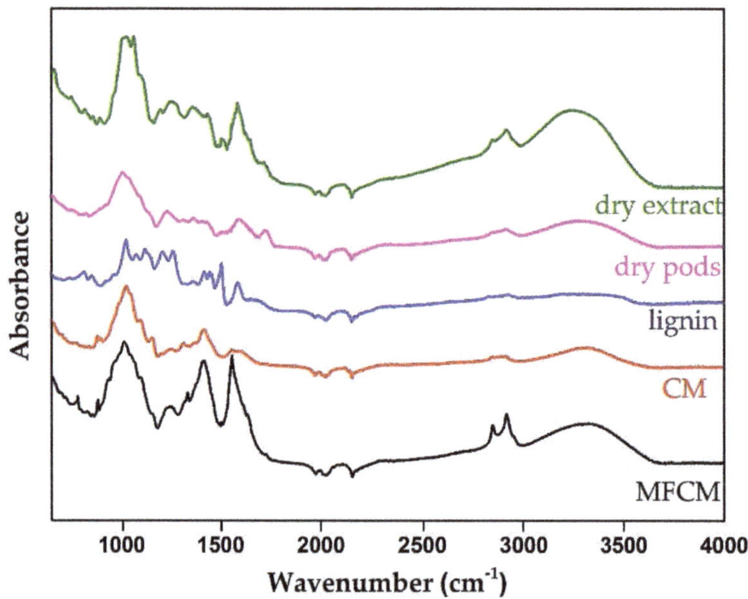

Figure 1. Attenuated Total Reflectance—Fourier Transform Infrared (ATR-FTIR) spectra of multi-functionalized cellulose microfibers (MFCM) (black line), CM (red line), lignin (blue line), initial dry pods (pink line), and dry extract (green line).

From the literature data, it is known that the bands in the region 1680–900 cm^{-1} may have come from phenolic compounds specific to the extract [14,15].

Among others, several specific aromatic C–H vibrations were present between the wavenumbers 670–900 cm^{-1} and 950–1225 cm^{-1} [16]. The band at the level of 1247 cm^{-1} corresponded to the extension of pyran rings, typical for flavonoid compounds [17].

Theoretically, the cellulose–PAs and cellulose–flavonoids bonds are the ester and ether ones, respectively, which were indeed highlighted at the wavenumbers 1338 cm^{-1} and 1260–1230 cm^{-1}, respectively. The bands specific to the ester groups were similar to those in the dry extract because PAs are found in the form of glycosidic esters according to Qian et al. [18]. By comparing the FTIR spectra of CM, MFCM, and the dry extract, we identified four specific bands for MFCM that were not found in the other spectra, at the wavenumbers 779 cm^{-1} (aromatic C–H out-of-plane bending), 1338 cm^{-1} (C–O stretching (phenyl), C–H (bending and CH$_2$), 1442 cm^{-1} (C–C stretching vibration in aromatic ring), and 1558 cm^{-1} (C=C–C aromatic ring stretch) [19].

2.3.2. Morphology of CM and MFCM

SEM (scanning electron microscopy) analysis revealed that the control sample (CM) used for multi-functionalization with phenolic compounds showed the presence of microfibers (Figure 2) with diameters between 5 and 10 μm. The surface morphology of the CM sample indicated a relatively smooth surface, while that of the MFCM highlighted a rougher surface due to the adherence of the phenolic compounds. In the case of MFCM (impregnated with the extract from *G. triachantos* leaves), it can be observed from Figure 3 that the chemical components specific to the extract adhered to the cellulose microfiber surface without requiring the presence of a fixing or crosslinking agent. These components (of MFCM) had a thickness between 40 and 120 nm and a length 0.2–1 μm.

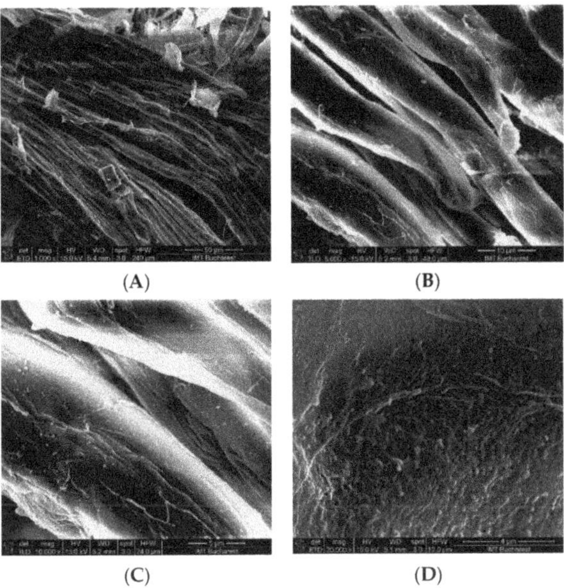

Figure 2. SEM morphology of the reference CM surface obtained from *Gleditsia triacanthos* pods, used for functionalization with active principles from *G. triacanthos* leaves: (**A**) 1000×, (**B**) 5000×, (**C**) 10,000×, and (**D**) 20,000×.

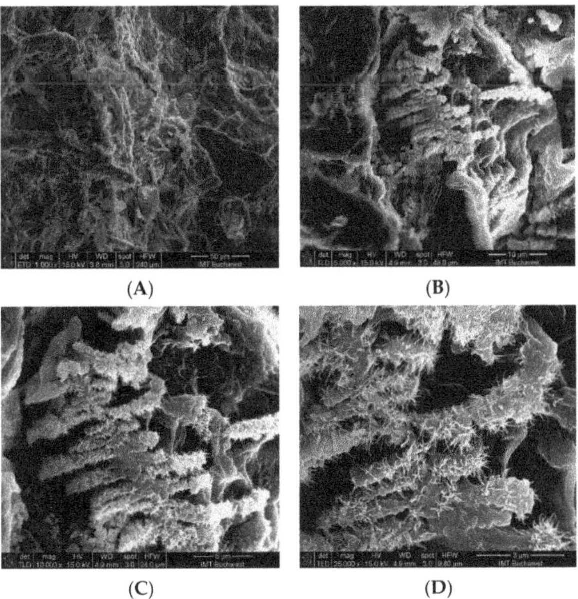

Figure 3. SEM morphology of the MFCM (multi-functionalized cellulose microfiber): (**A**) 1000×, (**B**) 5000×, (**C**) 10,000×, and (**D**) 20,000×.

2.3.3. Surface Wettability

The CA (contact angle) measurements revealed that CM and MFCM samples showed a superhydrophilic behavior (CA ≈ 0°), with the water drops being absorbed in the sample volume (Video Supplementary Materials).

2.3.4. Water Absorptivity and Retention Properties

The percentage of PBS (phosphate-buffered saline) uptake was slightly improved from 674.4% for the CM (control sample) to 721.51% for the MFCM. The freeze-dried phenolic compounds adhered to microfibers could be responsible for the increased absorption capacity, most probably due to the presence of the hydroxyl groups.

The difference in terms of water retention capacity (PBS) between the CM (control) and MFCM remained constant after 4 h, however, MFCM had better water retention properties (Figure 4).

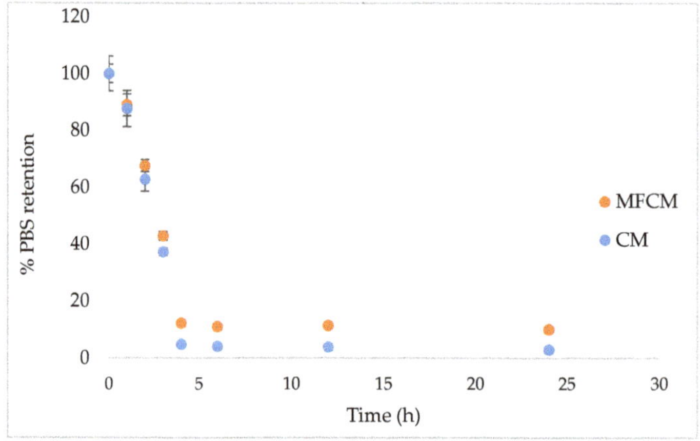

Figure 4. The water retention capacity of CM and MFCM matrices over time ($p < 0.05$).

2.3.5. In Vitro Phenolic Compound Release Studies and Antioxidant Activity

The release kinetics of the active principles were evaluated by determining the TPC content by the Folin–Ciocalteu method and through monitoring the antioxidant activity over time. The process of releasing the phenolic compounds from the MFCM varied considerably in time (Figure 5A)—in the first 6 h, the release was linear ($R^2 = 0.9935$), with a velocity of 3.1083 µg GAE/mL * h. The release of the phenolic compounds from the cellulose-based dressings was relatively short, reaching an almost complete release of the extract from MFCM within 120 h. A sudden release of a high concentration of TPC in PBS was observed after 2 h, with the rest being released in inverse proportion to the contact time. All experiments were performed in triplicate and the results have statistical significance ($p < 0.05$).

The TEAC (Trolox equivalent antioxidant capacity) results have shown that the highest concentration of compounds with antioxidant potential was released in the first 6 h [20], while in the DPPH (1,1-diphenyl-2-picrylhydrazyl radical) assay, the antioxidant activity was inversely proportional to the contact time, observing a constant release until the complete reduction of the phenolic compound content in the material (Figure 5B).

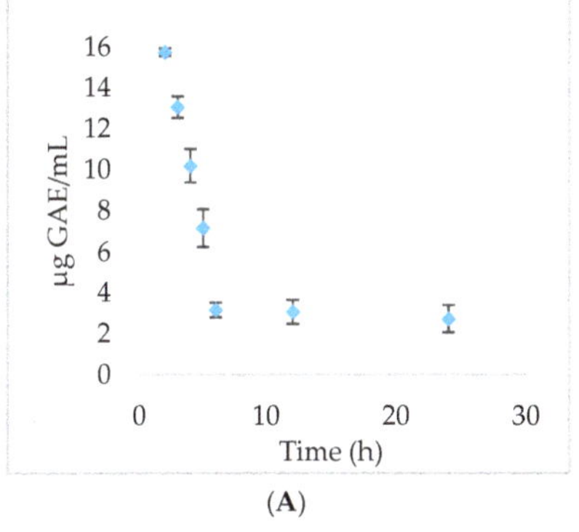
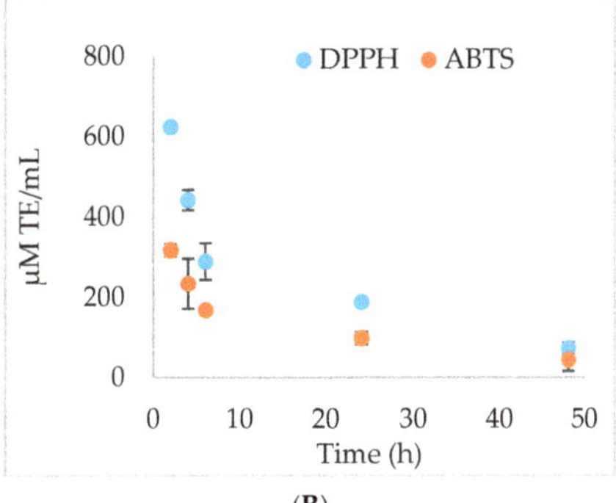

Figure 5. Kinetics of controlled release of phenolic compounds (**A**) and antioxidant effects (**B**) in temporal dynamics for multi-functionalized cellulose microfibers (MFCM), $p < 0.05$.

2.4. Antimicrobial Activity

The antimicrobial activity was performed by evaluating the viable cell counts (colony forming unit (CFU)/mL) and calculating the recovery factor initially and after 24 h. The data from Table 1 show that the active principles released from the cellulose matrix positively influenced the antimicrobial effect both compared to the strain and CM control. The MFCM showed a better antimicrobial activity than CM for all strains except *Enterococcus faecium* ATCC DMS 13590 after 24 h of incubation at 37 °C.

Table 1. Recovery rate of logarithmical colony forming unit/mL (lg CFU/mL) compared to microbial strain control (the closer it is to 1, the lower the antimicrobial effect).

Strains	Time	CM	MFCM
Staphylococcus aureus MRSA clinical strain	0	0.997 ± 0.045	0.995 ± 0.026
	24	0.522 ± 0.021	0.513 ± 0.005
Enterococcus faecium ATCC DMS 13590	0	0.778 ± 0.014	0.752 ± 0.029
	24	0.744 ± 0.009	0.747 ± 0.029
Pseudomonas aeruginosa 6 clinical strain	0	0.792 ± 0.085	0.681 ± 0.098
	24	0.690 ± 0.075	0.591 ± 0.075
Pseudomonas aeruginosa ATCC 27853	0	0.933 ± 0.077	0.725 ± 0.028
	24	0.848 ± 0.172	0.701 ± 0.172
Enterobacter cloacae clinical strain	0	0.983 ± 0.016	0.987 ± 0.016
	24	0.821 ± 0.003	0.732 ± 0.005
Acinetobacter baumannii clinical strain	0	0.890 ± 0.011	0.784 ± 0.021
	24	0.723 ± 0.004	0.652 ± 0.011
Escherichia coli ATCC 11229	0	0.761 ± 0.032	0.708 ± 0.035
	24	0.894 ± 0.006	0.802 ± 0.051
Candida albicans ATCC 10231	0	0.987 ± 0.035	0.969 ± 0.007
	24	0.875 ± 0.006	0.776 ± 0.141
Candida parapsilosis ATCC 22019	0	0.996 ± 0.021	0.996 ± 0.024
	24	0.807 ± 0.01	0.732 ± 0.022

MRSA: Methicillin-Resistant *Staphylococcus aureus*, ATCC: American Type Culture Collection.

2.5. Assessment of Cells Viability and Cytotoxicity

Cell viability and cell cytotoxicity were investigated by MTT ([3-(4,5-dimethylthiazol-2yl)]-2,5-diphenyltetrazolium bromide) assay—L929 cells, and by LDH (lactate dehydrogenase) assay—L929 cells, respectively, co-cultured with CM and MFCM (Figures 6 and 7).

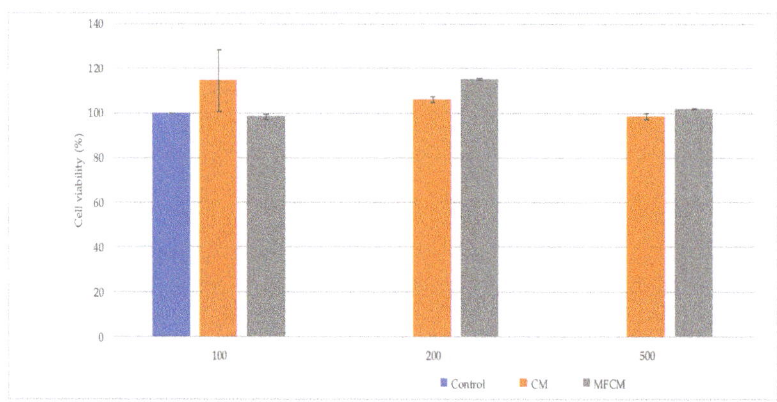

Figure 6. Cell viability (%) investigated by MTT ([3-(4,5-dimethylthiazol-2yl)]-2,5-diphenyltetrazolium bromide) assay—L929 cells co-cultured with CM and MFCM at various concentrations: 100, 200, and 500 µg/µL.

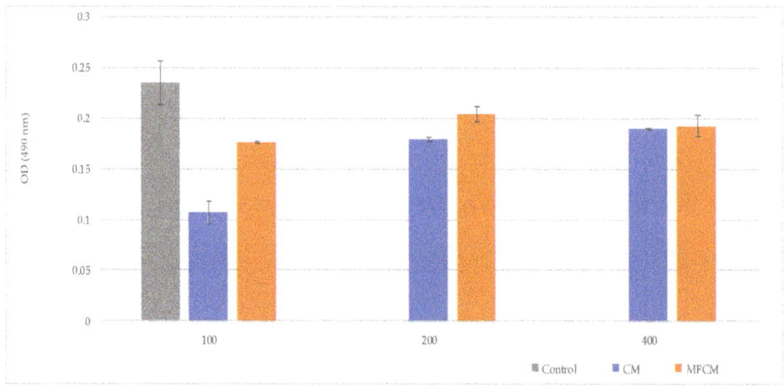

Figure 7. Cell cytotoxicity measured by LDH (lactate dehydrogenase) assay—L929 cells co-cultured with CM and MFCM at various concentrations: 100, 200, and 500 µg/µL.

Among the tested compounds, CM (100 µg/uL) harbored the highest level of biocompatibility with high cell proliferation rates and low LDH.

The high rates of cell viability in contact with the CM and MFCM demonstrated that these new cellulose microfibers obtained from *Gleditsia triachantos* are biocompatible materials.

3. Discussion

Phenolic compounds are plant secondary metabolites with a strong antioxidant effect, being known to initiate numerous processes involved in the wound healing process [21,22]. In this study, we aimed to extract the cellulosic material from the *G. triacanthos* pods, which was further multi-functionalized with a leaf extract of the same species, showing a significantly highphenolic compound content that in a first stage could exhibit both antimicrobial and antioxidant activities, contributing to skin regeneration. Data regarding

UHPLC-DAD-ESI/MS analysis of PA and flavonoids from *G. triachantos* leaf extract have not been previously reported. However, Mohammed (2014) isolated and identified eight flavone glycosides and two flavone aglycones such as luteolin-7-O-β-glucopyranosid, luteolin-7-O-β-galactopyranoside, apigenin-7-O-β-glucopyranoside, luteolin, and apigenin from the aqueous ethanol extract of *G. triacanthos* L. leaves. Vitexin, luteolin, isovitexin, and quercetin have also been previously identified in this plant extract [23]. In addition, some identified compounds from the alcoholic extract of *G. triachantos*, such as syringic acid [24], gallic acid [25], p-coumaric acid [26], apigenin [27], quercetin [28], and myricetin [29], were reported to exhibit anti-inflammatory effects. Thus, PAs promise to be multivalent, bioinspired molecules that either alone or in the form of biomaterials additives can be very useful to prevent and treat medical device-associated and drug-resistant infections [30].

On the other hand, Saleh (2016) [31] reported the analgesic activity of the fruit methanolic extract of *G. triacanthos* and the saponins fraction derived, but the presence of these compounds could not be considered in our paper because the foaming test (for saponins) was negative.

To the best of our knowledge, the microcellulose obtained from *G. triacanthos* pods is herein reported for the first time, although some studies have been carried out on the characterization of other polysaccharides such as galactomannan [32], which is a heterogenous, water-soluble, non-toxic, and biodegradable polysaccharide used as a stabilizing, thickening, and emulsifying agent in the biomedical industry. The cellulose was extracted by treatment of grounded pods with toluene/ethanol (2:1) solution and six cycles of delignification method adapted after Zhuo (2017) [33]. The cellulose obtained was divided quantitatively into two parts—one was treated with 50% ethanol (control) in accordance with the second part, which contained the extract from the leaves of the same species. After lyophilization, the two materials were evaluated by ATR-FTIR analysis and compared to the commercial lignin. SEM analysis revealed that the control sample (CM) used for multi-functionalization with phenolic compounds showed the presence of microfibers and that the extract adhered to the cellulose microfibers surface, without requiring the presence of a fixing or crosslinking agent. The data were in accordance with Khalil (2020) [34], which concluded that by acid hydrolysis, the fiber size was in the range of 9–16 μm. In the wound dressing case, it was shown that cellulose microfibers improved the hemostasis of the scaffolds without affecting its cytotoxicity and the strength and flexibility of the films, which could be used in drug delivery or active wound dressing.

SEM (scanning electron microscopy) analysis revealed that the presence of microfibers with diameters between 5 and 10 μm and CM sample indicated a relatively smooth surface while the MFCM surface morphology highlighted a rougher surface compared to the CM due to the adherence of the phenolic compounds.

Microscopic characteristics, such as surface roughness, surface energy of materials, and thin surface coatings, play an important role in determining the wettability of material as well as hydrophobicity/hydrophilicity characteristics. The CM and MFCM highlighted a high wettability surface by CA measurements. The high wettability surface could be attributed to hydroxyl groups located at the equatorial positions of glucopyranose rings corresponding to higher planar orientation. [35]. On the other hand, it is well known that hydrophilicity is one of the most important surface characteristics of wound dressings. The water absorption study is a gravimetric test that has the main purpose of determining the maximum amount of water absorption and retention expressed in percentages. This assay was designed to determine how a dressing manifests its properties in conditions as close as possible to what can happen on the surface of a wound. In essence, the weight gain of the dressing after the absorption of the liquid and its swelling over a certain period of time was measured and used as an indication of water absorption and retention. MFCM had better water retention properties than CM, possibly due to the presence of phenolic compounds and glycosides. In general, these compounds increased the rate of hydration, which represents an advantage for wound dressing materials.

It was observed that MFCM had a statistically significant better retention capacity than CM, which means that it will be more helpful for wounds with moderate exudate and for dry wounds. Water loss could allow exudate and edema fluid to be picked up from wounds in the dressing through an active process directed upwards, as has been reported for some commercially available dressings [36,37].

Another important aspect for MFCM is the modulation of antioxidant activity during the release of active principles from the material. Thus, in wound healing, in addition to the antimicrobial potential of dressings, an important aspect is the modulation of reactive oxygen species (ROS) release. It is well known that low levels of ROS can exhibit positive effects, including an antimicrobial effect, but their excessive production can lead to oxidative stress, which will further inhibit the wound healing process, thus favoring wound chronicity [38]. In addition, the plant polyphenols inhibit the release of pro-inflammatory mediators and have no toxicity effects on human tissues [39].

There are many studies that have tried to establish some correlations between the antioxidant and antimicrobial activity, but only a few involved clinical strains isolated from hospital-acquired wound infections, such as *Escherichia coli* and *Staphylococcus aureus*, although the results were, however, contradictory [30]. MRSA and *E. coli* seem to be sensitive to cinnamic acid [40], p-coumaric acid [41,42], caffeic acid, vanillic acid, protocatechuic acid [41,43], while ferulic acid was effective only against *E. coli* [43]. On the contrary, in other studies, *E. coli* was not sensitive to p-coumaric acid, vanillic acid, or proocatechuic acid [44]. The MFCM proved to be more active on Gram-negative strains. Previous studies have shown only a minimal link between the antioxidant activity and antimicrobial efficacy [30]. In addition, the CM and MFCM are biocompatible materials with high cell proliferation rates and low cytotoxicity on mouse murine fibroblast cells.

4. Materials and Methods

4.1. Plant Material

G. triachantos leaves and shells were collected from Ramnicu Valcea, Romania, in September 2019 from spontaneous flora. Their taxonomic affiliation was confirmed, and voucher specimens were deposited in the herbarium of the Botanical Garden "Dimitrie Brândză" from the University of Bucharest (no. 40065). The plant materials were manually sorted and dried at room temperature.

4.2. Phenolic Compounds Extraction from G. triachantos Leaves and Chemical Characterization

4.2.1. Hydro-Alcoholic Extraction from *G. triachantos* Leaves

An amount of 2.5 g of previously ground dry leaves was weighed and brought into 50 mL of ethanol solution 50%. The extract obtained by the ultrasound-assisted extraction method was maintained at room temperature for 20 days, and after that, it was sonicated again for 30 min. Then, it was filtered and stored at $-20\ °C$ until it was incorporated in cellulose microfibers.

4.2.2. Total Phenolic Compounds and Flavonoids Contents

TPC content was determined by the Folin–Ciocalteu method [45]. An aliquot was mixed with 0.1 mL Folin–Ciocalteu reagent, 1.8 mL distilled water, and 0.1 mL of saturated sodium carbonate. The tubes were vortexed for 15 s and allowed to stand at dark for 60 min for color development. Absorbance was then measured at 765 nm using a FlexStation 3 UV–VIS (Molecular Devices Company, Sunnyvale, CA, USA) spectrophotometer. A standard curve was prepared by using different concentrations of gallic acid in the same condition with samples ($R^2 = 0.9985$). TPC content was expressed as milligram gallic acid equivalent/g dry leaves (mg GAE/g DL). Analyses were performed in triplicate.

Flavonoid content was assessed through the $AlCl_3$ method [46]. Briefly, 0.1 mL sample/standard solution was mixed with 0.1 mL sodium acetate 10% and 0.12 mL $AlCl_3$ 2.5%, with the final volume being adjusted to 1 mL with ethanol 50%. The samples were then vortexed and incubated in the dark for 45 min. The absorbances were measured at

$\lambda = 430$ nm. A standard curve was prepared by using different concentrations of quercetin ($R^2 = 0.9958$). Total flavonoid content was expressed as milligram quercetin equivalent/g dry leaves (mg QE/g DL). Analyses were performed in triplicate.

4.2.3. Ultra-High-Performance Liquid Chromatography Diode Array Detector Electrospray Ionization Tandem Mass Spectrometry

All experiments were performed using an UltiMate 3000 UHPLC system (ThermoFisher Scientific, Bremen, Germany), consisting of a quaternary pump, diode array detector (DAD) set at 280 nm, column oven, and autosampler coupled to a Q Exactive Focus Hybrid Quadrupole-OrbiTrap mass spectrometer equipped with heated electrospray ionisation (HESI) probe (ThermoFisher Scientific, Bremen, Germany).

Separations were performed on Kinetex (C18, 100×2.1 mm, 1.7 µm, Phenomenex, California, USA) column (reverse-phase UHPLC column) and a binary solvent system consisting of solvent A (water with 0.1% formic acid) and solvent B (methanol with 0.1% formic acid). The UHPLC gradient for mass screening, the flow rate, and the operation conditions of the mass spectrometer were previously optimized [47]. PAs and flavonoids from the extract were also identified and quantified according to Ciucure [47] (Table S1), as well as data acquisition and data analysis.

4.3. CM and MFCM Obtaining and Characterization

4.3.1. CM and MFCM Obtaining

Cellulose extraction was performed according to Zhuo (2017) [33]. Briefly, an amount of 30 g ground shells of *G. triacanthos* was added over toluene/ethanol solution (2:1, *v/v*) and the mixture was further refluxed for 8 h at 90 °C. Then, the biomass was treated with sodium hypochlorite under the acidic condition (glacial acetic acid) at 70 °C for 6 h to remove the lignin, followed by treatment with 2% NaOH aqueous solution at 90 °C for 2 h to remove most of the hemicellulose. After that, the samples were again treated by sodium hypochlorite under acidic condition for 2 h, followed by final purification of 2% NaOH solution at 90 °C for 2 h. After each cycle, the presence of lignin was evaluated by hydroxycinnamic acid content according to British Pharmacopoeia [48] using a calibration curve with commercial lignin solubilized in DMSO (dimethylsulfoxide) ($\lambda = 524$ nm, $R^2 = 0.9996$). When lignin content was out of the detection limit, the cellulose was washed until it reached pH = 7.4, and each supernatant was evaluated by titration with AgNO₃, indicating the removal of chloride ions. The cellulose fibers were divided into 2 samples, one being dispersed in 50% ethanol (control), and the second in the extract obtained previously. Both products were dispersed 3 times using IKA ULTRA-TURRAX (IKA®-Werke GmbH & Co. KG, Staufen, Germany) for 10 min, followed by an ultrasound treatment for 60 min. Finally, the samples were lyophilized at −55 °C and 1 Pa to obtain the cellulose microfibers (CM) and the multi-functionalized cellulose microfibers (MFCM), respectively.

4.3.2. ATR-FTIR

The FTIR spectra for CM, MFCM, and lignin was recorded at room temperature using the Cary 630 FTIR Spectrometer in ATR mode (Agilent Technologies Inc., Santa Clara, CA, USA) and Agilent MicroLab Software FTIR System (Agilent Technologies, Inc., USA). The chosen measurement range was 4000–650 cm^{-1}, number of scans was 400, and resolution was 4 cm^{-1}. The FTIR spectra of dry extract were produced on the alcoholic extract of *G. tricanthos* leaves dried previously in an oven at 35 °C (for 24 h).

4.3.3. Morphology of CM and MFCM

The morphology of CM and MFCM were performed by Nova NanoSEM 630 Scanning Electron Microscope (FEI Company, Hillsboro, OR, USA) using UHR detector (Through-Lens-Detector-TLD) at an acceleration voltage of 15 kV and a working distance of about 5 mm. The samples were fixed with a double tape on a conductive support of aluminum, which, afterward, was covered with a few nanometers of gold layer using sputtering for 30 s.

4.3.4. Surface Wettability

Static (equilibrium) CAs were measured using a drop shape analysis instrument, model DSA100 (Krüss GmbH, Hamburg, Germany). The samples were positioned on a plane stage, under the tip of a stainless-steel needle, with a blunt end, with an outer diameter of 0.5 mm. The needle was attached to a syringe controlled by DSA3 software supplied with the instrument and was used to drop water with a controlled volume on the test surface as well as to evaluate the contact angle. The volume of water droplets was ≈ 1 µL with a pumping rate of 5 µL/min. The tests were carried out at room temperature [49].

4.3.5. Water Absorptivity and Retention Properties

An amount of 0.1 g CM was treated with methanol and dried (60 °C for 24 h), then immersed in PBS (pH 7.4). The excess buffer was removed with filter paper and the material's wet mass was determined using the method adapted after Ngadaonye (2013) [36]. The same procedure was applied to MFCM, while the percentage of absorbed water was calculated for CM and MFCM as follows:

$$\text{Swelling ratio (\%)} = (M_2 - M_1) \times 100/M_1$$

where M_1 and M_2 represent the weights of dry and wet samples, respectively.

Water retention properties were determined using water retention tests. Samples (CM and MFCM) with the same mass were dried and weighed. Samples were then immersed in PBS at 37 °C for 24 h, and the water on the surface was removed using filter paper. Samples were finally placed in plates at 37 °C and weighed after a specific incubation time [50]. The water retention capacity was calculated using the following equation:

$$\text{Water retention ratio (\%)} = M_i \times 100/M_0$$

where M_i and M_0 represent the percent of swelling at specific and initial time (maximum PBS swelling), respectively.

4.3.6. In Vitro Phenolic Compounds Release Studies and Antioxidant Activity

The MFCM were immersed in PBS at 37 °C for 5 days to determine the various compounds' kinetics release from these dressings. The release studies were conducted in closed glass vessels containing 1 mL PBS. The medium was removed (completely) periodically, at each sampling time (2 h, 4 h, 6 h, 24 h, 48 h, 72 h, 120 h), and fresh medium was introduced. Quantification of the active principle release at different time intervals was determined. The TPC content was determined by the Folin–Ciocalteu method. For each sampling time, the antioxidant activity was determined by DPPH and TEAC (Trolox equivalent antioxidant capacity) methods using a FlexStation 3 UV–VIS spectrophotometer (Molecular Devices Company, Sunnyvale, CA, USA).

The DPPH assay was performed according to the method of Madhu (2013) with some slight changes [51]. The standard curve was linear between 0 and 200 mM Trolox ($R^2 = 0.9998$). Results were expressed as millimolar TE (Trolox equivalent)/g material.

TEAC assay was performed according to Re (1999) with some modifications [52]. A stable stock solution of ABTS$\bullet^+$ (2,2'-azinobis (3-ethylbenzothiazoline-6-sulfonic acid)) was produced by reacting an aqueous solution of 7 mM ABTS with 2.45 mM potassium persulphate. The standard curve was linear between 0 and 200 mM Trolox ($R^2 = 0.9978$). Results are expressed as millimolar TE/g material.

4.4. Antimicrobial Activity

4.4.1. Sterilization of CM and MFCM

The samples were placed in an irradiation container. To measure the treatment doses, we placed dosimeters in the minimum dose positions (D min = 33.2 ± 1.6 kGy) and the maximum dose (D max = 35.7 ± 1.6 kGy). According to the dosimetry calculations performed, the minimum and maximum doses absorbed by the samples were within this range, with a confidence level of 95%. The dosimetry system used was Ethanol-Chlorobenzene (ECB). Gamma irradiation (Co-60) was performed with the SVST Co-60/B irradiator (Institute of Isotopes Co. Ltd. Budapest, Hungary) for 11 h 57 min at a dose ≥25 kGy. For sterility, verification/validation was performed at a concentration of 10% CM in peptone water and the doses were incubated at 37 °C for 24 h. After incubation, a volume of 0.1 mL was inoculated by being displayed on Trypton soy agar and Sabauroud agar in triplicates. The plates were incubated at 37 °C for 24 h for the enumeration of aerobic and mesophillic organisms and 7 days for fungi.

4.4.2. Evaluation of the Antimicrobial Activity

The bacterial strains were *Escherichia coli* ATCC 11229; *Staphylococcus aureus* ATCC 6538; *Pseudomonas aeruginosa* ATCC 27853; *Enterococcus faecium* DMS 13590; and multi-drug-resistant clinical isolates of *Staphylococcus aureus*, *Pseudomonas aeruginosa*, *Acinetobacter baumannii*, and *Enterobacter cloacae* from microbial collection of Microbiology Lab from Faculty of Biology, University of Bucharest. Two yeast strains were also tested: *Candida albicans* ATCC 10231 and *Candida parapsilosis* ATCC 22019 [53]. The CM and MFCM samples (10 mg) were placed in contact with a microbial suspension (10^4 CFU/mL—final density) and were subjected to vigorous shaking in order to assure the best contact between microbial cells and the samples. At 0 h and 24 h contact times, the viable cell counts (CFU/mL) of the microbial suspension were determined by spotting serial 10-fold dilutions on Muller–Hinton agar for bacteria and Sabauroud for yeasts. The microbial lg reduction of microbial growth in the presence of the CM and MFCM samples was calculated as follows: recovery rate = lg (CFU sample)/lg (CFU control strain).

4.5. Assessment of Cell Viability and Cytotoxicity by MTT and LDH Assay

MTT assay is a spectrophotometric method to analyze both cell viability and proliferation of L929 murine fibroblasts, which were co-cultivated with the CM and MFCM (at various concentrations: 100, 200, and 500 µg/µL) for 24 h and further incubated with 1 mg/mL MTT solution in the dark at 37 °C. After 4 h of incubation, formazan crystals were solubilized with isopropanol, resulting in a purple solution, and quantified at 550 nm using FlexStation 3 (Molecular Devices Company, Sunnyvale, CA, USA).

The level of cytotoxicity exerted on the L929 cells was quantified using the LDH test (Tox7 kit, Sigma-Aldrich, Burlington, MA, USA), following the manufacturer's instructions [54].

4.6. Statistical Analysis

All the analyses were performed in triplicate. The statistical analysis was performed by one-way analysis of variance (one-way ANOVA) using Microsoft Office Excel 365 (Redmond, WA, USA) at a 95% confidence interval.

5. Conclusions

Here, the microcellulose obtained from *G. triacanthos* pods is, to our knowledge, reported for the first time. The cellulose was extracted from *G. triacanthos* pods and submitted to six cycles of delignification for purification. The extract from the *G. triacanthos* leaves used for the functionalization of the cellulose microfibers had a content of TPC 113.62 ± 1.09 mg GAE/g dry plant, with the major compounds identified by UHPLC–DAD-ESI/MS being catechin (34,271.15 µg/L), chlorogenic acid (11,984.43 µg/L), quercetin (3007.89 µg/L), cinnamic acid (2736.42 µg/L), epicatechin (1431.47 µg/L), rutin (1302.31 µg/L), and vanillic acid (773.88 µg/L).

The multi-functionalized cellulose microfibers with the phenolic compounds extracted from the leaves of the same species manifested improved absorptivity and fluid retention properties and assured the controlled release of the active principles. The antioxidant effect correlated with the antimicrobial activity revealed that MFCM may represent a promising biocompatibility material for the design of high-performance wound dressings that could prevent wound infection and facilitate the wound healing process.

Supplementary Materials: The following are available online at https://www.mdpi.com/1422-006 7/22/1/33/s1.

Author Contributions: Conceptualization, I.C.M. and E.O.; methodology and investigation, I.C.M., E.-I.G., O.T., G.G.P., and I.Z.; formal analysis, I.C.M. and E.O.; writing—review and editing, I.C.M., M.C.C., and E.O.; supervision, M.C.C. All authors have read and agreed to the published version of the manuscript.

Funding: This research was funded by O.I.M., research project ICUB 29604/12.12.2019.

Institutional Review Board Statement: Not applicable.

Informed Consent Statement: Not applicable.

Data Availability Statement: Not Available.

Conflicts of Interest: The authors declare no conflict of interest.

Abbreviations

ABTS	2:2′-azinobis (3-ethylbenzothiazoline-6-sulfonic acid)
ATCC	American Type Culture Collection
CA	Contact Angle
CFU	Colony Forming Unit
CM	Cellulose Microfibers
DMSO	Dimethylsulfoxide
DPPH	1,1-diphenyl-2-picrylhydrazyl radical
ECB	Ethanol-Chlorobenzene
MFCM	Multi-functionalized cellulose microfibers (with a phenolic compound extract from *G. triachantos* leaves)
FTIR	Fourier Transform Infrared
ATR-FTIR	Attenuated Total Reflectance—Fourier Transform Infrared
GAE	Gallic acid equivalents
LDH	Lactate dehydrogenase
MF	Microcellulose fibers
MRSA	Methicillin-Resistant *Staphylococcus aureus*
MTT	[3-(4,5-dimethylthiazol-2yl)]-2,5-diphenyltetrazolium bromide
PAs	Phenolic acids
PBS	Phosphate-buffered saline
QE	Quercetin equivalents
ROS	Reactive oxygen species
SEM	Scanning electron microscopy
TE	Trolox equivalents
TEAC	Trolox equivalent antioxidant capacity
TEMPO	(2,2,6,6-tetramethyl-1-piperidinyl)oxidanyl
TPC	Total phenolic compounds
UHPLC–DAD-ESI/MS	Ultra-high-performance liquid chromatography diode array detector electrospray ionization tandem mass spectrometry

References

1. Bacakova, L.; Pajorova, J.; Bacakova, M.; Skogberg, A.; Kallio, P.; Kolarova, K.; Svorcik, V. Versatile Application of Nanocellulose: From Industry to Skin Tissue Engineering and Wound Healing. *Nanomaterials* **2019**, *9*, 164. [CrossRef] [PubMed]
2. Nie, K.; Song, Y.; Liu, S.; Han, G.; Ben, H.; Ragauskas, A.J.; Jiang, W. Preparation and Characterization of Microcellulose and Nanocellulose Fibers from Artemisia Vulgaris Bast. *Polymers* **2019**, *11*, 907. [CrossRef] [PubMed]
3. Sun, F.; Nordli, H.R.; Pukstad, B.; Kristofer Gamstedt, E.; Chinga-Carrasco, G. Mechanical characteristics of nanocellulose-PEG bionanocomposite wound dressings in wet conditions. *J. Mech. Behav. Biomed. Mater.* **2017**, *69*, 377. [CrossRef] [PubMed]
4. Xie, J.; Lia, J. Smart drug delivery system based on nanocelluloses. *J. Bioresour. Bioprod.* **2017**, *2*, 1.

5. Denet, A.-R.; Ucakar, B.; Préat, V. Transdermal Delivery of Timolol and Atenolol Using Electroporation and Iontophoresis in Combination: A Mechanistic Approach. *Pharm. Res.* **2003**, *20*, 1946. [CrossRef] [PubMed]

6. Rasheed, M.; Jawaid, M.; Karim, Z.; Abdullah, L.C. Morphological, Physiochemical and Thermal Properties of Microcrystalline Cellulose (MCC) Extracted from Bamboo Fiber. *Molecules* **2020**, *25*, 2824. [CrossRef]

7. Elanthikkal, S.; Gopalakrishnapanicker, U.; Varghese, S.; Guthrie, J.T. Cellulose microfibres produced from banana plant wastes: Isolation and characterization. *Carbohydr. Polym.* **2010**, *80*, 852. [CrossRef]

8. Ferus, P.; Barta, M.; Konôpková, J.; Turčeková, S.; Maňka, P.; Bibeň, T. Diversity in honey locust (*Gleditsia triacanthos* L.) seed traits across Danube basin. *Folia Oecologica* **2013**, *40*, 163.

9. Ferreras, A.E.; Funes, G.; Galetto, L. The role of seed germination in the invasion process of Honey locust (*Gleditsia triacanthos* L., *Fabaceae*): Comparison with a native confamilial. *Plant Species Biol.* **2015**, *30*, 126. [CrossRef]

10. Ferreras, A.E.; Galetto, L. From seed production to seedling establishment: Important steps in an invasive process. *Acta Oecologica* **2010**, *36*, 211. [CrossRef]

11. Svečnjak, L.; Marijanović, Z.; Okińczyc, P.; Marek Kuś, P.; Jerković, I. Mediterranean Propolis from the Adriatic Sea Islands as a Source of Natural Antioxidants: Comprehensive Chemical Biodiversity Determined by GC-MS, FTIR-ATR, UHPLC-DAD-QqTOF-MS, DPPH and FRAP Assay. *Antioxidants* **2020**, *9*, 337. [CrossRef] [PubMed]

12. Danial, W.H.; Abdul Majid, Z.; Mohd Muhid, M.N.; Triwahyono, S.; Bakar, M.B.; Ramli, Z. The reuse of wastepaper for the extraction of cellulose nanocrystals. *Carbohydr. Polym.* **2015**, *118*, 165. [CrossRef] [PubMed]

13. Garside, P.; Wyeth, P. Identification of Cellulosic Fibres by FTIR Spectroscopy—Thread and Single Fibre Analysis by Attenuated Total Reflectance. *Stud. Conserv.* **2003**, *48*, 269. [CrossRef]

14. Edelmann, A.; Diewok, J.; Schuster, K.C.; Lendl, B. Rapid Method for the Discrimination of Red Wine Cultivars Based on Mid-Infrared Spectroscopy of Phenolic Wine Extracts. *J. Agric. Food Chem.* **2001**, *49*, 1139. [CrossRef]

15. Fernández, K.; Agosin, E. Quantitative Analysis of Red Wine Tannins Using Fourier-Transform Mid-Infrared Spectrometry. *J. Agric. Food Chem.* **2007**, *55*, 7294. [CrossRef]

16. Laghi, L.; Versari, A.; Parpinello, G.P.; Nakaji, D.Y.; Boulton, R.B. FTIR Spectroscopy and Direct Orthogonal Signal Correction Preprocessing Applied to Selected Phenolic Compounds in Red Wines. *Food Anal. Methods* **2011**, *4*, 619. [CrossRef]

17. Espinosa-Acosta, G.; Ramos-Jacques, A.; Molina, G.; Maya-Cornejo, J.; Esparza, R.; Hernandez-Martinez, A.; Sánchez-González, I.; Estevez, M. Stability Analysis of Anthocyanins Using Alcoholic Extracts from Black Carrot (*Daucus Carota* ssp. *Sativus* Var. *Atrorubens* Alef.). *Molecules* **2018**, *23*, 2744. [CrossRef]

18. Qian, S.; Fang, X.; Dan, D.; Diao, E.; Lu, Z. Ultrasonic-assisted enzymatic extraction of a water soluble polysaccharide from dragon fruit peel and its antioxidant activity. *RSC Adv.* **2018**, *8*, 42145. [CrossRef]

19. Santos, D.I.; Neiva Correia, M.J.; Mateus, M.M.; Saraiva, J.A.; Vicente, A.A.; Moldão, M. Fourier Transform Infrared (FT-IR) Spectroscopy as a Possible Rapid Tool to Evaluate Abiotic Stress Effects on Pineapple By-Products. *Appl. Sci.* **2019**, *9*, 4141. [CrossRef]

20. Barbehenn, R.; Cheek, S.; Gasperut, A.; Lister, E.; Maben, R. Phenolic Compounds in Red Oak and Sugar Maple Leaves Have Prooxidant Activities in the Midgut Fluids of *Malacosoma disstria* and *Orgyia leucostigma* Caterpillars. *J. Chem. Ecol.* **2005**, *31*, 969. [CrossRef]

21. Shetty, B.S. Wound Healing and Indigenous Drugs: Role as Antioxidants: A Review. *Res. Rev. J. Med. Heal. Sci.* **2013**, *2*, 5.

22. Fikru, A.; Makonnen, E.; Eguale, T.; Debella, A.; Abie Mekonnen, G. Evaluation of in vivo wound healing activity of methanol extract of *Achyranthes aspera* L. *J. Ethnopharmacol.* **2012**, *143*, 469. [CrossRef] [PubMed]

23. Mohammed, R.S.; Abou Zeid, A.H.; El Hawary, S.S.; Sleem, A.A.; Ashour, W.E. Flavonoid constituents, cytotoxic and antioxidant activities of *Gleditsia triacanthos* L. leaves. *Saudi J. Biol. Sci.* **2014**, *21*, 547. [CrossRef] [PubMed]

24. Ham, J.R.; Lee, H.-I.; Choi, R.-Y.; Sim, M.-O.; Seo, K.-I.; Lee, M.-K. Anti-steatotic and anti-inflammatory roles of syringic acid in high-fat diet-induced obese mice. *Food Funct.* **2016**, *7*, 689. [CrossRef]

25. Karimi-Khouzani, O.; Heidarian, E.; Amini, S.A. Anti-inflammatory and ameliorative effects of gallic acid on fluoxetine-induced oxidative stress and liver damage in rats. *Pharmacol. Rep.* **2017**, *69*, 830. [CrossRef]

26. Pei, K.; Ou, J.; Huang, J.; Ou, S. p -Coumaric acid and its conjugates: Dietary sources, pharmacokinetic properties and biological activities. *J. Sci. Food Agric.* **2016**, *96*, 2952. [CrossRef]

27. Che, D.N.; Cho, B.O.; Shin, J.Y.; Kang, H.J.; Kim, J.-S.; Oh, H.; Kim, Y.-S.; Jang, S. Il Apigenin Inhibits IL-31 Cytokine in Human Mast Cell and Mouse Skin Tissues. *Molecules* **2019**, *24*, 1290. [CrossRef]

28. Xiong, G.; Ji, W.; Wang, F.; Zhang, F.; Xue, P.; Cheng, M.; Sun, Y.; Wang, X.; Zhang, T. Quercetin Inhibits Inflammatory Response Induced by LPS from *Porphyromonas gingivalis* in Human Gingival Fibroblasts via Suppressing NF-κB Signaling Pathway. *BioMed Res. Int.* **2019**, *2019*, 6282635. [CrossRef]

29. Semwal, D.; Semwal, R.; Combrinck, S.; Viljoen, A. Myricetin: A Dietary Molecule with Diverse Biological Activities. *Nutrients* **2016**, *8*, 90. [CrossRef]

30. Liu, J.; Du, C.; Beaman, H.T.; Monroe, M.B.B. Characterization of Phenolic Acid Antimicrobial and Antioxidant Structure–Property Relationships. *Pharmaceutics* **2020**, *12*, 419. [CrossRef]

31. Saleh, D.O.; Kassem, I.; Melek, F.R. Analgesic activity of *Gleditsia triacanthos* methanolic fruit extract and its saponin-containing fraction. *Pharm Biol.* **2016**, *54*, 576. [CrossRef] [PubMed]

32. Xu, W.; Liu, Y.; Zhang, F.; Lei, F.; Wang, K.; Jiang, J. Physicochemical characterization of Gleditsia triacanthos galactomannan during deposition and maturation. *Int. J. Biol. Macromol.* **2020**, *144*, 821. [CrossRef] [PubMed]

33. Zhuo, X.; Liu, C.; Pan, R.; Dong, X.; Li, Y. Nanocellulose Mechanically Isolated from Amorpha fruticosa Linn. *ACS Sustain. Chem. Eng.* **2017**, *5*, 4414. [CrossRef]

34. Khalil, H.P.S.A.; Jummaat, F.; Yahya, E.B.; Olaiya, N.G.; Adnan, A.S.; Abdat, M.; N. A. M., N.; Halim, A.S.; Kumar, U.S.U.; Bairwan, R.; et al. A Review on Micro- to Nanocellulose Biopolymer Scaffold Forming for Tissue Engineering Applications. *Polymers* **2020**, *12*, 2043. [CrossRef] [PubMed]

35. Yamane, C.; Aoyagi, T.; Ago, M.; Ago, M.; Sato, K.; Okajima, K.; Takahashi, T. Two Different Surface Properties of Regenerated Cellulose due to Structural Anisotropy. *Polym. J.* **2006**, *38*, 819. [CrossRef]

36. Ngadaonye, J.I.; Geever, L.M.; Killion, J.; Higginbotham, C.L. Development of novel chitosan-poly(N,N-diethylacrylamide) IPN films for potential wound dressing and biomedical applications. *J. Polym. Res.* **2013**, *20*, 161. [CrossRef]

37. Kickhöfen, B.; Wokalek, H.; Scheel, D.; Ruh, H. Chemical and physical properties of a hydrogel wound dressing. *Biomaterials* **1986**, *7*, 67. [CrossRef]

38. Schafer, M.; Werner, S. Oxidative stress in normal and impaired wound repair. *Pharmacol. Res.* **2008**, *58*, 165. [CrossRef]

39. Działo, M.; Mierziak, J.; Korzun, U.; Preisner, M.; Szopa, J.; Kulma, A. The Potential of Plant Phenolics in Prevention and Therapy of Skin Disorders. *Int. J. Mol. Sci.* **2016**, *17*, 160. [CrossRef]

40. Nascimento, G.G.F.; Locatelli, J.; Freitas, P.C.; Silva, G.L. Antibacterial activity of plant extracts and phytochemicals on antibiotic-resistant bacteria. *Brazilian J. Microbiol.* **2000**, *31*, 247. [CrossRef]

41. Salaheen, S.; Peng, M.; Joo, J.; Teramoto, H.; Biswas, D. Eradication and Sensitization of Methicillin Resistant *Staphylococcus aureus* to Methicillin with Bioactive Extracts of Berry Pomace. *Front. Microbiol.* **2017**, *8*, 253. [CrossRef] [PubMed]

42. Alves, M.J.; Ferreira, I.C.F.R.; Froufe, H.J.C.; Abreu, R.M.V.; Martins, A.; Pintado, M. Antimicrobial activity of phenolic compounds identified in wild mushrooms, SAR analysis and docking studies. *J. Appl. Microbiol.* **2013**, *115*, 346. [CrossRef] [PubMed]

43. Merkl, R.; Hrádková, I.; Filip, V.; Šmidrkal, J. Antimicrobial and antioxidant properties of phenolic acids alkyl esters. *Czech J. Food Sci.* **2010**, *28*, 275. [CrossRef]

44. Chatterjee, N.S.; Panda, S.K.; Navitha, M.; Asha, K.K.; Anandan, R.; Mathew, S. Vanillic acid and coumaric acid grafted chitosan derivatives: Improved grafting ratio and potential application in functional food. *J. Food Sci. Technol.* **2015**, *52*, 7153. [CrossRef]

45. Singleton, V.; Rossi, J. Colorimetry of Total Phenolics with Phosphomolybdic-Phosphotungstic Acid Reagents. *Am. J. Enol. Vitic.* **1965**, *16*, 144. [CrossRef]

46. Woisky, R.G.; Salatino, A. Analysis of propolis: Some parameters and procedures for chemical quality control. *J. Apic. Res.* **1998**, *37*, 99. [CrossRef]

47. Ciucure, C.T.; Geană, E. Phenolic compounds profile and biochemical properties of honeys in relationship to the honey floral sources. *Phytochem. Anal.* **2019**, *30*, 481. [CrossRef]

48. Gaur, R.; Azizi, M.; Gan, J.; Hansal, P.; Harper, K.; Mannan, R.; Panchal, A.; Patel, K.; Patel, M.; Patel, N.; et al. *British Pharmacopoeia*; Crown, Inc.: London, UK, 2009; Volume 3, p. 9983.

49. Stan, G.E.; Tite, T.; Popa, A.-C.; Chirica, I.M.; Negrila, C.C.; Besleaga, C.; Zgura, I.; Sergentu, A.C.; Popescu-Pelin, G.; Cristea, D.; et al. The Beneficial Mechanical and Biological Outcomes of Thin Copper-Gallium Doped Silica-Rich Bio-Active Glass Implant-Type Coatings. *Coatings* **2020**, *10*, 1119. [CrossRef]

50. Lu, B.; Wang, T.; Li, Z.; Dai, F.; Lv, L.; Tang, F.; Yu, K.; Liu, J.; Lan, G. Healing of skin wounds with a chitosan–gelatin sponge loaded with tannins and platelet-rich plasma. *Int. J. Biol. Macromol.* **2016**, *82*, 884. [CrossRef]

51. Madhu, G.; Bose, V.C.; Aiswaryaraj, A.S.; Maniammal, K.; Biju, V. Defect dependent antioxidant activity of nanostructured nickel oxide synthesized through a novel chemical method. *Colloids Surfaces A Physicochem. Eng. Asp.* **2013**, *429*, 44. [CrossRef]

52. Re, R.; Pellegrini, N.; Proteggente, A.; Pannala, A.; Yang, M.; Rice-Evans, C. Antioxidant activity applying an improved ABTS radical cation decolorization assay. *Free Radic. Biol. Med.* **1999**, *26*, 1231. [CrossRef]

53. Ong, S.Y.; Wu, J.; Moochhala, S.M.; Tan, M.H.; Lu, J. Development of a chitosan-based wound dressing with improved hemostatic and antimicrobial properties. *Biomaterials* **2008**, *29*, 4323. [CrossRef] [PubMed]

54. Serbezeanu, D.; Vlad-Bubulac, T.; Rusu, D.; Grădişteanu Pircalabioru, G.; Samoilă, I.; Dinescu, S.; Aflori, M. Functional Polyimide-Based Electrospun Fibers for Biomedical Application. *Material* **2019**, *12*, 3201. [CrossRef] [PubMed]

International Journal of
Molecular Sciences

Article

Rational Design of Ag/ZnO Hybrid Nanoparticles on Sericin/Agarose Composite Film for Enhanced Antimicrobial Applications

Wanting Li [1], Zixuan Huang [1], Rui Cai [1], Wan Yang [1], Huawei He [1,2,3] and Yejing Wang [1,2,*]

1   State Key Laboratory of Silkworm Genome Biology, Biological Science Research Center, Southwest University, Beibei, Chongqing 400715, China; lwt260@email.swu.edu.cn (W.L.); zixuan666@email.swu.edu.cn (Z.H.); cairui0330@email.swu.edu.cn (R.C.); yw961121@email.swu.edu.cn (W.Y.); hehuawei@swu.edu.cn (H.H.)

2   Chongqing Key Laboratory of Sericultural Science, Chongqing Engineering and Technology Research Center for Novel Silk Materials, Southwest University, Beibei, Chongqing 400715, China

3   Chongqing Key Laboratory of Soft-Matter Material Chemistry and Function Manufacturing, Chongqing 400715, China

*   Correspondence: yjwang@swu.edu.cn; Tel.: +86-23-6825-1575

Abstract: Silver-based hybrid nanomaterials are receiving increasing attention as potential alternatives for traditional antimicrobial agents. Here, we proposed a simple and eco-friendly strategy to efficiently assemble zinc oxide nanoparticles (ZnO) and silver nanoparticles (AgNPs) on sericin-agarose composite film to impart superior antimicrobial activity. Based on a layer-by-layer self-assembly strategy, AgNPs and ZnO were immobilized on sericin-agarose films using the adhesion property of polydopamine. Scanning electron microscopy, energy-dispersive X-ray spectroscopy, and X-ray powder diffraction spectroscopy were used to show the morphology of AgNPs and ZnO on the surface of the composite film and analyze the composition and structure of AgNPs and ZnO, respectively. Water contact angle, swelling ratio, and mechanical property were determined to characterize the hydrophilicity, water absorption ability, and mechanical properties of the composite films. In addition, the antibacterial activity of the composite film was evaluated against Gram-positive and Gram-negative bacteria. The results showed that the composite film not only has desirable hydrophilicity, high water absorption ability, and favorable mechanical properties but also exhibits excellent antimicrobial activity against both Gram-positive and Gram-negative bacteria. It has shown great potential as a novel antimicrobial biomaterial for wound dressing, artificial skin, and tissue engineering.

Keywords: sericin; AgNPs; ZnO; antimicrobial activity; green synthesis

Citation: Li, W.; Huang, Z.; Cai, R.; Yang, W.; He, H.; Wang, Y. Rational Design of Ag/ZnO Hybrid Nanoparticles on Sericin/Agarose Composite Film for Enhanced Antimicrobial Applications. *Int. J. Mol. Sci.* **2021**, 22, 105. https://dx.doi.org/10.3390/ijms22010105

Received: 3 November 2020
Accepted: 18 December 2020
Published: 24 December 2020

Publisher's Note: MDPI stays neutral with regard to jurisdictional claims in published maps and institutional affiliations.

1. Introduction

Sericin is a natural hydrophilic macromolecular protein produced by the silkworm, *Bombyx mori* [1,2]. Although sericin is a promising biomaterial for its good reactivity, hydrophilicity, biodegradability, and biocompatibility [3], its poor mechanical performance due to the amorphous structure restricts the biomaterial-related applications. Sericin has high levels of serine and aspartic acid. The polar side groups such as hydroxyl, carboxyl, and amino groups make it easy to interact with other polymers through blending, cross-linking, or copolymerization to yield improved biomaterials [1,4–6]. Agarose is a neutral polysaccharide derived from several species of red marine algae. It has a linear structure, consisting of D-galactose and 3,6-anhydrogalactose through β-1,4 and α-1,3 to alternately form repeating disaccharide units. It can form strong gels at low concentrations [7–9]. Considering its malleable mechanical properties, low cost, and good biocompatibility, agarose is often used to improve the performance of sericin as an auxiliary biomaterial [9–12].

The emergence of multidrug-resistant bacteria caused by the misuse of antibiotics has become a serious global health problem [13]. Considering its potential health risks and the incidence of cross-contamination, developing novel and effective fungicides is one of the most concerning issues worldwide [14–16]. Silver nanoparticles (AgNPs) are one of the most attractive inorganic antibacterial materials. It has exhibited excellent antimicrobial activity against a variety of microorganisms [17–21]. The antimicrobial activity is significantly affected by AgNPs size. The smaller the particles, the higher the antimicrobial efficiency [22–24]. However, with the decrease in particle size, AgNPs tend to aggregate, which decreases their antibacterial properties. Recently, AgNPs are incorporated in various materials such as TiO_2, SiO_2, ZnO, Fe_2O_3, and graphene, which have improved their dispersibility with enhanced antibacterial properties and long-term stability [25–29]. As an economic and non-toxic compound, zinc oxide nanoparticle (ZnO) is mostly used in cosmetics and biomedical materials. It can effectively prevent AgNPs aggregation with its good dispersion. The potential cytotoxicity of excessive AgNPs has attracted great attention. To ensure the biosafety and efficiency of AgNPs, incorporating metal oxides such as ZnO in AgNPs is an effective way to reduce the potential cytotoxicity of AgNPs without compromising their antimicrobial property [15,29,30]. In addition, AgNPs/ZnO hybrids have exhibited a synergistic antibacterial effect to inhibit bacterial growth or kill bacteria due to the strong interaction between AgNPs and ZnO [29,31–33].

Inspired by mussel secretory proteins, polydopamine (PDA) is produced by the self-polymerization of dopamine (3,4-dihydroxy-phenylalanine, DA) and has been widely applied as an effective adhesive agent [34,35]. PDA could effectively improve the adhesion ability and biocompatibility of materials. More importantly, it can promote the formation of metal nanoparticles with its reducing ability of the catechol group [35–38].

The purpose of this study is to develop sericin-based biomaterials loaded with Ag-NPs/ZnO for enhanced antimicrobial applications. Herein, agarose was blended with sericin to yield sericin/agarose (SS/AG, M) composite film with enhanced mechanical properties. Inspired by the adhesion and reduction properties of PDA, PDA was then coated on the surface of SS/AG film to capture ZnO and silver ions (Ag^+) one by one, respectively, yielding PDA-SS/AG (MP), ZnO-PDA-SS/AG (MPZ), AgNPs-PDA-SS/AG (MPA), and AgNPs-PDA-ZnO-PDA-SS/AG (MPAZ) composite films after Ag^+ was reduced to AgNPs by PDA. The morphology, structure, and properties of the films were well characterized, and the antimicrobial ability of these films against Gram-positive and Gram-negative bacteria was evaluated. This novel Ag@ZnO composite film has exhibited great potentials as a promising biomaterial for enhanced antimicrobial applications.

2. Results

To overcome the inherent brittleness of sericin and improve its antibacterial activity, we developed a simple, environment-friendly strategy to enhance the performance of sericin and expand its application in biomaterials. With the aid of agarose, we prepared a composite film by blending sericin and agarose, which greatly improved the mechanical properties of sericin. PDA was coated on the surface of the composite film to assist the assembly of ZnO and AgNPs layer by layer. Silver ions were adsorbed by PDA and then reduced to AgNPs by the catechol group of PDA [38–40]. The presence of ZnO not only improved the dispersibility and stability of AgNPs but also enhanced the antibacterial activity of AgNPs through a synergistic effect. The preparation and antibacterial effect of the films were shown in Figure 1.

2.1. SEM, EDX and XRD

Figure 2 presented the surface morphologies of M, MP, MPZ, MPA, and MPAZ films by scanning electron microscope (SEM) under different magnification. Sericin itself could not form a film due to its natural brittleness. SS/AG film (M) had a smooth surface (Figure 2a). After dopamine modification, MP exhibited a rough surface morphology (Figure 2b). With the addition of AgNPs and/or ZnO, some irregular particles appeared on the surface of the composite (Figure 2c–h). Higher magnification images showed the aggregation of AgNPs on the surface of the MPA film (Figure 2g), while AgNPs were relatively dispersed on the surface of MPAZ film due to the presence of ZnO (Figure 2h).

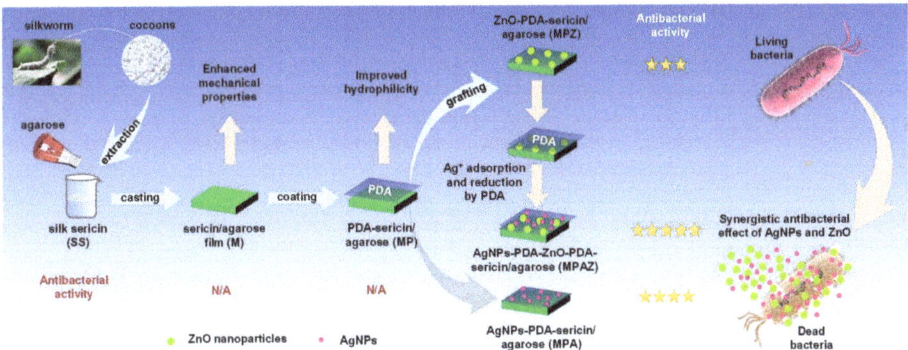

Figure 1. Preparation of AgNPs-PDA-ZnO-PDA-SS/AG (MPAZ) film with enhanced mechanical performance and antibacterial activity.

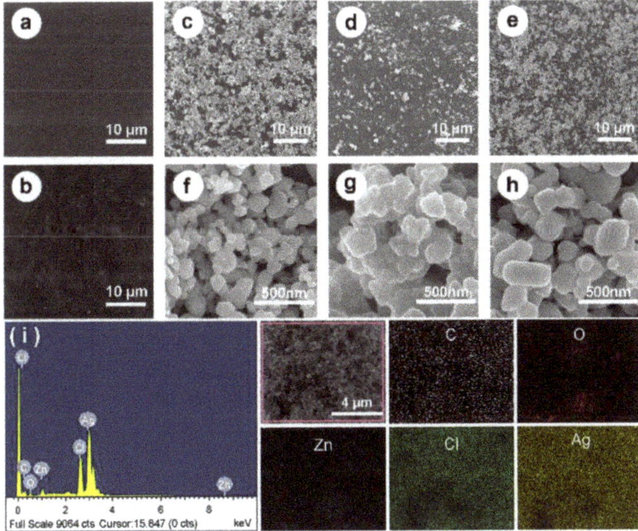

Figure 2. SEM and EDX. (**a**–**h**) The morphologies of different films are characterized by SEM. (**a**), M; (**b**), MP; (**c**–**e**), MPZ, MPA, and MPAZ, respectively; (**f**–**h**), higher magnification images of MPZ, MPA, and MPAZ, respectively; (**i**), EDX analysis of the elemental mapping of the selected region (indicated by a pink box) revealed the presence of zinc (Zn, blue), carbon (C, white), oxygen (O, red), silver (Ag, yellow) and chlorine (Cl, green), and the uniform distribution of ZnO and AgNPs on the surface of MPAZ film.

The composition of the MPAZ film was further characterized by energy-dispersive X-ray spectroscopy (EDX), as shown in Figure 2i. The elemental mapping of the selected region (indicated by a pink box) revealed the presence of zinc (Zn, cyan), carbon (C, white), oxygen (O, red), silver (Ag, yellow), and chlorine (Cl, green), and the uniform distribution of ZnO and AgNPs on the surface of MPAZ film. The characteristic peak of silver was at ~2.9 kV. The results further demonstrated that Ag^+ was effectively reduced to AgNPs by PDA.

To further identify the crystalline structure of ZnO and AgNPs, X-ray powder diffraction (XRD) was performed. All films showed the characteristic patterns of sericin and agarose (Figure 3), which appeared at 19.48° and 13.97° [41,42], respectively. The diffraction patterns of ZnO on the MPZ film could be assigned to the (100), (002), (101), (102), (110), (103), and (112) crystalline structure of ZnO (Figure 3) according to the JCPDS card of ZnO (No. 00-36-1451), which are located at 32.1°, 34.7°, 36.6°, 47.4°, 56.9°, 62.7°, and 68.2°, respectively. Two characteristic diffraction patterns were observed at 38.2° and 44.4° on the MPA film (Figure 3), corresponding to the (111) and (200) planes of AgNPs (JCPDS card No. 00-004-0783), respectively [43–45]. The XRD of MPAZ showed the specific patterns corresponding to the crystalline structure of ZnO and AgNPs [46–48], indicating that ZnO and AgNPs were successfully coated on the MPAZ film. The XRD pattern of ZnO did not change after AgNPs loading, indicating that AgNPs were not incorporated into the lattice of ZnO, but only deposited on the surface of ZnO [47]. The characteristic pattern (100) of ZnO was not observed on the MPAZ film, which may overlap with the pattern of Ag_2O ($2\theta = 32.2°$) as they are very close to each other [49]. Ag_2O may be derived from the oxidation of AgNPs on the surface of the film exposed to the air in aqueous solutions [50]. The combination of AgNPs and ZnO makes the crystal larger, which may lead to an increase in the diffraction intensity of ZnO on the MPAZ film. Meanwhile, the dispersion of ZnO resulted in an increase in the diffraction intensity of AgNPs on the MPAZ film [51,52].

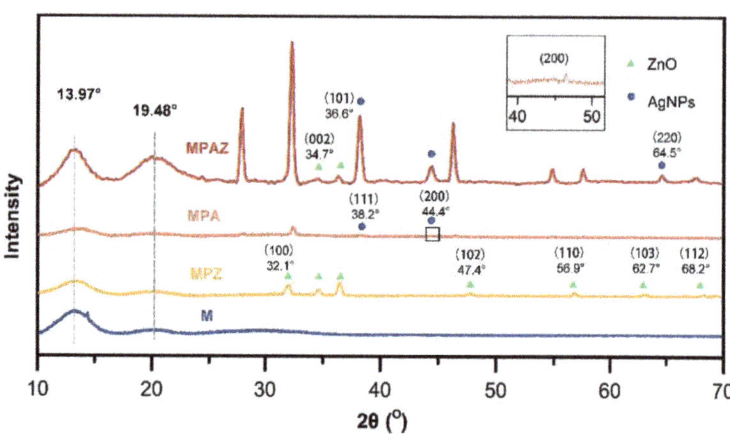

Figure 3. The characteristic XRD patterns of the films.

2.2. Swelling Ratio and Water Contact Angle

The swelling ratio is a feasible method to value the water absorption ability of materials. Figure 4a showed the swelling ratio of each film. The swelling ratio of SS/AG film was 382.73% at 12 h. The swelling ratio of the composite film coated with PDA and metal nanoparticles ranged from 157.93% to 269.84%. After two days, the swelling ratio of SS/AG film was almost unchanged, while the swelling ratios of other films slightly increased, which may be ascribed to the water absorption capability of PDA. The results suggested that composite films have excellent water absorption capability.

The water contact angle indicates the hydrophilicity of materials. Generally speaking, the smaller the contact angle, the better the hydrophilicity of a material [53]. The water contact angle of M, MP, MPZ, MPA, and MPAZ was 60.5°, 33.6°, 49.5°, 28.6°, and 33.5° (Figure 4b), respectively. All water contact angles were less than 90°, indicating the films were hydrophilic. The water contact angle decreased after AgNPs modification compared to that of MP, while the water contact angle increased after ZnO modification, which may be due to the difference of the hydrophilicity of AgNPs and ZnO [54,55].

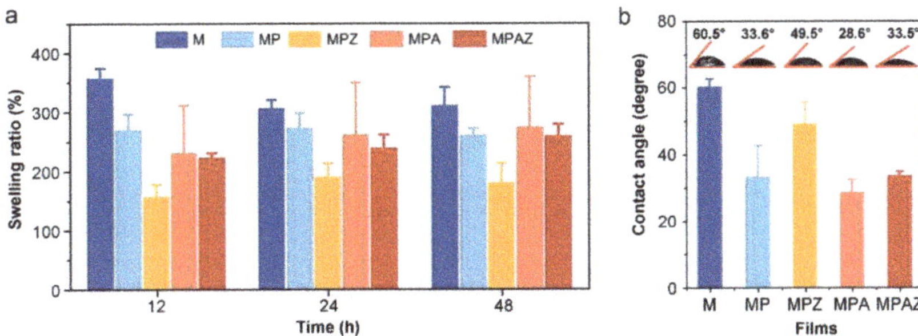

Figure 4. Swelling ratio and water contact angle of different films. a, Swelling ratio of the films; b, Water contact angle of the films.

2.3. Mechanical Properties

The molecular spatial structure of sericin was disordered, resulting in a lack of mechanical properties. Here, agarose was blended with sericin to improve its mechanical properties. A stereomicroscope was used to determine the thickness of different films (Figure S1). The statistical results from five independent tests were shown in Table S1. Figure 5 presented the stress-strain curves and Young's modulus of the films. The results showed that the mechanical properties of sericin had been greatly improved after the incorporation of agarose. As shown in Figure 5a–c, M had the tensile strength of 70.4 MPa, which was the highest among all films. However, its elongation at break (strain) was really the weakest, at only 4.3%. However, with the addition of metal nanoparticles, the thickness of the film increased, resulting in greatly improved mechanical properties of the film. Taking into account both the stress and strain, the mechanical properties of MPAZ were the best. It had a tensile stress of 57.37 MPa and a strain of 15.70%, respectively. Young's modulus also indicated the same result (Figure 5d). In general, the addition of ZnO and AgNPs improved the mechanical properties of the MPAZ film, which can meet the requirements of biomaterials to a certain extent [56–58].

2.4. Inhibition Zone Assay

Escherichia coli (*E. coli*) and *Staphylococcus aureus* (*S. aureus*) are typical Gram-negative and Gram-positive bacteria, respectively, which were used to evaluate the antibacterial properties of the films in this study. It was clear that M and MP did not form visible inhibition zones against *E. coli* or *S. aureus*. However, after the modification of ZnO or/and AgNPs, the films had formed obvious inhibition zones, both against *E. coli* and *S. aureus* (Figure 6), indicating the antibacterial activities of the films.

The diameters of the bacteriostatic zones were listed in Table 1. According to the diameters of the bacteriostatic zones, the antibacterial activity was ranked in the order of MPAZ, MPA, and MPZ. M and MP had no obvious antibacterial effect.

Table 1. The diameters of the bacteriostatic zones of different films.

	M (cm)	**MP (cm)**	**MPZ (cm)**	**MPA (cm)**	**MPAZ (cm)**
E. coli	0 ± 0.02	0 ± 0.02	0.25 ± 0.05	0.93 ± 0.03	1.05 ± 0.05
S. aureus	0 ± 0.02	0 ± 0.02	0.04 ± 0.01	0.34 ± 0.04	0.45 ± 0.05

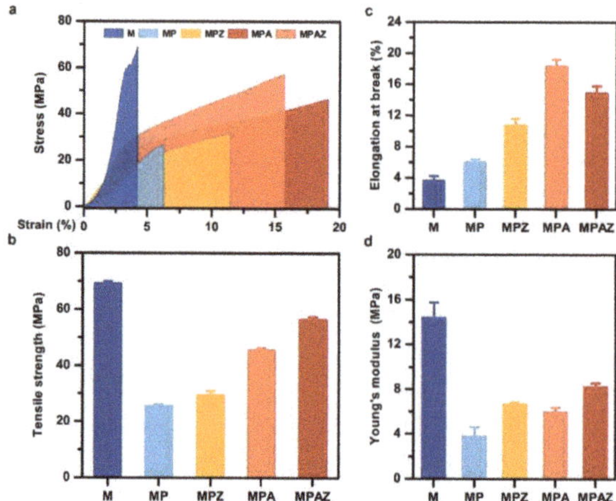

Figure 5. Mechanical properties of different films. (**a**), Stress-strain curves; (**b**), tensile strength; (**c**), elongation at break; (**d**), Young's modulus.

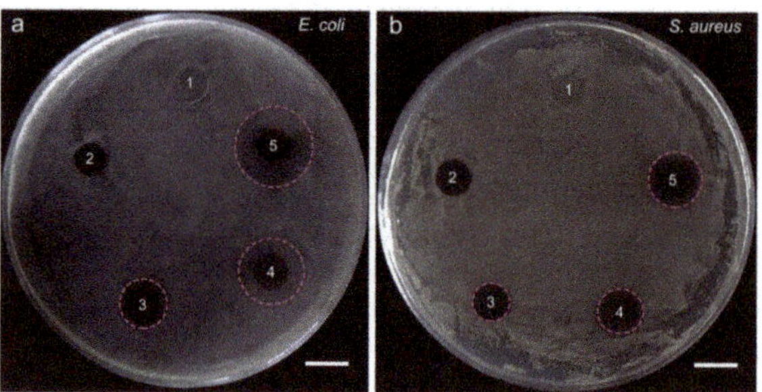

Figure 6. Inhibition zone assay of different films. (**a**), *E. coli*; (**b**), *S. aureus*. 1–5, M, MP, MPZ, MPA, and MPAZ, respectively. Scale bar, 1 cm.

2.5. Colony Counting Assay

After culture in the presence of different films for 24 h and 12 h, *E. coli* and *S. aureus* were spread on the plates to value the antibacterial activity of the films by counting the formed colony numbers, respectively, as shown in Figure 7a. Compared with other groups, the colony number in the MPAZ group were greatly reduced, both for *E. coli* and *S. aureus*, indicating that MPAZ had the best antibacterial effect among all tested films.

2.6. Growth Curve Assay

A bacterial growth curve assay was performed to compare the antibacterial effect of the films further. The result showed that M and MP did not affect bacterial growth. MPZ slightly repressed the bacterial growth compared with M and MP. Noticeably, both MPA and MPAZ effectively inhibited the growth of *E. coli* and *S. aureus*. In the presence of MPA, the growth of *E. coli* and *S. aureus* were delayed by 12 h and 8 h, respectively. However, in the presence of MPAZ, the growth of *E. coli* and *S. aureus* were arrested for 24 h and 12 h, respectively (Figure 7b,c). It was noted that after 24 h and 12 h, the OD of *E. coli* and *S. aureus* in the MPZ and MPA groups were almost comparable to those in the M and MP groups, respectively. Hence, the number of colonies on the plates seemed to be indistinguishable among MPZ, MPA, M, and MP groups (Figure 7a). The result was consistent with the above results, suggesting the antibacterial effect of MPAZ was the best among all films.

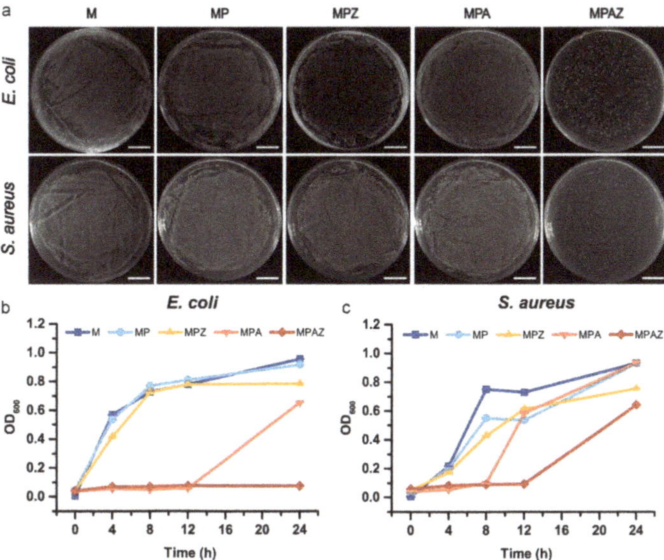

Figure 7. Colony counting and bacterial growth curve assays. (**a**), The colonies of *E. coli* and *S. aureus* after culture in the presence of different films for 24 h and 12 h, respectively; (**b,c**), bacterial growth curves in the presence of different films. Scale bar, 1 cm.

2.7. LIVE/DEAD BacLight Cell Viability Assay

Further, a LIVE/DEAD BacLight cell viability assay was performed to visualize the bactericidal capability of the films. In this assay, living cells are stained green, while dead cells are stained red [59,60]. The results showed that most bacteria were stained red in the presence of the MPA or MPAZ film, while most bacteria were dyed green in the presence of the M, MP, or MPZ film (Figure 8). The number of dead cells (red) indicated that the bactericidal activity of MPAZ was better than that of other films.

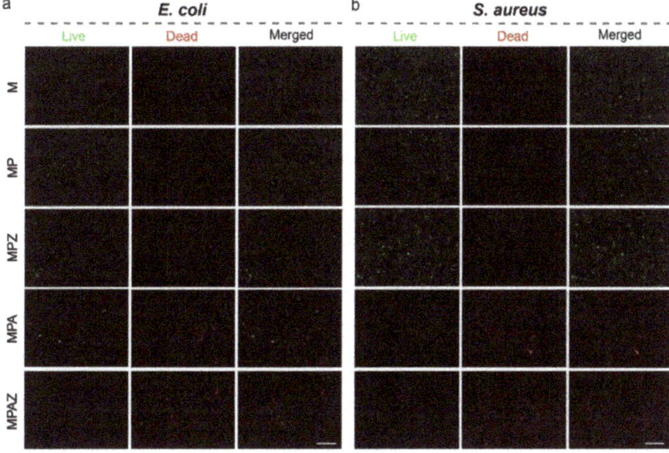

Figure 8. LIVE/DEAD BacLight cell viability assay. Fluorescence staining of *E. coli* (**a**) and *S. aureus* (**b**) cells after culture in the presence of different films. Scale bar, 50 μm.

2.8. SEM of Bacteria

Further, SEM was performed to visualize the interaction of MPAZ with the bacterial cells. In the presence of M or MP film, *E. coli* remained in their state in solution with intact cell walls. However, in the presence of MPZ, some bacterial cell walls remained intact but underwent deformation, indicating they were in a poor living state. MPA resulted in the significant shrink of *E. coli* cell walls under high magnification. In the presence of MPAZ, the bacterial cell wall suffered severe damage, resulting in cytoplasmic leakage (Figure 9). Similarly, *S. aureus* was spherical with a smooth and intact cell wall in the presence of M or MP film, but MPZ caused a small number of cell walls to shrink, and more cells with shrinking walls were found in the presence of MPA. MPAZ not only caused a large number of bacterial walls to shrink but also induced the deformation of cell walls (Figure 9). The results suggested that the bactericidal mechanism of MPAZ was partly due to the disruption of bacterial cell wall integrity and the resultant bacterial cells lysis, in which ZnO could promote the generation of reactive oxygen species and destroy the structure of cell membranes, and AgNPs could kill bacteria by releasing silver ions to destroy bacterial cell walls [61,62].

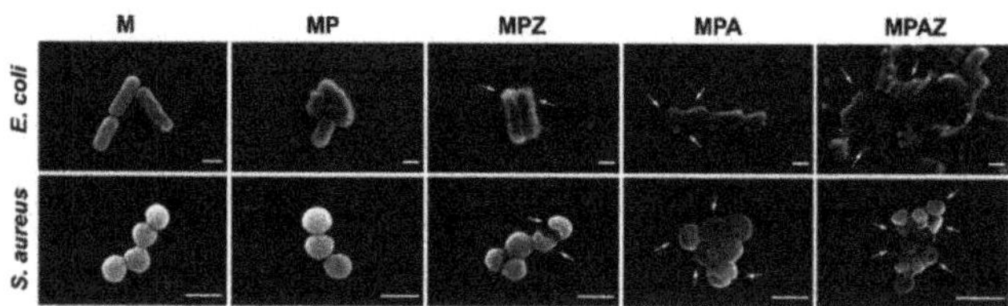

Figure 9. SEM of bacteria in the presence of different films. The arrows indicated the deformation, shrinking, and damage of bacterial cell walls. Scale bar, 1 μm.

S. aureus appeared to be more resistant to MPAZ than *E. coli*, which may be due to its special structure. The cell wall of *S. aureus* was significantly different from that of *E. coli*. The content of peptidoglycan in the cell wall of *S. aureus* is higher than that of *E. coli*, increasing the thickness of the cell wall, which helps *S. aureus* to protect its cells from the penetration of silver ions/AgNPs/ZnO into the cytoplasm [35]. In addition, our results suggested that MPAZ had the best antibacterial activity among all films, which may be due to the synergistic antibacterial effect of AgNPs and ZnO [15,27,30].

3. Materials and Methods

3.1. Materials and Chemicals

Silkworm cocoons were kindly provided by the State Key Laboratory of Silkworm Genome Biology, Southwest University (Beibei, Chongqing, China). Silver nitrate (AgNO$_3$), ZnO, and dopamine hydrochloride were from Aladdin (Shanghai, China). Agarose was provided by Biowest (Shanghai, China). Tris (hydroxymethyl) aminomethane (Tris) was from Sangon Biotech (Shanghai, China). *E. coli* (CICC 10389) and *S. aureus* (CICC 21600) were from the China Center of Industrial Culture Collection. LIVE/DEAD BacLight bacterial viability kit (L34856) was purchased from Thermo Fisher Scientific (Waltham, MA, USA). All other chemicals were of analytical grade and used directly.

3.2. Preparation of the Composite Films

The MPAZ film was prepared as per previous reports with slight modifications [38,63]. Agarose solution (2%, *w/v*) was adequately mixed with sericin solution (1%, *w/v*) as a volume ratio of 1:1 at 65 °C. Then, the mixture was dried at 37 °C overnight to become a SS/AG film. Dopamine (2 mg) was dissolved in 10 mL Tris-HCl, pH 8.5, and used immediately. Next, the SS/AG film was soaked in fresh dopamine solution at 25 °C for 6 h. After washing 3 times with water, the SS/AG film was dried at 25 °C for 12 h to yield a PDA-coated SS/AG (MP) film. Then, the MP films were immersed into ZnO solution (0.01 M) and AgNO$_3$ solutions (0.01 M) with shaking (50 rpm) at 25 °C for 2 h, respectively. After washing 3 times with water, the films were dried at 25 °C for 12 h to obtain ZnO-PDA-SS/AG (MPZ) film and AgNPs-PDA-SS/AG (MPA) film, respectively. The MPZ film was soaked into fresh dopamine solution at 25 °C for 15 min to coat PDA on its surface and then immersed into AgNO$_3$ solution (0.01 M) at 25 °C for 2 h after washing 3 times with water. After 3 rinses with water, the film was dried at 25 °C to yield AgNPs-PDA-ZnO-PDA-SS/AG film (MPAZ).

3.3. SEM, EDX, and XRD

The morphologies of different films were characterized by a field emission scanning electron microscope HITACHI SU8010 (Tokyo, Japan). Prior to SEM observation, the films with a dimension of 1 cm width and 1 cm height were precoated with Au for 90 s and then imaged on SU8010 with 5~10 kV acceleration voltage. The element mapping of MPAZ film was characterized by SU8010 equipped with an energy-dispersive X-ray spectroscopy mapping system at 15 kV. XRD spectra were recorded on a PANalytical x'pert (Almelo, The Netherlands) within 10–70° at a speed of 2°/min to identify the special crystalline structure of the composites.

3.4. Hydrophilicity and Swelling Ratio

The hydrophilicity of the film was measured by the sessile drip contact angle using a KRÜSS DSA100 contact angle analyzer (Hamburg, Germany) at 25 °C. The water contact angle was measured by dropping 4 μL water on the surface of the film at a time. Each sample was measured in triplicate. The result was an average of 3 tests.

The swelling ratio was used to characterize the water absorption capacity of the films. The dry films (1.5 cm × 1.5 cm, length × width) were immersed in water. Then, the films were removed out at different intervals and weighed after wiping off excess water with clean paper. The swelling ratio was defined as follows:

$$\text{Swelling ratio (\%)} = (m2 - m1)/m1 \times 100\% \tag{1}$$

where m1 and m2 were the mass of dry and swollen films, respectively. Three replications were performed for each film to ensure the accuracy of the test.

3.5. Mechanical Properties

The tensile strength (stress) and elongation at break (strain) of the films were measured on SHIMADZU AG-X plus (Tokyo, Japan). The thickness of the film was measured individually on a Stemi 2000C stereomicroscope (Shanghai, China), and the average value of 5 independent tests were applied. The samples (4 cm × 1 cm, length × width) were fixed on the mold and then stretched at a speed of 10 mm/min. The stress and strain values were recorded in real-time during operation. Each film was examined for at least 5 replicates to ensure the accuracy of the test, and the average value was calculated. Young's modulus was determined from the corresponding stress–strain curve.

3.6. Inhibition Zone Assay

E. coli and *S. aureus* were used to evaluate the antibacterial activity of the films. The bacteria were inoculated into Luria-bertani (LB) medium, and then cultured at 37 °C for 12 h with 220 rpm shaking speed until the optical density value at 600 nm (OD600) reached 1.0. Then, the bacterial suspension (200 μL) was collected and uniformly spread on an agarose plate. Next, the circular films (diameter, 0.7 cm) were sterilized by ultraviolet and then placed on the surface of the plate. After 12 h of incubation at 37 °C, the inhibition zones around the films were photographed, and the diameters were measured.

3.7. Colony Counting and Growth Curve Assays

The antibacterial properties of the films were further evaluated by colony counting and growth curve assays. Bacteria (1 × 10^7–10^8 colony-forming unit (CFU)/mL) were inoculated into LB medium and cultured at 37 °C in the presence of different films. Then, bacterial suspension (0.4 mL) was collected at different intervals to determine the bacterial growth curve by measuring the OD600 of the bacteria. *E. coli* and *S. aureus* were cultured in the presence of the films for 24 h and 12 h, respectively, and then spread on the plates after 500 times dilution with the media. The plates were photographed to compare the number of active bacterial colonies on the plates after culture at 37 °C for 12 h.

3.8. LIVE/DEAD BacLight Cell Viability Assay

The bacteria (10^9 CFU/mL) were cultured in LB medium in the presence of different films at 37 °C for 4 h. After removal of the films, the bacteria were collected by centrifugation (5000 rpm, 10 min), rinsed twice with phosphate-buffered saline (PBS, pH 7.4). Next, the bacteria were suspended in 100 μL PBS buffer (pH 7.4) and mixed with 10 μL staining reagent. After 15 min of incubation in the dark, as per the LIVE/DEAD BacLight cell viability guide kit, the living/dead bacteria cells were directly observed on a fluorescence microscopy EVOS FL auto cell imaging system (Waltham, MA, USA).

3.9. SEM of Bacteria

The bacteria were fixed in 2.5% glutaraldehyde after 4 h of culture in the presence of different films, rinsed twice with PBS buffer (pH 7.4), and then dehydrated by gradient ethanol concentration (30% for 15 min, 50% for 15 min, 70% for 15 min, 90% for 15 min, and 100% for 15 min). Finally, the morphologies of the bacteria in the presence of different films were characterized by SEM after air-drying overnight.

4. Conclusions

In summary, we developed a simple and eco-friendly strategy to efficiently assemble ZnO and AgNPs on sericin-agarose composite film with the assistance of PDA by layer-by-layer self-assembly. The resultant MPAZ film has not only desirable hydrophilicity, high water absorption ability, and favorable mechanical properties but also exhibits excellent antimicrobial activity against both Gram-positive and Gram-negative bacteria. Therefore, it is promising as a novel antimicrobial material for the applications such as antibacterial coatings, wound dressing, and tissue engineering.

Supplementary Materials: The following are available online at https://www.mdpi.com/1422-006 7/22/1/105/s1, Figure S1: Microscopic pictures of cross-sections of M (a), MP (b), MPZ (c), MPA (d), MPAZ (e) composite films, Table S1: Thickness of different films.

Author Contributions: W.L., Y.W., and H.H. conceived and designed the experiments; W.L., R.C., W.Y., and Z.H. performed the experiments; W.L., R.C., and W.Y. analyzed the data; Y.W. and H.H. contributed reagents/materials/analysis tools; W.L. wrote the draft; Y.W. and H.H. supervised the research and revised the manuscript. All authors have read and agreed to the published version of the manuscript.

Funding: This work was supported by the State Key Program of the National Natural Science of China (31530071), the National Natural Science of China (31972622), the Fundamental Research Funds for the Central Universities (XDJK2020TJ001, XDJK2020C049, XDJK2020D027).

Conflicts of Interest: All authors declare there are no conflict of interest, including specific financial interests and relationships and affiliations relevant to the subject of this manuscript.

References

1. Zhang, Y. Applications of natural silk protein sericin in biomaterials. *Biotechnol. Adv.* **2002**, *20*, 91–100. [CrossRef]
2. He, H.; Cai, R.; Wang, Y.; Tao, G.; Guo, P.; Zuo, H.; Chen, L.; Liu, X.; Zhao, P.; Xia, Q. Preparation and characterization of silk sericin/PVA blend film with silver nanoparticles for potential antimicrobial application. *Int. J. Biol. Macromol.* **2017**, *104*, 457–464. [CrossRef] [PubMed]
3. Tao, G.; Cai, R.; Wang, Y.; Zuo, H.; He, H. Fabrication of antibacterial sericin based hydrogel as an injectable and mouldable wound dressing. *Mater. Sci. Eng. C* **2021**, *119*, 111597. [CrossRef] [PubMed]

4. Lamboni, L.; Li, Y.; Liu, J.; Yang, G. Silk Sericin-Functionalized Bacterial Cellulose as a Potential Wound-Healing Biomaterial. *Biomacromolecules* **2016**, *17*, 3076–3084. [CrossRef]

5. Wang, P.; He, H.; Cai, R.; Tao, G.; Yang, M.; Zuo, H.; Umar, A.; Wang, Y. Cross-linking of dialdehyde carboxymethyl cellulose with silk sericin to reinforce sericin film for potential biomedical application. *Carbohydr. Polym.* **2019**, *212*, 403–411. [CrossRef]

6. Tao, G.; Wang, Y.; Cai, R.; Chang, H.; Song, K.; Zuo, H.; Zhao, P.; Xia, Q.; He, H. Design and performance of sericin/poly (vinyl alcohol) hydrogel as a drug delivery carrier for potential wound dressing application. *Mater. Sci. Eng. C* **2019**, *101*, 341–351. [CrossRef]

7. Shin, S.; Ikram, M.; Subhan, F.; Kang, H.Y.; Lim, Y.; Lee, R.; Jin, S.; Jeong, Y.H.; Kwak, J.-Y.; Na, Y.-J.; et al. Alginate–marine collagen–agarose composite hydrogels as matrices for biomimetic 3D cell spheroid formation. *RSC Adv.* **2016**, *6*, 46952–46965. [CrossRef]

8. Lee, P.Y.; Costumbrado, J.; Hsu, C.-Y.; Kim, Y.H. Agarose gel electrophoresis for the separation of DNA fragments. *Biology* **2012**, e3923. [CrossRef]

9. Zarrintaj, P.; Manouchehri, S.; Ahmadi, Z.; Saeb, M.R.; Urbanska, A.M.; Kaplan, D.L.; Mozafari, M.J. Agarose-based Biomaterials for Tissue Engineering. *Carbohydr. Polym.* **2018**, *187*, 66–84. [CrossRef]

10. Fernández-Cossío, S.; León-Mateos, A.; Sampedro, F.G.; Oreja, M.T.C. Biocompatibility of Agarose Gel as a Dermal Filler: Histologic Evaluation of Subcutaneous Implants. *Plasic Reconstr. Surg.* **2007**, *120*, 1161–1169. [CrossRef]

11. Sakai, S.; Kawabata, K.; Ono, T.; Ijima, H.; Kawakami, K. Development of mammalian cell-enclosing subsieve-size agarose capsules (<100 μm) for cell therapy. *Biomaterials* **2005**, *26*, 4786–4792. [CrossRef] [PubMed]

12. Yang, M.; Wang, Y.; Tao, G.; Cai, R.; Wang, P.; Liu, L.; Ai, L.; Zuo, H.; Zhao, P.; Umar, A.; et al. Fabrication of Sericin/Agarose Gel Loaded Lysozyme and Its Potential in Wound Dressing Application. *Nanomaterials* **2018**, *8*, 235. [CrossRef] [PubMed]

13. Ma, Z.; Kim, D.; Adesogan, A.T.; Ko, S.; Galvao, K.; Jeong, K.C. Chitosan Microparticles Exert Broad-Spectrum Antimicrobial Activity against Antibiotic-Resistant Micro-organisms without Increasing Resistance. *ACS Appl. Mater. Interfaces* **2016**, *8*, 10700–10709. [CrossRef] [PubMed]

14. Hajipour, M.J.; Fromm, K.M.; Akbar Ashkarran, A.; Jimenez de Aberasturi, D.; Larramendi, I.R.D.; Rojo, T.; Serpooshan, V.; Parak, W.J.; Mahmoudi, M. Antibacterial properties of nanoparticles. *Trends Biotechnol.* **2012**, *30*, 499–511. [CrossRef]

15. Agnihotri, S.; Bajaj, G.; Mukherji, S.; Mukherji, S. Arginine-assisted immobilization of silver nanoparticles on ZnO nanorods: An enhanced and reusable antibacterial substrate without human cell cytotoxicity. *Nanoscale* **2015**, *7*, 7415–7429. [CrossRef]

16. Chernousova, S.; Epple, M. Silver as Antibacterial Agent: Ion, Nanoparticle, and Metal. *Angew. Chemmie Int. Ed.* **2013**, *52*, 1636–1653. [CrossRef]

17. Sondi, I.; Salopek-Sondi, B. Silver nanoparticles as antimicrobial agent: A case study on E. coli as a model for Gram-negative bacteria. *J. Colloid Interface Sci.* **2004**, *275*, 177–182. [CrossRef]

18. Lara, H.H.; Ayala-Nuñez, N.V.; Ixtepan-Turrent, L.; Rodriguez-Padilla, C.J. Mode of antiviral action of silver nanoparticles against HIV-1. *J. Nanobiotechnol.* **2010**, *8*, 1–10. [CrossRef]

19. Senthilraja, A.; Krishnakumar, B.; Hariharan, R.; Sobral, A.J.; Surya, C.; John, N.A.; Shanthi, M. Synthesis and characterization of bimetallic nanocomposite and its photocatalytic, antifungal and antibacterial activity. *Sep. Purif. Technol.* **2018**, *202*, 373–384. [CrossRef]

20. Das, B.; Khan, M.I.; Jayabalan, R.; Behera, S.K.; Yun, S.-I.; Tripathy, S.K.; Mishra, A. Understanding the Antifungal Mechanism of Ag@ZnO Core-shell Nanocomposites against Candida krusei. *Sci. Rep.* **2016**, *6*, 36403. [CrossRef]

21. Tao, G.; Cai, R.; Wang, Y.; Liu, L.; Zuo, H.; Zhao, P.; Umar, A.; Mao, C.; Xia, Q.; He, H. Bioinspired design of AgNPs embedded silk sericin-based sponges for efficiently combating bacteria and promoting wound healing. *Mater. Des.* **2019**, *180*, 107940. [CrossRef]

22. Morones, J.R.; Elechiguerra, J.L.; Camacho, A.; Holt, K.; Kouri, J.B.; Ramírez, J.T.; Yacaman, M.J. The bactericidal effect of silver nanoparticles. *Nanotechnology* **2005**, *16*, 2346–2353. [CrossRef] [PubMed]

23. Elechiguerra, J.L.; Burt, J.L.; Morones, J.R.; Camacho-Bragado, A.; Gao, X.; Lara, H.H.; Yacaman, M.J. Interaction of silver nanoparticles with HIV-1. *J. Nanobiotechnol.* **2005**, *3*, 1–10. [CrossRef] [PubMed]

24. Samberg, M.E.; Orndorff, P.E.; Monteiro-Riviere, N.A. Antibacterial efficacy of silver nanoparticles of different sizes, surface conditions and synthesis methods. *Nanotoxicology* **2011**, *5*, 244–253. [CrossRef] [PubMed]

25. Yadav, H.M.; Kim, J.-S.; Pawar, S.H. Developments in photocatalytic antibacterial activity of nano TiO$_2$: A review. *Korean J. Chem. Eng.* **2016**, *33*, 1989–1998. [CrossRef]

26. Agnihotri, S.; Mukherji, S.; Mukherji, S. Immobilized silver nanoparticles enhance contact killing and show highest efficacy: Elucidation of the mechanism of bactericidal action of silver. *Nanoscale* **2013**, *5*, 7328–7340. [CrossRef] [PubMed]

27. Li, Z.; Tang, H.; Yuan, W.; Song, W.; Niu, Y.; Yan, L.; Yu, M.; Dai, M.; Feng, S.; Wang, M.; et al. Ag nanoparticle–ZnO nanowire hybrid nanostructures as enhanced and robust antimicrobial textiles via a green chemical approach. *Nanotechnology* **2014**, *25*, 145702. [CrossRef]

28. Dallas, P.; Tucek, J.; Jancik, D.; Kolar, M.; Panacek, A.; Zboril, R. Magnetically Controllable Silver Nanocomposite with Multifunctional Phosphotriazine Matrix and High Antimicrobial Activity. *Adv. Funct. Mater.* **2010**, *20*, 2347–2354. [CrossRef]

29. Ocsoy, I.; Paret, M.L.; Ocsoy, M.A.; Kunwar, S.; Chen, T.; You, M.; Tan, W. Nanotechnology in Plant Disease Management: DNA-Directed Silver Nanoparticles on Graphene Oxide as an Antibacterial against Xanthomonas perforans. *ACS Nano* **2013**, *7*, 8972–8980. [CrossRef]

30. Patil, S.S.; Patil, R.H.; Kale, S.B.; Tamboli, M.S.; Ambekar, J.D.; Gade, W.N.; Kolekar, S.S.; Kale, B.B. Nanostructured microspheres of silver@zinc oxide: An excellent impeder of bacterial growth and biofilm. *J. Nanoparticle Res.* **2014**, *16*, 2717. [CrossRef]
31. Lu, Z.; Gao, J.; He, Q.; Wu, J.; Liang, D.; Yang, H.; Chen, R. Enhanced antibacterial and wound healing activities of microporous chitosan-Ag/ZnO composite dressing. *Carbohydr. Polym.* **2017**, *156*, 460–469. [CrossRef] [PubMed]
32. Rajaboopathi, S.; Thambidurai, S. Synthesis of bio-surfactant based Ag/ZnO nanoparticles for better thermal, photocatalytic and antibacterial activity. *Mater. Chem. Phys.* **2019**, *223*, 512–522. [CrossRef]
33. Lu, W.; Liu, G.; Gao, S.; Xing, S.; Wang, J. Tyrosine-assisted preparation of Ag/ZnO nanocomposites with enhanced photocatalytic performance and synergistic antibacterial activities. *Nanotechnology* **2008**, *19*, 445711. [CrossRef] [PubMed]
34. Lee, H.; Dellatore, S.M.; Miller, W.M.; Messersmith, P.B. Mussel-Inspired Surface Chemistry for Multifunctional Coatings. *Science* **2007**, *318*, 426–430. [CrossRef] [PubMed]
35. Zhou, H.; Liu, Y.; Chi, W.; Yu, C.; Yu, Y. Preparation and antibacterial properties of Ag@polydopamine/graphene oxide sheet nanocomposite. *Appl. Surf. Sci.* **2013**, *282*, 181–185. [CrossRef]
36. Zhang, Z.; Zhang, J.; Zhang, B.; Tang, J. Mussel-inspired functionalization of graphene for synthesizing Ag-polydopamine-graphene nanosheets as antibacterial materials. *Nanoscale* **2013**, *5*, 118–123. [CrossRef]
37. Lu, Z.; Xiao, J.; Wang, Y.; Meng, M. In situ synthesis of silver nanoparticles uniformly distributed on polydopamine-coated silk fibers for antibacterial application. *J. Colloid Interface Sci.* **2015**, *452*, 8–14. [CrossRef]
38. Cai, R.; Tao, G.; He, H.; Song, K.; Zuo, H.; Jiang, W.; Wang, Y. One-Step Synthesis of Silver Nanoparticles on Polydopamine-Coated Sericin/Polyvinyl Alcohol Composite Films for Potential Antimicrobial Applications. *Molecules* **2017**, *22*, 721. [CrossRef]
39. Mao, B.; An, Q.; Zhai, B.; Xiao, Z.; Zhai, S.J. Multifunctional hollow polydopamine-based composites (Fe$_3$O$_4$/PDA@Ag) for efficient degradation of organic dyes. *RSC Adv.* **2016**, *6*, 47761–47770. [CrossRef]
40. Liu, L.; Cai, R.; Wang, Y.; Tao, G.; Ai, L.; Wang, P.; Yang, M.; Zuo, H.; Zhao, P.; He, H. Polydopamine-Assisted Silver Nanoparticle Self-Assembly on Sericin/Agar Film for Potential Wound Dressing Application. *Int. J. Mol. Sci.* **2018**, *19*, 2875. [CrossRef]
41. Kibar, G.; Dinç, D.Ş.Ö. In-situ growth of Ag on mussel-inspired polydopamine@poly(M-POSS) hybrid nanoparticles and their catalytic activity. *J. Environ. Chem. Eng.* **2019**, *7*, 103435. [CrossRef]
42. Ai, L.; He, H.; Wang, P.; Cai, R.; Tao, G.; Yang, M.; Liu, L.; Zuo, H.; Zhao, P.; Wang, Y. Rational Design and Fabrication of ZnONPs Functionalized Sericin/PVA Antimicrobial Sponge. *Int. J. Mol. Sci.* **2019**, *20*, 4796. [CrossRef] [PubMed]
43. Shamsuri, A.A.; Daik, R. Plasticizing effect of choline chloride/urea eutectic-based ionic liquid on physicochemical properties of agarose films. *BioResources* **2012**, *7*, 4760–4775. [CrossRef]
44. Sarma, B.; Sarma, B.K. Role of residual stress and texture of ZnO nanocrystals on electro-optical properties of ZnO/Ag/ZnO multilayer transparent conductors. *J. Alloy. Compd.* **2018**, *734*, 210–219. [CrossRef]
45. Satdeve, N.; Ugwekar, R.; Bhanvase, B.J. Ultrasound assisted preparation and characterization of Ag supported on ZnO nanoparticles for visible light degradation of methylene blue dye. *J. Mol. Liq.* **2019**, *291*, 111313. [CrossRef]
46. Kheirabadi, M.; Samadi, M.; Asadian, E.; Zhou, Y.; Dong, C.; Zhang, J.; Moshfegh, A.Z. Well-designed Ag/ZnO/3D graphene structure for dye removal: Adsorption, photocatalysis and physical separation capabilities. *J. Colloid Interface Sci.* **2019**, *537*, 66–78. [CrossRef]
47. Jaramillo-Páez, C.; Navío, J.A.; Hidalgo, M.C. Silver-modified ZnO highly UV-photoactive. *J. Photochem. Photobiol. A Chem.* **2018**, *356*, 112–122. [CrossRef]
48. Mahanti, M.; Basak, D. Highly enhanced UV emission due to surface plasmon resonance in Ag–ZnO nanorods. *Chem. Phys. Lett.* **2012**, *542*, 110–116. [CrossRef]
49. Murali, S.; Kumar, S.; Koh, J.; Seena, S.; Singh, P.; Ramalho, A.; Sobral, A.J. Bio-based chitosan/gelatin/Ag@ ZnO bionanocomposites: Synthesis and mechanical and antibacterial properties. *Cellulose* **2019**, *26*, 5347–5361. [CrossRef]
50. Bazant, P.; Kuritka, I.; Munster, L.; Kalina, L. Microwave solvothermal decoration of the cellulose surface by nanostructured hybrid Ag/ZnO particles: A joint XPS, XRD and SEM study. *Cellulose* **2015**, *22*, 1275–1293. [CrossRef]
51. Liu, J.; Hurt, R.H. Ion Release Kinetics and Particle Persistence in Aqueous Nano-Silver Colloids. *Environ. Sci. Technol.* **2010**, *44*, 2169–2175. [CrossRef] [PubMed]
52. Vaiano, V.; Matarangolo, M.; Murcia, J.J.; Rojas, H.; Navío, J.A.; Hidalgo, M.C. Enhanced photocatalytic removal of phenol from aqueous solutions using ZnO modified with Ag. *Appl. Catal. B Environ.* **2018**, *225*, 197–206. [CrossRef]
53. Ansari, S.A.; Khan, M.M.; Ansari, M.O.; Lee, J.; Cho, M.H. Biogenic Synthesis, Photocatalytic, and Photoelectrochemical Performance of Ag–ZnO Nanocomposite. *J. Phys. Chem. C* **2013**, *117*, 27023–27030. [CrossRef]
54. Lü, X.; Wang, X.; Guo, L.; Zhang, Q.; Guo, X.; Li, L. Preparation of PU modified PVDF antifouling membrane and its hydrophilic performance. *J. Membr. Sci.* **2016**, *520*, 933–940. [CrossRef]
55. Li, J.-H.; Shao, X.-S.; Zhou, Q.; Li, M.-Z.; Zhang, Q.-Q. The double effects of silver nanoparticles on the PVDF membrane: Surface hydrophilicity and antifouling performance. *Appl. Surf. Sci.* **2013**, *265*, 663–670. [CrossRef]
56. Li, K.; Chen, H.; Li, Y.; Li, J.; He, J. Endogenous Cu and Zn nanocluster-regulated soy protein isolate films: Excellent hydrophobicity and flexibility. *RSC Adv.* **2015**, *5*, 66543–66548. [CrossRef]
57. Liu, L.; Cai, R.; Wang, Y.; Tao, G.; Ai, L.; Wang, P.; Yang, M.; Zuo, H.; Zhao, P.; Shen, H.; et al. Preparation and characterization of AgNPs in situ synthesis on polyelectrolyte membrane coated sericin/agar film for antimicrobial applications. *Materials* **2018**, *11*, 1205. [CrossRef]

58. Qu, J.; Zhao, X.; Liang, Y.; Zhang, T.; Ma, P.X.; Guo, B.J. Antibacterial adhesive injectable hydrogels with rapid self-healing, extensibility and compressibility as wound dressing for joints skin wound healing. *Biomaterials* **2018**, *183*, 185–199. [CrossRef]
59. Lin, W.-C.; Lien, C.-C.; Yeh, H.-J.; Yu, C.-M.; Hsu, S.-H. Bacterial cellulose and bacterial cellulose–chitosan membranes for wound dressing applications. *Carbohydr. Polym.* **2013**, *94*, 603–611. [CrossRef]
60. Naskar, D.; Ghosh, A.K.; Mandal, M.; Das, P.; Nandi, S.K.; Kundu, S.C. Dual growth factor loaded nonmulberry silk fibroin/carbon nanofiber composite 3D scaffolds for in vitro and in vivo bone regeneration. *Biomaterials* **2017**, *136*, 67–85. [CrossRef]
61. Jalal, R.; Goharshadi, E.K.; Abareshi, M.; Moosavi, M.; Yousefi, A.; Nancarrow, P. ZnO nanofluids: Green synthesis, characterization, and antibacterial activity. *Mater. Chem. Phys.* **2010**, *121*, 198–201. [CrossRef]
62. Song, B.; Zhang, C.; Zeng, G.; Gong, J.; Chang, Y.; Jiang, Y. Antibacterial properties and mechanism of graphene oxide-silver nanocomposites as bactericidal agents for water disinfection. *Arch. Biochem. Biophys.* **2016**, *604*, 167–176. [CrossRef] [PubMed]
63. Wang, Y.; Cai, R.; Tao, G.; Wang, P.; Zuo, H.; Zhao, P.; Umar, A.; He, H. A Novel AgNPs/Sericin/Agar Film with Enhanced Mechanical Property and Antibacterial Capability. *Molecules* **2018**, *23*, 1821. [CrossRef] [PubMed]

International Journal of
Molecular Sciences

Article

Green Synthesis of Chromium Oxide Nanoparticles for Antibacterial, Antioxidant Anticancer, and Biocompatibility Activities

Shakeel Ahmad Khan [1,*], Sammia Shahid [2,*], Sadaf Hanif [2], Hesham S. Almoallim [3], Sulaiman Ali Alharbi [4] and Hanen Sellami [5]

1   Center of Super-Diamond and Advanced Films (COSDAF), Department of Chemistry, City University of Hong Kong, 83 Tat Chee Avenue, 999077 Kowloon, Hong Kong
2   Department of Chemistry, School of Science, University of Management and Technology, Lahore 54770, Pakistan; Sadafhanif59@gmail.com
3   Department of Oral and Maxillofacial Surgery, College of Dentistry, King Saud University, P.O. Box-60169, Riyadh 11545, Saudi Arabia; hkhalil@ksu.edu.sa
4   Department of Botany and Microbiology, College of Science, King Saud University, PO Box-2455, Riyadh 11545, Saudi Arabia; sharbi@ksu.edu.sa
5   Laboratory of Treatment and Valorization of Water Rejects, Water Research and Technologies Center (CERTE), Borj-Cedria Technopark, University of Carthage, Soliman 8020, Tunisia; sellami_hanen@yahoo.fr
*   Correspondence: shakilahmadkhan56@gmail.com (S.A.K.); sammia.shahid@umt.edu.pk (S.S.)

Citation: Khan, S.A.; Shahid, S.; Hanif, S.; Almoallim, H.S.; Alharbi, S.A.; Sellami, H. Green Synthesis of Chromium Oxide Nanoparticles for Antibacterial, Antioxidant Anticancer, and Biocompatibility Activities. *Int. J. Mol. Sci.* **2021**, 22, 502. https://doi.org/10.3390/ijms22020502

Received: 7 December 2020
Accepted: 4 January 2021
Published: 6 January 2021

Publisher's Note: MDPI stays neutral with regard to jurisdictional claims in published maps and institutional affiliations.

Abstract: This study deals with the green synthesis of chromium oxide (Cr_2O_3) nanoparticles using a leaf extract of *Abutilon indicum* (L.) Sweet as a reducing and capping agent. Different characterization techniques were used to characterize the synthesized nanoparticles such as X-ray diffraction (XRD), Scanning electron microscope (SEM), Transmission electron microscope (TEM), Energy-dispersive X-ray (EDX), Fourier transform infrared (FTIR), X-ray photoelectron spectroscopy (XPS), and ultraviolet-visible (UV-VIS) spectroscopy. The X-ray diffraction technique confirmed the purity and crystallinity of the Cr_2O_3 nanoparticles. The average size of the nanoparticles ranged from 17 to 42 nm. The antibacterial activity of the green synthesized nanoparticles was evaluated against four different bacterial strains, *E. coli*, *S. aureus*, *B. bronchiseptica*, and *B. subtilis* using agar well diffusion and a live/dead staining assay. The anticancer activities were determined against Michigan Cancer Foundation-7 (MCF-7) cancer cells using MTT and a live/dead staining assay. Antioxidant activity was investigated in the linoleic acid system. Moreover, the cytobiocompatibility was analyzed against the Vero cell lines using MTT and a live/dead staining assay. The results demonstrated that the green synthesized Cr_2O_3 nanoparticles exhibited superior antibacterial activity in terms of zones of inhibition (ZOIs) against Gram-positive and Gram-negative bacteria compared to plant extracts and chemically synthesized Cr_2O_3 nanoparticles (commercial), but comparable to the standard drug (Leflox). The green synthesized Cr_2O_3 nanoparticles exhibited significant anticancer and antioxidant activities against MCF-7 cancerous cells and the linoleic acid system, respectively, compared to chemically synthesized Cr_2O_3 nanoparticles. Moreover, cytobiocompatibility analysis displayed that they presented excellent biocompatibility with Vero cell lines than that of chemically synthesized Cr_2O_3 nanoparticles. These results suggest that the green synthesized Cr_2O_3 nanoparticles' enhanced biological activities might be attributed to a synergetic effect. Hence, green synthesized Cr_2O_3 nanoparticles could prove to be promising candidates for future biomedical applications.

Keywords: green synthesis; Cr_2O_3; *Abutilon indicum* (L.) Sweet; antibacterial; anticancer; antioxidant; biocompatibility

---

1. Introduction

Nanobiotechnology is the intersection of biology and nanotechnology that deals with nanotechnology's application in different biological systems. Nanobiotechnology further deals with the fabrication of biocompatible, ecofriendly, and biogenic nanomaterials and nanoparticles [1]. The nanoparticle Cr_2O_3, is of high significance and interest among various metal oxides-based nanoparticles because of its unique physicochemical properties such as a wide bandgap (~3.4 eV), high melting temperature, and increased stability [2]. The Cr_2O_3 nanoparticles have been widely utilized in different applications, including catalysis, photonics, coating materials, advanced colorants, etc. [3–6]. The trivalent Cr_2O_3 nanoparticles are considered the most stable compared to other chromium oxides [7]. Despite being a promising material, few studies have evaluated Cr_2O_3 nanoparticles for different biological applications because of their potential toxic effects that have been reported in many studies [8]. The biocompatibility of Cr_2O_3 nanoparticles is an essential parameter for their utilization in different biological systems. The poisonous effects of Cr_2O_3 nanoparticles can be reduced by coating or functionalization their surfaces with biogenic materials. One of the most promising ways of achieving this, is surface coating Cr_2O_3 with plants' biogenic phytomolecules [9].

The synthesis of nanoparticles using plants as a precursor has attracted much attention recently. As an alternative to conventional chemical and physical methods, the green synthesis of nanoparticles using biological sources (plants) is an economical, robust, ecofriendly, and easily scalable technique [10]. Most importantly, nanoparticles synthesized using plants appear to be more biocompatible than those prepared with chemical and physical methods. This is because of the fact that toxic chemicals are used in traditional chemical and physical techniques for synthesizing nanoparticles. After several rounds of washing, these toxic chemicals cannot easily be removed from the nanoparticle's surface. Therefore, poisonous chemicals present on the nanoparticle's surface making them less biocompatible and limiting their biological applications. Instead, plant based green synthesis of nanoparticles uses phytomolecules as the reducing and capping agents, and no additional chemicals are required. Moreover, plant biogenic phytomolecules have molecular functionalities that are biologically active and have antibacterial, antioxidant, anticancer, etc. properties. So, green synthesis using plants enhances the nanoparticle's biocompatibility and is responsible for the synergetic effect [9,10].

In this work, we synthesized Cr_2O_3 nanoparticles using leaf extracts of a medicinal plant (*Abutilon indicum (L.) Sweet*) for the first time, as per the author's best knowledge. *Abutilon indicum* (L.) Sweet has been widely employed for treating different kinds of diseases in Tamils, Siddha, Chinese, and traditional Ayurvedic medicine [10–12]. *Abutilon indicum (L.) Sweet* is a rich source of different biogenic phytomolecules such as terpenoids, alkaloids, saponins, polyphenols, tannins, etc., with various biological applications [13]. Many useful and biologically active compounds have been isolated from leaf extracts of *Abutilon indicum* (L.) Sweet [14]. Many reports are available that highlight the biological importance of this plant [15]. Till now, many plants have been utilized for the synthesis of nanoparticles. Among the plants used, some are either not biologically active or they are biologically active but have toxic effects. Therefore, nanoparticles for biological applications need to be synthesized with such plants that are biologically active with no toxic effects. In this regard, *Abutilon indicum (L.) Sweet* appeared as a more prominent plant that has both of these properties compared to other plants [10]. Many nanomaterials such as nanoparticles (gold, silver, ZnO, etc.) and nanorods (1D-MoO_3, etc.) have also been synthesized using leaf extracts of *Abutilon indicum* (L.) Sweet [11,12,16]. We have previously reported the green synthesis of MnO and CuO using leaf extracts of *Abutilon indicum* (L.) Sweet [10,17]. In this study, we have further utilized this plant for the green synthesis of Cr_2O_3 nanoparticles. The synthesized Cr_2O_3 nanoparticles using leaf extracts of *Abutilon indicum* (L.) Sweet were further evaluated for antibacterial, anticancer, biocompatibility, and antioxidant activities. They have presented excellent antioxidant and anticancer activities. The synthesized nanoparticles exhibited outstanding antibacterial activity by inhibiting the growth of both Gram-positive and Gram-negative bacterial strains. Moreover, the green synthesized Cr_2O_3 nanoparticles demonstrated excellent biocompatibility compared to chemically synthesized and already reported Cr_2O_3 nanoparticles.

2. Results and Discussion

2.1. Characterization

Abutilon indicum (L.) Sweet leaf extract was used as a reducing and capping agent for the synthesis of Cr_2O_3 nanoparticles. The Cr_2O_3 nanoparticles synthesis was monitored visually by detecting color change upon the addition of metal salt precursor in leaf extract. The color change of the reaction mixture from red to black indicated the formation of desired nanoparticles. This color transition occurred due to the surface plasmon resonance (SPR) phenomenon on the nanoparticle's surface [2,18]. *Abutilon indicum* (L.) Sweet leaf extract contains a rich source of biologically active phytomolecules (polyphenols, flavonoids, terpenoids, alkaloids, tannins, saponins, proteins, etc.) [10,13,19]. These phytomolecules can act as ligands and chelate with different metal ions to reduce and stabilize their ions to nano form [20,21]. The chromium sulfate salt ($Cr_2(SO_4)_3$), upon dissolution in water, becomes a freely moving ion. The freely moving Cr^{3+} ions due to electron-deficiency are attracted towards the plant's phytomolecules (polyphenols, etc.). As a result of this, chelate complex formation occurs between metal ions and the plant's phytomolecules upon transferring electrons (donor–acceptor mechanism) from oxygen to Cr^{3+} (Figure 1) [20,21]. This leads to the oxidation of polyphenols, flavonoids, etc., and converts them into keto form (Figure 1). On the other hand, Cr^{3+} is reduced to zero-valent specie Cr^0 and simultaneously stabilized by the other plant's phytomolecules (alkaloids, flavonoids, tannins, etc.) present in their vicinity. During air-drying and calcination, they are readily oxidized and converted into Cr_2O_3 nanoparticles capped with phytomolecules of *Abutilon indicum* (L.) Sweet leaf extract [20–22]. A similar green synthesis mechanism was also reported to synthesize ZnO, zinc oxide–silver, Fe_3O_4, and magnetite (Fe_3O_4) using different plants by Khalafi et al., Gurgur et al., López et al., and Yew et al., respectively [20–23].

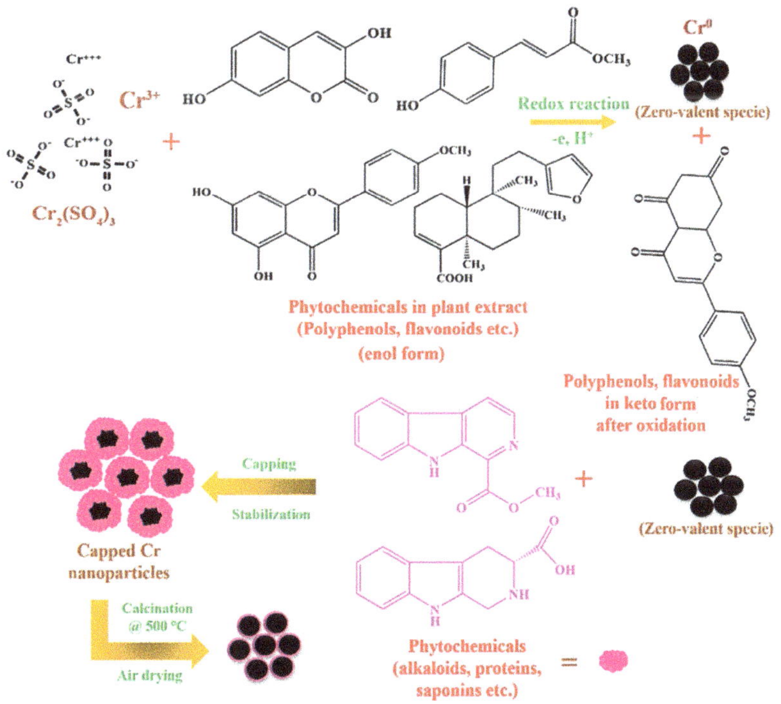

Figure 1. The schematic presentation for the green synthesis of Cr_2O_3 nanoparticles using *Abutilon indicum* (L.) Sweet leaf extract.

Further, the green synthesized Cr_2O_3 nanoparticles were analyzed using a UV–Visible spectrophotometer, and the results are presented in Figure 2a,b. UV–Visible spectrum results indicated the presence of two absorption peaks at 280 nm and 415 nm. The absorption band in the UV region is attributed to phytomolecules such as polyphenols and flavonoids, and these molecules absorb UV light because of the OH moieties [13,24,25]. The absorption band in the visible region corresponds to Cr_2O_3 [26–29]. Moreover, the plant leaf extracts presented the UV region's absorption band (200–390 nm) [10]. The FTIR analysis was further performed to determine the phytomolecules involved in synthesizing Cr_2O_3 nanoparticles as reducing and capping agents. Figure 2b,c presented the FTIR spectrum of plant leaf extract and nanoparticles. The results demonstrated that the synthesized nanoparticles displayed different FTIR peaks corresponding to O-H (3430 cm^{-1}), C-H (2921 cm^{-1}), C=O (1702 cm^{-1}), N-H (1646 cm^{-1}), C=C (1517 cm^{-1}), and C-O-C (1061 cm^{-1}). These peaks are matched with the FTIR signals of the leaf extracts with slight shifting. These results suggest that many biologically active phytomolecules are left adsorbed on the surface of the Cr_2O_3 nanoparticle [9,10]. Moreover, the FTIR signal at 612 cm^{-1}, corresponding to Cr-O, further validated the metal-oxygen bond formation [2,7,18].

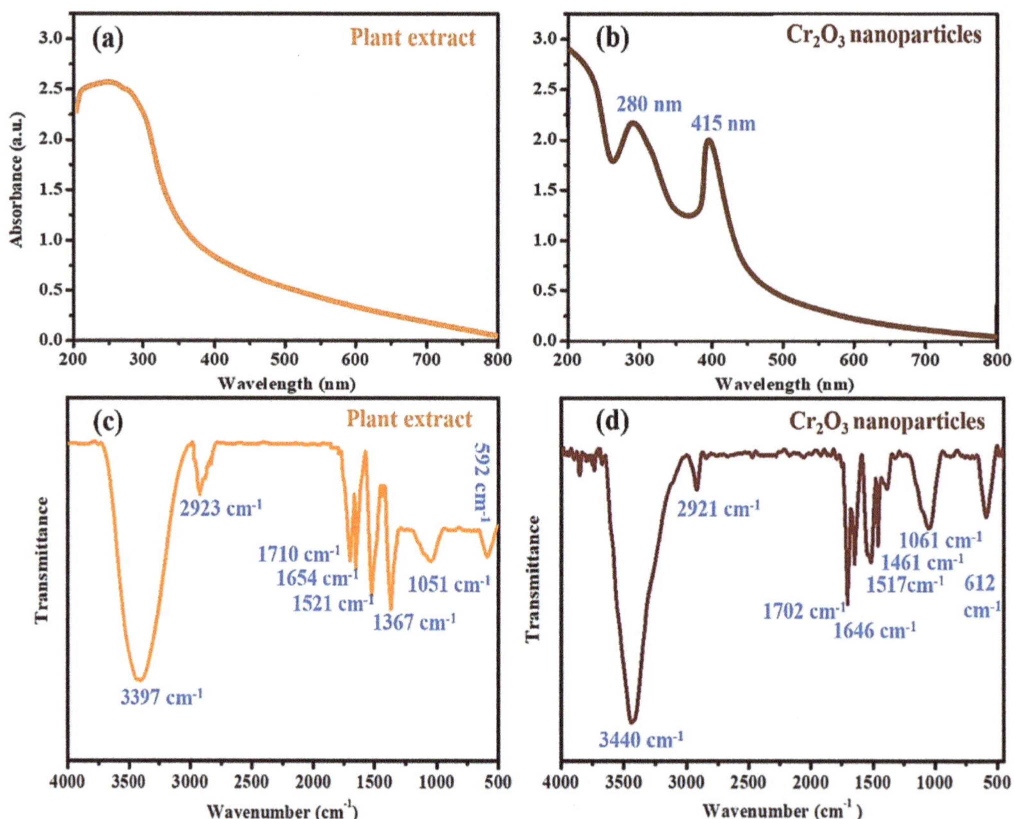

Figure 2. UV–Visible spectra of (**a**) plant leaf extract, (**b**) Cr_2O_3 nanoparticles. FTIR of (**c**) plant extract and (**d**) Cr_2O_3 nanoparticles.

The crystallinity of the green synthesized Cr_2O_3 nanoparticles was determined by XRD analysis, and the results are presented in Figure 3a. The XRD spectrum of synthesized nanoparticles revealed nine different Bragg's diffraction peaks, indexing to crystal planes of (012), (104), (110), (113), (024), (116), (214), (220), and (306) at 2θ = 24.5°, 33.6°, 36.2°, 41.5°,

50.2°, 54.9°, 63.4°, 76.8°, and 79.1°, respectively. The diffraction peaks of Cr_2O_3 nanoparticles are well-matched with Joint Committee on Powder Diffraction Standards (JCPDS) 38–1479 [30,31]. The peaks associated with impurities were not observed, indicating the purity of the nanoparticles. The peak's intensity further displayed the high crystalline nature of the nanoparticles. Figure 3b,c shows the SEM and TEM images of the synthesized Cr_2O_3 nanoparticles. SEM and TEM images displayed that the synthesized nanoparticles have spherical morphology. The Cr_2O_3 nanoparticles size determined by TEM ranged from 35–60 nm. The average nanoparticle size determined using DLS was 27.76 nm and ranged from 17–42 nm (Figure 3d). TEM and DLS particle size analysis results are consistent with each other.

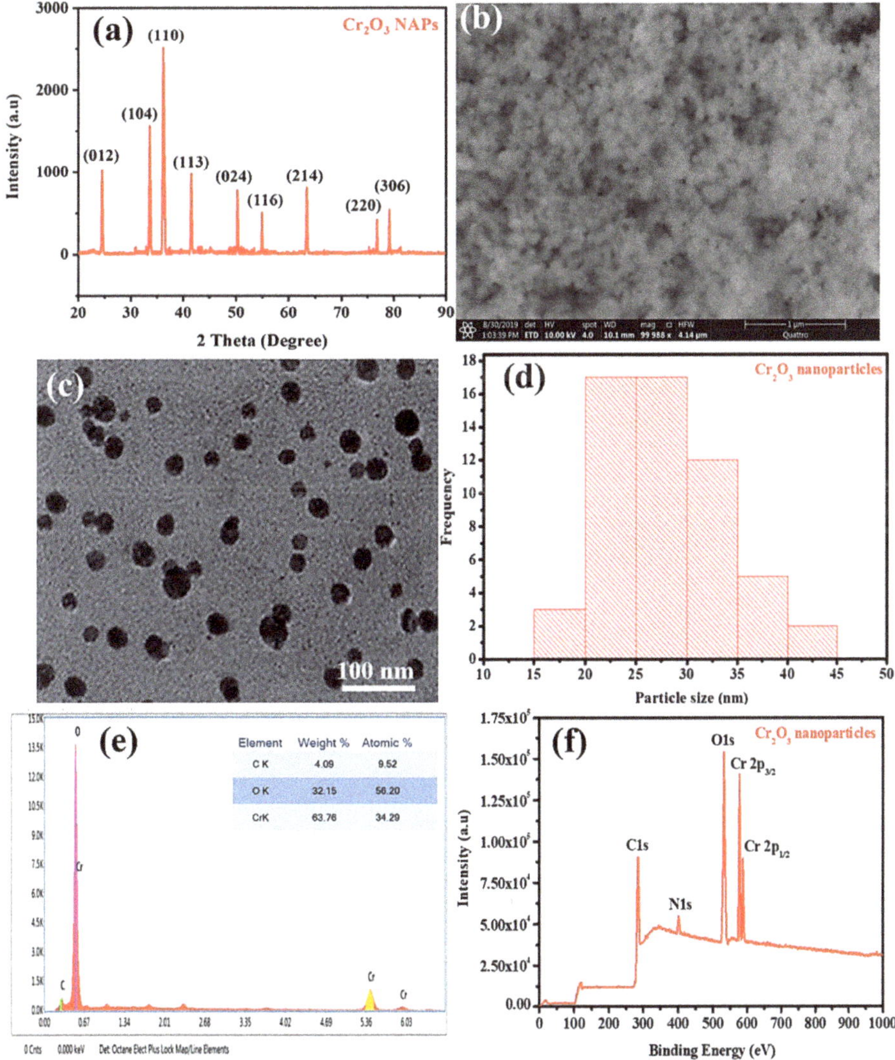

Figure 3. (**a**) XRD, (**b**) SEM, scale bar = 1 μm, (**c**) TEM, (**d**) DLS particle size distribution, (**e**) EDX and (**f**) XPS analysis of green synthesized Cr_2O_3 nanoparticles.

The compositional analysis of the synthesized nanoparticles was performed using Energy-dispersive X-ray spectroscopy. EDX spectra results showed that the nanoparticles were mainly composed of chromium (63.76%) and oxygen (32.15%), as shown in Figure 3e [2]. One extra peak associated with carbon (4.09%) is also evident in the EDX spectrum. The carbon peak could be attributed to the presence of phytomolecules (polyphenols, alkaloids, flavonoids, etc.) of leaf extract of *A. indicum* (L.) Sweet adsorbed on Cr_2O_3 nanoparticle's surface [9,10].

Further, the elemental analysis was also carried out using X-ray photoelectron spectroscopy (XPS), and the results are presented are shown in Figure 3f. XPS spectrum results indicated the presence of five peaks at binding energies of 284.5, 400.9, 530.9, 576.9, and 586.8 eV. These XPS peaks correspond to C1s, N1s, O1s, $Cr2p_{3/2}$, and $Cr2p_{1/2}$, respectively [32,33]. The carbon and nitrogen XPS peaks, other than oxygen and chromium, might be attributed to the adsorption of phytomolecules of *Abutilon indicum* (L.) Sweet leaf extract on the surface of nanoparticles. The phytomolecules of *Abutilon indicum* (L.) Sweet leaf extract has different molecular functionalities such as -OH, $-NH_2$, -CHO, $-CHO_2$, etc., in their molecules [13]. Both EDX and XPS analysis results are found to be consistent with each other. All these characterization results corroborated that the Cr_2O_3 nanoparticles of interest have been successfully green synthesized.

2.2. Antibacterial Propensity

The green synthesized Cr_2O_3 nanoparticles were evaluated for their antibacterial potential compared to the plant extract, chemically synthesized Cr_2O_3 nanoparticles, and the standard drug against four different pathogenic bacteria, including two Gram-positive and two Gram-negative. The results showed that all the samples presented concentration-dependent antibacterial activity, and the maximum inhibition in the bacteria's growth was observed with a 20 μg/mL concentration (Figure 4a–d). Moreover, the green synthesized Cr_2O_3 nanoparticles exhibited superior antibacterial activity in terms of ZOIs against Gram-positive and Gram-negative bacteria compared to the plant extract and chemically synthesized Cr_2O_3 nanoparticles. At all the concentration levels, they presented comparable inhibitory efficacy compared to the standard drug. The results further demonstrated that Gram-positive bacteria were found to be more susceptible than Gram-negative bacteria towards the green synthesized Cr_2O_3 nanoparticles. This might be due to the variances in the chemical structure and composition of both the bacteria's cell wall, and further, their different level of susceptibility towards metal oxide nanoparticles. The cell wall of Gram-negative bacteria is composed of lipopolysaccharides, lipoproteins, and phospholipids. In contrast, Gram-positive bacteria's cell walls include a thin layer of peptidoglycan and teichoic acid and large pores. Moreover, compared with Gram-negative bacteria, Gram-positive bacteria have a high negative charge on the cell wall surface, attracting nanoparticles more efficiently. Hence, the small size of nanoparticles at low temperatures can penetrate, spread, and damage the bacterial cell wall, which leads to bacteria demise [34].

The antibacterial activity of the green synthesized nanoparticles was further confirmed by CLSM, and the results are presented in Figure 5. SYTO-9 is a membrane-permeant dye which stains live/dead cells. In comparison, PI is an impermeant dye and can only stain dead cells upon its penetration. The PI penetrates the cells only via the dead cells' burst membrane and subsequently binds to the DNA, emitting a strong red fluorescence [35]. The results demonstrate that untreated bacterial cells (control) exhibited an intense green color, indicating that all the cells were alive and intact. On the other hand, bacterial cells treated with green synthesized Cr_2O_3 nanoparticles appeared red, which showed that the nanoparticles destroyed the bacterial cell's membrane's permeability and integrity, leading to cell demise.

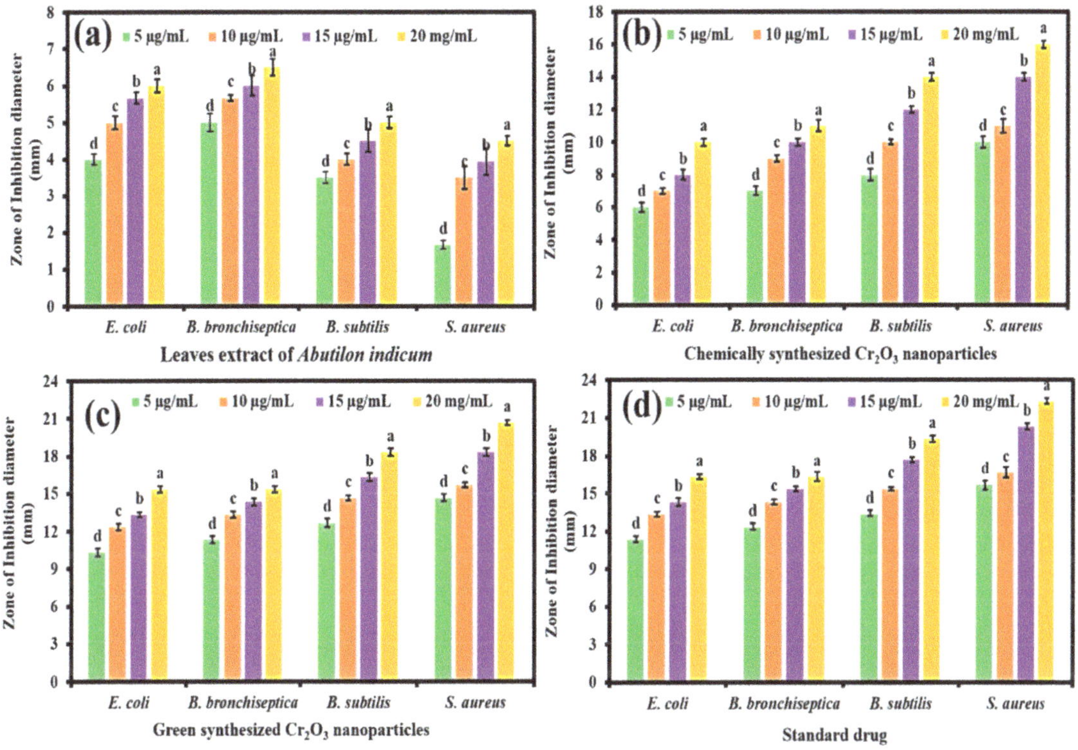

Figure 4. The antibacterial activity of (**a**) *Abutilon indicum* (L.) Sweet leaf extract, (**b**) chemically synthesized Cr_2O_3 nanoparticles, (**c**) green synthesized Cr_2O_3 nanoparticles in terms of zones of inhibition (ZOIs) at different concentration levels against different bacteria compared to standard drug (**d**). (Note: Tukey based heterogeneous lower-case letters represent significant pairs).

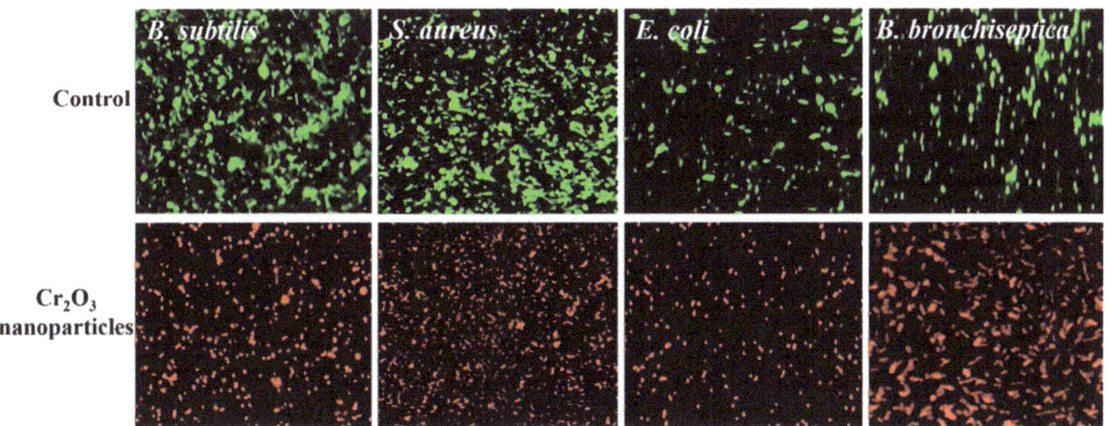

Figure 5. Live/dead bacterial cell images—live cells stained with green (SYTO-9) while dead cells stained with red (PI) (scale bar = 50 µm).

2.3. Anticancer Activity

The green synthesized Cr_2O_3 nanoparticles were evaluated for their anticancer potential in terms of cell viability percentage against MCF-7 cancerous cells compared to *Abutilon indicum* (L.) Sweet leaf extract, chemically synthesized Cr_2O_3 nanoparticles, and the standard drug. The results were shown that all the samples presented concentration-dependent anticancer activity. The maximum cytotoxic effect on MCF-7 cancer cells was observed with 120 µg/mL concentration of all the samples (Figure 6). The superior anticancer activity was demonstrated by the green synthesized Cr_2O_3 nanoparticles compared to *Abutilon indicum* (L.) Sweet leaf extract and chemically synthesized Cr_2O_3 nanoparticles at all tested concentrations. Moreover, the green synthesized Cr_2O_3 nanoparticles displayed slightly less anticancer activity against MCF-7 carcinoma cells than the standard drug. However, this difference was not sufficient, so we can suggest that they presented comparable levels of anticancer activity to the standard drug at all the tested concentrations. Our green synthesized Cr_2O_3 nanoparticles displayed better anticancer activity against human breast cancer cells at the 120 µg/mL concentration compared to (500 µg/mL) single-phase Cr_2O_3 nanoparticles synthesized using *Nephelium lappaceum* L. [36].

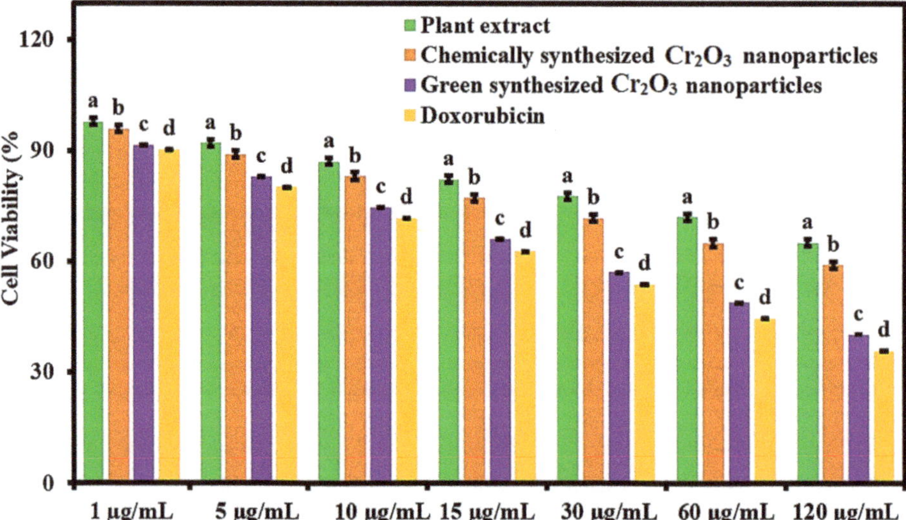

Figure 6. Anticancer activity of green synthesized Cr_2O_3 nanoparticles in terms of cell viability percentage against MCF-7 cancer cells compared to *Abutilon indicum* (L.) Sweet leaf extract, chemically synthesized Cr_2O_3 nanoparticles, and standard drug. (Note: Tukey based heterogeneous lower-case letters represent significant pairs).

Using an inverted microscope (Nikon Eclipse TE200), we further observed the morphological changes in MCF-7 carcinoma cells after their treatment with green synthesized Cr_2O_3 nanoparticles, plant extract, and chemically synthesized Cr_2O_3 nanoparticles at the concentration of 120 µg/mL. Figure 7a–d shows the inverted micrograph of MCF-7 cancerous cells. The images show that after treatment, drastic changes occurred in the morphology of MCF-7 cancer cells. The MCF-cells' volume and cytoplasm have been decreased, and the shape of the cells changed to round. All the samples induced toxicity, but green synthesized Cr_2O_3 nanoparticles were appeared to pose a significant and severe cytotoxic effect on MCF-7 cancer cells.

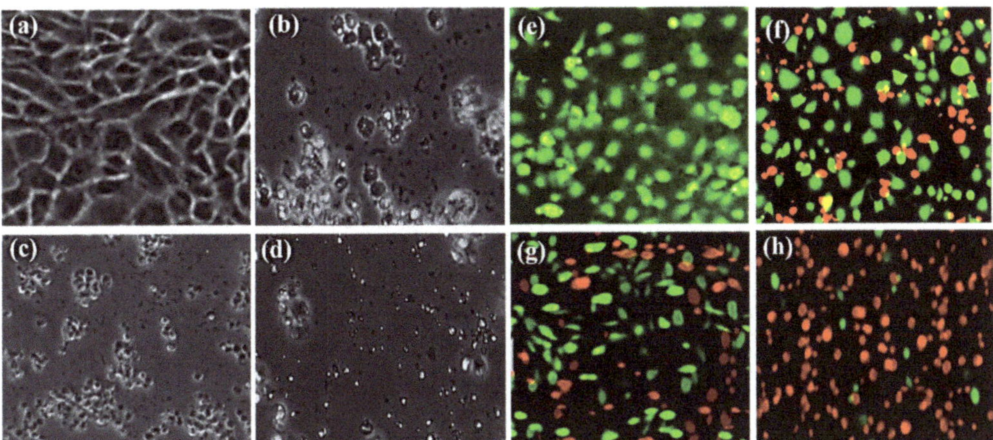

Figure 7. The morphological alterations in MCF-7 cancer cells after treatment with (**b**) plant extract, (**c**) chemically synthesized Cr_2O_3 nanoparticles, and (**d**) green synthesized Cr_2O_3 nanoparticles. The live/dead MCF-7 cancer cells stained with green and red fluorescent dye respectively after treatment with (**f**) plant extract, (**g**) chemically synthesized Cr_2O_3 nanoparticles, and (**h**) green synthesized Cr_2O_3 nanoparticles. (**a**) and (**e**) controls. (Scale bar = 100 μm).

To further affirm the anticancer activity against MCF-7 cancer cells, the live and dead fluorescence staining assay was employed using CLSM. Figure 7e–h shows the live/dead MCF-7 cancer cells stained with green and red dye, respectively. The results demonstrated that green synthesized Cr_2O_3 nanoparticles exhibited a maximum cytotoxic effect on MCF-7 carcinoma cells, and they had killed almost 90% cancerous cells (Figure 7h). On the other hand, chemically synthesized Cr_2O_3 nanoparticles induced a mild toxic effect on MCF-7 cancer cells and destroyed almost 50% of cancer cells. It is interesting to note that leaf extract also presented toxicity on MCF-7 cancer cells indicating that *Abutilon indicum* (L.) Sweet has biologically active phytomolecules. Henceforth, these results are consistent with the results of MTT and inverted microscopic analysis.

2.4. Antioxidant Activity

The antioxidant activity of green synthesized Cr_2O_3 nanoparticles was determined in the linoleic acid system and compared to plant leaf extract, chemically synthesized Cr_2O_3 nanoparticles, and standard (α-tocopherol). The results in the form of lipid peroxidation percentage are presented in Figure 8. The results demonstrate that maximum lipid peroxidation inhibition was observed with the standard, followed by plant extract and green synthesized Cr_2O_3 nanoparticles. Chemically synthesized Cr_2O_3 nanoparticles displayed the lowest level of antioxidant activity in terms of lipid peroxidation inhibition. It is interesting to note that the plant extract presented a comparable antioxidant activity compared to the standard. The enhanced antioxidant activity of green synthesized Cr_2O_3 nanoparticles might be attributed to the presence of phytomolecules of the plant leaf extract on the nanoparticle's surface, as evident from the FTIR, EDX, and XPS results. Hence, these results suggest that green synthesized Cr_2O_3 nanoparticles and plant extract can be used as powerful antioxidant agents in different applications. Moreover, our green synthesized Cr_2O_3 nanoparticles appeared to be more active in terms of antioxidant activity than previously reported for Cr_2O_3 nanoparticles synthesized using leaf extract of *Rhamnus virgate* [2].

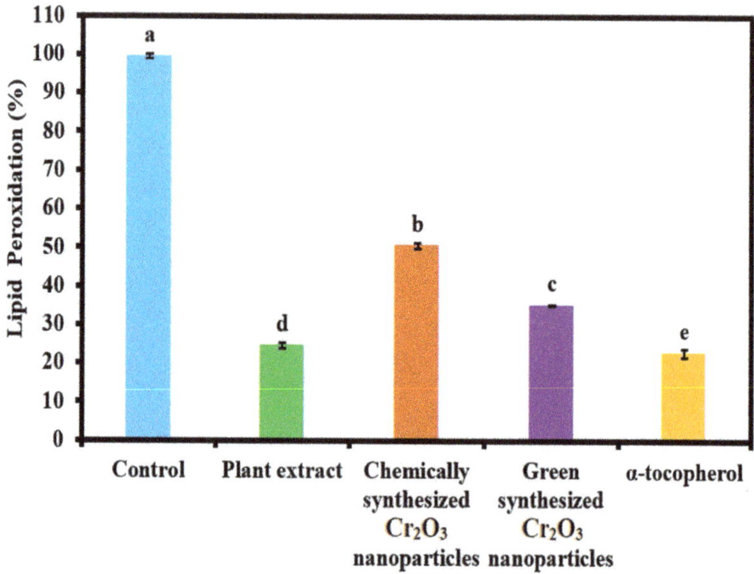

Figure 8. The antioxidant activity of green synthesized Cr_2O_3 nanoparticles in terms of linoleic acid peroxidation percentage compared to plant extract, chemically synthesized Cr_2O_3 nanoparticles, and standard (α-tocopherol). (Note: Tukey based heterogeneous lower-case letters represent significant pairs).

2.5. Cytobiocompatibility Analysis

The green synthesized Cr_2O_3 nanoparticles were further evaluated for their cytobiocompatibility analysis against the Vero cell lines (Kidney epithelial cells) in comparison to plant leaf extract and chemically synthesized Cr_2O_3 nanoparticles. As per International Organization for Standardization (ISO) 10993-5, a material could be considered toxic, moderately toxic, weak toxic, and cytobiocompatible if the cell viability (%) is less than 40%, 40 to 60%, 60 to 80%, and greater than 80% respectively. The results are presented in Figure 9a. The results of cell viability (%) demonstrated that the chemically synthesized Cr_2O_3 nanoparticles exhibited the least cytobiocompatibility (77.46 ± 0.31%). On the other hand, plant extract and green synthesized Cr_2O_3 nanoparticles presented excellent cytobiocompatibility (93.63 ± 0.24% and 88.50 ± 0.85%) with the Vero cell lines, respectively. Our green synthesized Cr_2O_3 nanoparticles exhibited good cytobiocompatibility with the normal cells compared to previous reports [2].

We further analyzed the cytobiocompatibility of the green synthesized Cr_2O_3 nanoparticles with Vero cell lines compared to the plant extract and chemically synthesized Cr_2O_3 nanoparticles using the live/dead staining technique. The results are displayed in Figure 9b–e. The results demonstrated that the plant leaf extract and green synthesized Cr_2O_3 nanoparticles exerted the lowest levels of cytotoxic effects on Vero cells as fewer cells died (Figure 9c,e). In contrast, chemically synthesized Cr_2O_3 nanoparticles exerted more cytotoxicity, and many cells appeared dead (Red) (Figure 9d). The good cytobiocompatibility of the green synthesized Cr_2O_3 nanoparticles with the Vero cell lines might be attributed to the presence of phytomolecules of plant leaf extract.

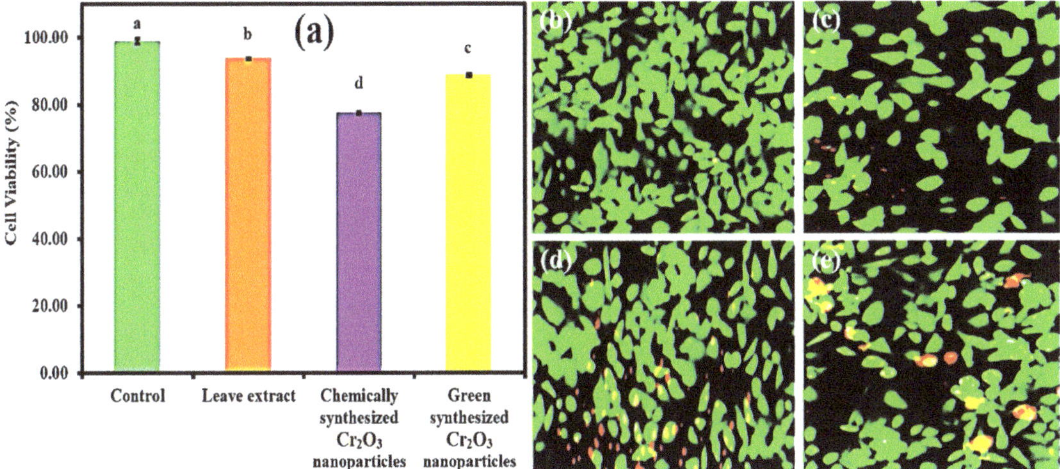

Figure 9. (**a**) The cytobiocompatibility analysis of green synthesized Cr_2O_3 nanoparticles against Vero cell lines compared to plant extract and chemically synthesized Cr_2O_3 nanoparticles. (Note: Tukey based heterogeneous lower-case letters represent significant pairs). CLSM images of (**b**) untreated Vero cell lines (control), and treated with (**c**) plant extract, (**d**) chemically synthesized Cr_2O_3, and (**e**) green synthesized Cr_2O_3 nanoparticles (Live cells with green and dead cells with red) (Scale bar = 50 μm).

3. Materials and Methods

The present research work was performed in the chemistry laboratory, Department of Chemistry, University of Management and Technology, Lahore. All the chemicals used were of analytical grade and available commercially. The chemicals used in the research work were purchased from Merck (Darmstadt, Germany) and Sigma Chemicals Co. (St. Louis, MS, USA). The commercially available Oleic acid-coated Cr_2O_3 nanoparticles were purchased with a 18 nm size for comparative biological analysis—these nanoparticles are named as chemically synthesized nanoparticles.

3.1. Collection of the Plant Material

Abutilon indicum (L.) Sweet plant was collected from the wild area native to tropical and subtropical regions. Its identification was made by Dr. Zaheer (Department of Botany, Punjab University, Lahore, Pakistan).

3.2. Preparation of Plant Extract

A total of 20 g of the plant's fresh leaves were taken. The leaves were washed with deionized (DI) water and dried in an oven at 80 °C. The dried leaves were crushed with a mortar and pestle. After fine crushing, the crushed leaves were mixed in 150 mL DI water and 100 mL methanol and heated at 50 °C for 1 h. After heating, the sample was kept for one day and then filtered, and the subsequent filtrate (plant extract) was stored in an air-tight bottle for further use (Figure 10).

Figure 10. The leaf extraction preparation and green synthesis of Cr_2O_3 nanoparticles using *Abutilon indicum* (L.) Sweet leaf extracts.

3.3. Green Synthesis of Chromium Oxide (Cr_2O_3) Nanoparticles

For the green synthesis of Cr_2O_3 nanoparticles, 10.20 g of $Cr_2(SO_4)_3$ was added to 100 mL of plant extract and stirred the resultant mixture for 60 minutes at 35 °C. After, a change in the color of the resulting mixture solution from red to black was observed, due to surface plasmon resonance indicating the formation of the required Cr_2O_3 nanoparticles. The nanoparticles were then centrifuged at 3000 rpm twice and then filtered and washed with deionized water/ethanol three times. Then, the Cr_2O_3 nanoparticles were dried in an oven at 40 °C and further calcinated at 500 °C in a muffle furnace for 3 hours (Figure 10). Finally, the obtained nanoparticles were stored in an air-tight container for characterization and biological applications.

3.4. Characterization

3.4.1. X-ray Diffraction

The crystallinity and purity of the green synthesized Cr_2O_3 nanoparticles in powder form were determined using the powder X-ray diffraction (XRD) (Bruker D2 PHASER with LYNXEYE XE-T detector, Haidian, Beijing, China) with a wavelength (λ) of 0.154 nm over the 2θ range 4–90°.

3.4.2. Scanning Electron Microscope (SEM) and Energy-Dispersive X-ray (EDX) Spectroscopy

The synthesized nanoparticles' morphology was characterized using an SEM (Quattro S) by placing the dried powder sample on the carbon tape. The compositional analysis was carried out with an energy-dispersive X-ray (EDX) spectroscopy using Thermo Fisher Scientific Ultradry (Madison, WI, USA) attached with SEM.

3.4.3. Transmission Electron Microscope (TEM)

The green synthesized Cr_2O_3 nanoparticles were dissolved in methanol, and sonication was performed at 25–30 °C and then they were transferred to a copper grid. The copper grid was set aside for drying for 5–10 min, then TEM (FEI/Philips Tecnai 12 BioTWIN, Baltimore, MD, USA) was used to acquire TEM images with an acceleration voltage of 200 kV [37].

3.4.4. Zetasizer Dynamic Light Scattering (DLS)

The green synthesized Cr_2O_3 nanoparticles were dissolved in DI water and sonicated for 5 min at 25–30 °C. About 10 mm sample solution was taken out and placed on glass cuvette. After that, the cuvette was placed in the cell holder, and scanning was performed using a dynamic light scattering particle size analyzer (Malvern Zetasizer Nano ZS, Worcestershire, WR14 1XZ, UK) from 1 to 100 nm at 25–30 °C [37].

3.4.5. X-ray Photoelectron Spectroscopy (XPS)

The XPS of green synthesized Cr_2O_3 nanoparticles was performed using ULVAC-PHI Quantera II (Ulvac-PHI Inc., Chigasaki. Kanagawa, Japan) with the following conditions monochromatic AlKα (hυ = 1486.6 eV) at 25.6 W with a beam diameter of 100 μm. Additionally, pass energy of 280 eV with 1 eV per step was used to perform a wide scan analysis.

3.4.6. UV-Visible Spectrophotometric Analysis

The green synthesized Cr_2O_3 nanoparticles were dissolved in DI water and sonicated for 5 min at 25–30 °C. The nanoparticles solution and plant extract were then transferred in a quartz cuvette, and after being placed in the cell, the absorption maxima were determined from 200 to 800 nm using a UV–Visible spectrophotometer (Shimadzu 1700, Columbia, Maryland, U.S.A.) at 25–30 °C [37].

3.4.7. Fourier Transform Infrared (FTIR)

The dried powder of green synthesized Cr_2O_3 and plant extract was placed on the quartz slide, and then the FTIR spectrum was measured from 450 to 4000 using Perkin Elmer Spectrum 100 spectrophotometer (Bridgeport Avenue Shelton, CT 06484-4794, USA) at 25–30 °C.

3.5. Antibacterial Propensity

The antibacterial propensity of the green synthesized Cr_2O_3 nanoparticles was determined using agar well diffusion assay against four different bacterial strains (*Staphylococcus aureus* ATCC[®] 23235™, *Bacillus subtilis* ATCC[®] 6051™, *Escherichia coli* ATCC[®] 25922™, and *Bordetella bronchiseptica* ATCC[®] 4617™), following the protocol previously reported by [9]. The antibiotic drug (Leflox) and dimethyl sulfoxide (DMSO) were used as a positive and negative control, respectively. Four concentrations (5, 10, 15, and 20 μg/mL) of each sample were prepared in DMSO. For antibacterial activity, washed petri-dishes and freshly synthesized media (nutrient agar) were sterilized by an autoclave for 15 min at 121 °C. The sterilized molten nutrient agar (30 mL) was poured into petri-dishes as a basal layer and set aside for a moment to form a solid gel, and subsequently, 3.5 mL of inoculum of each bacterium inoculated. The inoculum of each bacteria strain was prepared at 1×10^8 CFU/mL. The holes were then bored at four peripheral positions using a sterilized hollow iron rod. The holes were then filled with 20 μL of sample, positive and negative control dilutions. The petri-dishes were incubated for 24 h at 37 °C in an incubator. After 24 h, the clear zones of inhibition (ZOIs) were observed around the holes. The diameter of the ZOIs was recorded using a ruler in millimeters. The experiment was repeated three times.

Live/dead Bacteria Staining Assay

To further confirm the antibacterial activity of green synthesized Cr_2O_3 nanoparticles, a live and staining assay was performed using a confocal laser scanning microscope (CLSM, FV-1200, Olympus, Tokyo, Japan) following the protocol reported by [38]. The two fluorescent dyes SYTO-9 and propidium iodide (PI) were used for staining the live (green) and dead (red) bacteria, respectively. Each bacterium was cultured in a nutrient broth and incubated at 37 °C for 24 h to obtain the confluence of 10^5–10^6 colony forming units (CFU) per mL. After, bacteria were inoculated onto pasteurized cover glass coated with poly-L-lysine in 24-microtitre well plate and placed in an incubator for incubation for 1 hour to allow bacterial cells to attach to the cover glass. The suspended bacterial cells were then removed, and the cover glass was gently rinsed three times using a saline solution. Each bacterium on the cover glass was treated with 20 μg/mL concentration of green synthesized Cr_2O_3 nanoparticles and incubated again at 37 °C for 24 h. The bacteria on cover glass were then stained with a live/dead bacterial viability kit, as per the manufacturer's instructions. The bacterial cells were analyzed with CLSM at 485 nm excitation wavelength for SYTO-9 and PI and 530 nm emission wavelength for SYTO-9 while 630 nm for PI. We only considered green synthesized Cr_2O_3 nanoparticles for live/dead staining assay as they presented excellent antibacterial properties in terms of ZOIs.

3.6. Anticancer Activity

MTT (3-(4,5-dimethylthiazol-2-yl)-2,5-diphenyltetrazolium bromide) colorimetric assay was used for determining the anticancer activity of chromium oxide nanoparticles against MCF-7 (breast cancer cells) [10]. The MCF-7 carcinomatous cells were placed in Dulbecco's Modified Eagle's Medium (DMEM) in a humidified atmosphere consisting of 5% CO_2 and 95% air at 37 °C. To obtain the 5×10^8 cells/well, the MCF-7 cells were cultured in 96-well plates containing the 100 μL of DMEM for 24 h at 37 °C. The 50 μL of each sample (green synthesized Cr_2O_3 nanoparticles, plant extract, and pristine) at concentrations of 1, 5, 10, 15, 30, 60, and 120 μg/mL was added separately in each well, and the plate was then incubated for 24 h at 37 °C. After the plate was centrifuged to remove the supernatant and then washed with phosphate buffer saline. The 15 microliters of MTT reagent (0.5 mg/mL) was added to each well. The plate was placed in an incubator for incubation for 4 hours at 37 °C. To dissolve the crystals of formazan, a reduced product of MTT, 150 μL of DMSO were added in each well and stirred on a shaker for 10 min. The optical density (OD) of formazans products was measured at 570 nm using a spectrophotometer. The cell viability percentage was calculated using the following formula; with the help of the following equation;

$$\% \text{ Cell viability} = OD_{sample}/OD_{control} \times 100 \tag{1}$$

Live/Dead Cells Staining Assay

We further investigated the MCF-7 carcinomatous cell viability with the fluorescent staining technique to affirm the cytotoxicity using the live/dead double staining kit (viable cells stain with green and dead cells with red). The same experiment was repeated as described above till cancer cells treated with different samples (10 μL of 120 μg/mL) and subsequent incubation. After, the staining solution (4 μg/mL) was added to each well at 37 °C and incubated for 20 min. Photographs were taken with a fluorescence microscope (excitation wavelength 488/545 nm for viable/dead cells).

3.7. Antioxidant Activity in Terms of Linoleic Acid (%) Inhibition

The antioxidant activity of the green synthesized Cr_2O_3 nanoparticles in terms of linoleic acid (%) inhibition was determined compared to plant extracts and chemically synthesized Cr_2O_3 nanoparticles, following the protocol reported by [39]. In detail, 100 μg/mL concentration of each sample was added to the solution mixture of 0.2 M sodium phosphate buffer (pH 7.0, 10 mL), 99.99% ethanol (10 mL), and linoleic acid (0.13 mL). The resulting solution's total volume was made up to 25 mL with DI water and subsequently incubated for 360 hours at 40 °C. The extent of oxidation was measured using the thiocyanate method. Accordingly, 0.2 mL of the sample solution was taken and then added to 10 mL of ethanol (75%). Subsequently, 0.2 mL of aqueous ammonium thiocyanate solution (30%) and 0.2 mL $FeCl_2$ (20 mM in 3.5% HCl) was added. The reaction mixture was stirred for 3 min, and the absorption maxima were then measured at 500 nm wavelength. The percentage inhibition of linoleic acid was calculated using the following formula:

$$\% \text{ Inhibition} = [100 - (\text{Absorbance of sample})/(\text{absorbance of control})] \times 100 \tag{2}$$

The alpha-tocopherol was used as an external standard, and the control only consisted of linoleic acid without any treatment.

3.8. Cytobiocompatibility Analysis

The green synthesized Cr_2O_3 nanoparticles were further evaluated for their cytobiocompatibility analysis against the Vero cell line (Kidney epithelial cells) in comparison to chemically synthesized Cr_2O_3 nanoparticles. The MTT protocol was followed for determining the cytobiocompatibility analysis reported by [40].

3.9. Statistical Analysis

All the experiments (antibacterial, anticancer, antioxidant, and biocompatibility) were conducted three times, and the results are presented as mean $\pm$ standard deviation. One-way ANOVA at a fixed significance level (0.05) and the Tukey test were also applied to the results to determine the significance.

4. Conclusions

In this work, Cr_2O_3 nanoparticles have been successfully green synthesized using the leaf extract of *Abutilon indicum* (L.) Sweet as a reducing and capping agent. The green synthesized Cr_2O_3 nanoparticles were successfully characterized using XRD, SEM, TEM, EDX, FTIR, XPS, and UV-VIS spectroscopy. The green synthesized Cr_2O_3 nanoparticles displayed excellent antibacterial performance against all tested bacterial strains (*E. coli, S. aureus, B. bronchiseptica*, and *B. subtilis*) and were comparable to the standard available drugs. However, they showed better bacterial inhibition than plant leaf extract and chemically synthesized Cr_2O_3 nanoparticles. The green synthesized Cr_2O_3 nanoparticles also demonstrated significant anticancer and antioxidant activities against MCF-7 cancer cells and the linoleic acid system, respectively, comparable to the employed standard drug and external standard antioxidant, respectively. Moreover, green synthesized Cr_2O_3 nanoparticles presented excellent biocompatibility with Vero cell lines compared to chemically synthesized Cr_2O_3 nanoparticles. It is interesting to note that *Abutilon indicum* (L.) Sweet leaf extract was also found to be active towards antibacterial, antioxidant, and anticancer activities. These results suggest that the green synthesized Cr_2O_3 nanoparticles' enhanced biological activities might be attributed to the synergetic effect (physical properties and adsorbed phytomolecules on their surface). Thus, the antioxidant, antibacterial, biocompatibility, and anticancer activities results displayed the potential of green synthesized Cr_2O_3 nanoparticles for different future biomedical applications (antifungal, antilarvicidal, etc.). Hence, nanoparticles synthesis using leaf extracts of *Abutilon indicum* (L.) Sweet is an efficient, robust, economical, and green method that produces biocompatible and biological active nanoparticles. The use of leaf extracts of *Abutilon indicum* (L.) Sweet can be further extended for synthesizing various other biocompatible nanomaterials for biological applications.

Author Contributions: Conceptualization, S.A.K. and S.H.; methodology, S.A.K. and S.H.; software, S.A.K. and S.H.; validation, S.A.K., S.H. and S.S.; formal analysis, S.A.K., S.H. and H.S.; investigation, S.A.K. and S.H.; resources, S.S. and H.S.; data curation, S.A.K., S.H. and H.S.; writing—original draft preparation, S.A.K. and S.H.; writing—review and editing, S.S., H.S.A., S.A.A. and H.S.; visualization, S.S., H.S.A., S.A.A. and H.S.; supervision, S.S.; project administration, S.S., H.S.A., and S.A.A.; funding acquisition, H.S.A. and S.A.A. All authors have read and agreed to the published version of the manuscript.

Funding: This research was funded by King Saud University, Riyadh, Saudi Arabia, grant number RSP-2020/283. The APC was funded by RSP-2020/283.

Institutional Review Board Statement: Not applicable.

Informed Consent Statement: Not applicable.

Data Availability Statement: The data are not publicly available.

Acknowledgments: The authors are very thankful to the School of Science, Department of Chemistry, University of Management and Technology, Lahore Pakistan, for supporting this research work. The authors extend their appreciation to the Researchers Supporting Project number (RSP-2020/283) King Saud University, Riyadh, Saudi Arabia.

Conflicts of Interest: The authors declare no conflict of interest.

References

1. Singh, H.; Du, J.; Singh, P.; Yi, T.H. Ecofriendly synthesis of silver and gold nanoparticles by Euphrasia officinalis leaf extract and its biomedical applications. *Artif. Cells Nanomed. Biotechnol.* **2018**, *46*, 1163–1170. [CrossRef]
2. Iqbal, J.; Munir, A.; Uddin, S. Facile green synthesis approach for the production of chromium oxide nanoparticles and their different in vitro biological activities. *Microsc. Res. Tech.* **2020**, *83*, 706–719. [CrossRef] [PubMed]
3. Rao, T.V.M.; Zahidi, E.M.; Sayari, A. Ethane dehydrogenation over pore-expanded mesoporous silica-supported chromium oxide: 2. Catalytic properties and nature of active sites. *J. Mol. Catal. A Chem.* **2009**, *301*, 159–165. [CrossRef]
4. Patah, A.; Takasaki, A.; Szmyd, J.S. Influence of multiple oxide (Cr_2O_3/Nb_2O_5) addition on the sorption kinetics of MgH_2. *Int. J. Hydrog. Energy* **2009**, *34*, 3032–3037. [CrossRef]
5. Rakesh, S.; Ananda, S.; Gowda, N.M.M. Synthesis of Chromium(III) Oxide Nanoparticles by Electrochemical Method and Mukia Maderaspatana Plant Extract, Characterization, $KMnO_4$ Decomposition and Antibacterial Study. *Mod. Res. Catal.* **2013**, *2*, 127–135. [CrossRef]
6. El-Sheikh, S.M.; Mohamed, R.M.; Fouad, O.A. Synthesis and structure screening of nanostructured chromium oxide powders. *J. Alloys Compd.* **2009**, *482*, 302–307. [CrossRef]
7. Ahmed Mohamed, H.E.; Afridi, S.; Khalil, A.T.; Zohra, T.; Ali, M.; Alam, M.M.; Ikram, A.; Shinwari, Z.K.; Maaza, M. Phyto-fabricated Cr_2O_3 nanoparticle for multifunctional biomedical applications. *Nanomedicine* **2020**, *15*, 1653–1669. [CrossRef] [PubMed]
8. Alarifi, S.; Ali, D.; Alkahtani, S. Mechanistic investigation of toxicity of chromium oxide nanoparticles in murine fibrosarcoma cells. *Int. J. Nanomed.* **2016**, *11*, 1253–1259. [CrossRef]
9. Khan, S.A.; Shahid, S.; Lee, C.-S. Green Synthesis of Gold and Silver Nanoparticles Using Leaf Extract of Clerodendrum inerme; Characterization, Antimicrobial, and Antioxidant Activities. *Biomolecules* **2020**, *10*, 835. [CrossRef]
10. Khan, S.A.; Shahid, S.; Shahid, B.; Fatima, U.; Abbasi, S.A. Green Synthesis of MnO Nanoparticles Using Abutilon indicum Leaf Extract for Biological, Photocatalytic, and Adsorption Activities. *Biomolecules* **2020**, *10*, 785. [CrossRef]
11. Mata, R.; Nakkala, J.R.; Sadras, S.R. Polyphenol stabilized colloidal gold nanoparticles from Abutilon indicum leaf extract induce apoptosis in HT-29 colon cancer cells. *Colloids Surf. B* **2016**, *143*, 499–510. [CrossRef]
12. Mata, R.; Nakkala, J.R.; Sadras, S.R. Biogenic silver nanoparticles from Abutilon indicum: Their antioxidant, antibacterial and cytotoxic effects in vitro. *Colloids Surf. B* **2015**, *128*, 276–286. [CrossRef] [PubMed]
13. Saranya, D.; Sekar, J. GC-MS and FT-IR Analyses of Ethylacetate Leaf extract of *Abutilon indicum* (L.) Sweet. *Int. J. Adv. Res. Biol. Sci.* **2016**, *3*, 193–197.
14. Akbar, S. Abutilon indicum (Link) Sweet (Malvaceae). In *Handbook of 200 Medicinal Plants*; Springer: Cham, Switzerland, 2020. [CrossRef]
15. Surywanshi, V.S.; Umate, S.R. A Review on Phytochemical Constituents of Abutilon indicum (Link) Sweet–An Important Medicinal Plant in Ayurveda. *Plantae Scientia* **2020**, *3*, 15–19. [CrossRef]
16. Prashanth, G.K.; Prashanth, P.A.; Nagabhushana, B.M.; Ananda, S.; Krishnaiah, G.M.; Nagendra, H.G.; Sathyananda, H.M.; Rajendra Singh, C.; Yogisha, S.; Anand, S.; et al. Comparison of anticancer activity of biocompatible ZnO nanoparticles prepared by solution combustion synthesis using aqueous leaf extracts of Abutilon indicum, Melia azedarach and Indigofera tinctoria as biofuels. *Artif. Cells. Nanomed. Biotechnol.* **2018**, *46*, 968–979. [CrossRef] [PubMed]
17. Ijaz, F.; Shahid, S.; Khan, S.A.; Ahmad, W.; Zaman, S. Green synthesis of copper oxide nanoparticles using Abutilon indicum leaf extract: Antimicrobial, antioxidant and photocatalytic dye degradation activities. *Trop. J. Pharm. Res.* **2017**, *16*, 743–753. [CrossRef]
18. Hassan, D.; Khalil, A.T.; Solangi, A.R.; El-Mallul, A.; Shinwari, Z.K.; Maaza, M. Physiochemical properties and novel biological applications of Callistemon viminalis-mediated α-Cr_2O_3 nanoparticles. *Appl. Organomet. Chem.* **2019**, *33*, e5041. [CrossRef]
19. Kuo, P.C.; Yang, M.L.; Wu, P.L.; Shih, H.N.; Thang, T.D.; Dung, N.X.; Wu, T.S. Chemical constituents from Abutilon indicum. *J. Asian Nat. Prod. Res.* **2008**, *10*, 689–693. [CrossRef]
20. Khalafi, T.; Buazar, F.; Ghanemi, K. Phycosynthesis and enhanced photocatalytic activity of zinc oxide nanoparticles toward organosulfur pollutants. *Sci. Rep.* **2019**, *9*, 1. [CrossRef]
21. Gurgur, E.; Oluyamo, S.S.; Adetuyi, A.O.; Omotunde, O.I.; Okoronkwo, A.E. Green synthesis of zinc oxide nanoparticles and zinc oxide–silver, zinc oxide–copper nanocomposites using Bridelia ferruginea as biotemplate. *SN Appl. Sci.* **2020**, *2*, 1–2. [CrossRef]
22. López, Y.C.; Antuch, M. Morphology control in the plant-mediated synthesis of magnetite nanoparticles. *Curr. Opin. Green Sustain. Chem.* **2020**, *24*, 32–37. [CrossRef]
23. Yew, Y.P.; Shameli, K.; Miyake, M.; Kuwano, N.; Khairudin, N.B.; Mohamad, S.E.; Lee, K.X. Green synthesis of magnetite (Fe_3O_4) nanoparticles using seaweed (Kappaphycus alvarezii) extract. *Nanoscale Res. Lett.* **2016**, *11*, 1–7. [CrossRef] [PubMed]
24. Kasprzak, M.M.; Erxleben, A.; Ochocki, J. Properties and applications of flavonoid metal complexes. *RSC Adv.* **2015**, *5*, 45853–45877. [CrossRef]

25. Ajitha, B.; Ashok Kumar Reddy, Y.; Sreedhara Reddy, P. Green synthesis and characterization of silver nanoparticles using Lantana camara leaf extract. *Mater. Sci. Eng. C* **2015**, *49*, 373–381. [CrossRef] [PubMed]
26. Ahmad, Z.; Shamim, A.; Mahmood, S.; Mahmood, T. Biological synthesis and characterization of chromium (iii) oxide nanoparticles. *Eng. Appl. Sci. Lett.* **2018**, *1*, 23–29. [CrossRef]
27. Sackey, J.; Morad, R.; Bashir, A.K.; Kotsedi, L.; Kaonga, C.; Maaza, M. Bio-synthesised black α-Cr$_2$O$_3$ nanoparticles; experimental analysis and density function theory calculations. *J. Alloys Compd.* **2021**, *850*, 156671. [CrossRef]
28. Sharma, U.R.; Sharma, N. Green Synthesis, Anti-cancer and Corrosion Inhibition Activity of Cr$_2$O$_3$ Nanoparticles. *Biointerf. Res. Appl. Chem.* **2021**, *11*, 8402–8412.
29. Ramesh, C.; Mohan Kumar, K.T.; Latha, N.; Ragunathan, V. Green synthesis of Cr$_2$O$_3$ nanoparticles using Tridax procumbens leaf extract and its antibacterial activity on Escherichia coli. *Curr. Nanosci.* **2012**, *8*, 603–637. [CrossRef]
30. Chen, L.; Song, Z.; Wang, X.; Prikhodko, S.V.; Hu, J.; Kodambaka, S.; Richards, R. Three-dimensional morphology control during wet chemical synthesis of porous chromium oxide spheres. *ACS Appl. Mater. Interfaces* **2009**, *1*, 1931–1937. [CrossRef]
31. Hao, G.; Xiao, L.; Hu, Y.; Shao, F.; Ke, X.; Liu, J.; Li, F.; Jiang, W.; Zhao, F.; Gao, H. Facile preparation of Cr$_2$O$_3$ nanoparticles and their use as an active catalyst on the thermal decomposition of ammonium perchlorate. *J. Energ. Mater.* **2019**, *37*, 251–269. [CrossRef]
32. Sone, B.T.; Manikandan, E.; Gurib-Fakim, A.; Maaza, M. Single-phase α-Cr$_2$O$_3$ nanoparticles' green synthesis using Callistemon viminalis' red flower extract. *Green Chem. Lett. Rev.* **2016**, *9*, 85–90. [CrossRef]
33. Jin, C.H.; Si, P.Z.; Xiao, X.F.; Feng, H.; Wu, Q.; Ge, H.L.; Zhong, M. Structure and magnetic properties of Cr/Cr$_2$O$_3$/CrO$_2$ microspheres prepared by spark erosion and oxidation under high pressure of oxygen. *Mater. Lett.* **2013**, *92*, 213–215. [CrossRef]
34. Kamari, H.M.; Al-Hada, N.M.; Baqer, A.A.; Shaari, A.H.; Saion, E. Comprehensive study on morphological, structural and optical properties of Cr$_2$O$_3$ nanoparticle and its antibacterial activities. *J. Mater. Sci. Mater. Electron.* **2019**, *30*, 8035–8046. [CrossRef]
35. Arakha, M.; Saleem, M.; Mallick, B.C.; Jha, S. The effects of interfacial potential on antimicrobial propensity of ZnO nanoparticle. *Sci. Rep.* **2015**, *5*, 1–10. [CrossRef] [PubMed]
36. Isacfranklin, M.; Ameen, F.; Ravi, G.; Yuvakkumar, R.; Hong, S.I.; Velauthapillai, D.; Thambidurai, M.; Dang, C. Single-phase Cr$_2$O$_3$ nanoparticles for biomedical applications. *Ceram. Int.* **2020**, *46*, 19890–19895. [CrossRef]
37. Devi, H.S.; Boda, M.A.; Shah, M.A.; Parveen, S.; Wani, A.H. Green synthesis of iron oxide nanoparticles using Platanus orientalis leaf extract for antifungal activity. *Green Process. Synth.* **2019**, *8*, 38–45. [CrossRef]
38. Choi, K.-H.; Nam, K.; Lee, S.-Y.; Cho, G.; Jung, J.-S.; Kim, H.-J.; Park, B. Antioxidant Potential and Antibacterial Efficiency of Caffeic Acid-Functionalized ZnO Nanoparticles. *Nanomaterials* **2017**, *7*, 148. [CrossRef]
39. Iqbal, S.; Bhanger, M.I.; Anwar, F. Antioxidant properties and components of some commercially available varieties of rice bran in Pakistan. *Food Chem.* **2005**, *93*, 265–272. [CrossRef]
40. Emam, A.N.; Loutfy, S.A.; Mostafa, A.A.; Awad, H.; Mohamed, M.B. Cyto-toxicity, biocompatibility and cellular response of carbon dots-plasmonic based nano-hybrids for bioimaging. *RSC Adv.* **2017**, *7*, 23502–23514. [CrossRef]

International Journal of
Molecular Sciences

Article

Investigation of Nanoparticle Metallic Core Antibacterial Activity: Gold and Silver Nanoparticles against *Escherichia coli* and *Staphylococcus aureus*

Jimmy Gouyau [1], Raphaël E. Duval [1,2,*], Ariane Boudier [3] and Emmanuel Lamouroux [1,*]

1   Université de Lorraine, CNRS, L2CM, F-54000 Nancy, France; gouyau.jimmy@gmail.com
2   ABC Platform®, F-54505 Vandœuvre-lès-Nancy, France
3   Université de Lorraine, CITHEFOR, F-54000 Nancy, France; Ariane.Boudier@univ-lorraine.fr
*   Correspondence: Raphael.Duval@univ-lorraine.fr (R.E.D.); Emmanuel.Lamouroux@univ-lorraine.fr (E.L.);
    Tel.: +33-(0)372-747-218 (R.E.D.); +33-(0)372-745-668 (E.L.)

Abstract: Multidrug-resistant (MDR) bacteria constitute a global health issue. Over the past ten years, interest in nanoparticles, particularly metallic ones, has grown as potential antibacterial candidates. However, as there is no consensus about the procedure to characterize the metallic nanoparticles (MNPs; i.e., metallic aggregates) and evaluate their antibacterial activity, it is impossible to conclude about their real effectiveness as a new antibacterial agent. To give part of the answer to this question, 12 nm gold and silver nanoparticles have been prepared by a chemical approach. After their characterization by transmission electronic microscopy (TEM), Dynamic Light Scattering (DLS), and UltraViolet-visible (UV-vis) spectroscopy, their surface accessibility was tested through the catalytic reduction of the 4-nitrophenol, and their stability in bacterial culture medium was studied. Finally, the antibacterial activities of 12 nm gold and silver nanoparticles facing *Staphylococcus aureus* and *Escherichia coli* have been evaluated using the broth microdilution method. The results show that gold nanoparticles have a weak antibacterial activity (i.e., slight inhibition of bacterial growth) against the two bacteria tested. In contrast, silver nanoparticles have no activity on *S. aureus* but demonstrate a high antibacterial activity against *Escherichia coli*, with a minimum inhibitory concentration of 128 µmol/L. This high antibacterial activity is also maintained against two MDR-*E. coli* strains.

Keywords: metallic nanoparticles; silver; gold; synthesis; characterization; surface reactivity; stability; antibacterial activity; *Escherichia coli*; *Staphylococcus aureus*; multidrug resistant bacteria

Citation: Gouyau, J.; Duval, R.E.; Boudier, A.; Lamouroux, E. Investigation of Nanoparticle Metallic Core Antibacterial Activity: Gold and Silver Nanoparticles against *Escherichia coli* and *Staphylococcus aureus*. *Int. J. Mol. Sci.* **2021**, 22, 1905. https://doi.org/10.3390/ijms22041905

Academic Editors: Sotiris Hadjikakou, Christina N. Banti and Andreas K. Rossos

Received: 12 January 2021
Accepted: 10 February 2021
Published: 14 February 2021

1. Introduction

Antimicrobial resistance (AMR) is a vital public health issue [1]. In particular, antibiotic resistance and the spread of multidrug-resistant (MDR) bacteria are a global issue [2]. If nothing is done to combat AMR, the most pessimistic projections predict 10 million deaths per year by 2050 [3]. A recent Centers-for-Disease-Control-and-Prevention (CDC) report shows that more than 2.8 million antibiotic-resistant infections occur each year in the United States, and more than 35,000 people die [4]. In the same way, MDR-bacteria are also responsible for 33,000 deaths by year, in Europe [5]. Among these bacteria, Gram-positive bacteria, like *Staphylococcus aureus*, and Gram-negative ones, such as Enterobacteriaceae, *Pseudomonas aeruginosa*, and *Acinetobacter baumannii*, are the most frequently encountered bacteria in human infections [6]. Developing new and effective antibiotics against these bacteria is therefore a high priority emergency [7].

An exciting approach to address this issue can be metallic nanoparticles (MNPs) as new antibacterial agents. Indeed, metallic (e.g., Ag, Au, Cu . . .) nanoparticles' antibacterial activity facing numerous bacteria has been reported [8–13]. Interestingly, the concerned studies deal mainly with silver or gold nanoparticles. Among these, most studies are interested in NPs associated with antibiotics or any other potentially active molecules, such as chitosan; even if it is difficult to know what the antibacterial activity relates to, they attributed the activity to the MNPs. However, by definition, an MNP is an aggregate of metallic atoms with a specific size ranging from 1 to 100 nm [14]. To control their size and size distribution, it is necessary to stabilize their surface using surfactants. Consequently, the metallic aggregate (metallic core) and the stabilizers (organic shell) are, most of the time included in the term MNPs, unconsciously. Moreover, many studies do mention neither the origin of bacteria nor the corresponding antibiograms for a clinical isolate. All of this raises the question of the real effectiveness of MNPs (i.e., metallic aggregates) as a new antibacterial agent. To provide some elements of answer to this question, it appears necessary to have an adequate and complete description of both tested nano-objects and bacteria, allowing to attribute evaluation results to MNPs [15].

Hence, we have synthesized and characterized gold or silver nanoparticles and then determined their antibacterial activities. Silver nanoparticles were synthesized from aqueous silver nitrate, sodium tricitrate, and $NaBH_4$, whereas gold nanoparticles were prepared using the Turkevich procedure [16]. The full characterization of these nanoparticles performed by UV-visible spectroscopy, transmission electron microscopy (TEM), and dynamic light scattering (DLS) is detailed. The analyses indicate that the synthesis procedures are repeatable for the preparation of 12 nm spherical citrate-capped nanoparticles, exhibiting a narrow size distribution. The antibacterial activity of both nanoparticles was evaluated, using the broth microdilution method, against *Staphylococcus aureus* (ATCC 29213), *Escherichia coli* (ATCC 25922), and two antibiotic-resistant *E. coli* strains: EcR1 (penicillinase) and EcR2 (cephalosporinase overproduction).

2. Results

2.1. Characterization of Citrate-Capped Gold and Silver Nanoparticles

Citrate-capped gold nanoparticles (AuNPs) have been prepared following the Turkevich approach. Briefly, an aqueous solution of trisodium citrate is added to a refluxed aqueous solution of $HAuCl_4$. The color of the $HAuCl_4$ solution immediately changes from yellow to black before becoming red-wine. For citrate-capped silver nanoparticles (AgNPs), sodium borohydride and citrate have been used as reductant and stabilizer, respectively. They are added as an aqueous solution to $AgNO_3$ at room temperature. The resulting solution turns dark brown and then lightens. At least three independent experiments of these two kinds of nanoparticles have been prepared and characterized.

2.1.1. Transmission Electronic Microscopy

The detailed morphology of the citrate-capped gold and silver nanoparticles has been characterized by TEM. Figure 1 shows representative micrographs and statistical analysis of these samples. Both kinds of samples appear as spheroidal nanoparticles (Figure 1a,b). Few anisotropic shapes such as triangles are observed in the AuNPs sample, mainly constituted of spherical nanoparticles of similar size (Figure 1a). In contrast, the AgNPs sample exhibits spherical nanoparticles of different sizes (Figure 1b). Statistical analysis of the size distribution of the AuNPs and AgNPs highlights that both have a mean diameter close to 12 nm. Moreover, the AuNPs and AgNPs populations follow a Gaussian distribution (Figure 1c,d): 12.1 nm, $\sigma = 1.1$, and 12.2 nm, $\sigma = 2.7$, respectively.

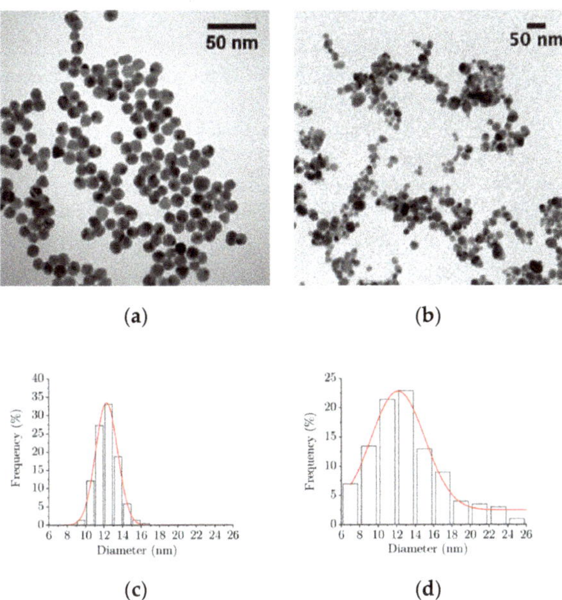

Figure 1. TEM micrographs and size distributions of (**a,c**) gold nanoparticles (AuNPs) and (**b,d**) silver nanoparticles (AgNPs).

2.1.2. Dynamic Light Scattering

The hydrodynamic diameters of nanoparticles suspended in water obtained by Dynamic light scattering (DLS) measurements are reported in Figure 2. For AuNPs (Figure 2a), only one peak is observed, whereas the AgNPs signal exhibits two peaks (Figure 2b).

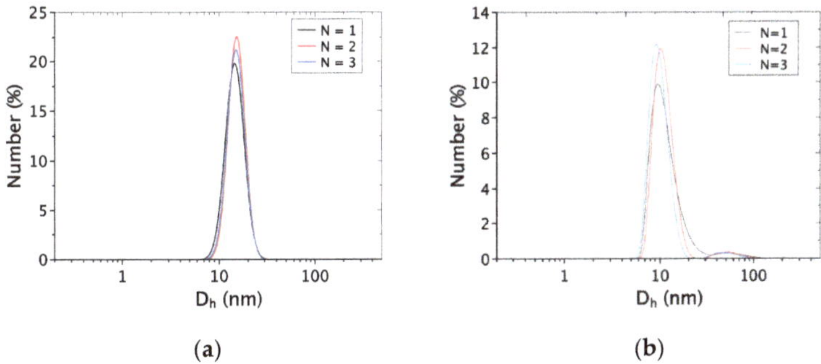

Figure 2. Dynamic Light Scattering (DLS) measurements for three independent experiments of (**a**) the AuNPs and (**b**) the AgNPs-aqueous solution (102.4 µM of metal). Number versus hydrodynamic diameter (D_h) in nanometers.

The average hydrodynamic diameter for AuNPs is 13.5 ± 3.0 nm (Figure 2a). For AgNPs, the main population has an average size of 13.6 ± 3.2 nm (Figure 2b). By considering TEM observations, the minority population with a size ranging from 20 to 100 nm could be attributed either to the aggregation of AgNPs in solution (weak interaction) or to the presence of bigger AgNPs not observed by TEM. The absence of such big AgNPs during TEM observations may be related to the difference between TEM sampling (few drops) and the one used for DLS measurement (≥ 1 mL).

2.1.3. UV-Visible Spectroscopy

Plasmonic nanoparticles, such as AuNPs and AgNPs, can also be characterized by UV-visible spectroscopy. One advantage is the sampling used, which is comparable to the one used for DLS measurements.

Figure 3 shows the UV-vis spectra for three independent experiments for AuNPs and AgNPs. It highlights the repeatability of the procedure used for NPs preparation, with a maximum wavelength at 518 ± 1 nm and 391 ± 2 nm, for AuNPs and AgNPs (Figure 3a,b), respectively. In the case of AuNPs, the fitting programs distributed by V. Amendola have been used to estimate the average size of the nanoparticles [17]. The average size obtained for the AuNPs samples is 11.6 ± 3 nm, which is in good agreement with TEM observations and DLS measurements.

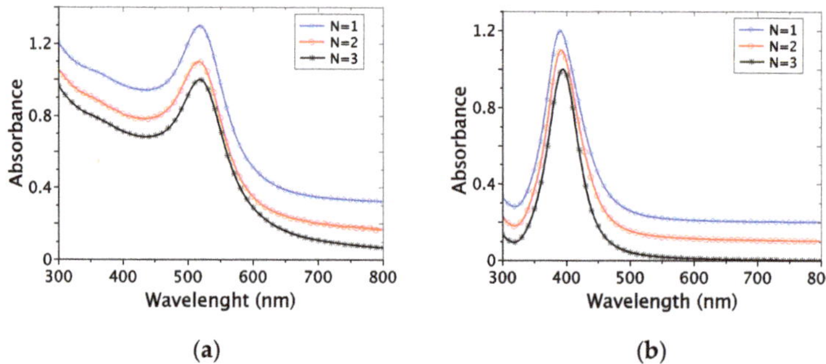

Figure 3. UV-visible spectra of (**a**) AuNPs and (**b**) AgNPs in water (for three independent experiments: $N = 1$ to 3). Stacked curves for more clarity.

Following Mie's theory [18], the AgNPs' average size does not exceed 20 nm (λmax = 391 ± 2 nm). Furthermore, the full width at half maximum of 57 ± 4 nm is characteristic of low size distribution. Moreover, there is no peak for a wavelength higher than 450–500 nm. These observations allow us to attribute the signal observed in DLS around 20–100 nm to the aggregation of AgNPs in solution (Figure 2b).

2.1.4. Surface Accessibility and Reactivity

The NPs antibacterial activity may imply NPs/bacteria interactions. Moreover, bacteria cell wall surfaces possess a negative zeta potential (i.e., -10 mV for *S. aureus* ATCC 12600 and -8 mV at 37 °C for *E. coli* ATCC 25922) [19,20], as well as the AuNPs and AgNPs (-43 mV and -40 mV, respectively). Hence, the catalytic reduction of 4-nitrophenol was chosen as a model reaction to study the NPs' surface accessibility and reactivity, involving reagent and reactants negatively charged.

The results obtained for the 4-nitrophenol catalytic reduction by $NaBH_4$ (35 °C and pH = 10) are shown in Figure 4 (a and b for AuNPs and AgNPs, respectively). In both cases, as t increases, the absorption peak around 400 nm decreases, while the one around 300 nm increases. These absorption variations are linked to the consumption of the 4-nitrophenol and the formation of the 4-aminophenol, respectively. As this reaction is from the first order, the corresponding reaction rate constants (k) can be determined from the slope of the linear plot Figure 4c,d): 5.3×10^{-3} s^{-1} for AuNPs, and 2.8×10^{-3} s^{-1} for AgNPs. As these k values are in good agreement with previously reported ones [21–23], we can conclude that both samples AuNPs and AgNPs exhibit good surface accessibility.

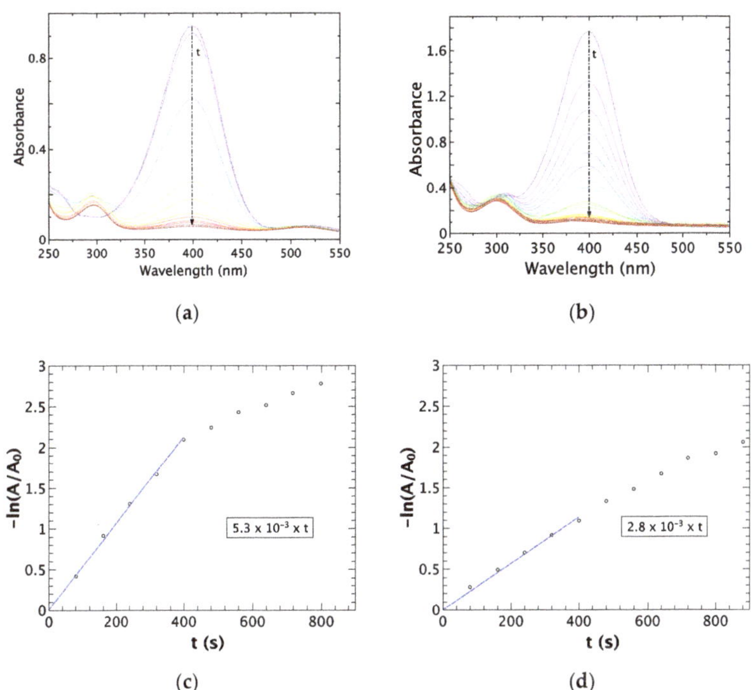

Figure 4. UV-vis spectra of 4-nitrophenol reduction as function of time and chemical reaction kinetics determination, respectively for (**a**,**c**) AuNPs and (**b**,**d**) AgNPs (pH = 10; T = 35 °C; 5 µg of metal; 10 µL of 4-nitrophenol at 1×10^{-2} µmol/L and 50 µL of NaBH$_4$ at 0.4 µmol/L).

2.2. Antibacterial Activity

Antibacterial activity evaluation was performed by the broth microdilution method using Cation-Adjusted Mueller Hinton Broth (CA-MHB) as a liquid culture medium.

However, before facing bacteria, the stability of the NPs in bacterial growth conditions (i.e., 24 h at 35 °C in the culture medium, which is a complex mixture) has to be checked. Indeed, proteins and cations, which are part of the culture medium, could interact with the NPs surface. Such kind of interactions could lead to either NPs surface destabilization/passivation or NPs dissolution. Hence, without stability checking, it would be challenging to attribute activity to metallic NPs.

2.2.1. Nanoparticles Stability in the Culture Medium

As the Surface Plasmon Resonance phenomenon is sensitive to NPs size and surface state, 256 µmol/L of NPs in water and then in CA-MHB (based on metal content) were analyzed by UV-vis spectroscopy (Figure 5). Replacing water with CA-MHB (black stars and red circles on Figure 5, respectively) leads to a slight shifting of the maximum absorbance: From 519 to 520 nm for AuNPs, and from 391 to 390 nm for AgNPs. The elevated absorbance observed in the UV region corresponds to CA-MHB's absorbance for both kinds of nanoparticles. After 24 h at 35 °C (blue rhombus in Figure 5), the characteristic absorbance peaks corresponding to AuNPs and AgNPs are no more observable.

According to Mie's theory, the observed absorbance signal modifications can be attributed to the change of the NPs surface, which could be induced by (i) an interaction between the NPs surface and compounds from the bacterial culture medium [24], (ii) the NPs aggregation/agglomeration or (iii) a change of NPs morphology.

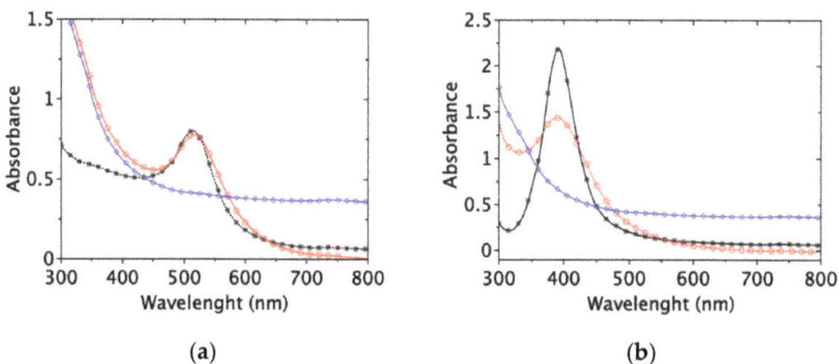

Figure 5. UV-visible spectra of 256 µmol/L of (**a**) AuNPs and (**b**) AgNPs in (*) water, in (○) CA-MHB, and in (◊) CA-MHB after 24 h at 35 °C.

To conclude about these modifications' origin, the NPs samples have been observed by TEM after the following treatment 24 h at 35 °C in CA-MHB (Figure 6). AuNPs are isolated spheroidal nanoparticles, whereas spheroidal AgNPs appear a little more agglomerated (Figure 6a,b). The observed nanoparticles' size distributions highlight mean sizes of 12.2 nm (σ = 1.2) and 11.9 nm (σ = 3.0) for AuNPs and AgNPs, respectively. A statistical test to compare them to the mean sizes obtained for NPs in water (i.e., 12.1 nm, σ = 1.1; and 12.2 nm, σ = 2.7) allows concluding that the mean sizes are identical with a confidence level of 95%. Consequently, neither the mean size of the NPs nor their morphology is altered by the conditions used for bacterial growth. It is worth noticing that due to the complex broth composition, DLS analysis did not provide interpretable data.

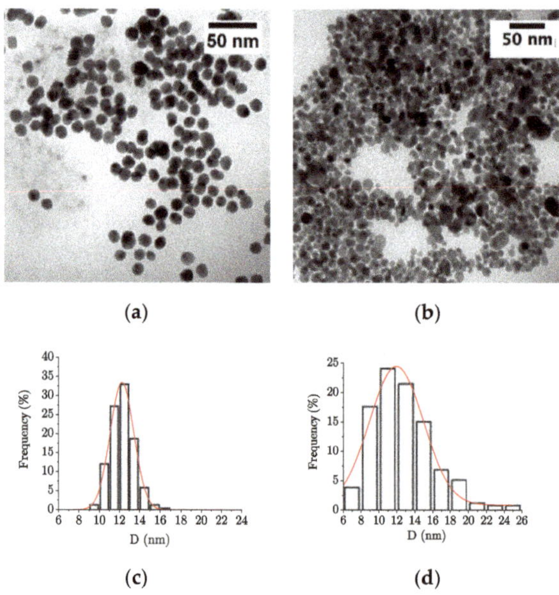

Figure 6. TEM micrographs and size distributions of (**a,c**) AuNPs and (**b,d**) AgNPs in CA-MHB.

2.2.2. Nanoparticles Antibacterial Activity Evaluation

NPs and bacteria (*S. aureus* ATCC 29213 or *E. coli* ATCC 25922) are incubated in CA-MHB. The inoculum concentration in bacteria is around 5.10^5 CFU/mL. A range of NP concentration from 256 to 1 µmol/L is obtained by two-fold dilutions. After incubation for 24 h at 35 °C, the absorbance at 540 nm, which depends on the bacteria concentration, allows the evaluation of the bacteria growth.

Figure 7 shows the obtained results for AuNPs (Figure 7a) and AgNPs (Figure 7b). The values obtained for the culture medium control (i.e., "CA-MHB") and the nanoparticle control (i.e., "NPs 256 µmol/L"), with absorbances of 0.05 ± 0.01 and 0.20 ± 0.10, respectively, compared to that obtained for the bacterial growth control, allow concluding that there is no measurable bacterial contamination of both samples. The controls of bacteria growth, with an absorbance equal or higher than 1.00, allow checking the growth of *S. aureus* and *E. coli* in the absence of NPs.

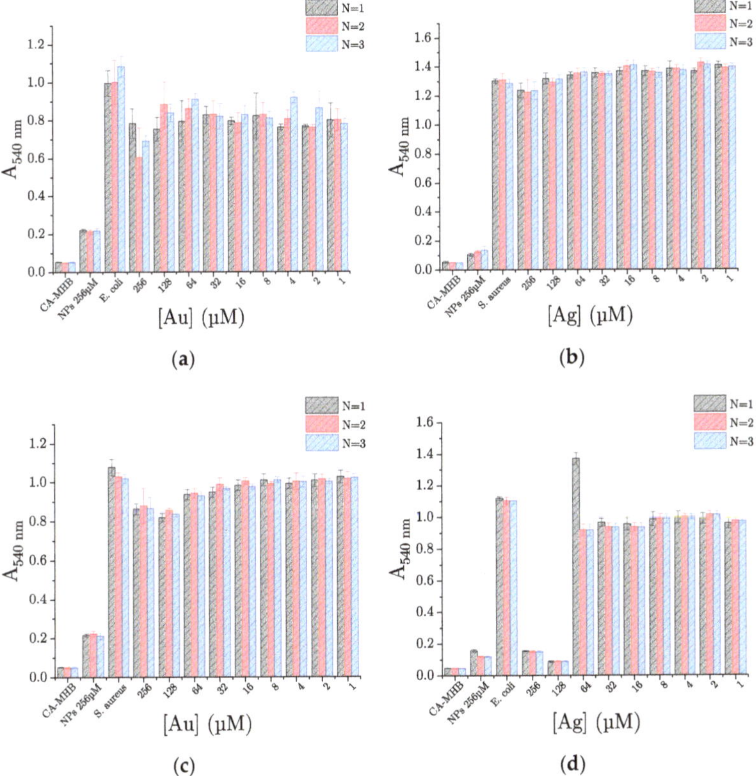

Figure 7. AgNPs and AuNPs antibacterial activity evaluation against (**a,b**) *Staphylococcus aureus* ATCC 29213 and (**c,d**) *Escherichia coli* ATCC 25922. "CA-MHB": Culture medium control (i.e., culture medium free of bacteria and nanoparticles); "NPs 256 µM": Nanoparticles control (i.e., bacteria-free culture medium, but with nanoparticles at 256 µmol/L); "*S. aureus* or *E. coli*": Bacterial growth control (i.e., growing bacteria free of nanoparticles).

Whatever the AuNPs concentrations tested against either *S. aureus* or *E. coli*, all measured absorbances are greater than those of NPs and culture medium controls (Figure 7a,c). Thus, in our experimental conditions (concentration ranging from 256 to 1 µmol/L), no minimum inhibitory concentration (MIC) is reached. However, if we look closely and compare the bacterial growth control absorbances to the value obtained for the highest AuNPs concentration tested (i.e., 256 µmol/L), a significant decrease is observed. Thus, at the concentration of 256 µmol/L, AuNPs at least partially inhibit bacterial growth, with an efficiency of around 12–14% and 20% against *S. aureus* and *E. coli*, respectively.

In the case of AgNPs, the antibacterial activity appears to depend on the tested bacteria. When *S. aureus* is incubated with AgNPs, the measured absorbances are equivalent to that of the bacterial growth control, regardless of the concentration tested (Figure 7b). So, in our experimental conditions, AgNPs demonstrate no antibacterial activity against *S. aureus*. In contrast, AgNPs exhibit an evident antibacterial activity against *E. coli*. Indeed, for concentration in metal higher or equal to 128 µmol/L, an absorbance lower than 0.2, equivalent to the NPs control, is measured (Figure 7d). Thus, the bacterial growth is totally inhibited at these concentrations, and obviously, we can determine a MIC = 128 µmol of metal per liter for AgNPs against *E. coli*. Besides, given our encouraging results on a wild-type strain of *E. coli* (i.e., with no acquired antibiotic resistance), we decided to evaluate the antibacterial activity of AgNPs against two clinical isolates of *E. coli* resistant to β-lactams: EcR1 (i.e., penicillinase) and EcR2 (i.e., cephalosporinase overproduction). Unexpectedly, our results show that AgNPs inhibit the bacterial growth of both human clinical strains in a comparable way, with a MIC of 64 and 128 µmol of metal per liter, against EcR1 and EcR2, respectively (Figure 8a,b). AgNPs would therefore have the same antibacterial activity against *Escherichia coli*, whether it is resistant or not to antibiotics (i.e., β-lactams).

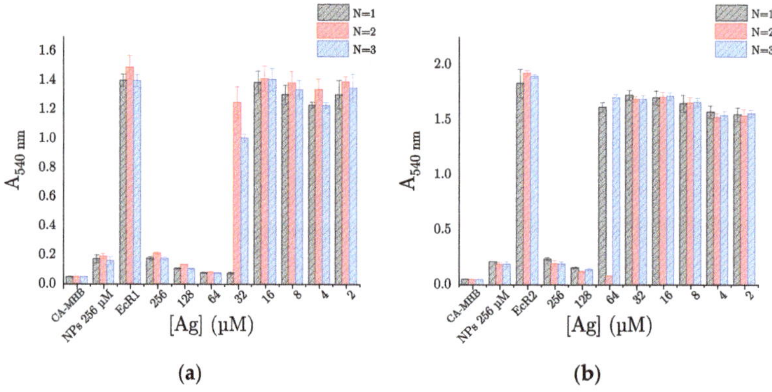

Figure 8. Evaluation of antibacterial activity by broth microdilution method results for AgNPs facing (**a**) EcR1 (ABC 23) penicillinase and (**b**) EcR2 (ABC 24) cephalosporinase. "CA-MHB": culture medium control (i.e., culture medium free of bacteria and nanoparticles); "NPs 256 µM": Nanoparticles control (i.e., bacteria-free culture medium, but with nanoparticles at 256 µmol/L); "EcR1 or EcR2": Bacterial growth control (i.e., growing bacteria free of nanoparticles).

3. Discussion

The investigation of MNPs (i.e., metallic aggregates) antibacterial activity implies to consider four key points. Firstly, it is necessary to adequately characterize the MNPs size and size distribution and the sample homogeneity (i.e., a significant sampling). Secondly, the stabilizer potential antibacterial activity should be known. Thirdly, the MNPs surface accessibility and their stability in culture medium have to be established. Fourthly, referenced bacteria strains or well-characterized clinical isolates (i.e., strains with their antibiograms) should be used for antibacterial activity evaluation. At this price, it is possible to correctly assess the antibacterial activity and attribute it only to the MNPs (i.e., "naked" metallic aggregates). Following this approach, we studied the MNPs (with M = Au or Ag) antibacterial activity against *E. coli* and *S. aureus*, two of the most frequent bacterial species encountered in human infections [25]. The stabilizer (i.e., citrate) do not exhibit antibacterial activity in the concentration range used [26]. The MNPs sample characterizations were consistent in that they consist of spherical 12 nm gold and silver NPs with a narrow size distribution. Moreover, both their size and morphology were not altered when introduced into the bacterial culture medium. According to the literature, four to seven different mechanisms of action for MNPs as an antibacterial agent are described [27–34]. For example, Lee and Jun described four main routes of antibacterial mechanism of AgNPs, namely (i) adhesion to the cell membrane, (ii) penetration onside the cell, (iii) ROS generation and cellular toxicity, and (iv) genotoxicity. In a more recent review, Joshi et al., listed seven different mechanisms of action: (i) Disruption of cell membrane, (ii) destabilization and disruption of membrane proteins, (iii) destabilization and disruption of cytoplasmic proteins, (iv) inactivation of enzymes and metabolic interference, (v) generation of ROS, (vi) damage to DNA and ribosomal assembly, and (vii) impairment in transmembrane electron transport system [34]. It is worth noticing they can be coupled with each other. Even if the precise antibacterial mechanism of MNPs is still not elucidated, all of the suspected mechanisms imply a direct interaction between metallic atoms and bacteria. However, 12 nm citrate-capped MNPs (M = Ag, Au) possess a negative zeta potential, as for *E. coli* and *S. aureus*; so, as high positive or negative zeta potential values induce high repulsive forces [35], interactions between MNPs and bacteria can be lowered. Using a model catalytic reaction, such as the nitrophenol reduction by $NaBH_4$ in the presence of MNPs, can remove such uncertainty. Indeed, as $NaBH_4$ (itself negatively charge) solution is stable at basic pH values, the nitrophenol with its pKa of 7.15 at 25 °C [36] will be present as nitrophenolate anions. So, performing this catalytic reaction at basic pH involves reactant and reagent negatively charged. The catalytic results obtained highlight the zeta potentials of both 12 nm citrate-capped MNPs (M = Ag, Au), and the bacteria membrane will not limit their interaction.

Then, the antibacterial activity of AuNPs and AgNPs was evaluated against two bacteria: *E. coli* (ATCC 25922) and *S. aureus* (ATCC 29213). Doing so, we were unable to determine a MIC for AuNPs on the two bacteria. At best, we were able to demonstrate a partial inhibition of bacterial growth at the highest concentration tested (i.e., 256 µmol/L). On the other hand, even though AgNPs demonstrated no antibacterial activity against *S. aureus*, we have clearly determined a MIC = 128 µmol/L for the same AgNPs, under the same experimental conditions, against *E. coli*. Our results clearly show an antibacterial activity (even low) of AuNPs against *E. coli* ATCC 25922 and *S. aureus* ATCC 25923, which strongly differs from recent studies that failed to report any significant antibacterial activity of AuNPs against the same bacterial strains [37,38]. *A contrario*, for AgNPs, our results seem to be consistent with other studies which found that citrate-capped silver NPs showed a higher activity against *E. coli* than *S. aureus* [39–42]. For example, minimal bactericidal concentrations (MBC) of 9.94 µg/mL (i.e., ≈92 µM of metal) and 19.88 µg/mL (i.e., ≈184 µM) were obtained by C. Quintero-Quiroz et al., for AgNPs (5–50 nm) against *E. coli* ATCC 25922 and *S. aureus* ATCC 29213, respectively [40]. However, with AgNPs of 95.5 nm in diameter (large size distribution) facing the same strains, M. Zarei et al., reported no antibacterial activity against *E. coli*, while they determined a MIC of 5 µg/mL (i.e., ≈46 µM of metal) against *S. aureus* [43]. Moreover, 12.9 ± 4.5 nm AgNPs, prepared from silver nitrate, sodium citrate and ascorbic acid, tested against *E. coli* (MG 1655) and *S. aureus* (ATCC 6538) led to the determination of MBC of 0.14 mg/mL (i.e., 1.3 µM) and 0.35 mg/mL (3.2 µM) respectively [41]; whereas AgNPs ranging from 1 to 20 nm facing *E. coli* (ATCC 25922) and *S. aureus* (ATCC 12600) resulted in MIC of 0.049 mg/mL (i.e., ≈454 µM) and 0.391 mg/mL (i.e., 3625 µM) [42]. The differences observed with our results can be attributed to the AgNPs large size distribution [40,43], the presence of an antibacterial agent in the NPs preparation procedure [41], or the selection of different bacteria strains and concentration range (more specifically when *S. aureus* is under consideration) [41–43]. Nevertheless, it is noteworthy that MIC values obtained for our AgNPs against *E. coli* are in good agreement with those from other studies [40,44].

At this stage, and based on our results, we can make two hypotheses: Either AgNPs or AuNPs do not have exactly the same mechanism of action (since the AgNPs completely inhibit the growth of *E. coli* while the AuNPs only partially inhibit the growth of the two bacterial species); or the antibacterial action mechanism of the AgNPs and AuNPs depends on the bacterium tested, and more specifically depends on the structure and composition of the bacterial wall (since AgNPs do not inhibit the growth of *S. aureus* at all, whereas we determined a MIC for these same nanoparticles on *E. coli*). Indeed, several studies demonstrated that AuNPs and AgNPs interacted electrostatically with cell wall lipopolysaccharides (in Gram-negative bacteria) and teichoic and lipoteichoic acids (in Gram-positive bacteria) [45–51]. Moreover, it has been postulated that according to Derjaguin–Landau–Verwey–Overbeek (DLVO) theory, the electrostatic repulsive forces and van der Waals attractive forces are involved in the interaction between nanoparticles and bacterial cells in an aqueous suspension [46]. But, several authors suggested that the "strength" of electrostatic interactions between MNPs and bacteria (which could correspond to first step of the antibacterial mechanism of action) were under the dependence of the polysaccharides cell wall composition, structure, density..., which vary between different bacterial strains and therefore could explain the difference of the antibacterial activity of MNPs in function of the tested bacteria and even the opposite results obtained [34].

Besides, as AMR is a major public health issue, we wanted to test the antibacterial activity of AgNPs against antibiotic-resistant human clinical isolates of *E. coli*: EcR1 (penicillinase) and EcR2 (cephalosporinase overproduction). These two resistance mechanisms are among the most frequently found in *E. coli* [25]. Surprisingly, we have also shown that AgNPs completely inhibit the growth of these two clinical isolates (i.e., EcR1 and EcR2), with MIC values identical to those obtained for the wild strain (i.e., *E. coli* ATCC 25922). Therefore, it would seem that the antibacterial mechanism of action of AgNPs is not counteracted by mechanisms of resistance to β-lactams, which inhibit the synthesis of peptidoglycan (i.e., major and essential constituent of the bacterial cell wall).

The absence of size and shape evolution of the AgNPs sample in CA-MHB indicates that it should be lite enough not to be observed if dissolution. Additionally, as 2.5 equivalent $NaBH_4$ versus $AgNO_3$ and AgNPs washing cycle, the antibacterial activity cannot be attributed to free Ag^+ ions present in colloidal solution. However, after the AgNPs/bacteria interaction or internalization, few Ag^+ ions should be released from the NPs and improve their antibacterial activity.

4. Materials and Methods

4.1. Materials

Chloroauric acid ($HAuCl_4 \cdot 3\,H_2O$) was obtained from Alfa-Aesar, Karlsruhe, Germany. Silver nitrate ($AgNO_3$), sodium citrate ($Na_3C_6H_5O_7$), sodium borohydride ($NaBH_4$), and absolute ethanol (CH_3CH_2OH) were purchased from Sigma-Aldrich, Steinheim, Germany. The standard silver TraceCERT® and gold TraceCERT® came from the same supplier.

For the antimicrobial test, Cation-Adjusted Mueller–Hinton Broth (CA-MHB) from BBL™ (Batch 7291628) was prepared following the manufacturer's instructions. Bacterial strains used, *Staphylococcus aureus* (ATCC 29213) and *Escherichia coli* (ATCC 25922), came from the American Type Culture Collection (ATCC); while EcR1 (ABC 23) and EcR2 (ABC 24) were *Escherichia coli* clinical isolates resistant to β-lactams (penicillinase and cephalosporinase overproduction, respectively) and came from ABC® Platform Bugs Bank. All strains were grown in CA-MHB.

Ultrapure water (15 MΩ cm filtered at 0.22 μm) was used in all procedures.

4.2. Citrate-Capped Nanoparticles Synthesis

4.2.1. Gold Nanoparticles

Sodium citrate was used to reduce chloroauric acid and NPs stabilization according to Turkevich procedure [16]. Briefly, chloroauric acid ($HAuCl_4$, $3H_2O$: 1×10^{-4} mol) in water (100 cm^3) was brought to a boil without refrigerant. A solution containing 5 molar equivalents of sodium citrate (i.e., 5×10^{-4} mol) in water (5 cm^3) was heated and added to the auric solution. The heating was switched off after 5 min. After return to room temperature, the nanoparticles were centrifuged for 30 min at 5 °C at 8500$\times$ *g*. The supernatant was removed. Further purification of the sample was made 3 times as follows the addition of water (10 cm^3), centrifugation (30 min, 5 °C, 8500$\times$ *g*) and removal of the supernatant. Finally, the colloidal solution was diluted to obtain the desired concentration.

4.2.2. Silver Nanoparticles

Silver nanoparticles (AgNPs) were prepared by reducing silver nitrate by sodium borohydride in the presence of sodium citrate. Silver nitrate (AgNO$_3$: 2×10^{-5} mol) in water (79.5 cm^3) was prepared at room temperature. A solution containing 1 molar equivalent of sodium citrate (i.e., 2×10^{-5} mol) in water (0.5 cm^3) was added to this solution of silver nitrate. Then, 0.5 mL of a sodium borohydride solution (NaBH$_4$: 5×10^{-5} mol) was added dropwise (60 mL/h) to the silver solution. After one hour under agitation, the solution was centrifuged for 60 min at 8500× g at 5 °C. The supernatant was removed. Further purification of the sample was made 3 times as follows the addition of water (10 cm^3), centrifugation (60 min, 5 °C, 8500× g) and removal of the supernatant. Finally, the colloidal solution was diluted to obtain the desired concentration.

4.3. Characterizations Methods

4.3.1. UV-Visible Absorbance Spectroscopy

To measured surface plasmonic resonance and check the repeatability of synthesis, aliquots of colloidal solution were analyzed by measuring the UV-Visible spectrum at a resolution of 1 nm at 20 °C using UV-Visible spectroscopy (Perkin Elmer, Shelton, CT, USA, series LAMBA 1050).

4.3.2. Transmission Electron Microscopy

The morphology and size of nanoparticles were determined using transmission electron microscopy (Philips, Tokyo, Japan, CM200). Colloidal solutions were diluted in absolute ethanol, then few drops of ethanol colloidal solution were loaded onto a carbon-coated copper grid. After evaporating the excess solvent, nanoparticles were visualized using TEM, which was operated at a 20 kV accelerating voltage. Images treatment was realized using ImageJ software (v1.51m9, NIH, Bethesda, MD, USA) [52].

4.3.3. Dynamic Light Scattering

Hydrodynamic sizes were determined using Multi-Angle Dynamic Light Scattering with Malvern Zetasizer Nano ZS by the combination of the signal at 13°, 90°, and 173° from source (λ = 633 nm) in plastic cuvettes.

4.3.4. Inductively Coupled Plasma Spectroscopy

Determination of metal (i.e., Au or Ag) concentration in aqueous colloid solution was performed using Inductively Coupled Plasma with detector by Absorption Emission Spectrometer (ICP-AES) with ICP Ultima (Jobin-Yvon Horiba, Kyoto, Japan). The concentration was determined according to TraceCERT® standard. The samples were prepared by digestion of nanoparticles in acid solution, aqua regia for AuNPs, and nitric acid for AgNPs.

4.4. Surface Accessibility

The surface accessibility was checked by nanoparticles' ability to catalyze the reduction reaction of 4-nitrophenol in 4-aminophenol in an aqueous solution at pH = 10. In a quartz cuvette, 40 µL of nanoparticles at 1 mmol/L of metal, 10 µL of 4-nitrophenol at 1×10^{-2} mol/L, and 50 µL of NaBH$_4$ at 0.4 mol/L at pH 10 were stirring. The reaction was monitored by recording UV-visible spectra between 250 and 800 nm every 10 s for 2 min, then each 30 s afterward.

4.5. Stability Study

Stability in CA-MHB was evaluated using %absorbance measurements, between 250 and 800 nm, for nanoparticles at 256 µmol/L of metal at t = 0 and t = 24 h at 35 °C. TEM images were recorded with these samples, and statistical tests have been performed to check the nanoparticle size evolution. For this purpose, with a population of 500 NPs, the bilateral student test and Fisher-Snedecor test were used to compare the mean diameters and the variances, respectively.

4.6. Antimicrobial Evaluation

Nanoparticles' antibacterial activities were determined using methods based on the broth microdilution method [53]. Briefly, twofold serial dilutions of drugs were prepared in CA-MHB in 96-well microtiter plates (Greiner, Bernolsheim, France, 650161), starting from a stock aqueous solution of 1024 µmol/L (number of mol of metal atom/L considered) to obtain a final concentration range from 256 to 1 µmol/L^1. Then, normed inoculum prepared from bacteria in the stationary phase, at 5.10^5 to 5.10^6 colony forming unit (CFU) by mL, was put in each well. Three controls were realized: Culture medium alone, nanoparticles in culture medium at 256 µmol/L, and bacteria without nanoparticles. After incubation for 24 h at 35 °C under agitation at 150 RPM, bacterial growth was evaluated with an ELISA plate reader (read at 540 nm, Multiskan EX, Thermo Electron Corporation, Saint-Herblain, France). All shown results are expressed as the means ± standard deviation of 8 wells at the same concentration of 96-wells microtiter plates. The three independent determinations are presented on the same graphic.

5. Conclusions

The purpose of the present study was to determine whether MNPs (i.e., metallic aggregates) can be considered new potential antibacterial agents. To provide some element of the answer, adequately characterized 12 nm citrate-capped gold and silver nanoparticles have been tested against *S. aureus* (ATCC 29213) and *E. coli* (ATCC 25922). In this study, we evidenced that both AuNPs and AgNPs are stable in CA-MHB at 35 °C over 24 h. Besides, the results clearly indicate that AuNPs exhibit only light antibacterial activity against *S. aureus* and *E. coli*. In contrast, AgNPs possess high antibacterial activity only against *E. coli*. We assume that the difference in the cell-wall structure of these bacteria (e.g., presence of an external membrane in Gram-negative bacteria, like *E. coli*) explains the difference in some bacteria sensitivity (depending on the composition/structure of their cell wall) to AgNPs. Moreover, AgNPs antibacterial activity appears to be maintained against antibiotic-resistant *E. coli* strains, presenting either a penicillinase or a cephalosporinase-overproduction. This study suggests that among MNPs (with M = Au or Ag), only AgNPs can be considered a new potential antibacterial agent. However, the scope of this study was limited in terms of NPs size and bacteria number. Hence, it would be interesting to assess the effects of NPs size and higher NPs concentrations on the antibacterial activity and the antibacterial spectra of the AgNPs, in particular other MDR-bacteria.

Author Contributions: Conceptualization, R.E.D. and E.L.; Investigation, J.G.; Supervision, R.E.D. and E.L.; Validation, R.E.D. and E.L.; Writing—original draft, E.L.; Writing—review and editing, R.E.D., A.B. and E.L. All authors have read and agreed to the published version of the manuscript.

Funding: This research received no external funding.

Institutional Review Board Statement: Not applicable.

Informed Consent Statement: Not applicable.

Data Availability Statement: The data that support the findings of this study are available from the corresponding author upon reasonable request.

Acknowledgments: The authors (J.G., R.E.D., A.B. and E.L.) thank Claire Genois, LCPME, UMR 7564, Université de Lorraine, for Inductively Coupled Plasma Spectroscopy measurements. J.G. acknowledges the French Ministry of Higher Education, Research and Innovation (MESRI) for a Ph.D. grant. The authors (J.G., R.E.D., and E.L.) are grateful to the French Ministry of Higher Education, Research and Innovation (MESRI), the French National Scientific Research Centre (CNRS), the Université de Lorraine, and Region Grand Est.

Conflicts of Interest: The authors declare no conflict of interest.

References

1. Duval, R.E.; Grare, M.; Demoré, B. Fight Against Antimicrobial Resistance: We Always Need New Antibacterials but for Right Bacteria. *Molecules* **2019**, *24*, 3152. [CrossRef] [PubMed]
2. Antimicrobial Resistance. Available online: www.who.int/news-room/fact-sheets/detail/antimicrobial-resistance (accessed on 12 October 2020).
3. Tackling Drug-Resistant Infections Globally: Final Report and Recommendations. Available online: https://amr-review.org/sites/default/files/160525_Final%20paper_with%20cover.pdf (accessed on 12 October 2020).
4. Centers for Disease Control and Prevention. *Antibiotic Resistance Threats in the United States*; CDC Organization: Atlanta, GA, USA, 2019. [CrossRef]
5. Cassini, A.; Högberg, L.D.; Plachouras, D.; Quattrocchi, A.; Hoxha, A.; Simonsen, G.S.; Colomb-Cotinat, M.; Kretzschmar, M.E.; Devleesschauwer, B.; Cecchini, M.; et al. Attributable Deaths and Disability-Adjusted Life-Years Caused by Infections with Antibiotic-Resistant Bacteria in the Eu and the European Economic Area in 2015: A Population-Level Modelling Analysis. *Lancet Infect. Dis.* **2019**, *19*, 56–66. [CrossRef]
6. Boucher, H.W.; Talbot, G.H.; Bradley, J.S.; Edwards, J.E.; Gilbert, D.; Rice, L.B.; Scheld, M.; Spellberg, B.; Bartlett, J. Bad Bugs, No Drugs: No Eskape! An Update from the Infectious Diseases Society of America. *Clin. Infect. Dis.* **2009**, *48*, 1–12. [CrossRef]
7. WHO Publishes List of Bacteria for Which New Antibiotics Are Urgently Needed. Available online: www.who.int/news-room/detail/27-02-2017-who-publishes-list-of-bacteria-for-which-new-antibiotics-are-urgently-needed (accessed on 12 October 2020).
8. Rudramurthy, G.R.; Swamy, M.K.; Sinniah, U.R.; Ghasemzadeh, A. Nanoparticles: Alternatives Against Drug-Resistant Pathogenic Microbes. *Molecules* **2016**, *21*, 836. [CrossRef]
9. Stanić, V.; Tanasković, S.B. Chapter 11—Antibacterial Activity of Metal Oxide Nanoparticles Nanotoxicity. In *Nanotoxicity: Micro and Nano Technologies*; Rajendran, S., Mukherjee, A., Nguyen, T.A., Godugu, C., Shukla, R.K., Eds.; Elsevier: Amsterdam, The Netherlands, 2020; pp. 241–274, ISBN 978-0-12-819943-5. [CrossRef]
10. Gold, K.; Slay, B.; Knackstedt, M.; Gaharwar, A.K. Antimicrobial Activity of Metal and Metal-Oxide Based Nanoparticles. Advanced Therapeutics. *Adv. Therap.* **2018**, *1*, 1700033. [CrossRef]
11. Brandelli, A.; Ritter, A.C.; Veras, F.F. Antimicrobial Activities of Metal Nanoparticles. In *Metal Nanoparticles in Pharma*; Rai, M., Shegokar, R., Eds.; Springer International Publishing: Cham, Switzerland, 2017; pp. 337–363, ISBN 978-3-319-63790-7. [CrossRef]
12. Wolny-Koładka, K.A.; Malina, D.K. Eco-Friendly Approach to the Synthesis of Silver Nanoparticles and Their Antibacterial Activity Against *Staphylococcus spp.* and *Escherichia coli*. *J. Environ. Sci. Health* **2018**, *53*, 1041–1047. [CrossRef] [PubMed]
13. Bastos, C.A.P.; Faria, N.; Wills, J.; Malmberg, P.; Scheers, N.; Rees, P.; Powell, J.J. Copper Nanoparticles Have Negligible Direct Antibacterial Impact. *NanoImpact* **2020**, *17*, 100192. [CrossRef]
14. Blackman, J.A.; Binns, C. Chapter 1 Introduction. In *Handbook of Metallic Nanoparticles*; Blackman, J.A., Ed.; Elsevier: Amsterdam, The Netherlands, 2008; pp. 1–16, ISBN 1570-002X. [CrossRef]
15. Duval, R.E.; Gouyau, J.; Lamouroux, E. Limitations of Recent Studies Dealing with the Antibacterial Properties of Silver Nanoparticles: Fact and Opinion. *Nanomaterials* **2019**, *9*, 1775. [CrossRef]
16. Turkevich, J.; Stevenson, P.C.; Hillier, J. A Study of the Nucleation and Growth Processes in the Synthesis of Colloidal Gold. *Discuss. Faraday Soc.* **1951**, *11*, 55–75. [CrossRef]
17. Amendola, V.; Meneghetti, M. Size Evaluation of Gold Nanoparticles by UV—Vis Spectroscopy. *J. Phys. Chem. C* **2009**, *113*, 4277–4285. [CrossRef]
18. Mie, G. Beiträge zur Optik Trüber Medien, Speziell Kolloidaler Metallösungen. *Ann. Physik* **1908**, *330*, 377–445. [CrossRef]
19. Gottenbos, B.; Grijpma, D.W.; van der Mei, H.C.; Feijen, J.; Busscher, H.J. Antimicrobial Effects of Positively Charged Surfaces on Adhering Gram-Positive and Gram-Negative Bacteria. *J. Antimicrob. Chemother.* **2001**, *48*, 7–13. [CrossRef] [PubMed]
20. Kim, J.K.; Harrison, M.A. Surrogate Selection for Escherichia Coli O157:H7 Based on Cryotolerance and Attachment to Romaine Lettuce. *J. Food Prot.* **2009**, *72*, 1385–1391. [CrossRef]
21. Corma, A.; Serna, P. Chemoselective Hydrogenation of Nitro Compounds with Supported Gold Catalysts. *Science* **2006**, *313*, 332–334. [CrossRef]
22. Satapathy, S.; Mohanta, J.; Si, S. Modulating the Catalytic Activity of Gold Nanoparticles through Surface Tailoring. *Chem. Sel.* **2016**, *1*, 4940–4948. [CrossRef]
23. Wu, K.-J.; Torrente-Murciano, L. Continuous Synthesis of Tuneable Sized Silver Nanoparticles via a Tandem Seed-Mediated Method in Coiled Flow Inverter Reactors. *React. Chem. Eng.* **2018**, *3*, 267–276. [CrossRef]
24. Pfeiffer, C.; Rehbock, C.; Hühn, D.; Carrillo-Carrion, C.; de Aberasturi, D.J.; Merk, V.; Barcikowski, S.; Parak, W.J. Interaction of Colloidal Nanoparticles with Their Local Environment: The (Ionic) Nanoenvironment around Nanoparticles is Different from Bulk and Determines the Physico-Chemical Properties of the Nanoparticles. *J. R. Soc. Interface* **2014**, *11*, 20130931. [CrossRef]
25. European Centre for Disease Prevention and Control. *Antimicrobial Resistance in the EU/EEA (EARS-Net). Annual Epidemiological Report 2019*; ECDC: Stockholm, Sweden, 2020.
26. Lee, Y.-l.; Cesario, T.; Owens, J.; Shanbrom, E.; Thrupp, L.D. Antibacterial Activity of Citrate and Acetate. *Nutrition* **2002**, *18*, 665–666. [CrossRef]
27. Taylor & Francis Group. *Silver Nanoparticles for Antibacterial Devices: Biocompatibility and Toxicity*; Cao, H., Ed.; CRC Press: Boca Raton, FL, USA, 2017; ISBN 978-1-4987-2532-3.

28. Franci, G.; Falanga, A.; Galdiero, S.; Palomba, L.; Rai, M.; Morelli, G.; Galdiero, M. Silver Nanoparticles as Potential Antibacterial Agents. *Molecules* **2015**, *20*, 8856–8874. [CrossRef] [PubMed]
29. Haider, A.; Kang, I.-K. Preparation of Silver Nanoparticles and Their Industrial and Biomedical Applications: A Comprehensive Review. *Adv. Mater. Sci. Eng.* **2015**, *2015*, 1–16. [CrossRef]
30. Khodashenas, B. The Influential Factors on Antibacterial Behaviour of Copper and Silver Nanoparticles. *Indian Chem. Eng.* **2016**, *58*, 224–239. [CrossRef]
31. Ullah Khan, S.; Saleh, T.A.; Wahab, A.; Ullah Khan, M.H.; Khan, D.; Ullah Khan, W.; Rahim, A.; Kamal, S.; Ullah Khan, F.; Fahad, S. Nanosilver: New Ageless and Versatile Biomedical Therapeutic Scaffold. *Int. J. Nanomed.* **2018**, *13*, 733–762. [CrossRef]
32. Yan, X.; He, B.; Liu, L.; Qu, G.; Shi, J.; Hu, L.; Jiang, G. Antibacterial Mechanism of Silver Nanoparticles in Pseudomonas Aeruginosa: Proteomics Approach. *Metallomics* **2018**, *10*, 557–564. [CrossRef]
33. Lee, S.H.; Jun, B.H. Silver Nanoparticles: Synthesis and Application for Nanomedicine. *Int. J. Mol. Sci.* **2019**, *20*, 865. [CrossRef] [PubMed]
34. Joshi, A.S.; Singh, P.; Mijakovic, I. Interactions of Gold and Silver Nanoparticles with Bacterial Biofilms: Molecular Interactions behind Inhibition and Resistance. *Int. J. Mol. Sci.* **2020**, *21*, 7658. [CrossRef] [PubMed]
35. Honary, S.; Zahir, F. Effect of Zeta Potential on the Properties of Nano-Drug Delivery Systems—A Review (Part 2). *Trop. J. Pharm. Res.* **2013**, *12*. [CrossRef]
36. Handbook, C.R.C. *CRC Handbook of Chemistry and Physics*, 88th ed.; CRC Press: Boca Raton, FL, USA, 2007; ISBN 0849304881.
37. Zhu, S.; Shen, Y.; Yu, Y.; Bai, X. Synthesis of Antibacterial Gold Nanoparticles with Different Particle Sizes Using Chlorogenic Acid. *R. Soc. Open Sci.* **2020**, *7*, 191141. [CrossRef] [PubMed]
38. López-Lorente, Á.I.; Cárdenas, S.; González-Sánchez, Z.I. Effect of Synthesis, Purification and Growth Determination Methods on the Antibacterial and Antifungal Activity of Gold Nanoparticles. *Mater. Sci. Eng. C* **2019**, *103*, 109805. [CrossRef] [PubMed]
39. Bilek, O.; Fialova, T.; Otahal, A.; Adam, V.; Smerkova, K.; Fohlerova, Z. Antibacterial Activity of AgNPs–TiO2 Nanotubes: Influence of Different Nanoparticle Stabilizers. *RSC Adv.* **2020**, *10*, 44601–44610. [CrossRef]
40. Quintero-Quiroz, C.; Acevedo, N.; Zapata-Giraldo, J.; Botero, L.E.; Quintero, J.; Zárate-Triviño, D.; Saldarriaga, J.; Pérez, V.Z. Optimization of Silver Nanoparticle Synthesis by Chemical Reduction and Evaluation of Its Antimicrobial and Toxic Activity. *Biomater. Res.* **2019**, *23*, 27. [CrossRef]
41. Kubo, A.L.; Capjak, I.; Vrček, I.V.; Bondarenko, O.M.; Kurvet, I.; Vija, H.; Ivask, A.; Kasemets, K.; Kahru, A. Antimicrobial Potency of Differently Coated 10 and 50 nm Silver Nanoparticles Against Clinically Relevant Bacteria *Escherichia coli* and *Staphylococcus aureus*. *Colloids Surf. B Biointerfaces* **2018**, *170*, 401–410. [CrossRef] [PubMed]
42. Hor, J.W.; Mohd, S.A.; Nor, D.; Daruliza, K.; Yazmin, B. Evaluation of Anti Bacteria Effects of Citrate-Reduced Silver Nanoparticles in Staphylococcus Aureus and Escherichia coli. *Int. J. Res. Pharm. Sci.* **2019**, *10*, 3636–3643. [CrossRef]
43. Zarei, M.; Karimi, E.; Oskoueian, E.; Es-Haghi, A.; Yazdi, M.E.T. Comparative Study on the Biological Effects of Sodium Citrate-Based and Apigenin-Based Synthesized Silver Nanoparticles. *Nutr. Cancer* **2020**, 1–9. [CrossRef]
44. Li, D.; Chen, S.; Zhang, K.; Gao, N.; Zhang, M.; Albasher, G.; Shi, J.; Wang, C. The Interaction of Ag2O Nanoparticles with Escherichia coli: Inhibition-Sterilization Process. *Sci. Rep.* **2021**, *11*, 1703. [CrossRef] [PubMed]
45. Abadeer, N.S.; Fülöp, G.; Chen, S.; Käll, M.; Murphy, C.J. Interactions of Bacterial Lipopolysaccharides with Gold Nanorod Surfaces Investigated by Refractometric Sensing. *ACS Appl. Mater. Interfaces* **2015**, *7*, 24915–24925. [CrossRef]
46. Pajerski, W.; Ochonska, D.; Brzychczy-Wloch, M.; Indyka, P.; Jarosz, M.; Golda-Cepa, M.; Sojka, Z.; Kotarba, A. Attachment Effciency of Gold Nanoparticles by Gram-Positive and Gram-Negative Bacterial Strains Governed by Surface Charges. *J. Nanopart. Res.* **2019**, *21*, 186. [CrossRef]
47. Jacobson, K.H.; Gunsolus, I.L.; Kuech, T.R.; Troiano, J.M.; Melby, E.S.; Lohse, S.E.; Hu, D.; Chrisler, W.B.; Murphy, C.J.; Orr, G.; et al. Lipopolysaccharide Density and Structure Govern the Extent and Distance of Nanoparticle Interaction with Actual and Model Bacterial Outer Membranes. *Environ. Sci. Technol.* **2015**, *49*, 10642–10650. [CrossRef] [PubMed]
48. Caudill, E.R.; Hernandez, R.T.; Johnson, K.P.; O'Rourke, J.T.; Zhu, L.; Haynes, C.L.; Feng, Z.V.; Pedersen, J.A. Wall Teichoic Acids Govern Cationic Gold Nanoparticle Interaction with Gram-Positive Bacterial Cell Walls. *Chem. Sci.* **2020**, *11*, 4106–4118. [CrossRef]
49. Bankier, C.; Matharu, R.K.; Cheong, Y.K.; Ren, G.G.; Cloutman-Green, E.; Ciric, L. Synergistic Antibacterial Effects of Metallic Nanoparticle Combinations. *Sci. Rep.* **2019**, *9*, 16074. [CrossRef]
50. Ivask, A.; ElBadawy, A.; Kaweeteerawat, C.; Boren, D.; Fischer, H.; Ji, Z.; Chang, C.H.; Liu, R.; Tolaymat, T.; Telesca, D.; et al. Toxicity Mechanisms in *Escherichia coli* Vary for Silver Nanoparticles and Differ from Ionic Silver. *ACS Nano* **2014**, *8*, 374–386. [CrossRef]
51. ElBadawy, A.M.; Silva, R.G.; Morris, B.; Scheckel, K.G.; Suidan, M.T.; Tolaymat, T.M. Surface Charge-Dependent Toxicity of Silver Nanoparticles. *Environ. Sci. Technol.* **2011**, *45*, 283–287. [CrossRef] [PubMed]
52. Schneider, C.A.; Rasband, W.S.; Eliceiri, K.W. NIH Image to ImageJ: 25 Years of Image Analysis. *Nat. Methods* **2012**, *9*, 671–675. [CrossRef] [PubMed]
53. CLSI. *Methods for Dilution Antimicrobial Susceptibility Tests for Bacteria That Grow Aerobically*; Clinical and Laboratory Standards Institute: Wayne, PA, USA, 2018; ISBN 1-56238-837-1.

International Journal of
Molecular Sciences

MDPI

Article

Antimicrobial Face Shield: Next Generation of Facial Protective Equipment against SARS-CoV-2 and Multidrug-Resistant Bacteria

Alberto Tuñón-Molina [1,†], Miguel Martí [1,†], Yukiko Muramoto [2], Takeshi Noda [2], Kazuo Takayama [3,*] and Ángel Serrano-Aroca [1,*]

1   Biomaterials and Bioengineering Lab., Centro de Investigación Traslacional San Alberto Magno, Universidad Católica de Valencia San Vicente Mártir, c/Guillem de Castro 94, 46001 Valencia, Spain; alberto.tunon@ucv.es (A.T.-M.); miguel.marti@ucv.es (M.M.)
2   Laboratory of Ultrastructural Virology, Institute for Frontier Life and Medical Sciences, Kyoto University, Kyoto 606-8507, Japan; muramo@infront.kyoto-u.ac.jp (Y.M.); t-noda@infront.kyoto-u.ac.jp (T.N.)
3   Center for iPS Cell Research and Application (CiRA), Kyoto University, Kyoto 606-8507, Japan
*   Correspondence: kazuo.takayama@cira.kyoto-u.ac.jp (K.T.); angel.serrano@ucv.es (Á.S.-A.)
†   These authors contributed equally to this work.

Citation: Tuñón-Molina, A.; Martí, M.; Muramoto, Y.; Noda, T.; Takayama, K.; Serrano-Aroca, Á. Antimicrobial Face Shield: Next Generation of Facial Protective Equipment against SARS-CoV-2 and Multidrug-Resistant Bacteria. *Int. J. Mol. Sci.* **2021**, *22*, 9518. https://doi.org/10.3390/ijms22179518

Academic Editor: Sotiris Hadjikakou

Received: 13 August 2021
Accepted: 24 August 2021
Published: 1 September 2021

Publisher's Note: MDPI stays neutral with regard to jurisdictional claims in published maps and institutional affiliations.

Abstract: Transparent materials used for facial protection equipment provide protection against microbial infections caused by viruses and bacteria, including multidrug-resistant strains. However, transparent materials used for this type of application are made of materials that do not possess antimicrobial activity. They just avoid direct contact between the person and the biological agent. Therefore, healthy people can become infected through contact of the contaminated material surfaces and this equipment constitute an increasing source of infectious biological waste. Furthermore, infected people can transmit microbial infections easily because the protective equipment do not inactivate the microbial load generated while breathing, sneezing or coughing. In this regard, the goal of this work consisted of fabricating a transparent face shield with intrinsic antimicrobial activity that could provide extra-protection against infectious agents and reduce the generation of infectious waste. Thus, a single-use transparent antimicrobial face shield composed of polyethylene terephthalate and an antimicrobial coating of benzalkonium chloride has been developed for the next generation of facial protective equipment. The antimicrobial coating was analyzed by atomic force microscopy and field emission scanning electron microscopy with elemental analysis. This is the first facial transparent protective material capable of inactivating enveloped viruses such as severe acute respiratory syndrome coronavirus 2 (SARS-CoV-2) in less than one minute of contact, and the methicillin-resistant *Staphylococcus aureus* and *Staphylococcus epidermidis*. Bacterial infections contribute to severe pneumonia associated with the SARS-CoV-2 infection, and their resistance to antibiotics is increasing. Our extra protective broad-spectrum antimicrobial composite material could also be applied for the fabrication of other facial protective tools such as such as goggles, helmets, plastic masks and space separation screens used for counters or vehicles. This low-cost technology would be very useful to combat the current pandemic and protect health care workers from multidrug-resistant infections in developed and underdeveloped countries.

Keywords: face shield; facial protective equipment; SARS-CoV-2; phage phi 6; MRSA; MRSE; polyethylene terephthalate; benzalkonium chloride; COVID-19; multidrug-resistant bacteria

1. Introduction

Even though the severe lockdowns carried out in many countries of the world, the coronavirus disease 2019 (COVID-19) pandemic is still increasing the number of global deaths in most countries [1–3]. The causative agent of this disease is the severe acute respiratory syndrome coronavirus 2 (SARS-CoV-2), which is an enveloped positive-sense single-stranded RNA virus [4] that belongs to the IV Baltimore group [5]. SARS-CoV-2 causes atypical viral pneumonia [6,7] which death risk increases by co-infection with bacteria such as *Streptococcus pneumoniae* [8–11].

The emergence of highly pathogenic viruses, such as SARS-CoV-2, that can co-infect with other viruses or bacteria [12], including antibiotic-resistant strains, constitutes one of the most current threatening to humans in this century. Additionally, bacterial resistance to antibiotics in pneumonia treatment is increasing at an alarming rate [13,14]. SARS-CoV-2 showed high stability in different material surfaces, including the surface of metals, plastics and cardboard [15–19]. Therefore, in addition to the aerosol transmission route, SARS-CoV-2 can be transmitted by contact with material surfaces contaminated with this pathogen [15–20]. In fact, it can spread faster than its two ancestors SARS-CoV and Middle East respiratory syndrome coronavirus [21] through coughing, sneezing, touching or breathing [22] and more easily through asymptomatic carriers [23,24].

Influenza virus (IFV) affects the nasal mucosa in the course of infections that simultaneously affect other sectors of the respiratory tract, including the lower tract [25]. IFV is also an enveloped single-stranded RNA virus-like SARS-CoV-2 [26]. Another important global risk is caused by respiratory infections caused by bacteria such as *S. pneumoniae* that is the most frequently isolated organism with the highest mortality [27]. This pathogen is the cause of many respiratory processes such as pneumonia, otitis, sinusitis, complicated with sepsis, meningitis and abscesses [28,29]. Apart from the therapeutic therapies aimed at combating these diseases and in those cases in which there are no effective therapies for the treatment of the infections caused by these pathogens, facial protection equipment acquires great importance. Facial protection equipment against infectious biological agents includes those with eye and/or respiratory protection (nose and mouth) to prevent the entry of microorganisms, splashes and biological aerosols through the respiratory or mucous tract.

The choice of a specific type of protection resides in the choice of equipment according to its application. Thus, there is protective equipment such as face masks that are made by porous fabric that filtrates the air and impede the pass of most of the microbial particles [30]. Another option of protective equipment is commonly called as face shields made of transparent plastic materials [31]. This type of protective equipment acts by forming a barrier between the wearer of the screen and the biological agent, thus avoiding, in the best of cases, the entry of the agent through the respiratory and mucosal tracts. Even though its effectiveness in combination with other protection measures is not questioned, by itself, this type of protection is not totally effective as many of the infectious biological agents are capable of surviving on its surface for a long time. Even though all the devices developed to date fulfill the function of acting as a barrier against direct exposure of the infectious biological agent, they may not be entirely effective, since the device has not been fabricated with antimicrobial materials capable of inactivating infectious agents when they are in contact with the material surface. Furthermore, this contaminated biological waste constitutes an environmental risk associated with the waste management of these protective systems.

Polyethylene terephthalate (PET) is a commercial low-cost transparent and recyclable polyester that is commonly used for the fabrication of facial protective equipment such as face shields [32]. However, this plastic material does not possess antimicrobial properties.

In this regard, quaternary ammonium compounds such as benzalkonium chloride (BAK) have been confirmed to be capable of inactivating enveloped RNA viruses [33] and Gram-positive multidrug-resistant bacteria [34]. In fact, this chemical compound is widely used as a disinfectant against bacteria, viruses, pathogenic fungi and mycobacteria [35].

The goal of this work consisted of producing a transparent face shield capable of providing extra-protection by acting as a physic barrier with intrinsic antimicrobial activity against enveloped viruses such as SARS-CoV-2 and bacteriophage phi 6, and multidrug-resistant bacteria.

2. Results and Discussion

2.1. Composite Material Morphology

Atomic force microscopy (AFM) and field emission scanning electron microscope (FESEM) with elemental analysis were performed in order to characterize the BAK micro-coating formed onto the PET surface. Figure 1 shows the AFM images of the treated and untreated PET plastics over a scan area of 10 µm × 10 µm.

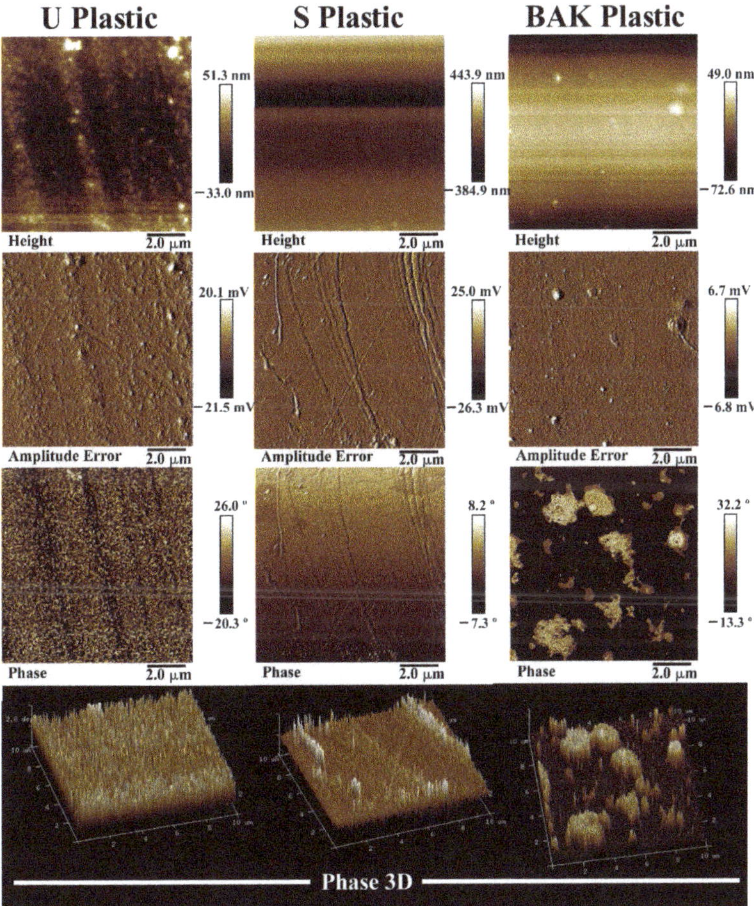

Figure 1. AFM topography images (height, amplitude error and phase images) and phase 3D representation recorded in tapping mode of the Untreated plastic (U Plastic), plastic treated by dip coating with the ethanol-based solvent (S Plastic) and filter with the biofunctional benzalkonium chloride (BAK) coating (BAK Plastic) scanning a 10 µm × 10 µm area. These images show how the BAK coating was formed on the polyethylene terephthalate surface after the dip-coating treatment. The U Plastic shows some impurities on its surface which disappear after immersion in ethanol (S Plastic). Thus, the clean S Plastic shows some surface imperfections (irregular stripes) which disappear after the BAK coating (BAK Plastic). The BAK plastic shows slightly higher zones observed as white zones in the phase image.

The 2D phase provides images whose contrast is produced by differences in the adhesion and viscoelastic properties of the sample surface [36]. Thus, the pictures of the topography and phase angle clearly indicated that a BAK coating was formed onto the

PET surface (BAK Plastic) after the dip-coating treatment with the solvent containing BAK (Figure 1). It can be clearly observed that the untreated plastic (U Plastic) possesses some impurities on its surface which disappear after immersing the disk in ethanol 70% for 1 min. Thus, the surface imperfections produced by the plastic fabrication procedure can be clearly observed in the 2D phase image of the PET plastic treated by dip coating with the absolute ethanol/distilled water (70/30 *v/v*) solvent mixture (S Plastic). However, the AFM images of the BAK Plastic clearly show that a BAK coating was formed covering all the imperfections observed in the 2D phase image of the S Plastic sample. This coating presents slightly higher zones that are observed white zones in the phase images. These results are in good agreement with the FESEM images shown in the following Figure 2.

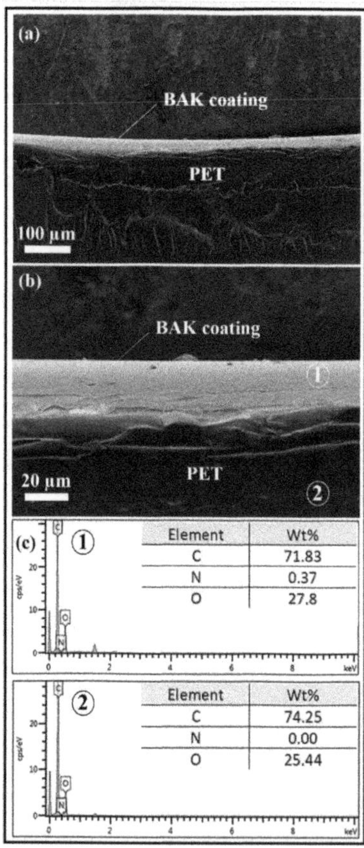

Figure 2. Morphology of the biofunctional coating of BAK onto the PET surface by Field Emission Scanning Electron Microscopy with energy-disperse X-ray spectroscopy (EDS) for elemental analysis: PET with 0.182 ± 0.034% *w/w* of biofunctional BAK coating (BAK Plastic) at two magnifications: (**a**) ×150 and (**b**) ×720, and (**c**) EDS elemental analysis of the coating and PET matrix.

The FESEM micrographs show clearly how a microcoating of BAK (light grey phase) is formed onto the PET surface (dark grey phase) with a thickness of approximately 25 µm (see Figure 2). Furthermore, the energy-disperse X-ray spectroscopy (EDS) analysis shows a nitrogen content of 0.37% weight on the BAK coating in good agreement with the nitrogen atom present in the BAK compound [34]. However, the EDS analysis on the PET matrix (polymer without nitrogen atoms) does not show any nitrogen content as expected.

2.2. Opacity

Figure 3 shows that there are no statistically significant differences of opacity (or transparency), calculated with Equation (1), of the PET disks before and after the treatments with solvent or the dip-coating treatment with BAK, which is essential to be used as transparent facial protective equipment (face shield screens, plastic masks, protective screens, protective glasses, etc.).

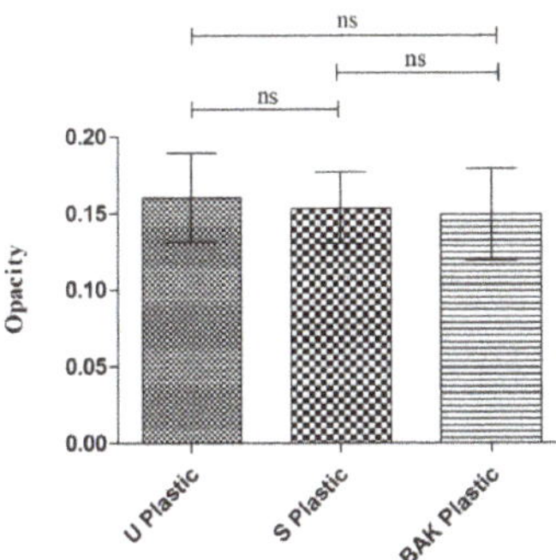

Figure 3. Opacity results of the Untreated PET (U Plastic), PET treated by dip coating with the ethanol-based solvent (S Plastic) and PET with the biofunctional BAK coating (BAK Plastic). The ANOVA results are indicated in this plot; ns: not significant.

2.3. Antibacterial Activity

Figure 4 shows the antibacterial results achieved with the U Plastic, the S Plastic and the BAK Plastic against two multidrug-resistant bacteria: methicillin-resistant *Staphylococcus aureus* (MRSA) and methicillin-resistant *Staphylococcus epidermidis* (MRSE).

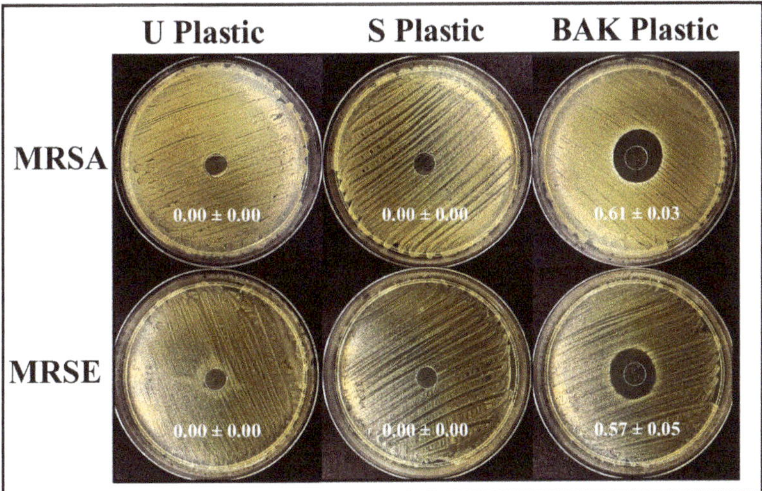

Figure 4. Antibacterial agar disk diffusion tests with two multidrug-resistant bacteria: methicillin-resistant *Staphylococcus aureus* (MRSA) and methicillin-resistant *Staphylococcus epidermidis* (MRSE). Untreated PET (U Plastic), PET treated by dip coating with the ethanol-based solvent (S Plastic) and the PET with the biofunctional BAK coating (BAK Plastic) after 24 h of incubation at 37 °C. The normalized widths of the antibacterial halos, expressed as mean ± standard deviation and calculated with equation (1), are shown in each image.

Therefore, we can observe that the plastic with the biofunctional coating of BAK showed potent antibacterial activity against MRSA and MRSE with similar normalized antibacterial halos of 0.61 ± 0.03 and 0.57 ± 0.05, respectively.

2.4. Antiviral Activity

The phage phi 6, which is an enveloped double-stranded RNA virus (group III of the Baltimore classification [5]), was used as biosafe viral model of SARS-CoV-2 and other enveloped viruses such as influenza due to safety reasons.

The BAK Plastic showed potent antiviral activity against phage phi 6 (100% of viral inhibition, see Figure 5).

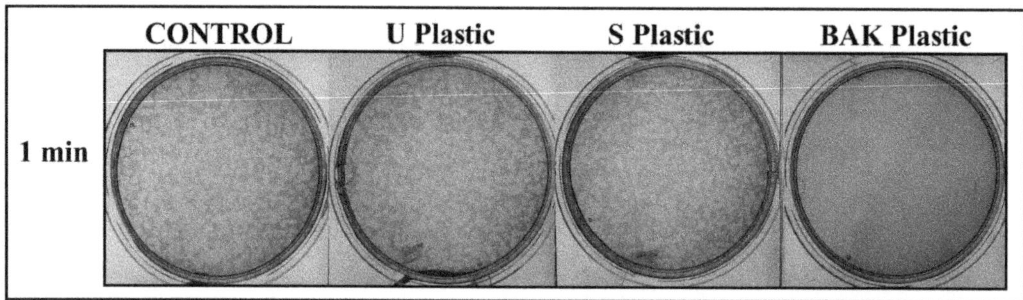

Figure 5. Loss of phage phi 6 viability measured by the double-layer method. Phage phi 6 titration images of undiluted samples for control, untreated PET (U Plastic), PET treated by dip coating with the ethanol-based solvent (S plastic) and PET with the biofunctional BAK coating (BAK plastic) at 1 min of viral contact.

Thus, no plaques were produced on the bacterial lawns after 1 min of contact between the BAK Plastic and the biosafe viral model. However, similar plaques to control can be observed on the bacterial lawns after 1 min of contact between the U Plastic or S Plastic and the biosafe viral model (see Figure 5). The phage titers of each type of sample were calculated and compared with the control (Table 1).

Table 1. Reduction of infection titers in plaque-forming units per mL (PFU/mL) determined by the double-layer assay for the phage phi 6. Logarithm of plaque-forming units per mL (log (PFU/mL)) of the control, untreated PET (U Plastic), PET treated by dip coating with the ethanol-based solvent (S Plastic) and PET with the biofunctional BAK coating (BAK Plastic) at 1 min of viral contact.

Sample	Phi 6 at 1 min (PFU/mL)
Control	$4.36 \times 10^6 \pm 2.92 \times 10^5$
U Plastic	$4.38 \times 10^6 \pm 1.98 \times 10^5$
S Plastic	$4.23 \times 10^6 \pm 1.36 \times 10^6$
BAK Plastic	0.00 ± 0.00

Table 1 shows that the titers obtained by contacting the phages with the U or S Plastic are similar to the control. However, the BAK plastic displayed a strong phage inactivation. The results achieved with SARS-CoV-2 after 1 min of contact with the U Plastic, the S Plastic and the BAK Plastic containing the biofunctional coating are shown in Figure 6.

These results clearly demonstrate that the BAK Plastic is very effective against SARS-CoV-2 even after 1 min of contact. This is also in good agreement with the antiviral results of the biosafe viral model used in this study (see Figure 5 and Table 1). However, since BAK is highly water-soluble and therefore could come out when the PET sheet with the BAK coating is in contact with water, the antiviral and antibacterial tests were performed again after washing with distilled water to analyze the antimicrobial durability of the BAK coating to water. The results of these experiments (results not shown) showed that the antimicrobial BAK coating dissolves really fast in distilled water and loses its antimicrobial activity as expected. However, the developed antimicrobial face shield is presented here as a single-use face protective equipment for the current and future microbial menaces.

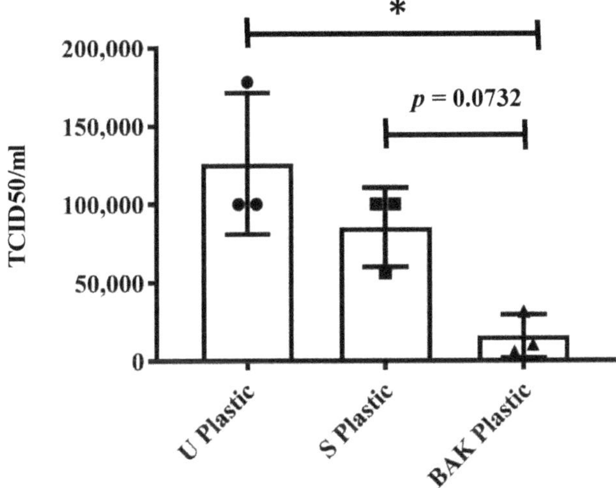

Figure 6. Reduction of infectious titers in PFU/mL of severe acute respiratory syndrome coronavirus 2 (SARS-CoV-2) after 1 min of contact determined by the median tissue culture infectious dose per mL (TCID50/mL) method. Untreated PET (U plastic), PET treated with the ethanol solvent (S Plastic) and the PET with the biofunctional BAK coating (BAK Plastic). A dot, square and triangle plot is a data set based on the value of each point. * $p < 0.05$.

These advanced face shields can provide superior protection to virologists working with highly infective pathogens in high-level biosafety labs, surgeons and healthcare workers in general. Furthermore, the proposed PET plastic is a recyclable material that can be reutilized [32] and thus contribute to decrease the increasing amount of this type of waste generated in the current pandemic. Furthermore, the antimicrobial coating can be produced easily as many times as necessary as a reusability characteristic of this technology. The BAK coating can be performed onto the outside side only for people not infected or also on the inside side if the protective equipment is going to be used for infected patients. It could also be performed on other types of transparent synthetic polymers such as polycarbonate, polymethyl methacrylate, poly (2-hydroxyethyl methacrylate), on non-transparent materials or on biodegradable polymers that would provide a solution to the need for bio-based protective tools for environmental reasons [37]. This next generation equipment will significantly reduce the increasing generation of infectious biological waste. These new technologies could revolutionize the face protective tool industry because other face protective equipment could be developed applying the same low-cost technology providing high antimicrobial activity (see Figure 7).

The antimicrobial mechanism of action of BAK against both bacterial and enveloped viruses is attributed to its positively charged nitrogen atoms that can eradicate the bacterial surface or disrupt the phospholipid bilayer membrane, the glycoproteinaceous envelope, and the spike glycoproteins of viruses such as phi6, SARS-CoV-2 and IFV [38,39]. BAK is a Food and Drug Administration-approved product for a broad-range of disinfecting applications such as additives in soaps and hand sanitizers [40–42]. We have demonstrated here that these transparent PET-based composites possess high antiviral and antibacterial activity to reduce the spread of COVID-19 and methicillin-resistant bacteria. This extra-protective composite material has been developed by a low-cost method of dip coating that let BAK to physically adsorbed [43] onto the surface of a commercial PET plastic commonly used for the fabrication of face shields and other protective equipment providing high antimicrobial activity. Nonetheless, further research is necessary to overcome the possible shortfalls on the applicability of this coating technology to other protective tools and to other types of transparent materials. The effect of packing, storage and transportation on the antimicrobial properties of the BAK coating need also to be studied before applying this technology at large-scale.

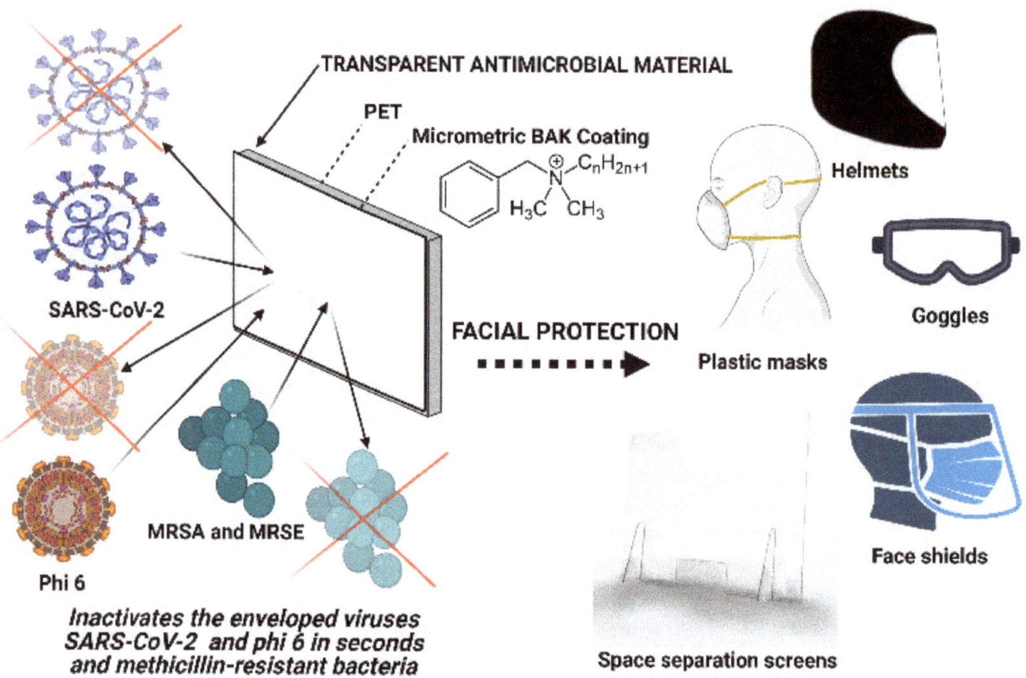

Figure 7. Applications of the coating technology of transparent polyethylene terephthalate (PET) with an antimicrobial coating of benzalkonium chloride (BAK) for the next generation of facial protective equipment: face shields, plastic masks, helmets, goggles, helmets and space separation screens.

3. Materials and Methods

3.1. Dip Coating Treatment with Benzalkonium Chloride

Sheets of PET with a thickness of 0.3167 ± 0.0408 mm used for the fabrication of commercial face shields were purchased from Plasticos Villamarchante S.L [44] (Valencia, Spain). Six disk specimens (n = 6) of approximately 10 mm in diameter of these transparent PET sheets (BAK Plastic) were treated with 70% ethyl alcohol with 0.1% *w/w* BAK (Montplet, Barcelona, Spain) by the dip-coating method [45] for 1 min at 25 °C to achieve a dry BAK content of 0.182 ± 0.034% *w/w* determined gravimetrically. Six more PET disks (n = 6) were subjected to the same dip-coating treatment but using only an absolute ethanol/distilled water solution (70/30% *v/v*) without BAK for 1 min at 25 °C (S Plastic). Untreated PET disks (U Plastic) disks (n = 6) were produced as reference material. The disks were subsequently dried at 60 °C for 48 h and sterilized under UV radiation for 1 h per disk side. The BAK used in this study was previously characterized by nuclear magnetic resonance on a BRUKER AVIIIHD 800 MHz (Bruker BioSpin AG, Fälladen, Switzerland) equipped with a 5 mm cryogenic CP-TCI [34].

3.2. Atomic Force Microscopy

Atomic force microscopy (AFM) was performed with a Bruker MultiMode 8 scanning probe microscope (SPM, Karlsruhe, Germany) operating in tapping mode in air and with the NanoScope V Controller and NanoScope 8.15 software version (Bruker, Karlsruhe, Germany). An antimony (n) doped silicon cantilever from Bruker was used with a scan rate of 0.500 Hz. The phase signal was set to zero at the resonance frequency of the tip. The tapping frequency was 5–10% lower than the resonance frequency. The drive amplitude and amplitude setpoints were 308.5 and 644.8 mV, respectively, and the aspect ratio was 1.00.

3.3. Electron Microscopy

A Zeiss Ultra 55 field emission scanning electron microscope (FESEM, Jena, Germany) was operated at an accelerating voltage of 10 kV to observe the biofunctional coating morphology of the treated PET surface at a magnification of $\times 150$ and $\times 720$. The plastic samples were prepared to be conductive by platinum coating with a sputter coating unit. This FESEM microscope is equipped with energy-disperse X-ray spectroscopy (EDS) for elemental ratio estimation at 2.00 kV.

3.4. Opacity

The opacity of the synthesized films was evaluated according to the spectrophotometric method utilized by Park and Zhao [46]. Thus, rectangular specimens (4 mm $\times$ 50 mm) of the dry films were directly placed in a spectrophotometer cell to measure the absorbance at 600 nm with a UV/VIS Nanocolor UV0245 spectrophotometer (Macherey-Nagel, Düren, Germany). An empty cell was utilized as a reference. After that, the opacity (O) of the films can be determined with Equation (1), in which Abs600 is the absorbance value at 600 nm and x is film thickness in mm.

$$O = \frac{\mathbf{Abs}600}{\mathbf{x}} \tag{1}$$

The measurements were performed with three specimens of each material and were reported as absorbance divided by film thickness (mean $\pm$ standard deviation).

3.5. Phage Host Culture

The phage phi 6 host is *Pseudomonas syringae* (DSM 21482). This Gram-negative bacterium was purchased from the Leibniz Institute DSMZ–German Collection of Microorganisms and Cell cultures GmbH (Braunschweig, Germany). *P. syringae* was cultured in solid tryptic soy agar (TSA, Liofilchem, Teramo, Italy). After that, the microorganism was cultured in liquid tryptic soy broth (TSB, Liofilchem, Roseto degli Abruzzi, Italy) incubated in an orbital shaker (CERTOMAT IS, Sartorius Stedim Biotech, Göttingen, Germany) at 25 °C and 120 rpm.

3.6. Phage Propagation

The specifications provided by the Leibniz Institute DSMZ–German Collection of Microorganisms and Cell Cultures GmbH were followed to propagate the phage phi 6 (DSM 21518) by using the double agar layer technique with top and bottom agar, which allows to perform lysis in bacterial host cultures [47].

3.7. Antiviral Test with the Biosafe Viral Model

50 µL of TSB with phages was placed onto each material disk at a titer of about 1×10^6 plaque-forming units per mL (PFU/mL) and allowed to incubate for 1 min. Each material disk was placed in a falcon tube with 10 mL TSB, and subsequently sonicated for 5 min and vortexed for 1 min at room temperature (24 ± 1 °C).

Phage titration was performed by serial dilutions of each falcon sample. 100 µL of each phage dilution was mixed with 100 µL of the host strain at $OD_{600\,nm} = 0.5$.

The infective activity of the phage phi 6 was measured based on the double-layer method [48]. Thus, 4 mL of top agar (TSB + 0.75% bacteriological agar, Scharlau) and 5 mM $CaCl_2$ were added to the mixture containing phages and bacteria, which was then poured on TSA plates. Incubation of the plates was performed for 24–48 h in a refrigerated oven at 25 °C. Phage titers of each sample were calculated in PFU/mL and compared with a control consisting of 50 µL of phage added directly to the bacterial culture without being in contact with any type of disk and without sonication or vortexing.

The antiviral activity of the material disks was estimated at 1 min of contact with the biosafe virus model in log reductions of titers. It was ensured that the residual disinfectants present in the titrated samples did not interfere with the titration process. It was also checked that sonication and vortexing did not affect the infectious activity of the phage phi 6. The antiviral assays were carried out three times during two different days (n = 6) to ensure reproducible results.

3.8. Antiviral Tests Using SARS-CoV-2

The SARS-CoV-2 strain (SARS-CoV-2/Hu/DP/Kng/19-027) was provided to us by Dr. Tomohiko Takasaki and Dr. Jun-Ichi Sakuragi from the Kanagawa Prefectural Institute of Public Health. SARS-CoV-2 was plaque-purified, propagated in Vero cells and stored at −80 °C. 50 µL of a virus suspension in phosphate-buffered saline (PBS) was placed onto each material disk at a titer of 1.3×10^5 median tissue culture infectious dose (TCID50) per disk, and then incubated for 1 min of contact at ambient temperature. After that, 1 mL PBS was added to each disk, and then vortexed for 5 min. After that, each tube was vortexed for 5 min at ambient temperature. Viral titers were determined by the TCID50 assay performed in a Biosafety Level 3 laboratory at Kyoto University. Thus, TMPRSS2/Vero cells [49] (JCRB1818, JCRB Cell Bank), cultured with the minimum essential media (MEM, Sigma-Aldrich, St. Louis, MO, USA) supplemented with 5% fetal bovine serum (FBS), 1% penicillin/streptomycin, were seeded into 96-well plates (Thermo Fisher Scientific, Waltham, MA, USA). Serial dilutions of 10-fold (from 10^{-1} to 10^{-8}) were performed in the culture medium. These dilutions were placed onto the TMPRSS2/Vero cells in triplicate and incubated at 37 °C for 96 h. Cytopathic effect was evaluated under a microscope and TCID50/mL was calculated using the Reed–Muench method [50].

3.9. Antibacterial Tests

The antibacterial activity was studied by the agar disk diffusion tests [51,52]. Thus, lawns of methicillin-resistant *Staphylococcus aureus* (MRSA), COL [53] and methicillin-resistant *Staphylococcus epidermidis* (MRSE), RP62A [54], in a concentration of approximately 1.5×10^8 colony forming units per mL (CFU/mL) in tryptic soy broth, were cultivated on trypticase soy agar plates. The lawns of bacteria were incubated aerobically at 37 °C for 24 h with the sterilized disks placed upon them. The antibacterial disks showed an inhibition zone (or *halo*) that can be normalized using Equation (1) [51].

$$nw_{halo} = \frac{\frac{d_{iz} - d}{2}}{d} \tag{2}$$

The term nw_{halo} represents the normalized width of the antibacterial inhibition zone, d_{iz} is the inhibition zone diameter and d indicates the material disk diameter. The material disk diameter was measured by image software analysis (Image J, Wayne Rasband (NIH), USA). The antibacterial tests were performed three times during two different days (n = 6) to ensure reproducible results.

3.10. Antimicrobial Durability of the BAK Coating to Water

The antiviral and antibacterial tests were performed again after washing 1 cm disks of the PET/BAK composite material (BAK Plastic) by immersion in 100 mL of distilled water during 1 min at 24 ± 1 °C to analyze the antimicrobial durability of the BAK coating to water.

3.11. Statistical Analysis

The statistical analyses were performed by ANOVA followed by Tukey's post hoc test ($* p > 0.05$, $*** p > 0.001$) using the GraphPad Prism software (GraphPad Prism 6, GraphPad Software Inc., San Diego, CA, USA).

4. Conclusions

A single-use antimicrobial face shield has been developed as the next generation of face protective equipment capable of inactivating enveloped viruses such as phage phi 6 and SARS-CoV-2 after 1 min of contact and multidrug-resistant bacteria such as MRSA and MRSE. This antimicrobial composite material was fabricated by a low-cost procedure consisting of dip-coating of polyethylene terephthalate with benzalkonium chloride. The formation of the antimicrobial coating was demonstrated by atomic force microscopy and field emission scanning electron microscopy with elemental analysis. This composite material avoids viral and bacterial inhalation and entry into the body through the respiratory tract or by splashing (in a surgical operation for example), providing an extra biosafety due to its capacity of inactivating the infectious microorganisms as soon as they are in contact with the protective element. Furthermore, this antimicrobial material is recyclable, and it reduces the generation of infectious biological waste. This antimicrobial material can be used for the fabrication of other face protective equipment such as goggles, helmets, plastic masks and space separation counter or vehicles screens and thus are very promising for the current and future microbial menaces. Nonetheless, further research is needed to contribute significantly on the up-scaling of this technology.

5. Patents

Facial protection element against risks of exposure to infectious biological agents (Utility model). U202130782. 15 April 2021.

Author Contributions: Á.S.-A. conceived the idea of this work; conceptualization, methodology, validation and formal analysis: M.M., K.T. and Á.S.-A.; software: K.T. and Á.S.-A.; investigation: A.T.-M., M.M., Y.M., T.N., K.T. and Á.S.-A.; resources: M.M., K.T. and Á.S.-A.; data curation, A.T.-M., K.T. and Á.S.-A.; visualization: K.T. and Á.S.-A.; writing—original draft preparation: Á.S.-A.; writing—review and editing: A.T.-M., M.M., K.T. and Á.S.-A.; supervision, M.M., K.T. and Á.S.-A.; project administration, K.T. and Á.S.-A.; funding acquisition, K.T. and Á.S.-A. All authors have read and agreed to the published version of the manuscript.

Funding: This research was supported by the Fundación Universidad Católica de Valencia San Vicente Mártir, Grant 2020-231-006UCV and by the Ministerio de Ciencia e Innovación (PID2020-119333RB-I00/AEI/10.13039/501100011033) (awarded to Á.S.-A). This research was also supported by grants from the Japan Agency for Medical Research and Development (AMED) (20fk0108533h0001), and the JST Core Research for Evolutional Science and Technology (JPMJCR20HA). This work was supported by Joint Usage/Research Center program of Institute for Frontier Life and Medical Sciences Kyoto University.

Institutional Review Board Statement: Not applicable.

Informed Consent Statement: Not applicable.

Data Availability Statement: Data is contained within the article.

Acknowledgments: The authors would like to express their gratitude to the Fundación Universidad Católica de Valencia San Vicente Mártir, the Ministerio de Ciencia e Innovación and the Japan Agency for Medical Research and Development for their financial support. We would like to thank Yoshio Koyanagi and Kazuya Shimura (Kyoto University) for setup and operation of the BSL-3 laboratory.

Conflicts of Interest: The authors declare no conflict of interest.

References

1. WHO Director-General's Opening Remarks at the Media Briefing on COVID-19. 11 March 2020. Available online: https://www.who.int/dg/speeches/detail/who-director-general-s-opening-remarks-at-the-media-briefing-on-COVID-19---11-march-2020 (accessed on 19 May 2021).
2. Briz-Redón, Á.; Serrano-Aroca, Á. The effect of climate on the spread of the COVID-19 pandemic: A review of findings, and statistical and modelling techniques. *Prog. Phys. Geogr. Earth Environ.* **2020**, *44*, 591–604. [CrossRef]
3. Briz-Redón, Á.; Serrano-Aroca, Á. A spatio-temporal analysis for exploring the effect of temperature on COVID-19 early evolution in Spain. *Sci. Total Environ.* **2020**, *728*, 138811. [CrossRef] [PubMed]
4. Wu, Y.; Guo, C.; Tang, L.; Hong, Z.; Zhou, J.; Dong, X.; Yin, H.; Xiao, Q.; Tang, Y.; Qu, X.; et al. Prolonged presence of SARS-CoV-2 viral RNA in faecal samples. *Lancet Gastroenterol. Hepatol.* **2020**, *5*, 434–435. [CrossRef]
5. Baltimore, D. Expression of animal virus genomes. *Bacteriol. Rev.* **1971**, *35*, 235–241. [CrossRef] [PubMed]
6. Walls, A.C.; Park, Y.J.; Tortorici, M.A.; Wall, A.; McGuire, A.T.; Veesler, D. Structure, Function, and Antigenicity of the SARS-CoV-2 Spike Glycoprotein. *Cell* **2020**, *181*, 281–292.e6. [CrossRef]
7. Yang, X.; Yu, Y.; Xu, J.; Shu, H.; Xia, J.; Liu, H.; Wu, Y.; Zhang, L.; Yu, Z.; Fang, M.; et al. Clinical course and outcomes of critically ill patients with SARS-CoV-2 pneumonia in Wuhan, China: A single-centered, retrospective, observational study. *Lancet Respir. Med.* **2020**, *8*, 475–481. [CrossRef]
8. Su, I.C.; Lee, K.L.; Liu, H.Y.; Chuang, H.C.; Chen, L.Y.; Lee, Y.J. Severe community-acquired pneumonia due to Pseudomonas aeruginosa coinfection in an influenza A(H1N1)pdm09 patient. *J. Microbiol. Immunol. Infect.* **2019**, *52*, 365–366. [CrossRef]
9. Chou, C.C.; Shen, C.F.; Chen, S.J.; Chen, H.M.; Wang, Y.C.; Chang, W.S.; Chang, Y.T.; Chen, W.Y.; Huang, C.Y.; Kuo, C.C.; et al. Recommendations and guidelines for the treatment of pneumonia in Taiwan. *J. Microbiol. Immunol. Infect.* **2019**, *52*, 172–199. [CrossRef]
10. Lee, J.Y.; Yang, P.C.; Chang, C.; Lin, I.T.; Ko, W.C.; Cia, C.T. Community-acquired adenoviral and pneumococcal pneumonia complicated by pulmonary aspergillosis in an immunocompetent adult. *J. Microbiol. Immunol. Infect.* **2019**, *52*, 838–839. [CrossRef]
11. Albrich, W.C.; Rassouli, F.; Waldeck, F.; Berger, C.; Baty, F. Influence of Older Age and Other Risk Factors on Pneumonia Hospitalization in Switzerland in the Pneumococcal Vaccine Era. *Front. Med.* **2019**, *6*, 286. [CrossRef]
12. Wu, X.; Cai, Y.; Huang, X.; Yu, X.; Zhao, L.; Wang, F.; Li, Q.; Gu, S.; Xu, T.; Li, Y.; et al. Co-infection with SARS-CoV-2 and influenza a virus in patient with pneumonia, China. *Emerg. Infect. Dis.* **2020**, *26*, 1324–1326. [CrossRef] [PubMed]
13. Feikin, D.R.; Schuchat, A.; Kolczak, M.; Barrett, N.L.; Harrison, L.H.; Lefkowitz, L.; McGeer, A.; Farley, M.M.; Vugia, D.J.; Lexau, C.; et al. Mortality from invasive pneumococcal pneumonia in the era of antibiotic resistance, 1995–1997. *Am. J. Public Health* **2000**, *90*, 223–229.
14. Huttner, B.; Cappello, B.; Cooke, G.; Gandra, S.; Harbarth, S.; Imi, M.; Loeb, M.; Mendelson, M.; Moja, L.; Pulcini, C.; et al. 2019 community-acquired pneumonia treatment guidelines: There is a need for a change toward more parsimonious antibiotic use. *Am. J. Respir. Crit. Care Med.* **2020**, *201*, 1315–1316. [CrossRef]
15. Bourouiba, L. Turbulent Gas Clouds and Respiratory Pathogen Emissions Potential Implications for Reducing Transmission of COVID-19. *JAMA* **2020**, *323*, 1837–1838. [PubMed]
16. Orenes-Piñero, E.; Baño, F.; Navas-Carrillo, D.; Moreno-Docón, A.; Marín, J.M.; Misiego, R.; Ramírez, P. Evidences of SARS-CoV-2 virus air transmission indoors using several untouched surfaces: A pilot study. *Sci. Total Environ.* **2021**, *751*, 142317. [CrossRef] [PubMed]
17. van Doremalen, N.; Bushmaker, T.; Morris, D.H.; Holbrook, M.G.; Gamble, A.; Williamson, B.N.; Tamin, A.; Harcourt, J.L.; Thornburg, N.J.; Gerber, S.I.; et al. Aerosol and Surface Stability of SARS-CoV-2 as Compared with SARS-CoV-1. *N. Engl. J. Med.* **2020**, *382*, 1564–1567. [CrossRef]
18. Ong, S.W.X.; Tan, Y.K.; Chia, P.Y.; Lee, T.H.; Ng, O.T.; Wong, M.S.Y.; Marimuthu, K. Air, Surface Environmental, and Personal Protective Equipment Contamination by Severe Acute Respiratory Syndrome Coronavirus 2 (SARS-CoV-2) from a Symptomatic Patient. *JAMA J. Am. Med. Assoc.* **2020**, *323*, 1610–1612. [CrossRef] [PubMed]
19. Chin, A.W.H.; Chu, J.T.S.; Perera, M.R.A.; Hui, K.P.Y.; Yen, H.-L.; Chan, M.C.W.; Peiris, M.; Poon, L.L.M. Stability of SARS-CoV-2 in different environmental conditions. *Lancet Microbe* **2020**, *1*, e10. [CrossRef]
20. USFDA. Q&A for Consumers: Hand Sanitizers and COVID-19. *United States Food Drug Adm..* 2020. Available online: https://www.fda.gov/drugs/information-drug-class/qa-consumers-hand-sanitizers-and-covid-19 (accessed on 19 May 2021).
21. Vellingiri, B.; Jayaramayya, K.; Iyer, M.; Narayanasamy, A.; Govindasamy, V.; Giridharan, B.; Ganesan, S.; Venugopal, A.; Venkatesan, D.; Ganesan, H.; et al. COVID-19: A promising cure for the global panic. *Sci. Total Environ.* **2020**, *725*, 138277. [CrossRef] [PubMed]
22. American Lung Association Learn About Pneumonia | American Lung Association. Available online: https://www.lung.org/lung-health-diseases/lung-disease-lookup/pneumonia (accessed on 19 May 2021).
23. Singhal, T. A Review of Coronavirus Disease-2019 (COVID-19). *Indian J. Pediatr.* **2020**, *87*, 281–286. [CrossRef] [PubMed]
24. Bai, Y.; Yao, L.; Wei, T.; Tian, F.; Jin, D.Y.; Chen, L.; Wang, M. Presumed Asymptomatic Carrier Transmission of COVID-19. *JAMA J. Am. Med. Assoc.* **2020**, *323*, 1406–1407. [CrossRef] [PubMed]
25. Skevaki, C.L.; Papadopoulos, N.G.; Tsakris, A.; Johnston, S.L. Microbiologic diagnosis of respiratory illness: Practical applications. In *Kendig and Chernick's Disorders of the Respiratory Tract in Children*; Elsevier Inc.: Amsterdam, The Netherlands, 2012; pp. 399–423.

26. Hodinka, R.L. Respiratory RNA Viruses. In *Diagnostic Microbiology of the Immunocompromised Host*; ASM Press: Washington, DC, USA, 2016; pp. 233–271.
27. Kadioglu, A.; Weiser, J.N.; Paton, J.C.; Andrew, P.W. The role of Streptococcus pneumoniae virulence factors in host respiratory colonization and disease. *Nat. Rev. Microbiol.* **2008**, *6*, 288–301. [CrossRef] [PubMed]
28. Weisfelt, M.; De Gans, J.; Van Der Poll, T.; Van De Beek, D. Pneumococcal meningitis in adults: New approaches to management and prevention. *Lancet Neurol.* **2006**, *5*, 332–342. [CrossRef]
29. Thorrington, D.; Andrews, N.; Stowe, J.; Miller, E.; van Hoek, A.J. Elucidating the impact of the pneumococcal conjugate vaccine programme on pneumonia, sepsis and otitis media hospital admissions in England using a composite control. *BMC Med.* **2018**, *16*, 1–14. [CrossRef]
30. Feng, S.; Shen, C.; Xia, N.; Song, W.; Fan, M.; Cowling, B.J. Rational use of face masks in the COVID-19 pandemic. *Lancet Respir. Med.* **2020**, *8*, 434–436. [CrossRef]
31. Amin, D.; Nguyen, N.; Roser, S.M.; Abramowicz, S. 3D Printing of Face Shields During COVID-19 Pandemic: A Technical Note. *J. Oral Maxillofac. Surg.* **2020**, *78*, 1275–1278. [CrossRef]
32. Shukla, S.R.; Palekar, V.; Pingale, N. Zeolite catalyzed glycolysis of polyethylene terephthalate bottle waste. *J. Appl. Polym. Sci.* **2008**, *110*, 501–506. [CrossRef]
33. Bélec, L.; Tevi-Benissan, C.; Bianchi, A.; Cotigny, S.; Beumont-Mauviel, M.; Si-Mohamed, A.; Malkin, J.E. In vitro inactivation of Chlamydia trachomatis and of a panel of DNA (HSV-2, CMV, adenovirus, BK virus) and RNA (RSV, enterovirus) viruses by the spermicide benzalkonium chloride. *J. Antimicrob. Chemother.* **2000**, *46*, 685–693. [CrossRef]
34. Martí, M.; Tuñón-Molina, A.; Aachmann, F.L.; Muramoto, Y.; Noda, T.; Takayama, K.; Serrano-Aroca, Á. Protective Face Mask Filter Capable of Inactivating SARS-CoV-2, and Methicillin-Resistant Staphylococcus aureus and Staphylococcus epidermidis. *Polymers* **2021**, *13*, 207. [CrossRef]
35. Mcdonnell, G.; Russell, A.D. Antiseptics and disinfectants: Activity, action, and resistance. *Clin. Microbiol. Rev.* **1999**, *12*, 147–179. [CrossRef] [PubMed]
36. Sanjeeva Murthy, N. Techniques for analyzing biomaterial surface structure, morphology and topography. In *Surface Modification of Biomaterials: Methods Analysis and Applications*; Elsevier Inc.: Amsterdam, The Netherlands, 2011; pp. 232–255.
37. Das, O.; Neisiany, R.E.; Capezza, A.J.; Hedenqvist, M.S.; Försth, M.; Xu, Q.; Jiang, L.; Ji, D.; Ramakrishna, S. The need for fully bio-based facemasks to counter coronavirus outbreaks: A perspective. *Sci. Total Environ.* **2020**, *736*, 139611. [CrossRef] [PubMed]
38. Schrank, C.L.; Minbiole, K.P.C.; Wuest, W.M. Are Quaternary Ammonium Compounds, the Workhorse Disinfectants, Effective against Severe Acute Respiratory Syndrome-Coronavirus-2? *ACS Infect. Dis.* **2020**, *6*, 1553–1557. [CrossRef] [PubMed]
39. Hora, P.I.; Pati, S.G.; McNamara, P.J.; Arnold, W.A. Increased Use of Quaternary Ammonium Compounds during the SARS-CoV-2 Pandemic and Beyond: Consideration of Environmental Implications. *Environ. Sci. Technol. Lett.* **2020**, *7*, 622–631. [CrossRef]
40. Yamanaka, T.; Bannai, H.; Tsujimura, K.; Nemoto, M.; Kondo, T.; Matsumura, T. Comparison of the virucidal effects of disinfectant agents against equine influenza a virus. *J. Equine Vet. Sci.* **2014**, *34*, 715–718. [CrossRef]
41. Gerba, C.P. Quaternary ammonium biocides: Efficacy in application. *Appl. Environ. Microbiol.* **2015**, *81*, 464–469. [CrossRef] [PubMed]
42. Tuladhar, E.; de Koning, M.C.; Fundeanu, I.; Beumer, R.; Duizer, E. Different virucidal activities of hyperbranched quaternary ammonium coatings on poliovirus and influenza virus. *Appl. Environ. Microbiol.* **2012**, *78*, 2456–2458. [CrossRef]
43. Neacşu, I.A.; Nicoară, A.I.; Vasile, O.R.; Vasile, B.Ş. Inorganic micro- and nanostructured implants for tissue engineering. In *Nanobiomaterials in Hard Tissue Engineering: Applications of Nanobiomaterials*; Elsevier Inc.: Amsterdam, The Netherlands, 2016; pp. 271–295.
44. Plásticos Villamarchante | Empresa de Moldeo por Inyección de Plástico | Fabricación de Piezas de Plástico. Available online: http://plasticosvillamarchante.com/ (accessed on 19 May 2021).
45. Zhang, J.; Li, B.; Wu, L.; Wang, A. Facile preparation of durable and robust superhydrophobic textiles by dip coating in nanocomposite solution of organosilanes. *Chem. Commun.* **2013**, *49*, 11509–11511. [CrossRef]
46. Llorens-Gámez, M.; Salesa, B.; Serrano-Aroca, Á. Physical and biological properties of alginate/carbon nanofibers hydrogel films. *Int. J. Biol. Macromol.* **2020**, *151*, 499–507. [CrossRef]
47. German Collection of Microorganisms and Cell Cultures. Available online: https://www.dsmz.de/collection/catalogue/details/culture/DSM-21518 (accessed on 23 August 2021).
48. Kropinski, A.M.; Mazzocco, A.; Waddell, T.E.; Lingohr, E.; Johnson, R.P. Enumeration of bacteriophages by double agar overlay plaque assay. *Methods Mol. Biol.* **2009**, *501*, 69–76. [PubMed]
49. Matsuyama, S.; Nao, N.; Shirato, K.; Kawase, M.; Saito, S.; Takayama, I.; Nagata, N.; Sekizuka, T.; Katoh, H.; Kato, F.; et al. Enhanced isolation of SARS-CoV-2 by TMPRSS2- expressing cells. *Proc. Natl. Acad. Sci. USA* **2020**, *117*, 7001–7003. [CrossRef]
50. Lin, H.T.; Tsai, H.Y.; Liu, C.P.; Yuan, T.T.T. Comparability of bovine virus titers obtained by TCID50/ml and FAID50/ml. *J. Virol. Methods* **2010**, *165*, 121–124. [CrossRef]
51. Martí, M.; Frígols, B.; Serrano-Aroca, Á. Antimicrobial Characterization of Advanced Materials for Bioengineering Applications. *J. Vis. Exp.* **2018**, 57710. [CrossRef] [PubMed]
52. Shao, W.; Liu, H.; Liu, X.; Wang, S.; Wu, J.; Zhang, R.; Min, H.; Huang, M. Development of silver sulfadiazine loaded bacterial cellulose/sodium alginate composite films with enhanced antibacterial property. *Carbohydr. Polym.* **2015**, *132*, 351–358. [CrossRef]

53. Gill, S.R.; Fouts, D.E.; Archer, G.L.; Mongodin, E.F.; DeBoy, R.T.; Ravel, J.; Paulsen, I.T.; Kolonay, J.F.; Brinkac, L.; Beanan, M.; et al. Insights on evolution of virulence and resistance from the complete genome analysis of an early methicillin-resistant Staphylococcus aureus strain and a biofilm-producing methicillin-resistant Staphylococcus epidermidis strain. *J. Bacteriol.* **2005**, *187*, 2426–2438. [CrossRef] [PubMed]
54. Christensen, G.D.; Bisno, A.L.; Parisi, J.T.; McLaughlin, B.; Hester, M.G.; Luther, R.W. Nosocomial septicemia due to multiply antibiotic-resistant Staphylococcus epidermidis. *Ann. Intern. Med.* **1982**, *96*, 1–10. [CrossRef] [PubMed]

MDPI

St. Alban-Anlage 66

4052 Basel

Switzerland

Tel. +41 61 683 77 34

Fax +41 61 302 89 18

www.mdpi.com

International Journal of Molecular Sciences Editorial Office

E-mail: ijms@mdpi.com

www.mdpi.com/journal/ijms